T0231329

Digital Imaging for Cultural Heritage Preservation

Heritage Preservation

Analysis, Restoration, and Reconstruction of Ancient Artworks

Digital Imaging and Computer Vision Series

Series Editor

Rastislav Lukac

Foveon, Inc./Sigma Corporation
San Jose, California, U.S.A.

Digital Imaging for Cultural Heritage Preservation

Analysis, Restoration, and Reconstruction of Ancient Artworks

Filippo Stanco
Sebastiano Battiato
Giovanni Gallo

CRC Press
Taylor & Francis Group
Boca Raton   London   New York

CRC Press is an imprint of the
Taylor & Francis Group, an **informa** business

CRC Press
Taylor & Francis Group
6000 Broken Sound Parkway NW, Suite 300
Boca Raton, FL 33487-2742

© 2011 by Taylor & Francis Group, LLC
CRC Press is an imprint of Taylor & Francis Group, an Informa business

No claim to original U.S. Government works

Printed in the United States of America on acid-free paper

International Standard Book Number: 978-1-4398-2173-2 (Hardback)

Visit the Taylor & Francis Web site at
http://www.taylorandfrancis.com

and the CRC Press Web site at
http://www.crcpress.com

Contents

Preface

Paintings, frescoes, antique photographic prints, incunabula, old books, handwritten documents, sculptures, ceramic fragments, and other ancient manufacts constitute the elements of an extremely valuable and immense historical patrimony. The digitalization of these treasures opens up the possibility of using image processing and analysis and computer graphics techniques to preserve this heritage for future generations and to augment it with accesory information or with new possibilities for its enjoyment and use. Digital imaging solutions can be used to generate virtually restored versions of the original artworks to be presented in online museums and/or for further development of historical studies; application of various feature extraction and image data analysis techniques are useful in addressing problems of authorship and artwork style categorization in the history of the arts; three-dimensional reconstruction of ancient artworks or entire archeological sites allows the creation of multidimensional models that incorporate information coming from excavations, archaeological know-how, and heterogeneous historical sources.

Pioneer works in this area have sprung up from the close, but too often occasional, cooperation of scientists with historians and archaeologists. Case studies in these kinds of interdisciplinary works are now abundant, yet the different worlds of information technology and art and history have not built a robust common language. This book wishes to respond to the growing demand of this area to bridge the existing gap between these different scientific communities. The urgency of providing a common ground, where technology may meet humanities, comes from the forthcoming generation of imaging devices that will provide better performances with lower costs, allowing the wide application of techniques nowadays considered experimental. This book comes as one of a few dedicated monographs dealing with digital imaging methods for cultural heritage preservation. It is a collection of contributions which cover the most prominent topics and applications of digital image processing, analysis, and computer graphics in the field of cultural heritage preservation. In this contributed volume, well-known experts present a collection of recent results covering a wide range of topics and related applications such as digitization, indexation of visual databases, automatic reassembly of ceramic fragments, restoration and analysis of the digital images of original artworks, digital archaeology using laser scanning, photogrammetry and 3D from 2D photo approaches, computer graphics by means of simulation for ancient mosaics reproduction, and copyright protection of artwork images using digital watermarking.

The book tries to cover the most important research and application directions in the area of digital imaging-driven cultural heritage preservation; hence, it can constitute an invaluable reference on the subject for both scholars and practitioners for many years to come.

This text is aimed at graduate students and advanced undergraduates, as well as researchers and practitioners, coming from both industry and academia. It assumes only some prior knowledge of digital image processing and computer graphics fundamentals. Each chapter provides the fundamentals for the topics under consideration and the detailed description of the relevant methods applied in concrete application. Case studies are also provided to demonstrate the effectiveness of the surveyed techniques. To make the material more accessible, each chapter has its own table of contents, drawings, and photographs, which will elucidate the presented concepts in detail, chapter summary, and bibliography for further reading.

The book opens with two contributions (Chapters 1, 2) that discuss and review how it is possible with present day technologies to realize digital projects that integrate the collected data with virtual reconstructions, taking into account the specialists' hypotheses. Both chapters illustrate, with plenty of examples of different complexities and scales, how different expertise converge into a new way of thinking, presenting, archiving, cultural heritage sites, artifacts, and knowledge.

Acquisition and processing of 3D data are at the core of this new field of study. In order to support the widespread diffusion and acceptance of these technologies a key factor is the availability of software tools. Chapter 3 is largely dedicated to reviewing the laser scanning techniques available and presenting proper software tools to manage 3D data. To complete the overview of the 3D-data acquisition methodologies available, Chapter 4 illustrates state-of-the-art techniques for image-based 3D-data acquisitions, and demonstrates how this approach can be effectively and economically adopted as an alternative to laser scanning.

Chapter 5 offers a methodological review of the steps that have to be performed in order to obtain a successful integration of information coming from archival sources, 3D scanning, aerial photography, and other image-based techniques to document a whole historical site. To be accepted by the general public, images and animations must satisfy high aesthetic standards. The chapter shows how it is possible to make use of advanced computer graphics techniques to enhance the experience of a monument or artifact. Chapter 6 illustrates how starting with 3D data and accurate photometric measurements, it is possible to recreate or synthesize atmospherical and light conditions around the Parthenon, offering to the observer a perceptive experience that would be impossible with a traditional visit to the site.

Colors, indeed, are a critical ingredient in any digitalization project of cultural heritage. Multispectral analysis and multispectral synthesis are the most advanced approaches in gathering colorimetric information from reality and are essential for a faithful reproduction. Mature computational methods are now available to support these operations, and Chapter 7 offers a complete review of them. Computational methods are addressing not only the issues of preservation, restoration, and presentation of artifacts; they offer powerful tools to investigate and confute historical hypotheses about art. Examples of how computer experiments and analysis may help to assess or disprove ideas in art history, when paired with more traditional historical competence and documental evidence, are reported in Chapters 8 and 9.

The application of computational methods to cultural heritage is not at all restricted to 3D-data collection and analysis. Several issues relative to genuinely bi-dimensional artifacts, like books and flat documents, or more recently archived video, have received considerable attention. Algorithms for automatic restoration of this class of artifacts are presented in Chapters 10 and 11.

Restoring artifacts from their fragmented remains is one of the most challenging tasks of archeologists. Fragment sets are frequently too large to be processed in an efficient way by a human expert. Alternative hypotheses about their original layout have to be explored and tested, and, sadly, fragments are rarely sufficient for a complete reconstruction of whole objects like, for example, vessels. As digitalization is made more efficient, the practice of virtual restoration from the 3D data has been the subject of many pioneering experiments. The challenge of reassembling a vase from the digital shapes of its fragments requires a sound orchestration of geometric data processing and of pattern recognition. Chapter 12 reports on an updated and detailed description of the state-of-the-art solutions available for this problem for the case of vessels and ceramics. Chapter 13 reports on similar efforts relative to other fragmented artifacts like frescoes.

Special techniques are required in the case of mosaics. Their digitalization for archival as well as for reconstructive purposes has to take into account the special structure of this artistic medium. Chapter 14 is devoted to the presentation of some case studies in digitalizing and virtual restoring of mosaics. The state-of-the-art techniques of computer-generated mosaics compared to more traditional photos and paintings, a well-investigated problem in non-photorealistic computer graphics renderings, is covered in Chapter 15.

Another interesting case study related to rock graffiti restoration and enhancement is reported in

Chapter 16. Finally, Chapter 17 reviews the watermarkings of digital images, addressing the crucial problems of copyright protection and controlled distribution of authored material.

Filippo Stanco, Sebastiano Battiato, Giovanni Gallo

Editors

Filippo Stanco

Filippo Stanco received a degree in computer science (summa cum laude) in 1999 and a Ph.D. in computer science in 2003, both from University of Catania, Italy. From 2003 to 2006 he was a research assistant at the Universities of Trieste and Catania, Italy. Since 2006, he has been an assistant professor with the Department of Mathematics and Computer Science, University of Catania. He coordinates the "Archeomatica Project" (www.archeomatica.unict.it) to develop new digital tools for the archaeological research and preservation of cultural heritage. His research interests include digital restoration, zooming, super-resolution, artifacts removal, interpolation, texture, and GIS. Dr. Stanco has been an organizer and associate editor of several conferences and symposiums, and he is a reviewer for several international journals. He is a member of IEEE, AIAr, and SIMAI.

Sebastiano Battiato

Sebastiano Battiato received his degree in computer science (summa cum laude) in 1995 and his Ph.D. in computer science and applied mathematics in 1999. From 1999 to 2003 he was the leader of the "Imaging" team at STMicroelectronics in Catania. He joined the Department of Mathematics and Computer Science at the University of Catania as assistant professor in 2004 and became associate professor in the same department in 2011. His research interests include image enhancement and processing, image coding, camera imaging technology and multimedia forensics. He has published more than 100 papers in international journals, conference proceedings and book chapters. He is a co-inventor of about 15 international patents, reviewer for several international journals, and he has been regularly a member of numerous international conference committees. Prof. Battiato is an associate editor of the IEEE Transactions on Circuits and System for Video Technology and of the SPIE Journal of Electronic Imaging. He is director (and co-founder) of the International Computer Vision Summer School, Sicily, Italy. He is a senior member of the IEEE.

Giovanni Gallo

Giovanni Gallo graduated from the University of Catania in 1984 and got his Ph.D. at New York University in 1992. Since then he has been professor of computer graphics and computer vision at Catania University where he coordinates a research group on image processing and graphics applied to medicine, archaeology, and data visualization. Prof. Gallo is also a member of the faculty of "Accademia di Belle Arti" of Catania and coordinates several interdisciplinary projects involving art and technologies. He has been an advisor of more than twelve Ph.D. students and authored several papers in his research field. He is a member of Eurographics and ACM-Siggraph. He is president of the Italian chapter of Eurographics for 2009-2012.

Contributors

Sebastiano Battiato
University of Catania
Catania, Italy

Lamia Benyoussef
École Centrale Marseille
Marseille, France

Simone Bianco
University of Milano Bicocca
Milan, Italy

Vittoria Bruni
Consiglio Nazionale delle Ricerche
Rome, Italy

Yang Cai
Carnegie Mellon University
Pittsburgh, PA

Roberto Caldelli
University of Florence
Florence, Italy

Marco Callieri
Consiglio Nazionale delle Ricerche
Pisa, Italy

Vito Cappellini
University of Florence
Florence, Italy

Paolo Cignoni
Consiglio Nazionale delle Ricerche
Pisa, Italy

Jacob Collins
c/o Adelson Galleries
New York

Alessandro Colombo
University of Milano Bicocca
Milan, Italy

David Corrigan
Trinity College
Dublin, Ireland

Massimiliano Corsini
Consiglio Nazionale delle Ricerche
Pisa, Italy

Andrew Crawford
Consiglio Nazionale delle Ricerche
Rome, Italy

Paul Debevec
University of Southern California
Playa Vista, CA

Andrea Del Mastio
University of Florence
Florence, Italy

Andrey Del Pozo
University of Illinois at Urbana-Champaign
Urbana, IL

Matteo Dellepiane
Consiglio Nazionale delle Ricerche
Pisa, Italy

Stéphane Derrode
École Centrale Marseille
Marseille, France

Gianpiero Di Blasi
University of Palermo
Palermo, Italy

Marco Duarte
Princeton University
Princeton, NJ

Per Einarsson
University of Southern California
Playa Vista, CA

Marcos Fajardo
University of Southern California
Playa Vista, CA

Yasuo Furuichi
Consultant
Kanagawa, Japan

Giovanni Gallo
University of Catania
Catania, Italy

Andrew Gardner
University of Southern California
Playa Vista, CA

Francesca Gasparini
University of Milano Bicocca
Milan, Italy

Tim Hawkins
University of Southern California
Playa Vista, CA

Andrew Jones
University of Southern California
Playa Vista, CA

Dave Kale
Stanford University
Stanford, CA

Anil Kokaram
Trinity College
Dublin, Ireland

Ashutosh Kulkarni
Stanford University
Stanford, CA

Therese Lundgren
University of Southern California
Playa Vista, CA

Philippe Martinez
Ecole Normale Superieure
Paris, France

Nikos Nikolaidis
Aristotle University of Thessaloniki
Thessaloniki, Greece

Ioannis Pitas
Aristotle University of Thessaloniki
Thessaloniki, Greece

Francois Pitie
Trinity College
Dublin, Ireland

Charis Poullis
University of Southern California
Playa Vista, CA

Giovanni Puglisi
University of Catania
Catania, Italy

Giovanni Ramponi
University of Trieste
Trieste, Italy

Fabio Remondino
B. Kessler Foundation (FBK)
Trento, Italy

Alfredo Restrepo Palacios
Universidad de los Andes
Bogotá, Colombia

M. Dirk Robinson
Ricoh Innovations
Menlo Park, CA

Silvio Savarese
University of Michigan
Ann Arbor, MI

Sara J. Schechner
Harvard University
Cambridge, MA

Raimondo Schettini
University of Milano Bicocca
Milan, Italy

Roberto Scopigno
Consiglio Nazionale delle Ricerche
Pisa, Italy

Ron Spronk
Queen's University
Kingston, Canada

Filippo Stanco
University of Catania
Catania, Italy

David G. Stork
Ricoh Innovations
Menlo Park, CA
Stanford University
Stanford, CA

Jessi Stumpfel
University of Southern California
Playa Vista, CA

Davide Tanasi
Arcadia University – Mediterranean Center for
   Arts and Sciences
Siracusa, Italy

Chris Tchou
University of Southern California
Playa Vista, CA

David Tingdahl
Katholieke Universiteit Leuven
Leuven, Belgium

Efthymia Tsamoura
Aristotle University of Thessaloniki
Thessaloniki, Greece

Christopher W. Tyler
Smith-Kettlewell Eye Research Institute
San Francisco, CA

Francesca Uccheddu
University of Florence
Florence, Italy

Luc Van Gool
Katholieke Universiteit Leuven
Leuven, Belgium

Maarten Vergauwen
GeoAutomation
Leuven, Belgium

Domenico Vitulano
Consiglio Nazionale delle Ricerche
Rome, Italy

Nicholas C. Williams
Towan Headland
Cornwall, UK

Andrew R. Willis
University of North Carolina at Charlotte
Charlotte, NC

Nathaniel Yun
University of Southern California
Playa Vista, CA

Silvia Zuffi
Consiglio Nazionale delle Ricerche
Milan, Italy

1

Experiencing the Past: Computer Graphics in Archaeology

Filippo Stanco

University of Catania
`Email: fstanco@dmi.unict.it`

Davide Tanasi

Arcadia University – Mediterranean Center for Arts and Sciences
`Email: dtanasi@mediterranencenter.it`

CONTENTS

1.1 The Past and the Future: Archaeology and Computer Science

In the last fifty years, the growing use of computer applications has become a main feature of the archaeological research [1]. Since the '90s, when computer science was oriented to the creation of work tools and solutions for the archive and management of quantitative data, to the development of virtual models and to the dissemination of knowledge, it quickly changed into a true theoretical

1

approach to the problems of archaeology. It is now, indeed, able to influence the interpretation proce- dures and to revolutionize the language and contents of the study of the past [2]. This new evidence introduced in several branches of the theoretical debate new scientific themes. There are different views about the integration of computers and archaeology. Digital archeology in the Anglo-Saxon cultural world [3] is considered as a computer approach to the modern cognitive archaeology. Archae- ological computing [4], on the other hand, is a methodology for the elaboration of archaeological data via computer. Archaeological computer science [5] is devoted to the representation with computer applets of the cognitive procedures behind the interpretation of the archaeological data, and the more popular virtual archaeology (VA) [6], is the analysis of the procedures of management and representation of the archaeological evidence through computer graphic 3D techniques.

From its first definition, by Reilly in 1991 [7], VA was intended as the use of digital reconstruction in archaeology. Recently to its research the development of new communicative approaches to archaeological contents through the use of interactive strategies has been added. The birth of the VA is not simply caused by the proliferation of 3D modeling techniques in many fields of the knowledge, but as a necessity to experience new systems to archive an overgrowing amount of data and to create the best medium to communicate those data with a visual language. From this point of view, the application of 3D reconstructions, obtained with different available techniques, became the core area of study of the VA in regard to the potential of cognitive interaction offered by a 3D model. In this way, virtuality turns into a a communication method even more effective if applied to particular fields, archaeological areas well preserved but not accessible [8], sites not preserved but known by traditional documentations [9], sites destroyed but depicted in iconographical repertoires [10], contextualization in progressive dimensional scale (object, context, site, landscape), and functional simulations repeating in virtual environment the processes of the experimental archaeology.

1.2 From the Field to the Screen: 3D Computer Graphics and the Archaeo- logical Heritage

The cognitive experiences of 3D computer graphics can essentially be divided into passive and active forms of interaction. The first case refers mainly to applications related to research and study, where the primary need is of documentary type, as the archaeological excavation or the monitoring of the degradation. In the second case, the interaction with the virtual recreated reality is further exploited in the enhancement of the archaeological heritage through the creation of a virtual museum, reachable on digital media or on the web, intended both as a virtual version of a proper museum and as a closer study of an archaeological site. Different is the case of the 3D reconstructions, developed within interdisciplinary research projects, made for the purpose of interpretation as a cognitive accessory available to the archaeologists.

1.2.1 3D Computer Graphics and the Archaeological Fieldwork

The other major field of application of 3D computer graphics in the world of archeology is the documentation of the excavation data in real time. Since archeology is the science of destruction par excellence, the need to document in a comprehensive and detailed way each item that is removed during excavation, imposed gradually the methods of graphic and photographic documentation in support of traditional 3D modeling [11]. This technique can be used both for recording singular evidence but also for the objects set inside a GIS system in which 3D data are fully integrated. From this point of view the combination of GIS systems in archaeology and the development of 3D laser scanning and image-based 3D modeling techniques determined the birth of experimental systems of

3D GIS. This system is able to visualize inside the geographic information system 3D data, such as point clouds from laser scanners, and it has already produced excellent results as demonstrated in the case of Miranduolo excavation (Siena) Italy [12] or in the investigations in the Jabal Hamrat Fidan region of the Faynan District in Southern Jordan [13], just to name a few recent examples. The point of the application of these techniques on the excavation activity is the possibility to perform analysis on multidimensional scale. At landscape scales, digital 3D modeling and data analysis allow archaeologists to integrate, without breaks, different archaeological features and physical context in order to better document the area. At monument/site scale, 3D techniques can give accurate measurements and objective documentation as well as a new aspect from a different point of view. At artifact scale, 3D modeling allows the reproduction of accurate digital/physical replicas of every artifact that can be studied, measured and displayed, as well as data for general public use, virtual restoration, and conservation.

1.2.2   Monitoring the Heritage

3D modeling could also be extremely useful for the identification, monitoring, conservation, restoration, and promotion of archaeological goods. The archaeological heritage is always under constant threat and danger. Architectural structures and cultural and natural sites are exposed to pollution, tourists, and wars, as well as environmental disasters such as earthquakes, floods, or climatic changes. Hidden aspects of our cultural heritage are also affected by agriculture, changes in agricultural regimes due to economic progress, mining, gravel extraction, construction of infrastructure, and the expansion of industrial areas. In this context, 3D computer graphics can support archaeology and the politics of cultural heritage by offering scholars a "sixth sense" for understanding the traces of the past, as it allow us to experience it [14]. 3D documentation of still existing archaeological remains or building elements is an important part of collecting the necessary sources for a virtual archaeology project. New developments allow this documentation phase, including the obtaining of correct measures and ground plans from photography using freely available tools. This is important when restoring archaeological remains, when older phases are reconstructed in a virtual way. The original state, the restored state, and eventual in between states can be recorded easily through this photo modeling technique [15]. Furthermore, the recent application of 3D computer graphics has proved crucial in planning strategies of restoration and conservation issues of monuments that are part of world cultural heritage, on which there is still an open debate, as in the case of the restoration of the Parthenon on the Acropolis of Athens [16].

1.2.3   The Virtual Museum

The rapid development of 3D visualization techniques, and the subsequent derivation in the promotion policies for the archaeological heritage as well as the creation of 3D models of monuments and artifacts of past civilizations has become the basis for the birth of the concept of virtual museum as a means of transmission of knowledge based on the use of multimedia [17, 18]. More recently, the seductive ability of visual communication, simplified and made more attractive by the opportunity to interact intuitively with multimedia content, has led to a huge proliferation of virtual museums on the web. The phenomenon in some cases has grown, losing sight of what are the ideological assumptions of the virtual museum itself. The virtual museum should not be considered as a transposition of a real museum in electronic form or on the web, nor can it be intended as a supplementary tool to complete the real museum, like a sort of exhibition space or additional digital catalog. Its nature is closely linked to its ultimate goal, namely to communicate knowledge to the wider audience possible, without stopping at generalizations between "scholars" and "public." This aim is achieved with the use of communication strategies based on visual narratives, interactive multimedia narratives that tell the story of each artifact, contextualizing it geographically, historically, and culturally and embedding it into a network of information that goes beyond the artifact itself and what the real museum contains.

In this perspective, the virtual museum has its best expression in the version of the "museum of territory" which virtually took out of the museum's closed walls the original cultural background of an entire territory. The virtual museum can also be considered as "a communicative projection of the real museum" which performs a dual function: educational, as it overlaps linguistical and physical barriers, and promotional, as it is also a tool of cultural exportation [19]. Furthermore, the use of virtual reality and augmented reality technologies may make even more immersive and involving the experience of decoding the information. Just to cite an example of excellence in the most recent virtual museum project, it can be mentioned the Iraq Virtual Museum [8, 9, 20–22], completely accessible from the web, where the most advanced digital visualization techniques have been applied. It was promoted between 2006 and 2007, by the Italian National Research Council and the Italian Ministry of Foreign Affairs, as a concrete action for the cultural development of the Iraqi people through a targeted intervention for recovery of the Baghdad Museum intended as a shared treasure. Closed to the public in 1991, during the first Gulf War, and subjected to severe devastation and looting in 2003, during the turmoil of the second Gulf War, the Museum of Baghdad and its archaeological treasures, which covers 8,000 years of Mesopotamian history, were returned to the community only in February of 2009, thanks to the efforts of many countries.

1.2.4   3D Modeling as a Cognitive Tool

Computer graphics can be applied to the reconstruction and visualization of several features of an archaeological site with the creation of a multidimensional model including every feature derived from the excavations. This process is fundamental for scholars of virtual archaeology, the goal of which is the complete reconstruction of an ancient landscape. Computer science has a primary role in this branch of cognitive archaeology, and 3D modeling is not considered to be an optional implement for the addition of aesthetic elements in reconstructions, but an indispensable tool for interpretation. The available technologies and methodologies for the digital recording of archaeological sites and objects are promising, and the scientific community is trying to adapt these approaches for detailed 3D documentations to go beyond the simple graphic and photographic data. The process of interpretation of archaeological evidence, often fragmentary and subject to many variables of alteration, finds in 3D computer graphics a valuable experimental environment in which to test the reliability of the assumptions. Often what makes sense during a study about the nature and function of artifacts poorly preserved and stored in store boxes, can be completely changed by a simple passive observation of a 3D reconstruction, if not completely disavowed by interactive or immersive virtual models in which the man-space-artifact-building relation is recreated. From this point of view, the 3D computer graphics become on the same level of the archeology itself, as a digital version of the experimental archaeology [23–25], characterized by the study of the "practice supporting the theory" [26]. It aims to the replication of experiments, the testing of methodological assumptions by applying them to known contexts, the experiments involving site formation processes. In the same way a similar research can be virtually conducted interacting with a 3D model replicating the reality. It is not a coincidence that in recent years several interdisciplinary projects of reconstructing the past have been completed thanks to the combination of the experiences both of experimental archaeology and 3D computer graphics [27]. In this sense of cognitive tool, the use of 3D models in archaeological research can be intended as a sort of benchmark of what the perceptual senses and the mind perceived in the first instance. A sort of "seeing causes believing" opposed to a simple and sometimes misleading "seeing is believing" which is often altered by the cultural superstructure of the archaeologists [28]. In this case digital technology is not only used to provide tools of discovery and communication but mostly interactive feedback [29]. Just this aspect of computer graphics in archaeology is that privileged by the multidisciplinary research projects of digital archeology, which aim to bridge the linguistic and cultural gap that still divides the researchers of archaeology and computer science and to contribute to the common reconstruction of the past through its virtual experiencing.

1.3 The Archeomatica Project

A digital archaeology research project, directed by the authors, the Archeomatica Project [30] was begun in late 2007 by a group of prehistoric archaeologists and researchers in image processing and computer graphics from the image processing Lab [31], both of them of the University of Catania. It aims to develop new implements for archaeological research in prehistory and protohistory within the field of 2D digital imaging and 3D graphics, mainly, to produce automatic systems of recognition and classification of graphic data, such as pottery figurative decoration, through the use of computer vision and pattern recognition techniques and to develop virtual models of prehistoric sites and items with a high degree of accuracy following the data obtained during excavation and study, through application of laser scanner and 3D modeling techniques. A cognitive process based on a peer-to-peer exchange of knowledge between experts of computer science and prehistory working side by side. The cooperative experience of the Archeomatica Project, which represents (through its scientific production) the most recent trends in digital archaeology and the modern politics of conservation of archaeological heritage, has also aimed to define a common multidisciplinary language to improve the quality of the message of this new discipline to the outside world. In these first years of research activity the Archeomatica Project has produced significant results in archaeological 3D modeling and for 3D digital restoration, in order to improve the cognitive capacities of the archaeologists.

1.4 Archaeological 3D Modeling

Archaeological 3D modeling is basically the recreation of landscapes, architecture, and objects by digital means based upon the current state of the salvaged monuments integrated with the data coming from historical and archaeological researches using software for developing 3D models [32], without the application of reverse engineering methodology.

It is probably the most popular computer-based technique applied to cultural heritage as it represents the core of the "serious games" used in many multimedia projects [33]. The archaeological 3D modeling is not just a simple cognitive tool to reproduce virtually aspects of the past, like objects of everyday life [34], to improve the knowledge and the comprehension. It is also, above all, a methodology of recording all the archaeological data in a much more complete way than the traditional photography and drawing and it is also an instrument of interpretation for the researchers who are involved in the theoretical reconstruction of the past itself. From this point of view, it is a kind of virtual benchmark of the archaeologists' theories where the hypothesis is tested and corrected in order to produce a truthful image of something buried by time. A kind of "solid modeling to illustrate the monument" becoming "solid modeling to analyze the monument" [35]. For this reason, the privileged application field for this technique is the prehistoric archaeological research, where, the scarcity of iconographical sources and the poor state of conservation of the findings, makes extremely complex both the process of decoding the information and of transmitting the knowledge to the public.

The researchers of the Archeomatica Project have chosen Blender [36] as a work tool, an open source cross-platform software for modeling, rendering, animation, post-production, creation, and playback of interactive 3D contents, extremely versatile, functional, and constantly open to implementations based on the research of its application in various fields. Where it was necessary for particular issues, image-based 3D modeling techniques, which consist in the elaboration of a 3D model from a set of high quality digital photos, have also been used [37, 38].

The study cases chosen for testing the archaeological 3D modeling are the site of Haghia

Triada (Crete, Greece) and Polizzello Mountain (Sicily, Italy), two pre-protohistoric archaeological complexes, which have been the subject of research of the archaeologists from the University of Catania for a long time, and which represent the full cultural evolution of Crete and Sicily between the second and first millennium BC. The decision was also supported by the small number of applications of 3D computer graphics on the Cretan and Sicilian archeology [39–41], a fact that greatly limits the interpretation study of evidences.

1.5   Haghia Triada, Crete

The site of Haghia Triada [42], in the Mesara plain in Crete (Figure 1.1), is one of the main sites of the Minoan civilization, constantly under study by the Centre for Cretan Archaeology of the University of Catania, under the permission of the Italian Archaeological School at Athens. The settlement of Haghia Triada has developed seamlessly over nearly two millennia, from the fourth to the second millennium BC, between the Early Bronze Age and the Early Iron Age. The flourishing of the site coincides with the Neopalatial period (from the seventeenth to the first half of the fifteenth century BC) and the Palatial Final period (from the second half of the fifteenth to the thirteenth century BC), that in terms of Minoan chronology correspond to Middle Minoan III, Late Minoan IB and Late Minoan IB, Late Minoan IIIB [43,44].

During the Neopalatial period was built the so-called Royal Villa, a monumental L-shaped complex, including districts and stately buildings, but also vast areas of warehouses, administrative offices, which represented the symbols of territorial power, probably under the control of Knossos, which at this time had acquired a leading role in central Crete. North of the Villa, a small village was located, perhaps demonstrating the special nature and specific function of "administrative" capital of the site. Set between the Villa complex and the village, an open area was enriched by the presence of a stoa and a propylon, which has the function of mediation space from the noble to the popular quarter. At the end of Late Minoan IB (1450-1430 BC) a serious seismic episode destroyed the Villa and the village, leading to a period of relative neglect in the area until the end of the fifteenth century BC, coinciding with a major political and cultural event in the history of Crete: the occupation of the palace of Knossos, the traditional power center island, by people coming from Mycenaean Greece, which extended its control over the west central part of the island, constituting a true unitary state of Mycenaean type [45,46]. During the Palatial Final period, the Mycenaean phase of Haghia Triada started, a process of architectural uprising aimed to create a unified and comprehensive urban planning, providing, in the Villa, at least five public buildings and a series of constructions with political and religious functions, as the Megaron, the Shrine, and the Stoa. In the village was instead created the commercial and administrative center dominated by large stoa and a series of large buildings not only with residential functions, as the so-called House of Razed Rooms (Casa delle Camere Decapitate), dated to Late Minoan IIIA2 and interpreted as a warehouse for the storage of grain. In fact, besides only small buildings that were short-lived, and the monumental complex of the so-called VAP House (Casa dei Vani Aggiunti Progressivamente), dating from several stages in the Late Minoan IIIA2-IIIB, which can be identified with the elite residence that had control of the site, Haghia Triada is characterized as a singular houseless town. In the new system of Mycenaean power, the site of Haghia Triada, indicated in Linear B texts with the name of pa-i-to (Phaistos) must have been an important administrative center, a sort of district capital or a second order center, probably controlled by a group in direct contact with the elites of Mycenaean Knossos and aimed to the exploitation of resources of the surrounding area [45,47]. The intense construction activity that characterized the site during about four centuries, marked by duplication, destruction and reconstruction, transformation, and reintegration, becomes the main problem for the process of interpreting the evidence and its spatial and cultural contextualization. The traditional

FIGURE 1.1
Diachronical plan of Haghia Triada.

(a)   (b)   (c)

FIGURE 1.2
Haghia Triada open area: (a) plan of the Propylon; (b) the Propylon in the current preservation state, from South; (c) the Stoa in the current preservation state, view from west.

graphic and photographic documentation of the excavation, which in many excavations is dating back to early 1900, in some cases may be insufficient to fully decode both the diachronic and synchronic architectural issues. From this point of view, the Haghia Triada site is an ideal benchmark for the application of 3D Modeling technologies and a perfect test for the real potential that this tool had to offer in the process of historical reconstruction.

Three monuments have been selected; these are considered the most problematic ones, both for the difficulty of interpretation and the architectural complexity. In a second phase of the study the case of the open area with the Propylon and the Stoa has been the focus of attention. It acts as a connecting element between the Villa and has been inhabited as early Neopalatial, and the House of Razed Rooms and VAP House, which was partially lived during the Palatial Final period. The decision to build the 3D models of these buildings, spatially adjacent but temporally successive, is part of the most ambitious project to create a 3D model of multi-plan site that can overcome the current phase plan [48], setting as the most advanced, modern, and expendable method of dissemination of the knowledge. In carrying out the work, two recurring problems were represented by the absence of a virtual model of the Haghia Triada ground: in which to locate the models, and the definition of light sources inside and outside of rebuilt buildings. The realization of the models was performed at different times and by different operators that followed, however, the same guidelines. In the initial phase of the project it was chosen to set the individual models in a virtual abstract ground model, waiting to start working with the project leaders of Crete Digital Satellite Remote Sensing Laboratory of Geophysical and Archaeo-environment of the Institute for Mediterranean Studies related to the Foundation of Research and Technology, Hellas (FORTH) [49], which already produced the Digital Terrain Model of the territory of Crete [50]. This will provide the location in scale of individual reconstructed buildings in the virtual version of the hill of Haghia Triada. As for the lighting following prior experiences [39, 51, 52], it was chosen to replicate the natural light of a summer morning for outdoor models, whereas for interior without windows a virtual source of artificial light, like the flame of an oil lamp, a candle, or a torch, has been introduced, developing the model with the help of Radiance software [53].

1.5.1 Propylon

The first case study [54, 55] is relative to the open area (so called Piazzale Inferiore 10) directly connected to the building of power itself, the Villa, and linked with the facilities of the village, in a period between the seventeenth and fifteenth century BC (Figure 1.2).

The excavations of the summer 2006 revealed on the north side the foundation of a very narrow and elongated rectangular room that served as a hub for the road from the village and therefore

FIGURE 1.3
3D reconstructive model of the open area between the village and the Villa with the virtual reconstruction of the Propylon, the façade of the Bastion and the Stoa.

constituted a real propylon, whose function of monumental vestibule is confirmed by the large threshold still recognizable in situ, according to the constructive concept input of the precinct with an elegant entrance, popular choice in the Neopalatial period [56, 57].

The need to consider the overall organization of this sector of the monumental complex has been addressed by the 3D modeling technology, with the application of a philological approach, based on a thorough investigation of the fundamental characteristics of Minoan architecture [58, 59] and of the repertoires represented on stone vessels [48] and on the frescoes [61, 62], all contemporary to the structures under study. In the middle of the north side of the open area stood the propylon which had its monumental face, probably, provided with a column, right-facing the open space and the Villa itself. The rear elevation provided a simple unaligned entrance, oriented to a gently sloping ramp that represented the main road of the village. The review of the archaeological evidence allowed reconstruction of the east side of the area, a sort of small stoa or porch [56, 57], with a second floor veranda facing the great paved road to the sea. In this case some artifacts documenting the characteristics of Minoan public and private buildings were fundamental (Figure 1.5.1).

1.5.2   House of the Razed Rooms

The other case study is represented by the so-called House of the Razed Rooms [42] (Figure 1.4). This house, built and used during the fourteenth century BC (late Minoan IIIA2), located in the north eastern part of the settlement, north of the Villa, was so deeply damaged by many subsequent activities that it became indispensable trying to create a virtual model to analyze its architectural phases [63]. Discovered by Federico Halbherr in 1911, the House of Razed Rooms, built on two superimposed terraces, takes its name by the razing of the walls of the upper terrace, and the covering and the re-qualification of the structures of the lower one, carried out when a new plan of this area was designed in the second half of the fourteenth century, after the fall of the Knossos palace. After the excavation of 1991, when only the southern part of the building was uncovered, in 1983 V. La Rosa [64] completed the exploration of the house, discovering the complete plan with a main L-shaped corridor (no. 2) with two groups of modular rooms (no. 5, 6, 7 in the western side and no. 1, 3, 4 in the northern part). The corridor 2 had a floor of 1.50 m higher than the other six rooms, that were accessible through trapdoor from the top. The entire building had a length of 20.10 m in north/east-south/west sense and a surface of 180 m^2; the walls had a thickness of 0.80-0.90 m with foundations on the rock. The plan, the thickness of the walls, and the absence of a direct connection

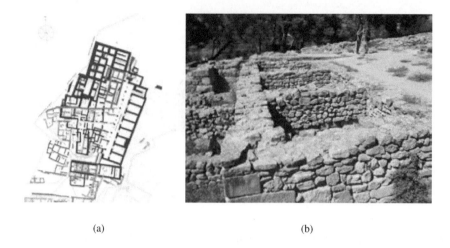

(a)   (b)

FIGURE 1.4
Haghia Triada, village: (a) multilayered plan; (b) House of the Razed Rooms in the current preservation state, view from south west.

between corridor 2 and the other rooms seem to suggest that probably this building did not have residential functions. Furthermore, the subdivision of the space in the same modular unities [65] is a feature, present in some contemporary buildings of Greece like Gla and Mycenae, that has to be interpreted as the mark of structures used as warehouses. In this context, it is possible to identify the two groups of rooms 1, 3, 4, and 5-7 of the House of the Razed Rooms as silos used probably for the long conservation of cereals and legumes [66].

The virtual model of the building (Figure 1.5) was very useful for many reasons [63], such as the possibility of producing two different models of the House of the Razed Rooms, one of the actual state of the building (Figure 1.5(a)), as to create a digital replica of the monument to be used for virtual museum politics, and a second one presenting a picture of the House at the moment of its use (Figure 1.5(b)).

1.5.3 VAP House

Another evidence, datable to the Late Palatial period, is the monumental mansion with several building phases, that has been identified with the house of the local authority, called by Italian archaeologists "Casa dei Vani Aggiunti Progressivamente" (VAP House, Figure 1.5.3) [47,67]. The main peculiarity of this building is how it was progressively enlarged by means of single rooms. This process involved the transformation of the original outer walls into partition walls. Seven subsequent building phases, with the construction of a second floor, the progressive adjunction of rooms and open spaces, can be distinguished between LM IIIA2 early and LM IIIB. At the time of the abandonment of the site in LM IIIB, VAP House was a substantial building, with twelve rooms and one small court on the ground floor, and at least five more rooms on the first floor. The ground floor measured about 320 m^2, and the total extension, including the upper floor, was about 450 m^2. The few finds from the House (mostly some pithoi and a few other vessels) suggest that it was abandoned in a nonviolent way and systematically emptied. A pyxis, a fragmentary pithos, and a clay stand of "snake tube" type were found in room A [68]. The latter was placed near the outer door of the room, beside the first step of the stair, and was probably used to hold cultic offerings.

The construction of room A, in the fifth building phase, involved a functional reassessment of

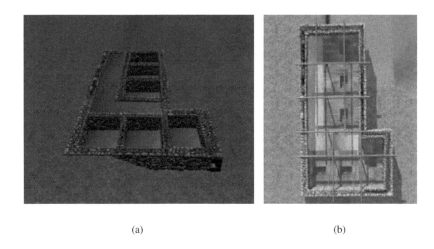

(a)   (b)

FIGURE 1.5
(a) Virtual replica of the House of Razed Rooms; (b) architectural study of the wooden floor with trapdoors and of the roof beams system on the virtual model.

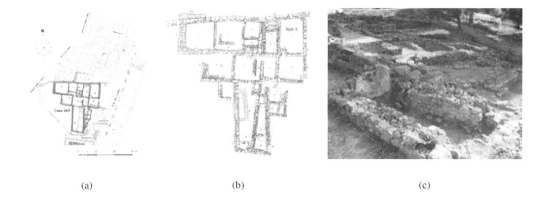

(a)   (b)   (c)

FIGURE 1.6
Haghia Triada, village: (a) Multilayered plan; (b) detailed plan of the VAP House; (c) VAP House in the current preservation state, view from south west.

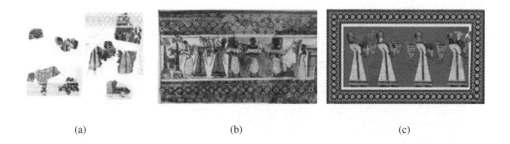

(a)   (b)   (c)

FIGURE 1.7
Vap House, Room A: (a) Fragments of the procession fresco; (b) procession scene on the Haghia Triada Sarcophagus; (c) virtual color reconstruction of the procession fresco.

the lower floor space that was related to the new layout of the principal North-South road of the settlement, and to the creation of the largest open area of Haghia Triada, the so-called Agora. It probably became the most important living room of the House, due to its large dimensions (6.80 × 4.90 m) and its proximity to the Agora. The ceremonial function of room A is confirmed by the frescoes, representing processions with people taking offerings and animals for sacrifice [67]. They belong to a figurative cycle, endowed with a strong symbolical character, which was probably intended to mark the beginning of an entirely new era in the political history of the settlement, when it became the capital of a small state in south-central Crete [43]. Virtual reconstruction is a valuable visualization tool when a complex and multiphased architecture has to be documented. It provides a graphical support to reason about the building and to visually access hypotheses about the appearance in the several construction phases. The reconstruction of the visible walls, derived from the graphic and photographic documentation available, offers a global picture of the entire house in its last phase before abandonment. To recreate a visual image of its ancient life explains the past by means of experiencing it [46]. The digital restoration of the procession fresco (Figure 1.7), and the re-creation of the ceremonial ambient of room A, with the location of the snake tube (Figure 1.8) in its original position, is another useful interpretative tool. The virtual model of the building, derived from the graphic and photographic documentation available, offers a global picture of the entire house as it is conserved now, including the fresco in room A and the snake tube. The virtual reconstruction (Figure 1.5.3) is intended to show the hypotheses about the state of the building in its several phases. To this aim the rooms have been modeled on different 3D layers: the user may choose the epoch and only the layer with the structures present in that epoch will be visualized. This interactive use greatly supports discussion and formulation of the alternative reconstruction choices. The building model has been completed with a minimal set of objects and decoration. In particular, a 3D model of a snake tube has been realized from photographic references of the snake tube found in room A (Figure 1.8. The modeling has been done generating the tube as a revolution solid, deformed and completed with snake-like handles (Figure 1.8(b)). In addition to the 3D reconstruction, a parallel virtual restoration activity has been performed on the very fragmentary remains of the fresco that was once on the wall of room A (Figure 1.7).

(a)　　　　(b)　　　　(c)　　　　(d)

FIGURE 1.8

(a) The snake tube from the room A; (b) 3D model of the snake tube obtained with Image Base 3D modeling technique; (c) Room A, to the right the stairway to room B and the location point of the snake tube, to the left the wall with the procession fresco, view from north. (d) virtual model of the Room A with the snake tube in its original position.

FIGURE 1.9 (SEE COLOR INSERT)

3D reconstructive model of the VAP House, view from south west.

1.6   Polizzello Mountain, Sicily

The indigenous settlement of Polizzello in the territory of Mussomeli (Caltanissetta) is situated on the top of an 877 m high, precipitous mountain that is almost completely encircled by the valleys created by the watercourse of two tributaries of the Platani River: the Fiumicello to the west and the Belici to the east. Strategically located in the very core of Sikania, the mountainous area of central Sicily described by Greek historians that corresponds to the modern territory of Caltanissetta and Agrigento, bordered to the east by the Salso River and to the west by the Platani river, this settlement has for the last 25 years been one of the key sites for the interpretation of the socio-political dynamics of the indigenous communities of central Sicily from the eighth to the late sixth century BC. The archaeological importance and richness of the area has always been well known to the local people who lived in the modern village of Polizzello. After several isolated cases of discovery of materials on the mountain, in 1926, P. Orsi and his collaborator R. Carta began an excavation that clarified the potential importance of this settlement within the region. In the course of these explorations a branch of the necropolis of the late eighth and seventh centuries BC, the settlement area and a shrine of Greek-indigenous type were found. After a long gap, the Superintedence of Agrigento restarted excavations between 1984 and 1990, focusing on the upper plateau, the so-called acropolis, still unknown. On the acropolis four buildings interpreted as open shrines (Sacelli), A, B, C, D, and enclosed by a kind of Temenos were identified (Figure 1.10(a)). Another larger circuit wall was also discovered, and interpreted as a fortification. Although the excavations were not all completed, the preliminary results gave the impression that the site was a very important cult place that had been active between the mid eighth and mid sixth centuries BC. The large assemblages of indigenous and Greek vessels, the bone, amber, ivory and faience items, the iron and bronze weapons, the bronze figurines of worshippers and their location in votive deposits inside the four buildings, defined the cultural richness and complexity of the indigenous community, its external interactions, and the great ritual relevance of the acropolis area. Between 2000 and 2006, the Superintendence of Caltanissetta and the University of Catania carried out a new systematic excavation project on the acropolis, with the aim of completing the previous explorations, definitively defining the different phases of occupation, principally the earliest, and restoring the archaeological area so that it could be opened to the public [69–71].

1.6.1   Buildings A, B, C, D, E

One of the most important aspects of the acropolis excavation was the completion of work on buildings A, B, C, D (Figures 1.11(a), 1.11(b)). They basically were circular or irregular precincts with entrance to the south, raised up with stone blocks in double curtain technique (emplecton) and paved with pressed earth floors. Inside several pieces of complement furniture were found as altars, hearths, and benches. Of great importance were the two buildings at the northern end of the acropolis, A and B, that performed different functions in the religious liturgy of the sanctuary activity. Building A, with a diameter of 8 m, probably kept the wooden statue of the deity, now disappeared, set on a rock base at the center, to which were offered votive gifts like vessels, iron weapons, and amber and bone beads, dated to the mid seventh century BC, accompanied by traces of animal sacrifices.

The new excavations were especially concerned with building B, where previous exploration had focused on the ground layer covering the structures. The building B, built shortly after A and partially leaning upon it, with a diameter of about 10 m, was the real treasure of the sanctuary, where all offers to the gods, the most valuable and varied, were kept at end of ceremonies that involved community meals and libations. Inside it, a pressed earth floor, a hearth, a circular bench running along the walls, a recess and an altar were revealed. A small statue of an ithyphallic warrior of indigenous type was found on the "altar" together with a set of astragaloi, nine of bone and one of lead. The circular bench

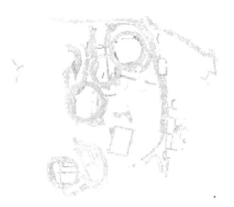 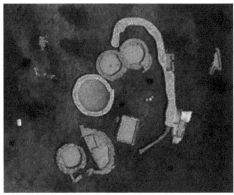

FIGURE 1.10
(a) Plan of the sacred area on the Acropolis of Polizzello Mountain; (b) virtual replica of sacred area on the Acropolis of Polizzello Mountain.

may have been used as seating for the participants in the rites, while the recess seems to have been used to isolate and hold specific depositions. Seventeen deposits of a total of 193 objects of the sixth century BC were found on the floor, simply placed and not hidden or covered, mostly in the northern part of the building. At least three groups of depositions can be distinguished. Pottery from the depositions includes a selected repertoire of the so-called Polizzello-Sant'Angelo Muxaro production: trefoil mouthed oinochoai, dippers, and cups. Large vessels such as pithoi, basins, and kraters imitate Greek prototypes. Some specimens of cups with terracotta figurines of a hut with animals inside are very significant on account of their rarity. Several Greek vessels were also found, mostly kylikes, ionic cups, kotylai, and a few kraters. In terms of metal items, the iron spears, sometimes of huge dimensions, are the most important items, followed by bronze arrows. Swords are entirely absent, and there are only a few daggers. Iron tools such as sickles and hoes are attested; axes are present only in bronze miniature versions. Bronze ornaments, such as bracelets, finger-rings, chains, and bronze foils are also represented as well as ivory, amber, and bone beads related to jewelery items. Furthermore, six regular-shaped lead ingots were recovered. Some of the most significant finds were an ivory and amber plate that was incised with a vegetal motif, and was probably part of the decoration of the lid of a wooden box; two sub-daedalic idols made of ivory; a couple of bronze dolphin laminae; and a bronze helmet of Cretan type decorated with a figure of hoplite. Although the previous excavations conducted in the area of buildings C and D had been very thorough, further investigation inside building D revealed a few vessels and several exotic items, mostly amber and ivory beads, dated to the seventh and sixth centuries BC. Furthermore, a second large building, called E (Sacello E), was found in the north western part of the acropolis, close to building A. Its 14.70 m diameter and megalithic construction technique is unique in Sicilian contexts, and it must therefore have been a structure of fundamental importance within the sacred area. This huge building may have been used at least through three different phases: in the second half of the eighth century BC, at the beginning of the seventh century BC, and between the end of the seventh and the beginning of the sixth century BC (Figure 1.12).

FIGURE 1.11
Polizzello Acropolis: (a) Buildings A, B, and E, view from south; (b) Building D view from north;
(c) Room III, view from south east; (d) Temenos House, view from east; (e) East House, digital
ortophotograph.

(a)   (b)   (c)

FIGURE 1.12
Building E: (a) phase I, second half of the eighth century BC; (b) phase II, beginning of the seventh century BC; (c) phase III, between the end of the seventh and the beginning of the sixth century BC.

1.6.2 Temenos and Room III

The sacred area represented by the central part of the acropolis, during the eighth century BC, was partially enclosed by a low and wide precinct wall of eccentric shape, a Temenos, well preserved along the eastern and northern slopes. It was made with stone blocks in double curtain technique (emplecton), with a constant width of 1.50 to 2.50 m. Totally obliterated on the western and southern sides, as a result of the slip of land and weather agency, it probably included a monumental entrance on the north side in front of building A, as shown by the particular bend of the wall and the construction technique of its northern limit. It is not excluded that a second main entrance could also be located to the south, where the circuit wall was damaged by building works and additional changes of later period. Another modification was the adjunction of a small rectangular ambient, later in chronology, Room III, partially built and excavated inside the Temenos, along its east side (Figures 1.11(c), 1.13(e), 1.13(f)).

1.6.3 Precinct F, East House, Temenos House

Research on the central part of the acropolis revealed a large and low rectangular precinct of 8.30 × 5.60 m, called Precint F (Recinto F) that contained traces of animal sacrifices, feasting activities, and pottery depositions that could be dated to the sixth century BC. The excavation of the Temenos and of the area east and north east of it, led to the discovery of a residential quarter represented by two houses, dated back to the end of the fifth century BC, namely the Temenos House (la Casa del Temenos), adjacent to the Temenos wall, and the East House (la Casa Est), in a smooth plain in eastern slope (Figure 1.13). The two buildings reflect the construction techniques of Greek Sicily and have such characteristics that they can be intended as part of a workshop district still partially unrevealed. The Temenos House (Figures 1.13(c), 1.13(d)), with a total length of 14.50 × 6 m, is adjacent to the exterior part the Temenos wall. The house consists of four rooms, with entrances to the east, and of an open area paved with cobblestones, on the north side. The construction technique shows large rectangular blocks with two courses and a flat roof with tiles. Within the rooms, rolled on the floor by the collapse of the ceiling, was found sets of pottery and objects, such as between transport amphorae, tablewares, cooking pots and storing jars, and especially some large mortars for grinding olives. Approximately 25 m northwest of the Temenos House was identified the second complex of the East House (Figures 1.11(e), 1.13(a), 1.13(b)), with an area of 10 × 2.50 m; it was built using the steep slope of the mountain in this area, using the rock as rear wall. It has the same kind of flat tiled roof and was accessible via a five step scale, located on the north side. Of the entire building just two connecting rooms were highlighted, the rectangular room A, to the south, 2.80 × 2.00, and a trapezoid room B 7.20 m long. Inside room A above the floor, many bronze arrowheads and large storing jars, amphoras, common wares, and mortars were found.

1.6.4   The Virtual Acropolis and the Multilayered 3D Model

The technique of 3D modeling in the case of the acropolis of Polizzello gave two completely innovative and extremely significant achievement, to augment the cognitive ability of the archaeologist: the promoting of an archaeological site and the monitoring of its conditions of degradation. Because of logistical problems, the Acropolis archaeological park will remain closed to the public for several years. Moreover some of the buildings (actual conditions of the site are shown in Figure 1.11), the oldest discovered in the exploration, were buried and not yet brought to light, so it was decided to create a virtual version of all the buildings of the acropolis in the same preservation conditions as at the time of the discovery. The "virtual acropolis," thus, allows the experience of visiting an interactive and immersive 3D environments.

The second result is to overcome all the traditional documentation methods of the many phases and chronological stages of a multilayered site like this, through color plant phase, with the creation of a multilayered 3D model, where all monumental evidences are organized into four temporal phases that summarize the history of the site between the eighth and the fourth century BC (Figure 1.14). In this way it's possible to focus better on the detailed planimetric evolution of the sanctuary between the eighth and sixth centuries BC and highlight the major functional difference of the last phase of the fourth century, that was only residential. With this methodology the overall development could be integrated with the architectural features of a single building, like building E, which itself has undergone many complex transformations. Finally, only for the last phase, for which the evidence revealed during the excavation was particularly significant, it has been possible to propose 3D models of how original buildings had to be, even on the basis of other evidence known from contemporary and historical sources, both for what concerns the structural data, the pottery, and objects that were used in everyday life.

1.7   Digital Restoration

The technique of 3D digital restoration of archaeological objects is perhaps the most common trend in interdisciplinary projects related to the interpretation and dissemination of archaeological knowledge. This is because of the potential that 3D has in subtracting the archaeological goods to the destructive effects of atmospheric agents, of pollution, of time, and, in some cases, of natural disasters and wars. The digital restoration is not only aimed to keep the archaeological goods from future risks, it is also suitable to return the conceptual and artistic integrity to monuments or complex objects, which in antiquity were considered as a unit and nowadays, for different reasons, are disassembled and divided between museums of different countries (a well known example is the complex of Parthenon sculptures [75]). It is also the chance to produce three-dimensional snapshots of specific phases of past life on the basis of the historical documentation of a monument [76]. The digital restoration is also the virtual version of the physical restoration, which for practical or economic reasons can not be made [77], which aims to give back a true image of a fragmented reality and, in some cases, wrongly recomposed [78]. In the last decade, in all the projects of digital archeology and also of 3D digital restoration, the reverse engineering approach had a large application in the policies of promotion of cultural goods [79–82, 82]. The high-definition 3D laser scanner is an instrument that collects 3D data from a given surface or object in a systematic, automated manner, at a relatively

FIGURE 1.13

(a) East House, virtual replica; (b) East House, 3D re-constructive model; (c) Temenos House, virtual replica (d) Temenos House, 3D re-constructive model; (e) Room III; (f) Room III 3D re-constructive model.

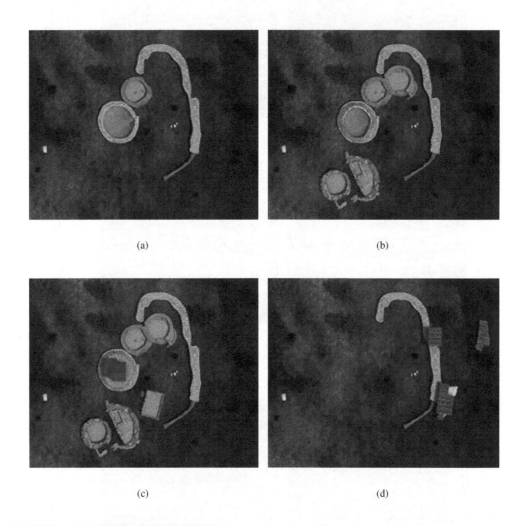

(a)   (b)

(c)   (d)

FIGURE 1.14 (SEE COLOR INSERT)
Multilayered 3D model of the Polizzello Acropolis: (a) phase I, half of the eighth century BC; (b) phase II, from the half of seventh to the beginning of sixth century BC; (c) phase III, end of sixth century BC; (d) phase IV, between the end of fifth and the beginning of fourth century BC.

high rate, in near real time using a laser ray to establish the surface coordinates. From a decade, this technology has been applied to archaeological research to construct geometric models with different characteristics [83, 84]. Most archaeological work has been carried out to digitize objects of an intermediate size, such as settlement structures, statues, and vessels. The most recent works have been focused on modeling structures during the excavation of archaeological sites, either of only one zone [11] or the complete ensemble [85]. These studies have been carried out from the ground surface or using helicopter and airplanes [86]. Traditional methods such as tapes and theodolites, and more modern technology such as total stations and GPS, provide accurate but relatively slow methods for gathering spatial data, and their use to scan small objects is not feasible.

The possibility to obtain in a limited amount of time a virtual exact replica of reality makes the laser scanning method ideal for studies of 3D digital restoration, where the virtual recomposition of fragmented elements, both physically and narratively, is fundamental [88]. The Archeomatica Project team of researchers in this field has proposed integrating the Blender-based 3D modeling and image-based 3D modeling with the laser scanning technique, in order to solve the possible data voids problems connected with complex scanning. The laser scanner hardware used is the relatively cheap Next Engine [89]. The choice was based essentially on the fact that it is very compact and handy, and then it proves to be very versatile especially when the objects to be scanned are placed in restricted spaces or cannot be removed. The case studies to test the 3D digital restoration, presented below, were chosen because they presented different levels of difficulty but also to demonstrate how the application of this technique has offered several new elements for the interpretation of objects and, sometimes, of their contexts.

1.7.1   Minoan Model

A good example of research in which 3D graphics techniques are applied to the problem of fragmentation is represented by an enigmatic small clay model, only partially preserved, from the site of Phaistos in Crete (Figure 1.15(a)), dating from the mid-fourth millennium BC (Early Minoan I-II in terms of Minoan chronology). The model was recovered in the south east room of the so-called Western Bastion, located on the northwestern side of Piazzale I [90]. A recent analysis of the piece [91], long known but ignored as incomprehensible at first glance because of its poor state, has led to its interpretation as a figurative scene of a religious nature. This interpretation was supported by its digital restoration the experts of Archeomatica Project have carried out. Basing on the conservation status of the object, it was possible to suggest the presence of two distinct anthropomorphic figurines, which body and long arms with large hands were stretched on the ground, and of a couple of cylindrical objects in front of the figurines (Figure 1.15(c)). The good level of knowledge achieved in the field of Minoan religion, and the several iconographical comparisons also present in other classes of artifacts, has led scholars to interpret the scene represented by the model as a scene of adoration of betili, large phallic stone symbols considered as bearers of good fortune, by two female figures in accordance with the custom of the Minoan religion [92]. This insight was made explicit through the creation of a 3D model, basing for the iconography of the female figures missing, the well known statue of the goddess of Myrtos [93], which is contemporary to the Phaistos model.

1.7.2   Asclepius

Another application of 3D digital laser scanning was the restoration of a colossal marble bust of the god Asclepius (Figure 1.16(a)) with an inscription [94], probably a copy of the I-II century AD of a late Hellenistic original and commonly considered one of the most representative pieces of the collection of Roman statuary of the Archaeological Museum "Paolo Orsi" of Syracuse [95–97]. The statue, 154 cm high, 90 cm wide, and 37 cm deep, kept for two fragments reassembled of torso and head, was found during the excavation of the foundations of the Spanish fortifications of Ortigia, around the mid-sixteenth century. Kept until 1810 in the Castle Maniace, the main Spanish fortress

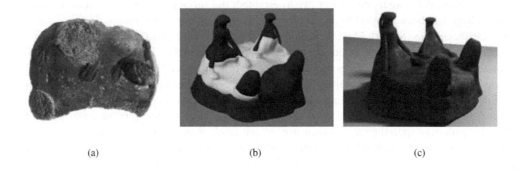

(a)   (b)   (c)

FIGURE 1.15
(a) Early Minoan clay model from Phaistos; (b) digital reintegration of the virtual model; (c) virtual
version of the clay model.

of Syracuse [98–100] and known also as "Don Marmoreo," the statue of the god was re-utilized
within the castle with radical changes, which altered the iconography, and probably it was placed in a
niche or fixed to a wall, as evidenced by the work of chipping and smoothing performed on the back.
These changes implied the engraving on his chest of a commemorative inscription in Spanish (Figure
1.17(a)), which remembers the granting of King Philip III to the noble ruler of the castle, John Rosa,
of dedicating, on 20 July 1618, the whole castle to St. James, and the four corner towers to patron
saints St. Peter, St. Catherine, St. Philip, St. Lucia, and the granting of the guns firing blanks at the
feast of the patron [95]. The epigraph is especially known for its transcripts made by scholars of the
last century. At present due to the deterioration of marble surfaces the inscription is almost illegible.

The statue of the Castle Maniace brings with it many issues of historical and archaeological
heritage, including the iconographical interpretation, the hypothesis of the presence of a cult of
Asclepius at Syracuse, the location of the temple which originally housed the statue, and the type of
alteration carried out in the Spanish ages. Although there was no recognizable signs on the statue
of the iconography of the god of medicine, son of Apollo, nor the caduceus and the serpent, and
although the beard and thick curly hair recall many canonical representations of divine brothers
Zeus, Hades, and Poseidon, a careful examination of certain stylistic features, such as facial lines
and strophion over his head, suggest a comparison with the statue of Asclepius of Mounichia, built
in the second century BC for the Asklepieion of Piraeus, and now kept in the National Museum
of Athens [96]. Chronologically closer is also the Roman Aesculapius of Villa Torlonia, of the
first imperial age, which shows the bearded god standing with caduceus, half naked, covered by an
himation supported by a prop [101]. The great reputation that the worship of God in the Greek world
had [102, 103], which devotion continued to the Roman Aesculapius, and the widespread presence
of his sanctuary in Sicily, often characterized by large architectural complexes as Agrigento [104]
and Eloro [105], all suggest that Syracuse, primate of the Western Greek city, was home for an
Asklepieion. Among the references of the literary sources, first of all is the quotation of Cicero in
the oration in Verrem (57.127-128) about the robbery made by Gaius Licinius Verres, propraetor of
Sicily from 73 to 71 BC, accused of the theft of a statue of Apollo Paian from the temple of Asclepius
at Syracuse. Moreover, the hypothesis that Asklepieion of Syracuse had its own place in Ortigia, as
the discovery of the statue Castle Maniace Asclepius would suggest, is further strengthened by the
discovery of a statue of Igea, daughter of Asclepius and associated for worship to the father as a god
of salvation, in the excavations of Piazza Pancali [106], and of an honorary inscription dedicated to
the city by a doctor found in the excavations of Corso Matteotti, probably, placed inside a sanctuary
area [107]. After the discovery by the Spanish, during his transfer to the Castle Maniace, because of

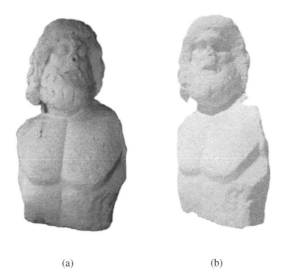

(a)   (b)

FIGURE 1.16

(a) Marble statue of Roman Republic period depicting the god Asclepius, from Syracuse Museum;
(b) 3D model of the statue obtained with laser scanning technique.

its large size, the statue probably had to be cut into separate blocks. In fact it is possible to see clearly the horizontal cuts at the neck, head, and half lower torso. On that occasion, were also likely to be selected only those parts necessary to build a herma, which suitably amended, was then relocated in the architecture of the castle. The most substantial change was the flattening of the chest to obtain a flat surface on which engrave the epigraph, evoking the idea of two pages of an open book.

The decision to produce a 3D model of the statue of Asclepius, using a laser scanner and improving the results with the 3D modeling during data processing, had two aims. First it was useful for increasing the knowledge about the artifact, investigating in detail the signs of ancient and modern working in order to correctly identify and distinguish the stylistic archetype, the changes made by the Spanish, and the restoring of the last century. Second, the creation of a virtual replica of the statue is a strong incentive for the promotion on multimedia, computer, and web totem, of this very fragile archaeological good, not suitable to be moved for exhibitions and cultural events. The 3D model of Asclepius was obtained by the technique of laser scanning, with a deliberate use, in this case, of an optical triangulation scanner rather than a time of flight (Figure 1.16(b)). The maneuverable and small scanner NextEngine has proved very useful for scanning the torso of Asclepius, as the statue, nontransferable elsewhere for the scan, was located very close to the wall of the exhibition hall of the museum, a fact that left little leeway to the operators of the laser scanner, allowing to capture correctly most of the details, leaving a small void area on the shoulder, extremely complex to acquire, and the back. Specific study of the result of 3D scanning and data processing has revealed some very significant changes to the original conditions of the statue. It is clear that the left deltoid has been removed as the upper left chest and lower abdominal part has been carved and smoothed. These operations, invisible to a simple analysis, should be interpreted as a conscious elimination of the drapery of his dress, which occurs constantly in the canonical representations of the god, in order to make the statue more symmetrical when, during the Spanish time, it became an inscribed herma (Figure 1.17(b)). It was also possible to observe how the reassembly of the head with the torso was a mistake that led to the anomalous downward inclination and slight twist to the left, unusual in ancient iconography, and probably due to incorrect restoration of the last century. The optimization

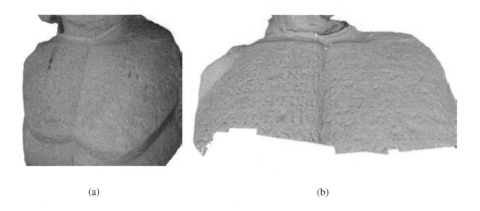

(a)   (b)

FIGURE 1.17
(a) Detail of the Spanish inscription on the chest of the Asclepius statue; (b) enhanced version of the inscription on the 3D model.

of the graphics gave a much more precise transcription of the engraved inscription. The work on the statue of the god Asclepius of Syracuse, however, is still a work in progress. In the future the research will include both the possibility of developing the model obtained to remove the erroneous restoration and attempts to bring back, based on stylistic evidence, the original iconography with which the god was represented before the Spanish era changes, also dealing with polychrome details, where these are documented.

1.7.3   Female Torso

The use of laser scanner and data processing of the data has significant applications in the digital restoration of individual archaeological artifacts, also for what concerns the clarification of certain technical or artistic features. A case study of this type was represented by the analysis on a portion of female torso in marble, dated between the middle and late Roman Empire (II-IV century AD), found in the excavations in the courtyard of the church of St. Agata la Vetere in Catania (Figure 1.18). The bust of $32 \times 18 \times 11$ cm, headless, retained only the left half, missing arms and the torso was preserved in the upper part. Near the sternum was evident that a sharp slit resulted in a lack of left side. Considering that the level of the excavation from which it came was related to a landfill and then the bust was not in primary position, its condition and typology seemed to be unusual. To better document the signs of working and to try through the modeling of 3D virtual models to propose a more precise interpretation, it was decided to acquire the bust with a laser scanner and then to operate on the model in Blender environment. The study and examination of the part under the cut, highlighted, in addition to signs of a progressive series of smoothing, the presence of a deep cut with a shallow central hole for housing a nail. The hypothesis was that the artifact was a kind of architectural appliquè in the form of semi-female bust; it would seem, however, refuted by the fact that the hole for the nail is very shallow and not sufficient to ensure a firm grip with a masonry wall. Furthermore, as the section is not vertical but curvilinear it is not possible to suggest any other hypotheses of reconstruction. A second hypothesis, more plausible, is that the bust was part of an acrolith of marble and wood, parts of which were made with small pieces of marble for reuse; this could justify the lack of depth and shape eccentric of the nail, which in this case would grip the wood. The analysis on the 3D model of the compatibility of the contact surfaces in the proximity of the neck and between the breasts, shows that probably the artifact was composed by two half parts. In

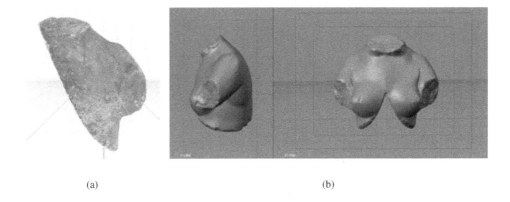

(a)   (b)

FIGURE 1.18

(a) Late Roman marble female torso from the excavation at St. Agata church at Catania; (b) phases of study and restoration analysis in virtual ambient of the torso acquired with laser scanning technique.

this case the hole for the plug would serve for a short nail that would clamp the two halves of the trunk between them. The peculiar shape of the chest can be justified suggesting the insertion of the wooden part of the acrolith, that also determined an uncommon chromatic effect for the Late Roman empire sculpture.

1.7.4   Hellenistic Thysia

This case is a true virtual replication of an architectural or archaeological reality aimed to preserve its condition before undergoing inevitable changes of time or human work. For these reasons, the technique of digital restoration finds wide application especially in the field of urban archeology where the stratification, the continuous building activity, and reduced chance to maintain accessible the discoveries become recurring features. In this perspective, the research carried out by the team of Archeomatica Project is set, in relation to some evidence emerged during archaeological excavations in Catania, between 2002 and 2005, in the complex adjoining the churches of St. Agata la Vetere [108] and St. Agata al Carcere [109], in the heart of downtown, where a continuous stratigraphy articulated from the Late Hellenistic period to the Middle Ages was identified [110]. Surveys held in conjunction with restoration of the two churches have revealed numerous monumental emergencies that were restored and preserved but, after the end of the works and the returning to the worship of the two buildings, cannot be showed to the public. During exploration [111], the most significant discovery was done in the sacristy of the Church of St. Agata al Carcere. Here, just below the modern pavement, in the midwest part, an accumulation of large rough blocks that seemed to follow a north-south alignment was unearthed (Figure 1.19(a)). The first hypothesis was that this accumulation was the result of the collapse of a great wall structure. It had an appreciable width of 2.25 m and a length of 2.80 m. The irregular blocks of lava stone of huge dimension were set as double curtains of a large wall filled with emplecton technique. It can be compared to other polygonal archaic walls dating from the first half of the sixth century BC, found in nearby excavations near the former Reclusorio della Puritá, inside the Roman theater, and in the cloister of the Benedictine monastery [112]. With the deepening of the investigation, in a small hole, filled with ashes and a few minute burnt animal bones, were found in situ, 2 two-handled olle and large two-handled stamnos with discoid lid, which suggest a chronological framework in the second half of the third century BC (Figure 1.7.4). This thysia finds comparisons in other part of the city, such as excavation yet unpublished at Palazzo Sangiuliano, or

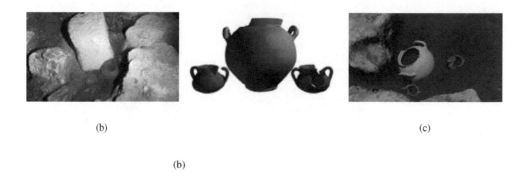

(b)   (c)

(b)

FIGURE 1.19
(a) Hellenistic votive deposit from the excavation at St. Agata church at Catania, at the moment of the discovery; (b) 3D models of three vessels of the votive deposit obtained with laser scanning technique; (c) virtual model of the excavation area depicting the discovery conditions of the votive deposit obtained integrating laser scanning and 3D modeling techniques.

more properly as in the above-mentioned complex of Reclusorio della Purità [113, 114], where there were found some votive pits dug into a pavement related to polygonal wall, containing two-handled olle and ashes. This Hellenistic depositional practice is well documented also in Syracuse in the sacred area near Altar of Hiero II [115]. For the Catania context, this means that in the Hellenistic age the ruins of the archaic city were still visible and considered sacred as to perform specific rituals related to them, rituals of remembering the ancient tradition of the city. To better analyze this significant evidence, extrapolating from the rest of the architectural complex, it was decided to operate the digital restoration of the environment of the sacristy, of the ruins of the ancient city wall, and of the Hellenistic deposition, using a hybrid technique, the 3D modeling in Blender environment for the walls and the laser scanning for the vessels of the deposit. The integration of the two techniques has allowed, on one hand, to assess the strengths and weaknesses and opportunities of integration and on the other provided a virtual model that offers a snapshot of the Hellenistic depositional area on the ruins of the Archaic wall, as it was when it was used.

1.8   Dealing with Image Data in Archaeology: New Perspectives

The future of virtual archaeology and of its applications in the field of research and promotion of archaeological goods seems to be linked in an unavoidable way to the evolution of the virtual reality (VR) environment [116]. The virtual reality allows the 3D visualization of concepts, objects, or spaces and their contextualization through the creation of a visual framework in which data is displayed. VR also enables interaction with data organized in 3D, facilitating the interaction between human, data, and information in order to enhance the sensorial perception [117]. It creates a virtual space that is a replica of the real space, where the information about every feature that constituted the different moments of life of the real space are "translated" in 3D data. The two crucial points of every project of VR are the selection of the informations (pictures, drawings, geometrical measures) and the choice of which facet of the original object's nature must be captured and reconstructed. "Visual computer models should make clear their sources and the criteria on which they are based" [118].

From this point of view, VR is not a copy of the "reality," but the representation of "one," or, several "instances," possibilities among others, various under different circumstances and contexts, even if it can offer a multilayered reality experience, a kind of digital surrogate of the real world [116]. It has the cohesive function of relating the raw data to the interpretation [35] and "the ability to get inside and walk around the reconstruction buildings gave a stronger feeling of enclosed space and volume, and enhance the sense of being there" [119]. The most important limit in the mimetic ability of the past and present reality of the VR is, in fact, represented by the cultural formation of the authors of the project, that in many cases are archaeologists. Diverse point of view can, in fact, affect the selection of the informations and produce very different results in terms of visual image of the past [120]. In order to understand archaeological systems, much more than a visually "realistic" geometric model is needed. "Dynamism and interaction" are essential. A dynamic model is a model that changes in position, size, material properties, lighting, and viewing specification. If those changes are not static but respond to user input, we enter into the proper world of virtual reality, whose key feature is real-time (RT) interaction. Here real-time means that the computer is able to detect input and modify the virtual world "instantaneously" at user commands. By selectively transforming an object, that is, by interpolating shape transformations, archaeologists may be able to form an object hypothesis more quickly [121]. The next step, on the same research direction, is the realization of systems of virtual experience of the recreated past, beyond the physical limits of the personal computer. "To make an archaeological excavation simulation compatible with the dynamics of the learning environment, the user must be able to navigate within it without being entirely bound to specified pathways, as typified by menu-driven routines or the buttons and links of a hyperspace environment" [122]. One field where the scholars in archaeology and computer science are recently getting involved is augmented reality (AR) or enhanced reality (ER) environments. Augmented reality has been defined as the simultaneous acquisition of supplemental virtual data about the real world while navigating around a physical reality [123]. In an AR environment the computer provides additional information that enhances or augments the real world, rather than replacing it with a completely virtual environment [124]. One of the objectives of AR is to bring the computer out of the desktop environment and into the world of the user working with a three-dimensional application. In contrast to VR, where the user is immersed in the world of the computer, AR incorporates the computer into the reality of the user. The user can then interact with the real world in a natural way, with the computer providing information and assistance. It is then a combination of the real scene viewed by the user and a virtual scene generated by the computer that augments the scene with additional information. The virtual world acts as an interface, which may not be used if it provides the same experience as face-to-face communication. AR enables users to go "beyond being there" and enhance the experience in order to achieve both the full interpretation of the traces of the past and the developing of the best tool for the dissemination of their message [125]. An example of the potential of this technology applied to archaeology is the augmented reality-based cultural heritage On-site Guide, acronym ARCHEOGUIDE, that provides new ways of information access at Greek archaeological sites in a compelling, user-friendly way through the use of 3D-visualization, mobile computing, and multi-modal interaction [126, 127]. The potential of this approach in the future could be enhanced by investing much more in the five fundamental elements of an AR environment, namely virtuality (objects that don't exist in the real world can be viewed and examined), augmentation (real objects can be augmented by virtual annotations), cooperation (multiple users can see each other and cooperate in a natural way), independence (each user controls his own independent viewpoint), and individuality (displayed data can be different for each viewer) [128]. In conclusion, the encouraging results of the application of the computer graphics 3D to the archaeological evidence has demonstrated that it is possible to use another "sense" do decrypt the traces of the past: three-dimensional recreation of ancient life and visual images are extremely effective in explaining the past because they allow us to experience it.

Acknowledgments

These Archeomatica Project members have participated: C. Falzone, A. Fiamingo, E. Greco, M. Marino, G. Mercadante, S. Provvidenza, E. Sangregorio, T. Scaffiddi, L. Truppia, R. Urso.

Bibliography

[1] E. B. W. Zubrow, *Digital Archaeology. A Historical Context.* in T. L. Evans, P. Daly (eds.), Digital Archaeology. Bridging Method and Theory, Routledge, London, 2006.

[2] G. Vannini, "Informatica per l'Archeologia o Archeologia per l'Informatica?," *Archeologia e Calcolatori*, vol. 11, pp. 311–315, 2000.

[3] P. Daly and T. L. Evans, *Archaeological Theory and Digital Pasts.* in T. L. Evans, P. Daly (eds.), Digital Archaeology. Bridging Method and Theory, Routledge, London, 2006.

[4] H. Eiteljorg, *Archaeological Computing.* II edition, Center for the Study of Architecture, Bryn Mawr, 2008.

[5] T. Orlandi, "Archeologia Teorica e Informatica Archeologica. Un Rapporto Difficile," *in Archeologia e Calcolatori*, vol. 15, pp. 41–50, 2004.

[6] M. Forte and R. Beltrami, "A Proposito di Virtual Archaeology: Disordini, Interazioni Cognitive e Virtualità," *in Archeologia e Calcolatori*, vol. 11, pp. 273–300, 2000.

[7] P. Reilly, *Towards a Virtual Archaeology.* in K. Lockyear, S. Rahtz (eds.), Computer Applications and Quantitative Methods in Archaeology 1990, BAR International Series 565, Oxford, 1990.

[8] M. Cultraro, F. Gabellone, and G. Scarrozzi, "The Virtual Musealization of Archaeological Sites: Between Documentation and Comunication," in F. Remondino, S. El-Hakim, L. Gonzo (eds.), *Proceedings of the 3rd ISPRS International Workshop 3D-ARCH 2009 "3D Virtual Reconstruction and Visualization of Complex Architectures," Trento, Italy, 25-28 February 2009, International Archives of Photogrammetry, Remote Sensing and Spatial Information Sciences*, vol. XXXVIII-5/W1, 2009.

[9] M. Cultraro, F. Gabellone, and G. Scardozzi, "Integrated Methodologies and Technologies for the Reconstructive Study of Dur-Sharrukin (Iraq)," in F. Remondino, S. El-Hakim, L. Gonzo (eds.), *Proceedings of the 3rd ISPRS International Workshop 3D-ARCH 2009 "3D Virtual Reconstruction and Visualization of Complex Architectures," Trento, Italy, 25-28 February 2009, International Archives of Photogrammetry, Remote Sensing and Spatial Information Sciences*, vol. XXXVIII-5/W1, 2009.

[10] V. Stojakovic and B. Tepavcevica, "Optimal Methods for 3D Modeling of Devastated Architectural Objects," in F. Remondino, S. El-Hakim, L. Gonzo (eds.), *Proceedings of the 3rd ISPRS International Workshop 3D-ARCH 2009 "3D Virtual Reconstruction and Visualization of Complex Architectures," Trento, Italy, 25-28 February 2009, International Archives of Photogrammetry, Remote Sensing and Spatial Information Sciences*, vol. XXXVIII-5/W1, 2009.

[11] M. Doneus and W. Neubauer, *Laser Scanners for 3D Documentation of Stratigraphic Excavations*. in M. Baltsavias, A. Gruen, L. Van Gool, M. Pateraki (eds.), Recording, Modeling and Visualization of Cultural Heritage, London, Taylor and Francis, 2006.

[12] M. Peripimeno, "Rilievo di Monumenti e Stratigrafie. L'Uso del Laser Scanner," in V. Fronza, A. Cardini, M. Valenti (eds.) *Informatica e Archeologia Medievale. L'esperienza senese, All'Insegna del Giglio, Siena*, pp. 111–129, 2009.

[13] T. E. Levy and N. G. Smith, "On-site Digital Archaeology: GIS-based Excavation Recording in Southern Jordan," in T. E. Levy, M. Daviau, R. Younker, M. M. Shaer (eds.), *Crossing Jordan. North American Contributions to the Archaeology of Jordan, Equinox, London*, pp. 47–58, 2007.

[14] S. Moser, *Archaeological Representation. The virtual Conventions for Constructing Knowledge about the Past*. in I. Hodder (ed.), Archaeological Theory Today, Polity Press, Malden, 2005.

[15] D. Pletinckx, "Virtual Archaeology as an Integrated Preservation Method," in *"Arqueologica 2.0, Proceedings of 1st International Meeting on Graphic Archaeology and Informatics, Cultural Heritage and Innovation," Seville 17-20 June 2009*, pp. 51–55, 2009.

[16] N. Toganidis, "Parthenon Restoration Project," in A. Georgopoulos, N. Agriantonis (eds.), *Anticipating the Future of the Cultural Past, Proceedings of the XXI International CIPA Symposium, Athens*, pp. 1–6, 2007.

[17] F. Niccolucci, "Virtual Museums and Archaeology: an International Perspective," in *Archeologia e Calcolatori, suppl. 1*, pp. 15–30, 2007.

[18] E. Bacci, S. Boni, T. Canonici, V. D. Pozzo, and A. Ribatti, "L'utilizzo della Ricostruzione nella Comunicazione del Patrimonio Archeologico. L'Approccio, il Metodo, le Finalità e Alcuni Spunti di Discussione," in *"Arqueologica 2.0, Proceedings of 1st International Meeting on Graphic Archaeology and Informatics, Cultural Heritage and Innovation," Seville*, pp. 405–409, 2009.

[19] F. Antinucci, "The Virtual Museum," *in Archeologia e Calcolatori, suppl. 1*, pp. 79–86, 2007.

[20] http://www.virtualmuseumiraq.cnr.it

[21] S. Chiodi, "Iraq Project: The Virtual Museum of Baghdad," in *Archeologia e Calcolatori, suppl. 1*, pp. 101–122, 2007.

[22] F. Gabellone and G. Scardozzi, "From Object to the Territory: Image-based Technologies and Remote Sensing for the Recostruction of Ancient Contexts," in *Archeologia e Calcolatori, suppl. 1*, pp. 123–142, 2007.

[23] P. Bellintani and L. Moser, *Archeologie Sperimentali – Metodologie ed Esperienze fra Verifica, Riproduzione, Comunicazione e Simulazione, Atti Convegno Comano Terme – Fiavé (Trento, Italy)*. 2003.

[24] G. Thomas, *Experimental Archaeology*, Routledge, London, 2009.

[25] L. Longo, "Archeologia Sperimentale, Esperimenti in Archeologia, Divulgazione. Osservazioni su Significato e Ruolo dell'Archeologia Sperimentale," in *Rivista Scienze Preistoriche*, no. LIII, pp. 549–568, 2003.

[26] J. Coles, *Archeologia sperimentale*, Longanesi, Milano, 1981.

[27] M. Moser, S. Hye, G. Goldenberg, K. Hanke, and K. Kovacs, "Digital Documentation and Visualization of Archaeological Excavations and Finds Using 3D Scanning Technology," in *Arqueologica 2.0, Proceedings of 1st International Meeting on Graphic Archaeology and Informatics, Cultural Heritage and Innovation*, pp. 351–355, June 2009.

[28] D. C. Dennett, "Seeing Is Believing," in K. Akins ed., *Perception, Vancouver Studies in Cognitive Science,* Oxford University Press, vol. 5, pp. 158–172, 1996.

[29] B. Frischer, "Art and Science in the Age of Digital Reproduction: From Mimetic Representation to Interactive Virtual Reality," in *Arqueologica 2.0, Proceedings of 1st International Meeting on Graphic Archaeology and Informatics, Cultural Heritage and Innovation*, pp. 35–48, June 2009.

[30] http://www.archeomatica.unict.it

[31] http://iplab.dmi.unict.it

[32] D. Margounakis, *Virtual Reconstructions in Archaeology.* in D. Politis ed., E-Learning Methodologies and Computer Applications in Archaeology, 2008.

[33] E. F. Anderson, L. McLoughlin, F. Liarokapis, C. Peters, P. Petridis, and S. de Freitas, "Serious Games in Cultural Heritage," in *Procedings of the 10th International Symposium on Virtual Reality, Archaeology and Cultural Heritage VAST,* M. Ashley, F. Liarokapis (eds.), State of Arts and Reports, pp. 29–48, 2009.

[34] F. Salvadori, "Modellazione dei Reperti," in V. Fronza, A. Cardini, M. Valenti (eds.) *Informatica e Archeologia Medievale. L'esperienza senese,* All'Insegna del Giglio, pp. 131–147, 2009.

[35] P. Reilly, *Three Dimensional Modeling and Primary Archaeological Data.* in P. Reilly, S. Rahtz (eds.), Archaeology and the Information Age: A Global Perspective, 1992.

[36] http://www.blender.org

[37] F. Remondino and S. El-Hakim, *A Critical Overview of Image-based 3D Modeling.* In M. Baltsavias, A. Gruen, L. Van Gool, M. Pateraki (eds.), Recording, Modeling and Visualization of Cultural Heritage, London, Taylor and Francis, 2005.

[38] F. Verbiest, G. Willems, D. Pletincky, and L. Van Gool, *Image-based Rendering of Cultural Heritage.* in M. Baltsavias, A. Gruen, L. Van Gool, M. Pateraki (eds.), Recording, Modeling and Visualization of Cultural Heritage, London, Taylor and Francis, 2005.

[39] I. Roussos and A. Chalmers, "High Fidelity Lighting of Knossos," in D. Arnold, A. Chalmers, F. Niccolucci (eds.), *4th International Symposium on Virtual Reality, Archaeology and Intelligent Cultural Heritage, VAST*, pp. 195–201, 2003.

[40] R. Ercek, D. Viviers, and N. Warzée, "3D Reconstruction and Digitalization of an Archaeological Site, Itanos, Crete," in *Arqueologica 2.0, Proceedings of 1st International Meeting on Graphic Archaeology and Informatics, Cultural Heritage and Innovation*, pp. 289–294, 2009.

[41] K. Papadopoulos and G. P. Earl, "Structural and Lighting Models for the Minoan Cemetery at Phourni, Crete," in Perlingieri, C. and Pitzalis, D. (eds) *Proceedings of the 10th International Symposium on Virtual Reality, Archaeology and Cultural Heritage VAST*, pp. 57–64, 2009.

[42] V. La Rosa, *La "Villa Royale" d'Haghia Triada.* in R. Hgg (ed.), The Function of the "Minoan Villa," Astrom editions, 1997.

[43] V. La Rosa, *Haghia Triada à l'Époque Mycénienne: l'Utopie d'une Ville Capitale.* in J. Driessen, A. Farnoux (eds.), La Crète Mycénienne (BCH Suppl. 30), De Boccard, 1997.

[44] P. Rehak and J. Younger, *Review of Aegean Prehistory VII : Neopalatial, Final Palatial, and Postpalatial Crete.* in T. Cullen (ed.), Aegean Prehistory, A Review, Boston, Archaeological Institute of America, 2001.

[45] J. Bennet, "The Structure of the Linear B Administration at Knossos," *in American Journal of Archaeology*, vol. 89, pp. 231–249, 1985.

[46] J. Driessen, *Centre and Periphery: Some Observations on the Administration of the Kingdom of Knossos.* in S. Voutsaki, J. Killen (ed.), Economy and Politics in the Mycenaean Palace States, Cambridge, Cambridge Philological Society, 2001.

[47] S. Privitera, "Looking for a Home in a Houseless Town: Domestic Architecture at Final Palatial Hagia Triada," in K. Glowacki, N. Vogeikoff-Brogan (eds.), *The Archaeology of Houses and Households in Ancient Crete*, 2005.

[48] V. La Rosa, "Il Colle sul quale Sorge la Chiesa ad Ovest è tutto Seminato di Cocci." Vicende e Temi di uno Scavo di Lungo Corso, *in Creta Antica*, vol. 4, pp. 11–68, 2003.

[49] http://digitalcrete.ims.forth.gr

[50] A. Sarris, V. Trigkas, G. Papadakis, M. Papazoglou, E. Peraki, N. Chetzogiannaki, M. El-vanidou, E. Karimali, K. Kouriati, M. Katifori, G. Kakoulaki, E. Kappa, K. Athanasaki, and N. Papadopoulos, "A web GIS Approach for the Cultural Resources Management of Crete: The Digital Archaeological Atlas of Crete," in A. Posluschn, K. Lambers, I. Herzog (eds.), *Layers of Perception. Proceedings of the 35th International Conference on Computer Applications and Quantitative Methods in Archaeology (CAA)*, 2008.

[51] D. Gutierrez, V. Sundstedt, F. Gomez, and A. Chalmers, "Dust and Light: Predictive Virtual Archaeology," in *Journal of Cultural Heritage*, no. 8, pp. 209–214, 2007.

[52] J. Happa, M. Mudge, K. Debattista, A. Artusi, A. Gonalves, and A. Chalmers, "Illuminating the Past - State of Art," in *Procedings of the 10th International Symposium on Virtual Reality, Archaeology and Cultural Heritage VAST*, M. Ashley, F. Liarokapis (eds.), State of Arts and Reports, pp. 9–28, 2009.

[53] http://radsite.lbl.gov/radiance

[54] E. Sangregorio, F. Stanco, and D. Tanasi, "The Archeomatica Project: Towards a New Application of Computer Graphics in Archaeology," in *Proceedings of 6th Eurographics Italian Chapter Conference*, pp. 1–5, July 2008.

[55] F. Stanco and D. Tanasi, "La Computergrafica nella Ricerca Archeologica. Dal 3D Modeling alla Digital Archaeology," in *Proceedings of V Congresso Nazionale di Archeometria*, pp. 605–617, 2009.

[56] V. La Rosa, "Considerazioni sull'Area ad Ovest del c.d. Bastione ad Haghia Triada," *in ASAtene LXXXIV, s. III, 6, tomo II*, pp. 819–877, 2006.

[57] P. Militello, "Un nuovo Propylon ad Hhaghia Triada e gli Spazi Recintati nella Creta Neopalaziale," *Creta Antica*, vol. 9, pp. 11–18, 2008.

[58] L. A. Hitchcock, *Minoan Architecture: A Contextual Analysis.* Astrom Editions, 2000.

[59] J. W. Shaw, "Minoan Architecture: Materials and Techniques," *Studi di Archeologia Cretese, Aldo Ausilio Editore*, vol. VII, 2009.

[60] P. Warren, *Minoan Stone Vases*, Cambridge University Press, 1969.

[61] M. Hue, *Les Représentations Architecturales Dans la Peinture Murale Égéenne de l'Age du Bronze, l'Exemple d'Akrotiri (Thera)*. In "Architecture et poésie dans le monde grec: hommage à George Roux," Étienne R., Le Dinahet M., Yon M., (eds.), Maison de l'Orient, 1989.

[62] C. Boulotis, *Villes et Palais dans l'Art Égéen du IIe Millénaire Av. J.-C*. In "L'habitat égéen préhistorique," Darcque P., Treuil R., De Boccard, (eds.) 1990.

[63] F. Stanco and D. Tanasi, "Reconstructing the Past: il 3D Modeling nella Ricerca Archeologica," in L. Bezzi, D. Francisci, P. Grossi, D. Lotto (eds.), *Atti del 3 Workshop Open Source, Free Software e Open Format nei processi di ricerca archeologica*, 2008.

[64] V. La Rosa, "Recenti Acquisizioni nel Settore Nord dell'Abitato di Haghia Triada," in *Atti VI Congr. Intern. Studi Cretesi*, pp. 411–418, 1990.

[65] P. Darcque, *L'Habitat Mycénien. Formes et Fonctions de l'Espace Bti en Grèce Continentale à la Fin du IIe Millénaire Avant J.-C*. De Boccard, 2006.

[66] F. Sigaut, "A Method for Identifying Grain Storage Techniques and its Application for European Agricultural History," in *Tools and Tillage*, vol. 6, pp. 3–32, 1998.

[67] S. Privitera, "The LM III Frescoes from the Villaggio at Hagia Triada: New Observations on Context and Chronology," *Creta Antica*, vol. 9, pp. 111–137, 2008.

[68] G. Gesell, "The Minoan Snake Tube. A Survey and Catalogue," *American Journal of Archaeology*, vol. 80, pp. 247–259, 1976.

[69] D. Tanasi, "Chapter 9. A Late Bronze Age Upland Sanctuary in the Core of Sikania?," in M. Fitzjohn (ed.), *Uplands of Ancient Sicily and Calabria. The archaeology of landscape revisited, Accordia Specialist Study on Italy*, vol. 13, pp. 157–170, 2007.

[70] D. Palermo, E. Pappalardo, and D. Tanasi, "Le Origini di un Santuario," in *Atti del convegno di studi EIS EKRA. Insediamenti d'altura in Sicilia dalla preistoria al III secolo a.C.*, pp. 47–78, 2008.

[71] R. P. C. Guzzone, D. Palermo, *Montagna di Polizzello. Campagna di Scavo 2004*. Betagamma Editrice, Viterbo.

[72] M. R. Matini, A. Einifar, A. Kitamoto, and K. Ono, "Digital 3D Reconstruction Based on Analytic Interpretation of Relics: Case Study of Bam Citadel," in Y. Takase (ed.), *Digital Documentation, Interpretation & Presentation of Cultural Heritage, Proceedings of the XXII International CIPA Symposium*, pp. 1–6, 2009.

[73] R. Guglielmini, "Poggioreale Old Town in Sicily: Strategies, Memory and Planning Place," in A. Georgopoulos, N. Agriantonis (eds.), *AntiCIPAting the Future of the Cultural Past, Proceedings of the XXI International CIPA Symposium*, pp. 1–6, 2007.

[74] G. Toubekis, I. Mayerb, M. Doering-Williams, K. Maeda, K. Yamauchi, Y. Taniguchi, S. Morimoto, M.Petzet, M.Jarke, and M. Jansen, "Preservation and Management of the Unesco World Heritage Site of Bamiyan: Laser Scan Documentation and Virtual Recostruction of the Destroyed Buddha Figures and the Archaeological Remains," in Y. Takase (ed.), Digital Documentation, Interpretation & Presentation of Cultural Heritage, Proceedings of the XXII International Cipa Symposium, pp. 1–6, 2009.

[75] J. Stumpfel, C. Tchou, T. Hawkins, P. M. B. Emerson, M. Brownlow, A. Jones, N. Yun, and P. Debevec, "Digital Reunification of the Parthenon and its Sculptures," in D. Arnold, A. Chalmers, F. Niccolucci (eds.), *4th International Symposium on Virtual Reality, Archaeology and Intelligent Cultural Heritage, VAST*, pp. 1–10, 2003.

[76] G. Bitelli, V. A. Girelli, M. Marziali, and A. Zanutta, "Use of Historical Images for the Documentation and the Metrical Study of Cultural Heritage by Means of Digital Photogrammetric Techniques," in A. Georgopoulos, N. Agriantonis (eds.), *Anticipating the Future of the Cultural Past, Proceedings of the XXI International CIPA Symposium*, pp. 1–6, 2007.

[77] A. C. R. Eppich, "Recording and Documenting Cultural Heritage 3D Modeling for Conservation in Developing Regions," in M. Baltsavias, A. Gruen, L. Van Gool, M. Pateraki (eds.), *Recording, Modeling and Visualization of Cultural Heritage*, London, Taylor and Francis, pp. 11–20.

[78] http://formaurbis.stanford.edu, "Stanford Digital Forma Urbis Romae Project."

[79] M. Callieri, P. Cignoni, F. Ganovelli, G. Impoco, C. Montani, P. Pingi, and R. S. F. Ponchio, "Visualization and 3D Data Processing in David's Restoration," vol. 24, pp. 16–21, March/April 2004.

[80] V. Valzano, A. Bandiera, and J. A. Beraldin, "Realistic Representations of Cultural Heritage Sites and Objects Through Laser Scanner Information," in *The 10th International Congress "Cultural Heritage and New Technologies."* Vienna, Austria, November 2005.

[81] J. Ristevski, "Feature: Laser Scanning for Cultural Heritage Applications," vol. 26, November 2006.

[82] T. Abmayr, F. Hartl, M. Reinkoster, and C. Frohlich, "Terrestrial Laser Scanning–Applications in Cultural Heritage Conservation and Civil Engineering," in *Proceedings of ISPRS Congress 2008*, vol. XXXVII, 2008.

[83] D. Peloso, "Tecniche Laser Scanner per il Rilievo dei Beni Culturali," *in Archeologia e Calcolatori*, vol. 16, pp. 199–224, 2005.

[84] W. Boeheler, "Comparison of 3D Laser Scanning and other 3D Measurement Techniques," in M. Baltsavias, A. Gruen, L. Van Gool, M. Pateraki (eds.), *Recording, Modeling and Visualization of Cultural Heritage*, London, Taylor and Francis, pp. 89–100.

[85] T. Gaisecker, *Pinchango Alto. 3D Archaeology Documentation Using the Hybrid 3D Laser Scan System of RIEGL.* in M. Baltsavias, A. Gruen, L. Van Gool, M. Pateraki (eds.), Recording, Modeling and Visualization of Cultural Heritage, London, Taylor and Francis, 2006.

[86] M. Doneus, C. Brieseb, M. Feraa, U. Fornwagnera, M. Griebla, M. Jannera, and M. C. Zingerlea, "Documentation and Analysis of Archaeological Sites Using Aerial Reconnaissance and Airborne Laser Scanning," in A. Georgopoulos, N. Agriantonis (eds.), *Anticipating the Future of the Cultural Past, Proceedings of the XXI International CIPA Symposium*, pp. 1–6, 2007.

[87] W. Boeheler and A. M. G. Heinz, "The Potential of Non-contact Close Range Laser Scanners for Cultural Heritage Recording," in J. Albertz, M. Petzet, J. Haspel eds., *Surveying and Documentation of Historic Buildings - Monuments - Sites Traditional and Modern Methods, in Proceedings of XVIII International CIPA Symposium*, pp. 1–8, September 2001.

[88] K. Cain, C. Sobieralski, and P. Martinez, "Reconstructing a Colossus of Ramesses II from Laser Scan Data," in *SIGGRAPH '03: ACM SIGGRAPH 2003 Sketches & Applications*, (New York), pp. 1–1, ACM, 2003.

[89] http://www.nextengine.com

[90] F. Carinci and V. L. Rosa, "Revisioni Festie II," in *Creta Antica*, vol. 10/I, pp. 147–300, 2009.

[91] V. La Rosa, *A New EM Clay Model from Phaistos*. in O. Krzyszkowska ed., Cretan Offerings, Studies in Honour of Peter Warren (BSA Suppl. Studies).

[92] M. Gallo, "Per una Riconsiderazione del Betilo in Ambito Minoico," *in Creta Antica*, vol. 6, pp. 47–58, 2005.

[93] P. Warren, *Myrtos: An Early Bronze Age Settlement in Crete*. 1972.

[94] L. Arcifa, D. Calì, A. Patanè, F. Stanco, D. Tanasi, and L. Truppia, "Laserscanning e 3D Modeling nell'Archeologia Urbana: lo Scavo della Chiesa di Sant'Agata al Carcere a Catania (Italia)," in *Arqueologica 2.0, Proceedings of 1st International Meeting on Graphic Archaeology and Informatics, Cultural Heritage and Innovation*, pp. 301–305, June 2009.

[95] G. Libertini, *Il Regio Museo Archeologico di Siracusa*. La Libreria dello Stato, Roma, 1929.

[96] G. Caputo, "Note alle Sculture del Museo Siracusano," *in Bollettino d'Arte*, vol. 29, pp. 420–423, 1939.

[97] A. Crispino and A. Musumeci, *Musei nascosti. Collezioni e Raccolte Archeologiche a Siracusa dal XVIII al XX secolo*, Electa, Napoli, 2009.

[98] G. Giarrizzo, *Dal Cinquecento all'Unità d'Italia*. in D'Alessandro V., Giarrizzo G. (eds.), La Sicilia dal Vespro all'Unità d'Italia, Utet, Torino, 1989.

[99] L. Cassataro, *Il Castello di Federico II a Siracusa: Guida al Monumento*, Arnaldo Lombardi Editore, Palermo, 1997.

[100] F. Maurici, *Castelli Medievali di Sicilia*, Regione Sicilia, Palermo, 2001.

[101] C. Gasparri, "Materiali per Servire allo Studio del Museo Torlonia di Scultura Antica," in *Atti dell'Accademia Nazionale dei Lincei*, vol. 24, pp. 35–238, 1980.

[102] G.D. Hart, and M.S.J. Forrest, *Asclepius: The God of Medicine*, RSM Press, London, 2000.

[103] M. Melfi, *I Santuari di Asclepio in Grecia*, Roma L'Erma di Bretschneider, 2007.

[104] E. De Miro, *Agrigento 2. I Santuari extraurbani. L'Asklepieion*. Rubbettino, Soveria Mannelli, 2003.

[105] G. Voza, *Nel Segno dell'Antico*, Arnaldo Lombardi Editore, Palermo, 1999.

[106] P. Orsi, "Siracusa. II. Scoperta di Due Statue nella Città," in *Notizie degli Scavi di Antichità*, pp. 338–343, 1901.

[107] G. Cultrera, "Gli Antichi Ruderi di Via del Littorio," in *Notizie degli Scavi di Antichità*, pp. 225–226, 1940.

[108] A. Patanè and D. Tanasi, "Ceramiche Fini dai Livelli Tardo Romani degli Scavi 2003-2004 a Sant'Agata la Vetere (Catania),"in Malfitana, D., Poblome, J., Lund, J. (eds.), *Old Pottery in a New Century. Innovating Perspectives on Roman Pottery Studies. Consiglio Nazionale delle Ricerche, Catania*, pp. 465–475, 2006.

[109] D. Tanasi and D. Calì, "Sicilia. Catania, S. Agata la Vetere e S. Agata al Carcere," in *Archeologia Medievale*, vol. 23, pp. 429–431, 2006.

[110] L. Arcifa, "Dalla Città Bizantina alla Città Normanna: Ipotesi sullo Sviluppo Urbanistico di Catania in Età Medievale," in Casamento, A., Guidoni E. (eds.), *Storia dell'urbanistica/Sicilia IV. Le città medievali dell'Italia meridionale e insulare. Kappa. Roma*, vol. 23, pp. 279–291, 2004.

[111] A. Patanè, D. Tanasi, and D. Calì, "Indagini Archeologiche a S. Agata la Vetere e S. Agata al Carcere," in V. La Rosa (ed.), Tra lava e mare. Contributi all'archaiologhia di Catania, 2007.

[112] E. Tortorici, "Osservazioni e Ipotesi sulla Topografia di Catania Antica," in *Edilizia Pubblica e Privata nelle Città Romane* in Atlante tematico di topografia antica, Atta 17, 2008.

[113] M. G. Branciforti, *Gli Scavi Archeologici nell'ex Reclusorio della Purità di Catania.* in Megalai Nesoi (ed.). Studi dedicati a Giovanni Rizza per il suo ottantesimo compleanno. Consiglio Nazionale delle Ricerche, Catania, 2005.

[114] S. Amari, *Materiali per la Datazione dello Scavo Condotto all'Interno dell'ex Reclusorio della Purità a Catania* in Megalai Nesoi. Studi dedicati a Giovanni Rizza per il suo ottantesimo compleanno. Consiglio Nazionale delle Ricerche, Catania, 2005.

[115] B. V. Gentili, *Siracusa, Ara di Ierone. Campagna di scavo 1950-1951* in Notizie degli Scavi di Antichità, 1954.

[116] M. Mudge, M. Ashley, and C. Schroer, "A Digital Future for Cultural Heritage," in A. Georgopoulos, N. Agriantonis (eds.), *Anticipating the Future of the Cultural Past, Proceedings of the XXI International CIPA Symposium, Athens*, pp. 1–6, 2007.

[117] S. Hermon and L. Kalisperis, "Between the Real and the Virtual: 3D Visualization in the Cultural Heritage Domain - Expectations and Prospects," in *Arqueologica 2.0, Proceedings of 1st International Meeting on Graphic Archaeology and Informatics, Cultural Heritage and Innovation*, pp. 99–103, June 2009.

[118] F. Niccolucci, "Virtual Archaology: an Introduction," in F. Niccolucci (ed.), *Virtual Archaeology, Proceedings of the VAST Euroconference, Arezzo 24-25 November 2000*, BAR I.S. 1075, Oxford, Archaeopress, pp. 3–6, 2002.

[119] J. Wood and G. Chapman, *Three Dimensional Computer Visualization of Historic Buildings.* in P. Reilly, S. Rahtz (eds.), Archaeology and the Information Age. A Global Perspective, London-New York, 1992.

[120] A. Coralini and E. Vecchietti, "L'archeologia Attraverso un Virtual Model," in A. Coralini, D. Scagliarini Corlàita (eds.), *Ut Natura Ars. Virtual Reality ed Archeologia, Atti della Giornata di Studi*, pp. 17–40, 2002.

[121] J. Barcel, "Virtual Reality for Archaeological Explanation Beyond "Picturesque" Reconstruction," in *Archeologia e Calcolatori*, vol. 12, pp. 221–244, 2001.

[122] B. Molyneaux, *From Virtuality to Actuality: the Archaeological Site Simulation Environment.* in P. Reilly, S. Rahtz (eds.), Archaeology and the Information Age. A Global Perspective, London-New York, 1992.

[123] P. Milgram and S. Yin, "An Augmented Reality Based Teleoperation Interface for Unstructured Environments," in *ANS 7th Meeting on Robotics and Remote Systems, Augusta*, pp. 101–123, 1997.

[124] N. Magnenat-Thalmann and G. Papagiannakis, *Virtual Worlds and Augmented Reality in Cultural Heritage Applications.* in M. Baltsavias, A. Gruen, L. Van Gool, M. Pateraki (eds.), Recording, Modeling and Visualization of Cultural Heritage, London, Taylor and Francis.

[125] M. Billinghurst and H. Kato, "Collaborative Mixed Reality," pp. 261–284, in Proceedings of the First International Symposium on Mixed Reality (ISMR '99). Mixed Reality - Merging Real and Virtual Worlds, Berlin, Springer Verlag, 1999.

[126] V. Vlahakis, N. Ioannidis, J. Karigiannis, M. Tsotros, M. Gounaris, D. Stricker, T. Gleue, P. Daehne, and L. Almeida, "Archeoguide: Challenges and Solutions of a Personalized Augmented Reality Guide for Archaeological Sites," *IEEE Computer Graphics and Applications*, vol. 22, no. 5, pp. 52–60, 2002.

[127] D. Stricker, A. Pagani, and M. Zoellner, "In-situ Visualization for Cultural Heritage Sites Using Novel, Augmented Reality Technologies," in *Arqueologica 2.0, Proceedings of 1st International Meeting on Graphic Archaeology and Informatics, Cultural Heritage and Innovation*, pp. 141–145, June 2009.

[128] D. Schmalsteig, A. Fujrmann, Z. SZalavari, M. Gervautz, and E. Studierstube, "An Environment for Collaboration in Augmented Reality," in *CVE '96 Workshop Proceedings, Nottingham*, September 1996.

2

Using Digital 3D Models for Study and Restoration of Cultural Heritage Artifacts

Matteo Dellepiane, Marco Callieri, Massimiliano Corsini, Roberto Scopigno

Consiglio Nazionale delle Ricerche
Email: m.dellepiane@isti.cnr.it, m.callieri@isti.cnr.it,
m.corsini@isti.cnr.it, r.scopigno@isti.cnr.it

CONTENTS

2.1 Introduction

The introduction of new technologies in the context of cultural heritage (CH) and Archeology has often been a difficult issue. This is probably related to the lack in confidence in replacing consolidated approaches with experimental methods heavily based on innovative hardware or software systems. This already happened for a number of revolutionary technologies: for example, the advent of photography, color images and digital cameras took some time before changing the reference methods for archival and studies in the context of archeological excavation or of restoration actions. The same considerations hold for the use of digital 3D models in CH applications. One basic issue is the need to switch from a two-dimensional visualization and reasoning approach (essentially based on photos and drawings) to the possibility to explore and visualize the object in its full three-dimensional

(3D) nature. Nevertheless, in the last few years both 3D modeling and 3D scanning have become a valued way to present and analyze CH artifacts. Several interesting practical applications have been made available to the public in museums, or virtually on the Web. Here we just cite a single sample paper [56], which deals with the problem of interactive inspection and rendering of complex 3D models; many other experiences are presented in this chapter.

The availability of tools for the reconstruction and the virtual interaction with accurate digital 3D models, supporting accurate and flexible visualization features, is a crucial issue for many applications, such as architectural design, graphics simulation, and scientific visualization. The users of these systems need accurate, realistic and/or interpretative visual representations, real-time navigation, and flexible interaction tools. Ease of use is also an important issue in the design of visualization tools for CH applications, because most CH users (such as museum curators, art historians, restorers, and museum visitors) have limited skills in managing 3D graphics technology. The graphics user interface (GUI) design and the overall usability of the rendering tools play a critical role in determining whether the tool will be just a nice toy or a useful technical instrument. Another problem is that the complexity (the huge number of graphics primitives required to represent a shape) of accurate, realistic-looking models exceeds the interactive capabilities of most graphics workstations. We need real-time visualization of those huge meshes without sacrificing the high-quality achievable with 3D scanning technologies. We have several examples of use of 3D models in virtual presentation of CH artworks to the wide public (just to cite a few, see [2–6]), most of them very successful in their field. On the other hand, visualization to the public is only one of the possible goals. Another important objective is to design applications or tools to help CH experts in their everyday work, being it the study or research on CH, or the conservation/restoration. Until now, the use of 3D graphics and 3D scanning in CH applications has been limited by the costs of the acquisition devices and the slowness and complexity of the processing phase required to transform raw sampled data into accurate and usable digital 3D models. Technological issues play a role, but this is not the only (or major) reason for the scarce use of 3D graphics in CH applications. Another key factor is the lack of some killer applications in the realm of the daily needs of CH experts. Visualization is the most diffuse use of digital 3D models, but unfortunately in many cases virtual 3D presentation is still considered by CH practitioners as a medium to reach the public, rather than a basic instrument for CH research and conservation. This chapter aims to show that there are wide possibilities for the use or development of software tools oriented to the CH domain. This activity is often perceived by the academic community as an applied research or a research transfer activity (designing tools according to specific needs of the CH users community). This is partially true, since in some cases academic research and development is playing the role that should usually be covered by commercial companies; but the small budget that characterizes most of the ICT application in the CH domain can drastically change in the near future, raising the level of interest of the industry. On the other hand, in many cases the specific CH needs motivated interesting basic research on CH technology and had an impact on the advance of our research domain. Some of the possible applications of digital 3D models in the CH domain together with a description of some tools specifically designed for CH applications are presented in the following, trying to cover several potential interests of CH experts.

2.2 Visual Communication of Art

As briefly mentioned in the introduction, visual communication is by large the most common utilization of digital 3D models in the CH domain. Just to make an example, the 3D media is becoming more and more a common resource in the production of broadcasted programs on history and art. The different visual presentation modalities can be divided in those using still images and those adopting video streams. The latter can be further divided into *passive* modalities (i.e., videos

FIGURE 2.1
Some snapshots from "The Parthenon" movie, by P. Debevec (see Chapter 6).

or computer-generated animations) and *active* modalities (i.e., interactive multimedia installations or virtual navigation systems.) Let us briefly present the potential of those technologies, and some major experiences and issues.

2.2.1 Computer-Generated Animations

Computer-generated animations are by far the more common channel to present digital 3D models or hypothetical reconstructions of CH artworks to the large public (Figures 2.1 and 2.2). Nowadays, any broadcasted program on art or history contains at least some animation clips; high-budget movies recreate the past using the available technology at its best. The Hollywood blockbuster "The Gladiator" is just one example of the many *historical* movies which allow to visually present highly veridical representations of our past. There is a long standing discussion on how far these representations are also a good medium to convey historically or culturally veridical representations of the past (i.e., up to what extent they comply with art history knowledge). Technology is nowadays extremely high quality and the images/videos produced are so compelling (photorealistic), that the resulting video sequences can become *the historical reality* in the mind of the public, maybe much more than that reached/obtained by years of CH research and dissemination. But, as usual, the problem is not in the medium, but in its use. One rising issue is that the level of quality reached by the current movie production is paired by an extreme complexity of the production process and of the related costs. Computer animation systems are now very complex, flexible, and expensive. They also raised the bar of the public perception of what is a good-quality virtual representation or experience. Anyone attempting to produce a video or a visual presentation (for example in the CH domain) should expect the potential users will make their personal evaluation by comparing it to what they are used to seeing (i.e., Hollywood movies). It is often very hard to run against such tough competition.

Nevertheless, there have been very nice examples of videos produced in the academic/research domain, based on a low budget and giving very good examples of what can be accomplished when we pair CH dissemination needs and new visual technologies. One of the best videos ever produced in this domain is "The Parthenon" movie, produced by Paul Debevec

FIGURE 2.2
Snapshot from the movie clips presented on "The Virtual Museum of Iraq" web; videos produced by
F. Gabellone (CNR-IBAM, Italy).

(http://www.debevec.org/Parthenon/film.html) in 2004. The video is a very good example of how the
story of a monument can be presented; it visually reunites the Parthenon and its sculptural decorations,
separated since the early 1800s. The film used a number of cutting-edge technologies (3D scanning,
photometric stereo, inverse global illumination, photogrammetric modeling, image-based rendering,
BRDF measurement, and Monte-Carlo global illumination) in order to create the twenty-some shots
used in the film (see Chapter 6 for more details). Another more recent example of intense utilization
of computer animation to present CH artworks is the video content at the base of "The Virtual
Museum of Iraq" (http://www.virtualmuseumiraq.cnr.it/). Most of the animations presented there
have been produced by an Italian CNR group, led by Francesco Gabellone, using mostly image-based
reconstruction approaches and commercial video production technology. Those animations can
also be considered as a nice example of what can be produced by a small multidisciplinary group
(computer scientists, architects, and art historians) working under a small budget.

Computer animation can be an ideal instrument as well for visualizing the CH scholars' research
results, if the production complexity and cost will reduce. We have indicators that the diffusion of
the knowledge on computer animation technology is considerably improving in young generations.
Another important factor for a wider diffusion is the impressive improvement of low-cost open source
software platforms (e.g., Blender [7]).

2.2.2 Interactive Visualization

An advantage of interactive visualization is to insert the user in the loop. Conversely to passive media
such as computer animation, it is now the user that drives the navigation and the inspection of the
digital artifact. An interactive system allows the user to follow his specific interest while choosing
the exploration path, focusing on the details that hit personal interest and giving the possibility to
choose the duration of the visualization session on the base of the specific insight experience and
needs. Virtual Inspector (see Figure 2.3) [56] is an example of a CH-oriented visualization system

FIGURE 2.3
A snapshot of the Virtual Inspector interface.

FIGURE 2.4
The Arrigo Kiosk, with Virtual Inspector integrated in a website-like kiosk.

which aims to fulfill some of the issues sketched before. The system allows naive users to inspect a very dense 3D model at interactive frame rates on off-the-shelf PCs, presenting the 3D model and all the multimedia data that has been linked to selected points of its surface. A main goal in the design of the system is to provide the user with a very easy and natural interaction approach, based on a straightforward point and click metaphor. Visualization efficiency is obtained, without sacrificing quality, by adopting a state-of-the-art continuous level-of-detail (LOD) representation [68]. Finally, the adoption of the XML encoding for the specification of the GUI structure and behavior makes Virtual Inspector a very flexible and configurable system.

Virtual Inspector has been used for the implementation of several multimedia presentation, including: the Arrigo VII kiosk [9] (Museo Opera Primaziale Pisana, on show since the end of 2004, see Figures 2.3 and 2.4), a system for documenting the results of the restoration of Michelangelo's David [10], an interactive kiosk to present the excavations results in the area of Luni (La Spezia, Italy), etc. But Virtual Inspector is just an example of those systems devoted to the inspection and manipulation of a single artifact. The virtual navigation of large scenes, representing a complex of artifacts such as for example a complete archaeological site or a historical town, is also of great interest. The need to represent and navigate a large scene opens a number of issues [11] that have to be considered, namely:

- Need to sample a large extent, which has to be processed to build up the global model (processing huge 3D sampled domains is not as easy as processing a single statue);

- Representing the 3D model in a format which is adequate for interactive visualization (the

FIGURE 2.5
A snapshot of the VR installation in Ename, Belgium [12].

adoption of multiresolution methodologies for encoding and rendering the 3D models is here mandatory);

• Devising intuitive GUI approaches for driving the navigation in a complex scenario; usability is a must, to prevent that the user would be concentrated on how to drive the navigation rather than enjoying and getting insight from the virtual visit.

Moreover, the use and combination of cutting-edge technologies can boost the user experience also in the visit of large areas: the example of Ename [12] shows the combination of several approaches for the comprehensive presentation and exploration of an interesting historical settlement (Figure 2.5).

2.2.3   Geographic Web Browsers Deploying 3D Models

Web-based systems such as Google Earth [13] or Microsoft Live Local [14] could boost considerably the usage of 3D reconstructions and the availability of 3D models of CH artifacts. These systems allow the addition of small-resolution 3D models on top of terrain models, and link those representations with high-resolution 3D models of the architectures and artworks which characterize a given territory or urban context. We could easily forecast that in the near future local authorities will be active sponsors for massive data gathering to populate those systems, aiming at increasing the tourism industry and providing a wide dissemination of local CH masterpieces. This could give an impressing impulse to the use of both image-based approaches, for the reconstruction of low-polygon models, and 3D scanning, for the production of higher accuracy models.

Open source platforms are also available for large-scale representations of CH, supporting virtual navigation of CH resources, either on site or on the web. Some applications have been experimented by the Virtual Heritage lab of CNR-ITABC, focusing on the reconstruction of archeological sites. An example is the "Virtual Museum of the Ancient via Flaminia" [15] (http://www.vhlab.itabc.cnr.it/flaminia/). The system has been designed as a Virtual Reality system to be installed inside an important museum in Rome. One of the design goals was to keep the social aspects which are typical of the visit in a museum, therefore it was designed to support a multi-user navigation and interaction. Through the VR application, visitors can explore the landscape that characterized the Roman road, Via Flaminia, and have an intense experience visiting four different sites (an old bridge over the Tevere River, a Roman necropolis, the Villa of Livia, wife of emperor Augusto, and finally a Roman arch), all of them reconstructed in details using different 3D technologies. Authors performed a topographical, archaeological, and architectonic survey, whose goal was data acquisition at different levels of detail (DGPS, laser station, 3D laser scanner, photogrammetry, cartography, remote sensing, aerial and satellite photos processing). Data have been integrated in the system, to build up models of the landscape and of the CH artifacts, which could be navigated in interactive time with the selected VR platform. A similar navigation experience is also available (in single-user mode) on the web via the support of the VR Web GIS platform [16].

2.3   Art Catalogs and Digital Repositories

CH cataloguing efforts are still based on just textual data enriched by (a few, often B/W) photographs. It is very easy to forecast a future where catalogs will adopt digital 3D models as the main representation resource for characterizing the shape and the appearance of the artifacts. Unfortunately, whether this will happen in a near or a far future depends more on the availability of funds and improved technological skills of the CH governing bodies than on technological reasons.

The wide availability of 3D catalogs is the prerequisite for devising interesting applications concerning 3D search and retrieval methodologies, able to find shape-similar artifacts in large digital repositories, given a sample object. A considerable research effort is now devoted to this topic; current solutions are too naive to work well in a field so demanding such as CH, but technology might improve fast. Moreover, the availability of 3D models of similar artworks can be very valuable to support further study, such as, for example, understanding how representation of God changed as the religion moved to different parts of the world [17] or supporting the assessment of attribution hypothesis [18]. While the current 3D scanning technology is considered mature for cataloguing efforts (even if the data encoding issue is still not completely solved, with so many alternative available 3D formats), some technical issues are still open, especially considering the problem of data preservation on long time frames. The usual problem (possible deterioration of hardware supports) is made more complex in the case of 3D graphics by the very fast renovation cycle of the software tools. This problem is less critical for repositories of basic 3D models than for the case of repositories for 3D interactive applications. In the first case, the adoption of publicly known and documented data format makes data obsolescence a more tractable problem. But this issue has to be considered and solutions have to be planned well in advance.

Finally, it is immediately evident how a CH repository focusing only on the storage and retrieval of 3D data is not enough [19]. CH applications require a joint management of the 3D model and of all the metadata which characterize the represented artifact. Moreover, it is strategic also to know how the digital 3D model has been acquired and how the raw sampled data have been processed to derive the final model. All this information is crucial to be able to assess the accuracy/quality of the digital model. These topics have been a major focus of the EU projects CASPAR (http://www.casparpreserves.eu/) and 3D-COFORM (http://www.3d-coform.eu/).

2.4   Digital 3D as a Tool for Art Scholars

Accurate 3D models and visualization tools enable very powerful means of presentation, inspection, or navigation of cultural heritage artifacts. But visual communication and insight is used also in several other applications or for other purposes. Important activities like proposing or validating the attribution of an artwork, planning and documenting restoration, and identification of fake artworks are based not only on material analysis [20–22] and historic documentation studies, but also on visual cues and comparisons between different artworks. Visual comparisons and localized annotation are

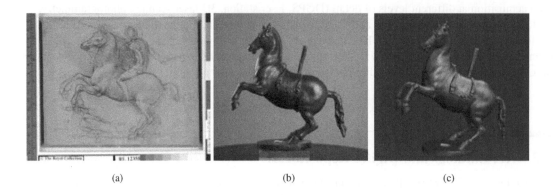

(a)   (b)   (c)

FIGURE 2.6
(a) The Windsor silverpoint drawing (WRL #12358: The Royal Collection © 2007 Her Majesty Queen Elizabeth II). (b) The small bronze horse (Archeological Museum, Florence). (c) A snapshot of the model produced with 3D scanning.

also very useful for art historians. Therefore, the digital 3D medium can become a powerful asset also for the work of the art scholars. The consolidated approach is to study an artwork by performing direct manipulation and visual inspection (which is a rare privilege of a very restricted number of experts, with heavy limitation in time and space). In many other cases, the work of the scholar is based on the analysis of 2D images, which present just a partial representation of the artwork (the information conveyed depends on the specific view depicted in the image and on the specific illumination used to make the photo). The digital 3D medium can offer a very similar experience to the personal manipulation to a much wider number of scholars (with no danger or deterioration for the original), from any location in space and at any time. Once an accurate digital 3D model (which should encompass a representation of both the shape and of the surface reflection characteristics, as discussed in Chapter 3) is at disposal, there is the possibility to create arbitrary renderings from any point of view, giving to the scholars the possibility to locate and examine the important details as if they were manipulating the original artifact. In the next section, we show some interesting examples in this domain: the use of image registration techniques to support and attribution theory, the creation of a collaborative study environment for Romanesque Art, and the use of 3D data to classify faces sculptures in a Cambodian temple.

2.4.1 Using 3D Scanning to Analyze an Attribution Proposal

The work described here is a good example of how 3D technologies could be adopted to help the work of art scholars. It is also an example of application of techniques designed to solve a different problem: an *image-to-3D model* registration technique was applied to find a relation between a hand drawing and a bronze statuette [18]. The hypothesis proposed by Mark Fondersmith suggested that a Leonardo metalpoint (RLW #12358, Windsor Royal Collection, see Figure 2.6) could be an optically-traced drawing of a small bronze horse (inv. #19446, Archeological Museum, Florence, see Figure 2.6). According to this hypothesis, the small bronze horse should have been modeled much earlier than the official attribution (it is currently recognized as an artwork by B. Cellini). Fondersmith's hypothesis went even farther, proposing that the small horse could have been cast by Leonardo, while an apprentice of Verrocchio (1470-1480). Later, Leonardo might have used the small bronze horse as a reference for the Windsor metalpoint (1480). According to this hypothesis, Leonardo could have used a camera obscura (or some other similar optical device) to create the

FIGURE 2.7
Images from the first alignment (main body).

FIGURE 2.8
Images from the second alignment (head).

Windsor metalpoint. The small bronze horse could have been traced in two stages: first, the body (see Figure 2.7) and then the head and one leg, drafted after rotating the bronze to the left (see Figure 2.8). The two sets of front hooves would be a true evidence of the rotation. The redrawn neckline might show how Leonardo reconciled the second angle of the head with the original profile. The precisely crosshatched areas of the drawing describe the sculptural form of the small bronze horse. The authors were contacted by Fondersmith for executing an accurate digital acquisition of the bronze, with the purpose of performing further study on a physical replica produced from the digital 3D model via rapid reproduction technology. After having acquired the digital model with 3D scanning technology (a screenshot of the colored model is shown in Figure 2.6), it was proposed to do some tests using a semi-automatic image registration approach [54].

In order to try to produce some scientific evidence of the asserted shape similarity, to help assess the attribution proposal, it was decided to treat the Leonardo's drawing as if it was a photo. The alignment of the drawing to the geometric 3D model was performed in the usual way: some correspondences (points pairs) between the 3D model and the drawing were selected manually [54]. The correspondences on the drawing were set by following the indications provided by the attribution hypothesis. The two proposed points of view were considered: the first one for the body silhouette and the second, rotated one for the head and the right leg. Clearly, the points provided for the registration were different for the two cases. The focal length and perspective deformation was fixed to a plausible value for the use of a camera obscura. The result of the first alignment is shown in Figure 2.7, where the overlapping of the 3D model and the drawing is shown, progressively, with different transparency levels. As it can be noted, the alignment of some parts of the drawing is surprisingly precise: the lines of the neck and the back are coincident with the silhouette of the model, and the profiles of both fore and hind right legs are quite similar. Moreover, the position of the associated camera (denoted by a cone in the right-most section of Figure 2.9) is clearly compatible with the point of

FIGURE 2.9
The relative position of the camera for the first alignment is presented in the left image, while the relative position of the camera for the second alignment is in the right image.

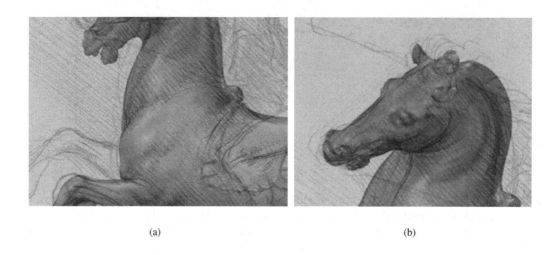

(a)   (b)

FIGURE 2.10
Two detailed views: on the neck and saddle region (a) and on the head (b).

view of a person using a camera obscura. The second alignment was performed to validate the second proposed position, from which the head would have been sketched. Results are shown in Figure 2.8: not only the profile of the head is very similar to the drawing, but also the line of the neck is overlapped to an internal line in the drawing. The associated camera position (Figure 2.9) shows that the point of view is not only rotated but also raised with respect to the first alignment. The results of the alignment provided a visually significant proof and confirmed the compatibility of the use of a camera obscura to sketch the drawing from the bronze statue. It's important to stress that no deformation was introduced while performing the image-to-3D mapping. Two images of zoomed-in details are presented in Figure 2.10. Unfortunately, it is very hard to provide numerical proofs of the quality/accuracy of an image-to-3D model alignment; this can be done with photos, which exactly reproduce the object, but it would be of no utility in this case, since the drawing is clearly not a mere numeric tracing of the silhouette but it is the hand of the artist that performs the tracing even when a camera obscura is used. Anyway, the surprising alignment obtained seemed to be too precise to be casual, especially considering the results obtained on similar objects that were tested.

In conclusion, this peculiar application showed that it was possible to provide some empirical

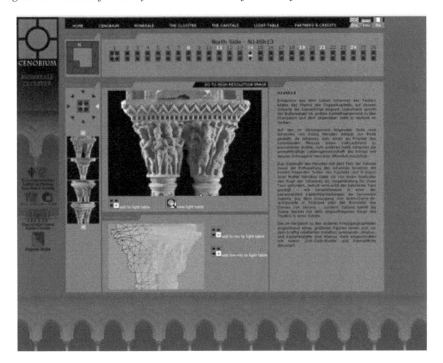

FIGURE 2.11
CENOBIUM: integrated information about one selected capital.

proof to sustain the attribution proposal. Obviously, this is not the last word in the long lasting dispute relative to the attribution of the bronze horse, but it could help raise some serious concerns on the current attribution (i.e., to Benvenuto Cellini). The debate should be solved by further study or analysis of the constituent material and by a more in-depth study of the historical documentation.

2.4.2 The CENOBIUM Project: An Integrated Visual Comparison of Historiated Capitals

The CENOBIUM (Cultural Electronic Network Online: Binding up Interoperable Usable Multimedia) project [24] regards the visual comparison (for art history studies) of the Romanesque cloister capitals of the Mediterranean region. One of the goals of this project is to illustrate cultural exchange in the twelfth and thirteenth centuries through the example of architectural decorations. In this domain, three cloisters have been considered and digitized so far: the cloister of Monreale (Sicily, Italy), the cloister of Sant'Orso in Aosta (Italy), and the cloister of Cefalù (Sicily, Italy). Currently, the CENOBIUM website (http://cenobium.isti.cnr.it) includes the data about these three cloisters; we are planning to extend it in the near future. For most of the capitals of these cloisters several high-resolution digital images and highly detailed colored 3D models have been produced with different technologies, including laser scanning and image-based 3D reconstruction. The CENOBIUM website is a web application which attempts to integrate the different type of information about the acquired capitals. All the information related to a specific capital (text, images, and 3D models) are shown in a single web page (see Figure 2.11) to improve readability. Such information is handled through a database which inter-connects the different kinds of media (text, images, and 3D models). From the Capitals section the user has the possibility to explore, in an integrated manner, all the multimedia information related to each capital. By clicking on the selected view of the capital, we can visualize each full-resolution image. Since the resolution of the acquired images is very high, a dedicated

FIGURE 2.12
The visual comparison tool called *LightTable* [24]: on the left are the models downloaded locally, on the right the screenshot of the browser that allows inspection of the 3D model and three images of the same capital.

FIGURE 2.13
Three examples of carved faces in the Bayon temple. These models were used to classify the 173 acquired stone faces in three groups.

image server (*Digilib*, developed by Max Planck Istitut) gives the possibility to navigate them interactively. A textual description containing historical information is given in the right part. One of the most interesting features of the CENOBIUM website is the possibility to visually compare 3D models and images by selecting them during the navigation. The selected items are handled by the *LightTable* [24]: when an item is selected, the LightTable downloads it on the local PC; then, for those 2D or 3D items that are ready-to-use, the user has the possibility to visualize them simultaneously for the inspection. The maximum number of items that can be visually compared is four. Figure 2.12 shows the LightTable in use. The CENOBIUM system will be extended further, adding more data and functionalities; it could become a very useful resource for the work of art historians or art scholars. The most interesting goal of the whole project stands in the creation of a collaborative environment which is mainly devoted to art historians: the users will be able to compare 3D models and images from places which are disseminated around Europe. Moreover, the possibility to add annotations could boost the work of the small but active community of experts.

2.4.3   Classifying and Archiving Carved Faces: The Bayon Digital Archival Project

The Bayon digital archival project [25] is a very complex acquisition experience, which used several technologies [26] to acquire the extremely big (100 m length on each side by 43 m height at the most) architecture of the Bayon temple in Cambodia.

FIGURE 2.14

An example of a rapid prototyping project, developed by CNR-ISTI in collaboration with the company *Scienzia Machinale* (www.smrobotica.it). Left to right: the original artifact, the 3D model obtained using laser triangulation, the prototyping machine in action, the reproduction during an intermediate stage, and the final results.

Besides the importance of the acquisition campaign, we point out a peculiar use of the 3D data for the work of art historians. Throughout the temple, there are 173 stone faces, carved on the towers. They were classified in three groups (see Figure 2.13) and an automatic method assigned each of the acquired stone faces to one of the groups. In this simple example, the analysis of data helped the experts in classifying a complex set of data, without the need of working on the real environment, which is set in a very difficult to reach area.

2.5 Physical Reproduction from the Digital Model

An extremely interesting feature of 3D scanning and modern photogrammetry (and also of standard 3D modeling) is that the digital 3D model is obtained without any direct contact with the artifact. This is extremely important in the CH context, since the manipulation of the originals can easily produce damages. Once we have an accurate digital 3D model, another technology, *rapid prototyping*, can find many interesting uses in the CH domain. This is another case of a technology inherited from the industrial field. *Rapid prototyping* devices are able to create accurate reproductions of an object starting from the digital 3D model. The process is usually automatic and can be applied on a variety of materials (chalk, plastic, plastic with metallic coating, stones, etc.). As a major difference from the consolidated reproduction approach (production of rubber molds for the subsequent production of gypsum copies), digital reproduction allows one to obtain copies in any reproduction scale, or of just portions of the object. Like the other technologies in this field, the improvement in hardware and software components led to a rapid decrease in the costs.

An example of a practical application is shown in Figure 2.14. The subject of the work was a marble head of Mecenate (Archaeological Museum, Arezzo, Italy). CNR-ISTI was commissioned by the German Research Ministry for the production of an accurate marble copy, to be used in the context of the German "Mecenate" research project. The scanning was performed on the original object using a laser triangulation scanner, so that an extremely accurate model was produced. The 3D model was the input for driving the carving path of a robotized drilling system, which is able to sculpt a marble block with great accuracy and repeatability. After a final manual intervention for carving finer details and polishing the surface, a very detailed 3D reproduction of the original artifact was obtained.

Other applications, which do not use typical prototyping devices, have also been considered in the context of the reproduction from digital models. For example, laser cutting machines have been used to reproduce ancient astrolabes [27]. Hence, physical reproductions from 3D models can be used in several ways:

- **Temporary or permanent replacement of originals.** If an artifacts has to be removed from its original position, it is possible to replace it with an accurate copy. The replacement can be temporary, for example when a museum lends an object to an external exhibition. In the case of severely damaged or fragile objects, it is possible to put them in a more protected environment (museum, controlled environments) by replacing them permanently with a copy. In this way, the visitor can come in contact with the environment in which the artifact was posed (note that from a medium distance, the difference between the original and the replica becomes not perceptible) and at the same time the original artifact can be protected and conserved.

- **Reconstruction and coloring hypotheses.** An accurate copy of the object can be very useful if the restorers want to experiment and propose hypotheses about the original shape of missing parts, or about the original colors of polychromatic statues or decorations. While rapid prototyping devices are able to produce colored objects, and obtain also realistic and visually pleasing results [28], the re-coloring is usually made by hand to obtain full accuracy (accurate selection of color tints and layout of color). This peculiar application helps the restorers in their practical work, with the possibility to produce and compare several hypotheses.

- **Wide-scale accurate physical copies.** A more commercial application is the possibility to obtain accurate small-scale copies of an artifact, for commercial purposes. This raises several issues about copyright and opportunity, but it is also an interesting option for funding the activities of a museum or a CH governing institution.

2.6 Virtual Reconstruction and Reassembly

2.6.1 Virtual Reconstruction

Virtual reconstruction has always been one of the most straightforward CH applications of 3D graphics. Besides the use of pure 3D modeling, which creates geometry starting from images or historical documents, we will focus mainly on some examples of automatic or semi-automatic reconstruction of 3D models from incomplete or peculiar data. A fascinating application is the reconstruction of artifacts which are not existent anymore, using the historical material that can be found, for example, on the web. A nice example is the work aimed at the reconstruction of the Great Buddha of Bamiyan [29], which was destroyed by the Taliban in 2001. This model was automatically obtained from a set of images which were partly found in web repositories. The resulting 3D model was proposed as a starting point for the project to rebuild the huge statues. Figure 2.15 shows a sample set of the images used for reconstruction, a rendering of the niche without the statue and a rendering of the niche with the reconstructed model. This kind of automatic reconstruction, which is based on photogrammetry and stereo-matching techniques, is able to obtain a sufficiently accurate reconstruction of an object which cannot be scanned anymore. A similar but more simple approach is based on the information which can be inferred from plans of the remains of ancient buildings. Semi-automatic image processing techniques are able, for example, to extract the wall lines, so that it is possible to extrude them in order to rebuild the original architecture. Other drawings can be used to add the peculiar shapes of the decorations. A recent and interesting project [30] aimed at combining some of the already mentioned reconstruction techniques with *procedural reconstruction* methods to obtain a realistic and navigable model of Ancient Rome. A different approach stands in the reconstruction of the original shape of uncomplete or damaged artifacts. As an example, the

FIGURE 2.15
Digital reconstruction of the Bamiyan Buddhas. Top: a sample set of the images used for reconstruction. Bottom: the model of the niche without and with the reconstructed model.

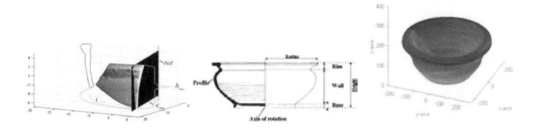

FIGURE 2.16
Some images from works by R. Sablatnig (TU Wien) left ro right: an example of profile computed from the digital model of a fragment; profile-based characterization of vases; an example of reconstruction of a vase from profiles.

cranium shape of a human prehistoric race was reconstructed by comparing and integrating the 3D acquisition of the few remains of the original skull and the radiographies of similar races [31]. A final example of virtual reconstructions are all those systems that start from a fragment of an object and try to reconstruct it. An interesting application field is pottery [32–35], where the artifacts are usually obtained by rotation and thus their shape possesses sufficient regularity and symmetry to allow easy reconstruction from fragments (see Figure 2.16). Hence, after scanning one or more fragments of a vase, it is possible to infer its diameter, and produce a digital model of its entire shape.

2.6.2   Virtual Reassembly

Besides the reconstruction of missing geometry, another field of application is the digital reconstruction of disassembled or fragmented artworks. Physical reassembly is a process done manually by archeologists. The adoption of a computer-aided approach can be justified either in case of extreme fragility of the artifact or of complicated manipulation (e.g., the fragments either are too heavy or too numerous to be manipulated easily by an archeologist). Early methods have been proposed for special cases, such as the reassembly of sherds of ancient pottery, where some hypothesize that regularity and symmetry of shape can simplify the reassembly task [33]. A recent result has shown that the generic process can be solved in a robust manner by taking into account also the non-precise and eroded fractures of archeological remains [36]. The joint improvement of 3D scanning and automatic reassembling methods can open new insights in very complex reconstruction problems. In the case of extremely eroded or very lacking sets of fragments, the mathematical approaches can encounter severe issues: an alternative solution is to provide the restorer with a tool which helps in recomposing the original structure. For example, the user can propose a possible recomposition to the system, which would then try to validate it by comparing the adjoining fracture surfaces (i.e., by using a shape-based matching approach). Even the interactive adjoining of the fragments can be aided by applying constraints to the possible relative and absolute movements of the 3D models (as proposed in [37]).

A peculiar example of virtual restoration is the work on the church St. Maria di Cerrate (Lecce, Italy) [38]. Here the problem was to produce a virtual restoration starting from the erroneous results of a real restoration. One of the walls of the nave of the church fell to the ground and was rebuilt using the fallen stones (see Figure 2.17), without taking into account the (apparently unknown) right order of the stone blocks. As a result, those blocks were shuffled and the remains of an old fresco were no longer legible. The problem was solved at the virtual level, first by moving again all stone blocks in the right location (see Figure 2.17(b)) and second by drawing a global virtual restoration,

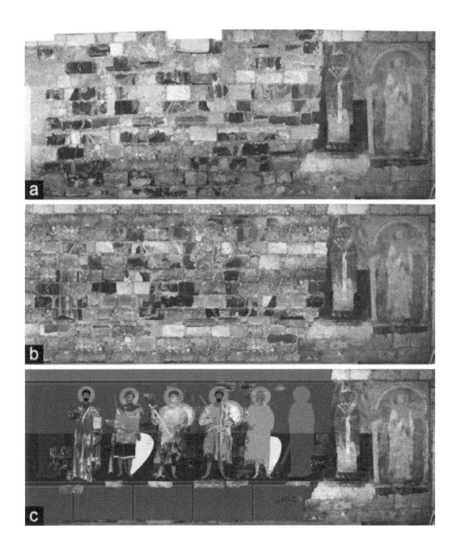

FIGURE 2.17
Two snapshots from a video produced by F. Gabellone (IBAM-CNR) to show the results of a virtual restoration project concerning a wrongly reconstructed wall with remains of an old fresco in St. Maria di Cerrate (Lecce, Italy); actual status is in (a); results of the virtual recomposition of the fragments in the correct position are presented in (b); final virtual restoration in (c).

       (a)   (b)   (c)

FIGURE 2.18
(a) and (b) A physical recomposition of a small subset of the fragments of the Frontone of Luni [37]; at the same time, many sherds still wait to be reassembled. (c) A rendering of the digital model of one of the statues, its colored version according to the current color of the statue, and a painted model reconstructed according to evidence of its possible original colors.

according to the knowledge available on the fresco (see Figure 2.17(c)). This work is presented to the public by means of a prerecorded video and an interactive system.

Moreover, creating assemblies with 3D CG or VR technologies helps to better understand the past. Digital 3D models of artifacts can be used as building blocks in interactive CG/VR applications (or to produce passive animations), to give a better understanding of how different materials or components were used in the past to build architectural structures [39] or complex instruments (which could range from the simple compound tool to the big industrial machine).

2.6.3   Virtual Repainting

The availability of accurate 3D models opens interesting capabilities for the dissemination of the original aspect of ancient sculptures or architectures. In many cases we have statues which either completely lost their original painting (this is the case of many archeological masterpieces) or present severe deteriorations. A seminal work has been performed by professor Vinzenz Brinkmann (see [40] for an overview of several different experiences on this topic, including the description of several projects he coordinated on this subject). Brinkmann adopted an approach based on the reproduction of gypsum or marble replicas which were then painted, to give them the look and feel of the original painted artifacts.

With the aid of user-friendly tools, it is possible to easily produce hypotheses about the original color of the statues, based for example on the analysis of the original pigments found on the surface. Since there are usually several possible proposals, the use of digital 3D models avoids the use of physical replicas or 2D drawings. A simple example of 3D model repainting is described in [37]. Figure 2.18 summarizes this project, where we have first a reassembly problem (many sherds have still to be included in the reconstruction; some original historical reassembly hypothesis have still to be validated) and, second, a problem of investigating and repainting the original color from residual traces still visible on the surfaces.

A very interesting practical deployment of the results of virtual repainting can be proposed by adopting digital video-projection rather than repainting a physical replica. Using cheap video projectors it is possible to virtually restore color to the surface of these artworks by repainting the digital 3D model and projecting back the model on the surface of the original artifact [41]. The same approach can be used also to present a different reconstruction hypothesis for the original painting, or to digitally repaint solid copies produced with rapid reproduction technologies. An issue in this type of application is how to register the original (or the physical copy) with the projected digital image. Manual registration is a slow and complicated process, while the same action could be transformed into a semi-automatic process by coupling the video projector with a video acquisition channel and adopting image-based solutions which could iteratively improve the mapping of the rendered image on the original.

2.7 Supporting the Restoration Process

Restoration of CH artifacts can be positively affected by the use of accurate digital 3D models. Restoration is nowadays a very complex task, where multidisciplinary skills and knowledge are required. A complex set of investigations usually precedes the restoration of a valuable artwork: visual inspection, chemical analysis, different type of image-based analysis (RGB or colorimetric, UV light reflection, X-ray, etc.), structural analysis, historical/archival search, etc. These analyses might also be repeated to monitor the status of the artwork and the effects of the restoration actions. An emerging issue is how to manage all the resulting multimedia data (text/annotations, historical documents, 2D/3D images, vectorial reliefs, numeric data coming from the analysis, etc.) in a common and integrated framework, making all information accessible to the restoration staff (and, possibly, to experts and ordinary people as well). The final goal is to help the restorer in the selection of the proper restoration procedure by giving full access to the analysis performed, and to assess in an objective manner the results of the restoration (to compare the pre- and post-restoration status of the artwork, to document the restoration process). Since most of the information gathered is directly related to a multitude of spatial locations on the artwork surface, digital 3D models can be an ideal media to index, store, cross-correlate, and obviously visualize all this information. 3D models can also be a valuable instrument in the final assessment phase, supporting the interactive inspection of the multiple digital models (depicting pre- and post-restoration status) to check the eventual variations in shape and/or color.

Moreover, a number of investigations can be performed directly on the digital 3D model by adopting computer-based simulations or computations. This has been done in the past to assess the static and structural status of buildings or sculptures, or to detect risky conditions due to an exaggerated stress of the materials. Deterioration is another effect that can be simulated, to give a preview of the future conditions of the artworks subject to corrosion or deterioration (e.g., the erosion of sculpted stone decorations in our polluted historical towns). Very few works focused on this subject, which involves an accurate simulation of both shape and reflection properties and modification/evolution of the inspected surface [42]. A similar task, with a different goal, is the virtual presentation of the forecasted effects of a restoration action; the goal here is to allow the restorers to show to decision bodies or to the public, before the execution of the restoration, a plausible model of the expected results. Many post-restoration discussions and harsh polemics could be prevented by a preliminary presentation of the planned results and of the visual changes that will be brought to the work of art. Therefore, a future goal for computer-aided restoration technologies would be the possibility to simulate the geometric and appearance effects of degradation or of the inverse restoration on the different materials. This is a highly challenging task, since it's necessary to couple high-quality geometry acquisition techniques with accurate models of the physical and chemical properties of materials. Being able to simulate how a given metallic artifact would oxidize, or a marble stone degrade/erode under the attack of pollution, acid rain, and other effects, would be a valuable instrument for CH management and conservation. In this context, the ideal goals are both to guess the extent and locations of future damages or to bring back an endangered artwork to its plausible original status.

FIGURE 2.19 (SEE COLOR INSERT)
Exposure of David's surface to dust, mist, or other contaminations. This visualization shows, using a false-color ramp, the different classes of exposition produced by the simulation (red: absence of fall, blue: high density of fall), under a maximal angle of random fall of 5 degrees (on the left) and 15 degrees (on the right).

The two following subsections present some results obtained while using digital 3D models and CG tools in the framework of CH restoration.

2.7.1 Tools for Investigation and Diagnostics

As stated before, specific *scientific investigations* can be conducted directly on the digital 3D model. In the David restoration project (2003-2004) [43], the authors performed two main "digital" investigations: the characterization of the *surface exposure* with respect to the fall of contaminants, and the computation of a number of *physical measures* [44]. In both cases, ad-hoc software tools were implemented to process the data produced and to present the results to the users.

Surface Exposure Characterization

A tool to evaluate the exposure of the David's surface to the *fall of contaminants* (e.g., fall of rain, mist or dust) was designed and implemented. The phenomenon depends on: the direction of fall of the contaminant, the surface slope, the self-occlusion, and the accessibility of the different surface parcels. The tool produces several qualitative and quantitative results, useful to characterize the artwork surface. The falling directions of the contaminant agents were modeled by assuming a *random fall direction*, uniformly distributed around the vertical axis of the statue within an angle α which defines the maximum fall inclination. Figure 2.19 shows some of the results obtained on the David. The different exposures are visualized using a false-color ramp; the digital 3D model is therefore used both to compute the simulation and to present visually the results. Numeric data have also been produced (tables and graphs) [43, 44].

Physical Measures

Physical measures can be computed directly on the digital 3D model (e.g., David's *surface*: 19.47 squared meters; or its *volume*: 2.098 cubic meters). Knowing the unit weight of the artwork material, the total *weight* can be immediately computed from the volume measure. *Point-to-point distances* are also often needed, and can be simply computed on the 3D model by adding a *linear measuring* feature to the browser used to visualize the digital model. A linear measuring feature is included in several geometric modeling tools. One of the most specific for the treatment of geometric data

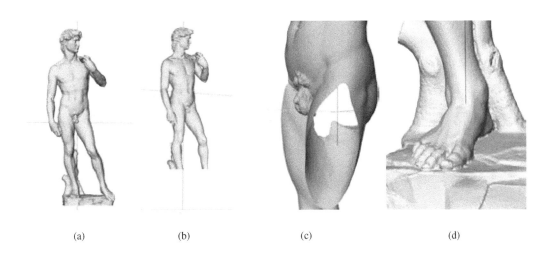

(a)   (b)   (c)   (d)

FIGURE 2.20
Spatial location of David's barycentre: (a) with and (b) without basement and feet; zoomed images of the former in (c) and (d).

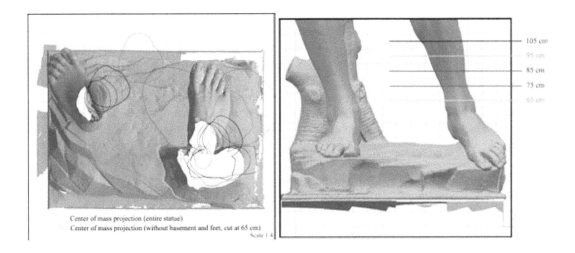

FIGURE 2.21 (SEE COLOR INSERT)
Visualization of the projection of the center of mass (marked by a yellow circle) and of the profiles of some cut-through sections (ankles, knees, and groin; see the respective height of those cut-through sections in the right-most image).

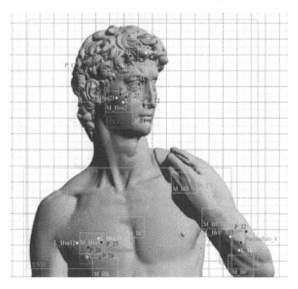

FIGURE 2.22 (SEE COLOR INSERT)
The digital model is used as an index to the scientific investigations performed on selected points or
on subregions of the statue's surface.

coming from 3D scanning is the *MeshLab* [19,41] tool (developed by the authors and freely available
under the GPL license). In this tool the user simply selects two points on David's surface to compute
the linear distance between those two points.

One of the issues evaluated in David's restoration was the *static* of the statue, since some cracks
on the back of the ankles worried the curators. These cracks could have been generated by a wrong
distribution of the mass of the statue, since there are historical papers which sustain that the original
basement (up to the mid nineteenth century) was not properly planar and that the statue was slanting
forward. Therefore, an investigation on the statics of the statue was done during the last restoration
project. The basic data for the static investigation are the mass properties (volume, center of mass,
and the moments and products of inertia of the center of mass), which were computed directly on
the digital 3D model using an algorithm that exploits an integration of the whole volume assuming
constant density of the mass [47]. From this computation, the obtained center of mass of the statue
was placed in the interior of the groin, approximately in the pelvis (see a visualization in Figure 2.20).
The vertical projection of the center of mass on the base of the statue (i.e., the sculptured rocky base
where the David stands) is the blue line, which exits from the marble on the high posterior part of
the left thigh and enters again in the marble on the right foot. The center of mass was estimated also
after the *digital removal of the basement* (cutting the statue at the height of the main cracks); the new
position is shown again in Figure 2.20. The projection of the center of mass on the statue base was
documented with a large size plot (see Figure 2.21) produced by an application [48] designed to
support the easy production of large format prints (orthographic drawings and cut-through sections,
rendered according to the user-selected reproduction scale) from very high resolution digital models
produced with 3D scanning technology.

It's important to underline here that the results presented in this subsection were produced around
six years ago by writing specific small software components. Nowadays, most of those computations
can be easily performed by using in a coordinated manner one or multiple features or filters of the
above-mentioned MeshLab system.

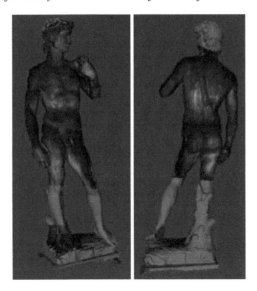

FIGURE 2.23
Mapping multiple UV images on the digital 3D model.

2.7.2 Tools Supporting Knowledge Management

A second important use for digital 3D models is to consider them as an instrument to document, organize, and present the restoration data. During the David restoration campaign, a number of *scientific investigations* were performed; some of them will be repeated periodically, in order to monitor the status of the statue. These investigations include: different chemical analysis (to find evidence of organic and inorganic substances on the surface of the statue), petrographic and colorimetric characterization of the marble, UV imaging, X-ray, etc. All the results produced by the scientific investigations can be organized and made accessible using the 3D models as the integration medium. The 3D model of the David can be used to build different spatial indexes to those data (see Figure 2.22), pointing out their location on the surface of the statue and supporting hyperlinks to web pages describing the corresponding investigation and the results obtained. Specific features for linking or mapping multimedia data on a digital 3D model should be therefore supported by a 3D browser oriented to CH applications. Some works have been done in the CH domain to adopt or configure GIS technology to the specific needs of CH management. This has produced some specific systems, based on a 2D or 2.5D representation, that can represent the domain of interest (e.g., a 2D map or a 2D photo representing a portion of an architecture) and map it selectively on those space all the information available [49], following the GIS approach. Another option is to design new 3D browsers, by giving to the user the ability to create all the links required to connect the 3D representation with the multimedia info and the documents of interest. This entails being able to: store in a repository the set of links introduced on the 3D model and all the related documents; support selective visualization of the 3D model and of the set of links (which can be point-wise or based on the selection of a region). Coming back to the David diagnostic analysis, some of them produced image-based results, which can be directly mapped on the statue surface and presented in an integrated manner. An example is the case of the ultraviolet (UV) imaging investigation. Images produced under UV light are very important to give visual evidence to organic deposits on the marble surface (e.g., wax), which have to be removed with proper solvents. The UV investigation performed by the *Opificio delle Pietre Dure* (a renowned Italian public restoration institution) produced many 2D images taken from different viewpoints. These images can be mapped onto the 3D surface

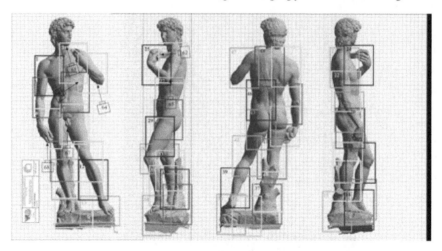

FIGURE 2.24
Schema of the photographic campaign, which divides the David surface in 68 photos.

using an approach which computes the inverse projection and the camera specification from each single photograph and combines all the available photographs in a single texture map which is wrapped around the 3D geometry [48]. In this manner, image-based information is mapped on the corresponding location of the 3D object surface and the information content of the input UV images can be inspected with the help of an interactive browser (Figure 2.23). Other important sources of data are the two high-resolution photographic surveys of the David, performed by a professional photographer with digital technology and according to specific guidelines of the restoration team. The photographic sampling was planned as shown in Figure 2.24 with the aim to document the status of the statue before and after the restoration. These RGB images can be mapped as well to the 3D mesh (see Figure 2.25) with the same methodology used for the UV images. At the time of the restoration (2003-2004) it was possible to map only a subset of the color data on a subset of the mesh (see Figure 2.25(a)); later on, the progress of multiresolution encoding schemes and of color mapping solutions allowed us to build a complete mapping of the photographic sampling on a full resolution 3D model of the David [10], and to support interactive access to that dataset by means of the *Virtual Inspector* tool (see Figure 2.25(b)). Moreover, the curators asked the restorers to produce detailed drawings reporting the results of a precise survey on the status of David's surface. It was decided to manage this phase by drafting very accurate graphic annotations on the set of high-resolution photos depicting the pre-restoration status, which cover all the surface of the statue. These annotations describe in a very detailed manner:

- the imperfections of the marble (small holes or veins);

- the presence of deposits and stains (e.g., brown spots or the traces of staining rain);

- the surface consumption;

- the remaining traces of Michelangelo's workmanship.

The restorers drew these annotations on transparent acetate layers positioned onto each printed photo (in A3 format). Therefore, four different graphic layers were created for each one of the sixty-eight high-resolution photos. These graphic reliefs have been scanned, registered (roto-translation+scaling) on the corresponding digital RGB image, and saved at the same resolution of the corresponding RGB image. Mapping all this image detail on the 3D model is not a good choice, since each relief has been

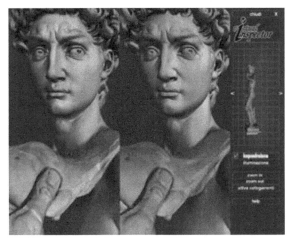

(a) (b)

FIGURE 2.25
Mapping of photographic sampling on the David's digital model; (a) images rendered with *MeshLab* on the head portion, or (b) with *Virtual Inspector* on two complete David models with color coming from the pre- and post-restoration sampling.

drafted not taking into account the region of overlap with the nearby photos. Hence, after mapping all the photos on the 3D model, the overlapping regions would show different reliefs which would not be coherent on the surface of the model. Therefore, it was decided to visualize those data by following a more common 2D approach. A web-based system was implemented to browse the RGB images and to plot (in overlay) any relief layer selected by the user (see Figure 2.26). It was decided to use a 2D-based visualization approach, instead of trying to map reliefs and RGB images on the 3D surface, again due to performances reasons, ease of deployment, and data accuracy constraints. The amount of information contained in those 2D layers (each of them is a 5M pixels image) is impressive. A specific requirement of the curator was that the data (color and reliefs) should be presented at full resolution (in this case, the pixel resolution is much larger than geometric resolution of the 3D model).

In the near future, the availability of new technologies will allow one to draw similar reliefs directly on the digital skin of the 3D model. The solution, granted by the increased performances of GPUs and 3D scanning technology, is to allow the restorer to directly draw the relief on the surface of the digital model. A testbed is under implementation and evaluation during the writing of this chapter in the framework of the diagnostic and pre-restoration investigation phase for the Nettuno statue (Florence, Italy). Restorers received a copy of a high-resolution model of the statue [51] in order to experiment with a pipeline where the reliefs are directly executed on the digital statue, by using the painting feature recently introduced in the *MeshLab* system.

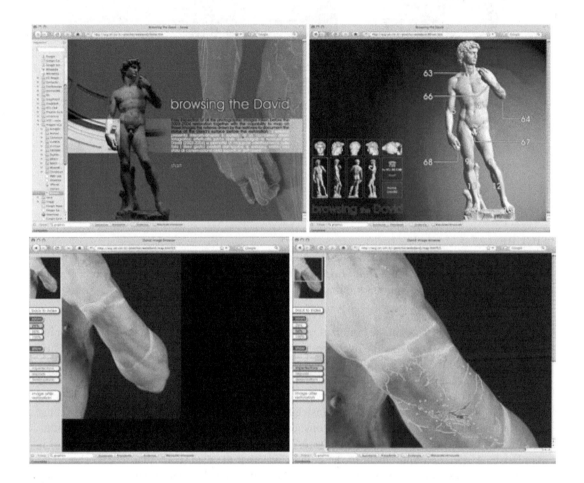

FIGURE 2.26

A few snapshots from the web-based system to browse the David's photographic campaign (pre- and post-restoration) and the restorers' reliefs (access to the web browser is unfortunately restricted to authorized users only). The four images present (the order in which the figure are referred is Top-Left , Top-Right , Bottom-Left , Bottom-Right): the home page; one of the selection panels (the one to select the front-view images); snapshot from the presentation interface, with the image depicting the left arm shown at 25% resolution; finally, the same image, presented at 50% of its resolution and with one of the reliefs in overlay.

2.8 Conclusions

An overview of several applications of digital 3D technology in the cultural heritage framework has been presented in this chapter. Even if the range of the possible uses of digital 3D models is extremely wide, according to the authors' opinion only part of the potential of this medium and related technology have been explored until now. There is still ample space for the design of innovative tools, new processing and visualization methodologies, and, finally, for the consolidation of these new approaches in the CH domain.

As shown throughout this chapter, several ongoing projects are facing the issue of designing 3D technology which could be easily adopted by CH experts, focusing on ease of use, ease of delivery, and reduced costs; one can easily predict that other innovative ideas will be surely presented in the next few years. It is possible that digital 3D models acquisition and visualization will assume, in the medium term, the same importance held nowadays by the digital photography medium in all CH contexts, from excavation to restoration, from study to musealization. Current years might be referred to, in the future, as the beginning of the *digital 3D revolution* in the CH domain. But reaching this goal depends not only on substantial technological progress, it will also depend on the capability to communicate and collaborate of the various communities (computer science, archaeology, history of art, restoration, chemistry, etc.) which are involved in this complex and important domain.

Acknowledgments

The research leading to these results has received funding from the European Community's Seventh Framework Programme (FP7/2007-2013) under grant agreement no. 231809 (EC IST IP project *3D-COFORM*) and from Tuscany Region under the *ST@RT* project.

Bibliography

[1] M. Callieri, F. Ponchio, P. Cignoni, and R. Scopigno, "Virtual inspector: A flexible visualizer for dense 3D scanned models," *IEEE Computer Graphics and Applications*, vol. 28, no. 1, 2008.

[2] J. Stumpfel, C. Tchou, T. Hawkins, P. Debevec, J. Cohen, A. Jones, and B. Emerson, "Assembling the sculptures of the Parthenon," in *VAST 2003* (A. C. D. Arnold and F. Niccolucci, eds.), (Brighton, UK), pp. 41–50, Eurographics, Nov. 5-7, 2003.

[3] A. Gaitatzes, D. Christopoulos, and M. Roussou, "Reviving the past: Cultural Heritage meets Virtual Reality," in *VAST '01: Proceedings of the 2001 Conference on Virtual Reality, Archeology, and Cultural Heritage* (New York), pp. 103–110, ACM, 2001.

[4] M. Carrozzino, C. Evangelista, and M. Bergamasco, "The immersive time-machine: A virtual exploration of the history of Livorno," in Archives of Photogrammetry, Remote Sensing and Spatial Information Systems, Vol. XXXVIII-5/W1, pp. 1682-1777, 2009.

[5] Y. Takase, K. Yano, T. Nakaya, Y. Isoda, T. Kawasumi, K. Matsuoka, T. Seto, D. Kawahara, A. Tsukamoto, M. Inoue, and T. Kirimura, "Virtual Kyoto: Visualization of historical city with 4D-GIS, virtual reality and web technologies," in *ISPRS 2008*, p. B5: 975, 2008.

[6] I. Besora, P. Brunet, M. Callieri, A. Chica, M. Corsini, M. Dellepiane, D. Morales, J. Moyés, G. Ranzuglia, and R. Scopigno, "Portalada: A virtual reconstruction of the entrance of the Ripoll monastery," in *3DPVT08: Fourth International Symposium on 3D Data Processing, Visualization and Transmission*, pp. 89–96, June 2008.

[7] Blender Foundation, "Blender - The free open source 3D content creation suite, available for all major operating systems under the GNU General Public License." More info on: http://www.blender.org/, 2010.

[8] P. Cignoni, F. Ganovelli, E. Gobbetti, F. Marton, F. Ponchio, and R. Scopigno, "Batched multi triangulation," in *IEEE Visualization 2005*, pp. 27–35, 2005.

[9] C. Baracchini, A. Brogi, M. Callieri, L. Capitani, P. Cignoni, A. Fasano, C. Montani, C. Nenci, R. P. Novello, P. Pingi, F. Ponchio, and R. Scopigno, "Digital reconstruction of the Arrigo VII funerary complex," in *VAST 2004*, pp. 145–154, 2004.

[10] M. Dellepiane, M. Callieri, F. Ponchio, and R. Scopigno, "Mapping highly detailed color information on extremely dense 3D models: The case of David's restoration," *Computer Graphics Forum*, vol. 27, no. 8, pp. 2178–2187, 2008.

[11] J.-A. Beraldin, M. Picard, S. F. El-Hakim, G. Godin, L. Borgeat, F. Blais, E. Paquet, M. Rioux, V. Valzano, and A. Bandiera, "Virtual reconstruction of heritage sites: Opportunities and challenges created by 3D technologies," in *International Workshop on Recording, Modeling and Visualization of Cultural Heritage*, 2005.

[12] D. Pletinckx, D. Callebaut, A. Killebrew, and N. Silberman, "Virtual-reality heritage presentation at Ename," *Multimedia, IEEE*, vol. 7, pp. 45–48, Apr-Jun 2000.

[13] Google Inc., "GoogleEarth - the world's geographic information at your fingertips." More info on: http://earth.google.com/, 2010.

[14] Microsoft Inc., "Microsoft Live Local." More info on: http://preview.local.live.com/, 2010.

[15] M. Forte, S. Pescarin, E. Pietroni, and C. Rufa, "Multiuser interaction in an archaeological landscape: The Flaminia project," in *Proceedings of the 2nd International Conference on Remote Sensing in Archaeology From Space to Place* (Rome 4-7 Dec. 2006), pp. 189–196, BAR Int., 2006.

[16] S. Pescarin, M. Forte, L. Calori, C. Camporesi, A. Guidazzoli, and S. Imboden, "Open Heritage: An Open Source approach to 3D real-time and web-based landscape reconstruction," in *VSMM Virtual Reality at Work in the 21th Century*, Belgium, 2005.

[17] K. Ikeuchi, A. Nakazawa, K. Hasegawa, and T. Ohishi, "The Great Buddha Project: Modeling Cultural Heritage for VR Systems through Observation," in *The Second IEEE and ACM International Symposium on Mixed and Augmented Reality* (Los Alamitos, CA, USA), p. 7, IEEE Computer Society, 2003.

[18] M. Dellepiane, M. Callieri, M. Fondersmith, P. Cignoni, and R. Scopigno, "Using 3D scanning to analyze a proposal for the attribution of a bronze horse to Leonardo da Vinci," in *The 8th International Symposium on Virtual Reality, Archaeology and Cultural Heritage (VAST 07)*, pp. 117–124, Eurographics, Nov 2007.

[19] M. Doerr and D. Iorizzo, "The dream of a global knowledge network: A new approach," *ACM Journal on Computing and Cultural Heritage*, vol. 1, no. 1, p. Article No.: 5, 2008.

[20] C. Andalo, M. Bicchieri, P. Bocchini, G. Casu, G. Galletti, P. Mando, M. Nardone, A. Sodo, and M. P. Zappala, "The beautiful "Trionfo d'Amore" attributed to Botticelli: A chemical characterisation by proton-induced X-ray emission and micro-Raman spectroscopy," *Analytica Chimica Acta*, pp. 279–286, Feb. 2001.

[21] A. S. Serebryakov, E. L. Demchenkoa, V. I. Koudryashova, and A. D. Sokolovb, "Energy dispersive X-ray fluorescent (ED XRF) analyzer X-Art for investigation of artworks," in *Proceedings of 5TH Topical Meeting on Industrial Radiation and Radioisotope Measurement Applications*, pp. 699–702, 2004.

[22] D. Manoogian, "Technical analysis of three paintings attributed to Jackson Pollock," *Harvard University Art Museums, Technical report*, 2007.

[23] T. Franken, M. Dellepiane, F. Ganovelli, P. Cignoni, C. Montani, and R. Scopigno, "Minimizing user intervention in registering 2D images to 3D models," *The Visual Computer*, vol. 21, pp. 619–628, Sept. 2005.

[24] M. Corsini, M. Dellepiane, U. Dercks, F. Ponchio, M. Callieri, D. Keultjes, A. Marinello, R. Sigismondi, R. Scopigno, and G. Wolf, "Cenobium - putting together the romanesque cloister capitals of the mediterranean region," *Bar International Series (Proceedings of III International Conference on Remote Sensing in Archaelogy)*, 2010.

[25] K. Ikeuchi, K. Hasegawa, A. Nakazawa, J. Takamatsu, T. Oishi, and T. Masuda, "Bayon digital archival project," in *Proceedings of Virtual Systems and Multimedia*, pp. 334–343, 2004.

[26] A. Banno, T. Masuda, T. Oishi, and K. Ikeuchi, "Flying laser range sensor for large-scale site-modeling and its applications in bayon digital archival project," *Internationa Journal on Computer Vision*, vol. 78, no. 2-3, pp. 207–222, 2008.

[27] G. Zotti, "Tangible heritage: Production of astrolabes on a laser engraver," *Computer Graphics Forum*, vol. 27, no. 8, pp. 2169–2177, 2008.

[28] P. Cignoni, E. Gobbetti, R. Pintus, and R. Scopigno, "Color enhancement for rapid prototyping," in *The 9th International Symposium on VAST International Symposium on Virtual Reality, Archaeology and Cultural Heritage*, Eurographics, 2008.

[29] A. Gruen, F. Remondino, and L. Zhang, "Image-based Automated Reconstruction of the Great Buddha of Bamiyan, Afghanistan," *Computer Vision and Pattern Recognition Workshop*, vol. 1, p. 13, 2003.

[30] B. Frischer, D. Abernathy, G. Guidi, J. Myers, C. Thibodeau, A. Salvemini, P. Müller, P. Hofstee, and B. Minor, "Rome reborn," in *SIGGRAPH '08: ACM SIGGRAPH 2008 new tech demos* (New York), ACM, 2008.

[31] C. Zollikofer, M. Ponce de Len, D. Lieberman, F. Guy, D. Pilbeam, A. Likius, H. Mackaye, P. Vignaud, and M. Brunet, "Virtual cranial reconstruction of Sahelanthropus Tchadensis," *Nature*, vol. 434, pp. 755–759, 2005.

[32] A. Willis, D. B. Cooper, et. al., "Bayesian pot-assembly from fragments as problems in perceptual-grouping and geometric-learning," in *International Conference on Pattern Recognition (ICPR), Vol. III*, pp. 297-302, 2002.

[33] M. Kampel and R. Sablatnig, "Virtual reconstruction of broken and unbroken pottery," in *International Conference on 3D Digital Imaging and Modeling (3DIM)*, (Los Alamitos, CA, USA), pp. 318–326, IEEE Computer Society, 2003.

[34] F. Melero, A. Leon, F. Contreras, and J. Torres, "A new system for interactive vessel reconstruction and drawing," *Bar International Series (Proceedings of Computer Applications in Archaeology, CAA 2003)*, no. 1227, pp. 78–81, 2003.

[35] A. Willis and D. B. Cooper, "Bayesian assembly of 3D axially symmetric shapes from fragments," in *Conference on Computer Vision and Pattern Recognition (CVPR), Vol. 1*, p. 8289, 2004.

[36] Q.-X. Huang, S. Flry, N. Gelfand, M. Hofer, and H. Pottmann, "Reassembling fractured objects by geometric matching," *ACM Transactions on Graphics*, vol. 25, no. 3, pp. 569–578, 2006.

[37] M. Dellepiane, M. Callieri, R. Scopigno, E. Paribeni, E. Sorge, N. S. Faro, and V. Marianell, "Multiple uses of 3D scanning for the valorization of an artistic site: The case of Luni terracottas," in *Eurographics Italian Charter Conference (Salerno, IT)*, pp. 7–14, Eurographics, 2008.

[38] F. Gabellone, "Virtual Cerrate: A DVR-based knowledge platform for an archaeological complex of the Byzantine age," in *CAA 2008, Computer applications and quantitative methods in Archaeology* (Budapest, Hungary), 2008.

[39] R. Levy and P. Dawson, "Reconstructing a Thule whalebone house using 3D imaging," *IEEE MultiMedia*, vol. 13, no. 2, pp. 78–83, 2006.

[40] P. Liverani, *I colori del bianco - Policromia nella scultura antica*. De Luca Editori d'Arte, 2004.

[41] R. Raskar, G. Welch, K. Lim Low, and D. Bandyopadhyay, "Shader lamps: Animating real objects with image-based illumination," in *Rendering Techniques (Proceedings of the 12th EG Workshop on Rendering Techniques, London, UK)* (S. J. Gortler and K. Myszkowski, eds.), pp. 89–102, Eurographics, June 25-27, 2001.

[42] J. Dorsey, H. Rushmeier, and F. Sillion, "Advanced material appearance modeling," in *SIGGRAPH '08: ACM SIGGRAPH 2008 classes* (New York), pp. 1–145, ACM, 2008.

[43] S. Bracci, F. Falletti, M. Matteini, and R. Scopigno, *Exploring David: Diagnostic tests and state of conservation*, Giunti Editore, 2004.

[44] M. Callieri, P. Cignoni, F. Ganovelli, G. Impoco, C. Montani, P. Pingi, F. Ponchio, and R. Scopigno, "Visualization and 3D data processing in David restoration," *IEEE Computer Graphics & Applications*, vol. 24, pp. 16–21, Mar.-Apr. 2004.

[45] P. Cignoni, "MeshLab: An open source, portable, and extensible system for the processing and editing of unstructured 3D triangular meshes." More info on: http://meshlab.sourceforge.net/, 2010.

[46] P. Cignoni, M. Callieri, M. Corsini, M. Dellepiane, F. Ganovelli, and G. Ranzuglia, "Meshlab: An open-source mesh processing tool," in *Sixth Eurographics Italian Chapter Conference*, pp. 129–136, 2008.

[47] B. Mirtich, "Fast and accurate computation of polyhedral mass properties," *Journal of Graphics Tools*, vol. 1, no. 2, pp. 31–50, 1996.

[48] P. Cignoni, M. Callieri, R. Scopigno, G. Gori, and M. Risaliti, "Beyond manual drafting: A restoration-oriented system," *Journal of Cultural Heritage*, vol. 7, no. 3, pp. 214–226, 2006.

[49] C. Baracchini, P. Lanari, R. Scopigno, and F. Tecchia, "Sicar: Geographic information system for the documentation of restoration analyses and intervention," in *Proceedings of SPIE*, Vol. 5146, p. 149, 2003.

[50] M. Callieri, P. Cignoni, and R. Scopigno, "Reconstructing textured meshes from multiple range RGB maps," in *7th International Fall Workshop on Vision, Modeling, and Visualization 2002*, (Erlangen), pp. 419–426, IOS Press, Nov. 20 - 22, 2002.

[51] M. Callieri, P. Cignoni, M. Dellepiane, and R. Scopigno, "Pushing time-of-flight scanners to the limit," in *The 10th International Symposium on Virtual Reality, Archaeology and Cultural Heritage VAST (2009)*, pp. 85–92, Eurographics, 2009.

3

Processing Sampled 3D Data: Reconstruction and Visualization Technologies

Marco Callieri, Matteo Dellepiane, Paolo Cignoni, Roberto Scopigno

Consiglio Nazionale delle Ricerche
Email: m.callieri@isti.cnr.it, m.dellepiane@isti.cnr.it,
p.cignoni@isti.cnr.it, r.scopigno@isti.cnr.it

CONTENTS

3.1 Introduction

This chapter describes the so-called 3D scanning pipeline (i.e., how raw sampled 3D data have to be processed to obtain a complete 3D model of a real-world object). This kind of raw data may be the result of the sampling of a real-world object by a 3D scanning device [1] or by one of the recent image-based approaches (which returns raw 3D data by processing a set of images) [2]. Thanks to the improvement of the 3D scanning devices (and the development of software tools), it is now quite

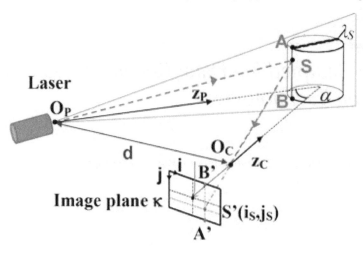

FIGURE 3.1
Schema of the *geometric triangulation* principle adopted by *active optical* scanners. (Image courtesy of Giovanna Sansoni et al. [5].)

easy to obtain high-quality, high-resolution three-dimensional sampling in relatively short times. Conversely, processing this fragmented/raw data to generate a complete and usable 3D model is still a complex task, requiring the use of several algorithms and tools; the knowledge of the processing tasks and solutions required is still the realm of well-informed practitioners and it often appears as a set of obscure black boxes to most of the users. Therefore, we focus the chapter on: (a) the geometric processing tasks that have to be applied to raw 3D scanned data to transform them into a clean and complete 3D model, and (b) how 3D scanning technology should be used for the acquisition of real artifacts. The sources of sample 3D data (i.e., the different hardware systems used in 3D scanning) are briefly presented in the following subsection. Please note that the domains of 3D scanning, geometric processing, visualization, and applications to CH are too wide to provide a complete and exhaustive bibliography in a single chapter; we decided to describe and cite here just a few representative references to the literature. Our goal is to describe the software processing, the pitfalls of current solutions (trying to cope both with existing commercial systems and academic tools/results), and to highlight some topics of interest for future research, according to our experience and sensibility. The presentation follows in part the structure of a recently published paper on the same subject [3].

3.1.1 Sources of Sampled 3D Data

Automatic 3D reconstruction technologies have evolved significantly in the last decade; an overview of 3D scanning technologies is presented in [1, 4, 5]. The technological progress in its early stages has been driven mostly by industrial applications (quality control, industrial metrology). Cultural heritage (CH) is a more recent applications field, but its specific requirements (accuracy, portability, and also the pressing need to sample and integrate the color information) were often an important testbed for the assessment of new, general-purpose technologies.

Among various 3D scanning systems, the more frequently used for 3D digitization are the so-called *active optical* devices. These systems shoot some sort of controlled, structured illumination over the surface of the artifact and reconstruct its geometry by checking how the light is reflected by the surface. Examples of this approach are the many systems based on the *geometric triangulation* principle (see Figure 3.1). These systems project light patterns on the surface, and measure the position of the reflected light by a CCD device located in a known calibrated position with respect

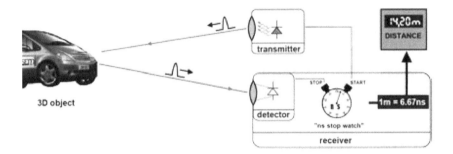

FIGURE 3.2
Basic principle of a *time-of-flight* scanning system. (Image courtesy of R. Lange et al. [6].)

to the light emitter. Light is either *coherent*, usually a laser stripe swept over the object surface, or *incoherent*, such as the more complex fringe patterns produced with video- or slide-projectors. Due to the required fixed distance between the light emitter and the CCD sensor (to be accurate, the apex angle of the triangle that interconnects the light emitter, the CCD sensor and the sampled surface point should be not too small, i.e., 10-30 degrees), those systems usually support working volumes that range from a few centimeters to around one meter.

The acquisition in a single shot of a much larger extent is supported by devices that employ the so-called *time-of-flight* (TOF) approach (see Figure 3.2). TOF systems send a pulsed light signal towards the surface and measure the time elapsed until the reflection of the same signal is sensed by an imaging device (e.g., a photo-diode). We can compute, given the light speed, the distance to the sampled surface; since we know the direction of the emitted light and the distance over this line, the XYZ locations of the sampled point can be easily derived. An advantage of this technology is the very wide working volume (we can scan an entire building facade or a city square with a single shot); when compared with the capabilities of triangulation-based systems, disadvantages of TOF devices are the lower accuracy and lower sampling density (i.e., the inter-sampling distance over the measured surface is usually in the range of the centimeter). The capabilities of TOF devices have been improved recently with the introduction of variations of the basic technology, based on modulation of the sensing probe, that allowed to increase significantly the sampling speed and maintain very good accuracies (in the order of a few millimeters).

Very promising but still not very common are the *passive optical* devices, where usually a large number of images of the artifact are taken and a complete model is reconstructed from these images [2,7,8]. These approaches, mostly based on consumer digital photography and sophisticated software processing, are presented in Chapters 4 and 5.

The quality of the contemporary commercial scanning systems is quite good if we take into account accuracy and speed of the devices; unfortunately, cost is still high, especially for CH applications, which are usually characterized by very low budgets. The introduction of inexpensive low-end laser-based systems (e.g., the Next Engine device [9]), together with the impressive improvement of image-based 3D reconstruction technologies (e.g., the Arc3D web-based reconstruction tool [10,11]) are very beneficial for the field. The availability of low-cost options is strategical to increase the diffusion of 3D scanning technology and to rise awareness and competence in the application domain. Moreover, a wider user community could ultimately drive to significant price reduction of the high-end scanning systems as well.

Finally, most of the existing systems consider just the acquisition of the external shape (geometric information), while a very important aspect in many applications is *color sampling*. This limitation is caused mainly by the fact that most 3D sensors come from the industrial world and therefore

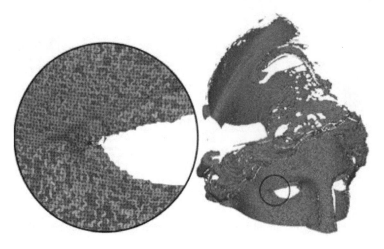

FIGURE 3.3
An example of a range map (from the 3D scanning of the Minerva of Arezzo, National Archeological Museum in Florence, Italy.)

are designed to focus more on geometry acquisition than color. Color is the weakest feature of contemporary technology, since the color-enabled scanners produce usually just a low-quality sampling of the surface color (with a notable exception of the technology based on multiple laser wavelengths [12], unfortunately characterized by a very high price that made their market share nearly negligible). Moreover, existing devices sample only the *apparent* color of the surface and not its *reflectance properties*, that constitute the characterizing aspect of the surface appearance. There is wide potential for improving current technology to cope with more sophisticated surface reflection sampling.

3.2   Basic Geometric Processing of Scanned Data

Unfortunately, almost every 3D scanning system does not produce a final, complete 3D model but a large collection of raw data which have to be post-processed. This is especially the case of all *active optical* devices, since only the portion of the surface directly visible from the device is captured in a single shot. The main problem of a 3D acquisition is indeed this fragmentation of the starting data: scanners do not produce a complete model by simply *pushing a button* (unless the application field is very restricted, such as some devices designed specifically for scanning small objects, e.g., teeth). On the other hand, the good news is that most of the scanning technologies produce raw data which are very similar: the so-called *range maps* (see Figure 3.3). This homogeneity in the input data makes the 3D scan processing quite independent from the specific sampling device adopted. Of course, depending on: the technology and specifications of the employed sensor; the scale, material, and nature of the target object; and the kind of 3D model needed, the various processing phases might slightly vary, but the main phases and the workflow will generally still fall in the scheme we present here. The result of a single scan, a range map, is the counterpart of a digital image: while a digital image encodes in each pixel the specific local color of the sampled surface, the range map contains for each sample the geometrical data (a point in XYZ space) which characterize the location in the 3D space of the corresponding small parcel of sampled surface. The range map therefore encodes the geometry of just the surface portion which can be seen and sampled from a selected

FIGURE 3.4
Two examples of results obtained with: a time-of-flight (TOF) scanner, depicting the Dome square at S. Gimignano (Italy); laser-line triangulation scanner, depicting a capitol from the cloister of the Dome at Cefalú (Italy) .

viewpoint (according to the specific features of the scanning device). The complete scan of an artifact requires usually the acquisition of many shots taken from different viewpoints, to gather complete information on its shape. Each shot produces a range map. The number of range maps required to sample the entire surface depends on: the working volume of the specific scanner; the surface extent of the object; and its shape complexity. Usually, we sample from a few tens up to a few hundred range maps for each artifact. Range maps have to be processed to convert the data encoded into a *single*, *complete*, *non-redundant* and *optimal* 3D representation (usually, a triangulated surface or, in some cases, an optimized point cloud). Examples of digital models produced with 3D scanning technology are presented in Figure 3.4 .

3.2.1   The 3D Scanning Pipeline

As previously stated, the homogeneous nature of the raw data makes the processing quite independent from the scanning device used and the processing software adopted. The processing of the raw data coming from the 3D scanning device is divided in subsequential steps, each one working on the data produced in the previous one; hence the name *3D scanning pipeline*. These steps are normally well recognizable in the various software tools, even if they are implemented with very different algorithms. Before or after each step, there may be a *cleaning stage*, aimed at the elimination of small defects of the intermediate data that would make the processing more difficult or longer. The overall structure of the 3D scanning pipeline is presented in an excellent overview paper [13], which describes the various algorithms involved in each step. Some new algorithms have been proposed since the date of publication of that review paper, but the overall organization of the pipeline is not changed.

The processing phases that compose the 3D scanning pipeline are:

- *Alignment* of the range maps. By definition the range map geometry is relative to the current sensor location (position in space, direction of view); each range maps is then in a different reference frame. The aim of this phase is to place the range maps into a common coordinate space where all the range maps lie in the correct position with respect to each another. Most alignment methods do require user input and rely on the *overlapping regions* between adjacent range maps.

- *Merge* of the aligned range maps to produce a single digital 3D surface (this phase is also called *reconstruction*). After the alignment, all the range maps are in the correct position, but the object is still composed by multiple overlapping surfaces with lot of redundant data. A single, non-redundant representation (usually, a triangulated mesh) has to be reconstructed out of the many partially overlapping range maps. This processing phase, generally completely automatically, exploits the redundancy of the range maps data in order to produce (often) a more correct surface where the sampling noise is less evident than in the input range maps.

- Mesh *editing*. The goal of this step is to improve (if possible) the quality of the reconstructed mesh. For example, usual actions are to remove or reduce impact of noisy data or to fix the unsampled regions (hole filling).

- Mesh *simplification* and *multiresolution encoding*. 3D scanning devices produce huge amount of data; it is quite easy to produce 3D models so complex that it is impossible to fit them in RAM or to display them in real time. This complexity has usually to be reduced in a controlled manner, by producing discrete level of details (LOD) or multiresolution representations.

- *Color mapping*. The information content is enriched by adding color information (an important component of the visual appearance) to the geometry representation.

All these phases are supported either by commercial [14–16] or academic tools [17–19]. Unfortunately, given the somehow limited diffusion of 3D scanning devices (mainly due to their cost) and the still restricted user base of this kind of technology in non-industrial environments, it is not so easy to find free software tools to process raw 3D data. The availability of free tools would be extremely beneficial to the field, given the considerable cost of the major commercial tools, to increase awareness and dissemination to a wider user community. These instruments could be very valuable for the CH institutions that want to experiment with the use of this technology on a low budget. Most of the non-commercial development of processing tools has been carried out in the academic domain; some of those efforts are sponsored by European projects [20,21]. The very recent availability of low-cost scanning devices or of image-based 3D acquisition approaches, paired with availability of free processing tools, could boost the adoption of 3D scanning technologies in many different application domains. The different phases of the scanning pipeline are presented in a more detailed manner in the following subsections.

Alignment
The *alignment* task converts all the acquired range maps in a common reference system. This process is usually partially manual and partially automatic. The user has to find an initial, raw registration between any pair of overlapping range maps; the alignment tool then uses this approximate placement to compute a very accurate registration of the two meshes (*local registration*). The precise pair-wise alignment is usually performed by adopting the Iterated Closest Point (IPC) algorithm [22, 23], or variations of the same. This pairwise registration process (repeated on all pairs of adjacent and overlapping range maps) is then used to automatically build a *global registration* of all the meshes, and this last alignment is enforced among all the maps in order to move everything in a unique reference system [24]. Range map alignment has been the focus of a stream of recent papers; we cite here only one single result [25], the interested reader can find references to other recent approaches in the previous work.

Quality of the alignment is crucial for the overall quality of the merged model, since errors introduced in alignment may be the cause of wrong interpolation while we merge the range maps. Since small errors could add one to the other, the resulting global error can be macroscopic. An example of 3D acquisition where this potential effect has been monitored and accurately measured is the acquisition of Donatello's Maddalena [26]. A methodology for quality control is proposed in that paper, where potential incorrect alignment produced by ICP registration is monitored and

FIGURE 3.5

An example of a pair-wise alignment step: the selection of a few corresponding point pairs on the two range maps allows one to find an initial alignment, which is then refined by the ICP algorithm.

corrected by means of data coming from close-range digital photogrammetry. The alignment phase is usually the most time-consuming phase of the entire 3D scanning pipeline, due to the substantial user contribution required by current commercial systems and the large number of scans sampled in real scanning campaigns. The initial placement is heavily user-assisted in most of the commercial and academic systems (requiring the interactive selection and manipulation of the range maps). Moreover, this kernel action has to be repeated for all the possible *overlapping range map pairs* (i.e., in average 6-8 times the number of range maps). Consider that the scanning of a 2 meters tall statue generally requires from 100 up to 500 range maps, depending on shape complexity and sampling rate required. If the set of range maps is thus composed by hundreds of elements, then the user has a very complex task to perform: for each range map, find which are the partially-overlapping ones; given this set of overlapping range maps, determine which one to consider in pair-wise alignment (either all of them or a subset); finally, process all those pair-wise initial alignments. If not assisted, this becomes the major bottleneck of the entire process: a slow, boring activity that is also really crucial for the accuracy of the overall results (since an inaccurate alignment may lead to very poor results after the range maps merging). An improved management of large sets of range maps (from 100 up to 1,000) can be obtained by both providing a hierarchical organization of the data (range maps divided into groups, with atomic alignment operations applied to an entire group rather than to the single scan) and by using a multiresolution representation of the raw data to increase efficiency of the interactive rendering/manipulation process over the range maps. Moreover, since the standard approach (user-assisted selection of each overlapping pair and creation of the correspondent alignment arc) becomes very impractical on large sets of range maps, tools for the automatic setup of most of the required alignment actions have to be provided (see subsection 3.2.2).

Merging

Reconstruction/merging from point samples or range maps has been one of the more active fields on scanning-related research in the last few years. A very important feature of a reconstruction code is to perform a *weighted integration* of the range maps (or point set) and not just joining them. Since we usually have a high degree of overlap and sampled data are noisy, a weighted integration can significantly improve the accuracy of the final result by reducing the impact of most of the noisy samples. Another important feature of a reconstruction code is the capability to fill up small holes (i.e., region not sampled by the scanner, see subsection 3.2.3). Finally, since reconstruction algorithms require a very large memory footprint on a big dataset, they have to be designed to work locally on subsections of the data, loading only the data subset involved in the generation of a single portion of

FIGURE 3.6
Range maps are often taken according to a regular order: an example of *circular* radial acquisition performed around a statue's head (left); an example of a *raster-scan* scanning order adopted for the acquisition of a bas-relief (right).

the final results (*out-of-core* reconstruction).

Fit Results to Application Requirements
Simplification/multiresolution is another key ingredient in scanned data processing. When we choose a reconstruction kernel region approximately equal to the inter-sampling distance used in scanning, the reconstructed models may become huge in size (i.e., many millions of triangles). Most applications require significant complexity reduction in order to manage these models interactively. Two problems arise when we try to simplify such models: we need solutions working on external memory to cope with these big models [27]; and simplification has to be accurate [28, 29] if we want to produce high-quality models and accurate interactive visualization (possibly, based on multiresolution data representation schemes). These issues are presented in Section 3.5.

Color Mapping
Color is usually the final stage of the pipeline. Here we have to support the reconstruction of textured meshes from a sampling of the object's surface reflection properties. The easiest and most common approach is the acquisition and mapping of the *apparent color* (reflected color, illumination-dependent) using a digital photo camera. But we can also immerse the artifact into a controlled lighting setup, to produce a more sophisticated acquisition of the reflection properties of the object's surface (see for example the methodology for the acquisition of bi-directional reflectance distribution function - BRDF proposed in [30]). This latter approach is unfortunately often impractical for those applications, such as acquisition of CH artifacts, which require one to perform the acquisition in real-life conditions (i.e., when we cannot move the object to be scanned in an acquisition lab). The issues involved in color acquisition and management are discussed in detail in Section 3.3.

3.2.2 Implementing Range Maps Alignment as an Automatic Process

We have already introduced that range maps alignment is usually the most time-consuming phase in the 3D scanning pipeline. But the goal is obviously to make scanning as much as possible an automatic process. Therefore, completely automatic scanning systems have been proposed. Some of those systems are based on the use of complex and costly positioning/tracking machinery (e.g., see [31] or the several commercial scanners based on optical or magnetic tracking of the scanning head location); others adopt passive silhouette-based approaches which do not extend well to the acquisition of

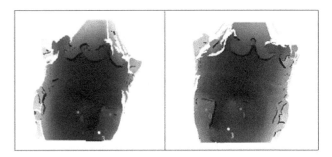

FIGURE 3.7
An example of four matching point pairs selected on two range maps by an automatic matching algorithm.

medium- or large-scale artifacts. An alternative approach is to design new algorithms for processing the data produced by standard scanning systems, which are able to transform the classical scanning pipeline into a mostly unattended process. In particular, the range maps alignment phase is the only task where a considerable human intervention is required. Several papers proposed methods for *automatic alignment*, usually based on some form of shape analysis and characterization [32].

The general alignment problem can be made more easy to manage by considering some assumptions which usually hold in practical use. While designing a new automatic registration solution [33], we started from a few initial conditions directly gathered by our experience in 3D scanning. First, the *detection of the pairs of overlapping range maps* can be reduced to a simpler task, once we notice that 3D acquisition is usually done by following simple scanning pose paths. Users usually acquire range maps in *sequences*, following either a *vertical*, *horizontal*, *raster-scan*, or *circular* translation of the scanning system (see Figure 3.6). The different types of sequences share a common property: they contain an ordered set of n range maps, such that range map R_i holds a significant overlapping with at least R_{i-1} and R_{i+1}. Vertical, horizontal, or raster-scan stripes are often produced when acquiring objects like bas-reliefs, walls, or nearly planar items. Circular stripes are indeed more useful when acquiring objects like statues, columns, or objects with an axial symmetry.

If we can assume that the acquisition has been performed using one of these stripe-based patterns, then we may reduce the search for overlapping and coarse registration to each pair of consecutive range maps (R_i, R_{i+1}). An automatic registration module can process each couple (R_i, R_{i+1}), to produce in output the roto-translation matrix M_i that aligns R_{i+1} to R_i. Matrix M_i can be computed by applying some basic geometric processing to the two range maps: *feature points* can be detected by evaluating a shape descriptor on the two meshes (or point set); potential *corresponding feature points pairs* can be detected by adopting RANSAC-like approaches: from these possible point pairs, we can select the one which produces, after ICP alignment, the matrix M_i which performs the best alignment (see [33] as an example of this type of solution). Many other approaches for the design of shape feature characterization and matching are also possible [34, 35]. As an alternative to geometry-based solutions, it is also possible to work with an image-based approach [36]: correspondent point pairs can be also retrieved by looking to image features in the RGB channel associated to the XYZ samples (under the assumption that the scanner produces self-registered shape and color samples).

The subset of registration arcs is not complete nor sufficient by itself when we restrict the search to consecutive pairs in linear sequences, since we usually have many other potential overlaps between range maps. On the other hand, information on those consecutive pairs is sufficient for the application of an intelligent ICP-based solution that allows one to complete the graph. Designing a smart alignment tool able to complete the needed arcs (interconnecting R_i with all the overlapping range maps, not just R_{i-1} and R_{i+1}) is easy. We can represent the space of the set of range maps

FIGURE 3.8
The coarse alignment obtained over a set of range maps representing a bas-relief *(top)* and the final model *(bottom)* obtained after automatic completion of all overlapping pairs; alignment has been performed in non-attended mode using the solution presented in [33].

with a *spatial indexing* data structure, a regular grid which encodes for each 3D cell the set of range maps passing through that sub-volume. This data structure allows an easy automatic detection of the potential overlaps between generic pairs of range maps. We can then iterate automatic ICP-based alignment on all those overlapping range map pairs which have not been already processed according to the linear ordering. The alignment tool can therefore introduce all needed arcs (in a completely unattended manner), by selecting and processing only those arcs which satisfy a minimum-overlap factor.

The automatic registration approach sketched above [33] has been tested on many complex scanning campaigns (where each range map is usually affected by noise, artifacts, and holes). An example concerning a bas-relief is shown in Figure 3.8, whose approximate length is 2.5 meters. In this case two raster-scan (snake-like) stripes were acquired and processed, for a total of 117 meshes (about 45.5M vertices). The overall automatic alignment requires usually much less than the raw scanning time and it is therefore sufficiently fast to run in background during the acquisition, processing all the scans in sequence as soon as they are produced.

Solutions similar to the ones presented in this subsection are unfortunately still not provided by commercial software; inclusion of automatic alignment would speed up the processing of scanned data, reducing the manpower required and the overall costs.

3.2.3 Enhancing Incomplete Surface Sampling

According to 3D scanning experience, obtaining a *complete sampling* of a complex artifact surface is often impossible [31]. Various are the reasons why we usually end up with an incomplete scan: presence of self-obstructing surfaces; presence of small cavities or folds; sections of the surface which are not cooperative with respect to the optical scanning technology adopted (highly reflective materials like polished metals, transparent components such as glass or gems, surfaces which do not reflect the emitted light probe, etc). In all those cases, we have to decide if the model has to be completed or if we have to keep it incomplete. In the CH domain we are usually asked to produce models which should contain only sampled geometry, i.e., the use of software solutions which fill up the gaps is not allowed. On the other hand, incomplete digital models perform very poor in visualization, since the holes usually attract the observer's attention much more than the other clean parts. *Clean* sampled surfaces are therefore needed for visualization, obtained by closing all the gaps with plausible surface patches. A very nice extension to available data formats would be an attribute that could allow us to differentiate between sampled and interpolated geometry (a sort of *confidence value* assigned to each geometric element), making it possible to make visually evident those two different data components. This would be an interesting addition to provenance data encoding.

Gaps filling can be obtained by two orthogonal approaches: *volumetric* and *surface oriented*. In the first case the holes can be filled:

- At reconstruction time, for example by enhancing volumetric reconstruction approaches based on a discrete distance field with a diffusion process which extends the distance field in regions not covered by scanned samples [37] or by adopting methods based on the Poisson reconstruction formulation [38];

- After the reconstruction, by putting the model in a volumetric grid and devising solutions able to produce a watertight surface [39,40].

Unfortunately, these approaches make it very hard to differentiate sampled and interpolated geometry. Moreover, methods based on volumetric diffusion are very complicated to use because steering the diffusion process to obtain a plausible completion surface is not easy; a time-consuming trial and error process is usually needed to find the parameters best fitting a given dataset. Poisson-based reconstruction is more frequently used, due to ease of use and free availability (both as source code distributed by authors and as a feature available in open source tools, e.g., MeshLab [41]).

On the other hand, *geometric processing* solutions can be devised to detect and fill unsampled regions. Accordingly, *surface-oriented* approaches try to detect and close holes preserving triangle shape and curvature [42,43]. The problem is not simple because we need geometrically robust solutions, able to close any gap with a surface patch which should share curvature continuity with respect to the surface regions adjacent to the open border. Some issues are the necessity to deal with self-intersections and isles, the algorithm speed and robustness, and the difficulty in creating fully automatic but reliable methods. Moreover, in some cases a basic concept of curvature continuity is not enough, since a missing region can contain surface features (such as a given texture or a carved detail) that we may want to reproduce in a way conforming with adjacent regions. So called *surface inpainting* methods have been proposed to support intelligent cut and paste of surface detail from one completely sampled region to a partially sampled region [44,45].

3.3   Color Sampling and Processing

There are many application domains which require not just shape sampling, but also accurate acquisition and management of color data. Sophisticated approaches for sampling the surface reflection characteristics have been proposed (e.g., generic BRFD sampling [30] or technologies for the acquisition of the reflectance of human faces [46,47]). In some cases, those approaches are unfortunately too complicated to be massively applied in all those fields where we do not have the pleasure to work in controlled lab conditions, as it is the case of CH digitizations (usually performed in crowded museums and under a not proper illumination setup). Let's describe first the easier approach to gathering color data and then how those data can be encoded in digital models. At the end of the section, we will review the other more sophisticated approaches to gathering samples of the surface reflection properties.

3.3.1 Basic Acquisition of Color Data

The basic approach, acquiring just the so-called *apparent color* and mapping those samples to the 3D model, is still widely used in most of the practical cases. A series of pictures can be taken with a digital camera, trying to avoid shadows and highlights by taking them under a favorable lighting setup; these photographs are then stitched onto the surface of the object. However, even in this simpler case, the processing needed to build a plausible texture is not straightforward [48]. Naive mapping of apparent color on the mesh can produce severe discontinuities that are due to the varying illumination conditions of the surface sampled by the photos (perceived color depends on specific illumination and direction of the viewer in the instant the photo is shot). Many different approaches have been proposed to reduce the aliasing and to produce seamless color mapping; we cite here only some representative papers: using the range intensities produced by some active optical scanning devices to correct the color information [49]; detecting and removing cast shadows [50], which are usually a major problem in color acquisition in outdoor scenes; devising methods for computing the inverse illumination (i.e., recovering approximate surface reflectance from a sampling of the real surface under known illumination conditions [51, 52].)

3.3.2 Recovering Camera Parameters

A basic problem in managing color information is how to register the images with the geometric data in a time-efficient way. Once intrinsic (focal length and distortion of the camera lenses) and extrinsic parameters (view specifications) have been computed for each image by registering it onto the geometric model, many approaches exist to map the color info on the 3D model, based on mesh parameterization or color-per-vertex encoding. The bottleneck in the color pipeline is the *image-to-geometry registration* phase, a complicated time-consuming phase which requires substantial intervention of a human operator since the current approach is based on the selection of several corresponding point pairs which link each 2D image to the 3D mesh [53].

We designed a tool to support image registration, *TexAlign* [54], which solves the image-to-geometry registration problem by constructing a *graph of correspondences*, where: the 3D model and all the images are represented as nodes; a link is created for any correspondence defined between two nodes (implementing either an *image-to-geometry* or an *image-to-image* correspondence). This graph of correspondences is used: (a) to keep track of the work done by the user; (b) to infer automatically new correspondences from the instantiated ones; and (c) to find the shortest path, in terms of the number of correspondences that must be provided by the user, to complete the registration of all the images. The goal is to assist the user in the management of large sets of images. This system has been used to map several complex photographic samplings (e.g., in the framework of the Michelangelo's David restoration we mapped on the digital 3D model 61 images showing the pre-restoration status and 68 images depicting the post-restoration condition [55], see Figure 3.9).

Analogously to the range map alignment problem, this image registration phase should be as much as possible solved automatically. We need fully automatic and robust approaches able to solve the general problem (i.e., a large and complex object, where each image covers only a subset of its overall extent). Finding automatically the correspondences between a set of 2D images and a 3D mesh is not an easy task, also because the geometry features are usually less dense than the image features we can retrieve in the photographs. A possible research direction could be to move from the usual search for *image-to-geometry* correspondences to a context where we use both *image-to-geometry* and *image-to-image* correspondences. As shown in [36] and other recent works, finding correspondences in images is simpler than detecting *image-to-geometry* correspondences. Since a large number of *image-to-image* correspondence pairs can be detected in an automatic manner, we can deploy that information to speed up the overall image registration process, or to solve those cases where a single image covers a region where the surface has insufficient shape features to allow an accurate selection of *image-to-geometry* correspondences. We have recently proposed an approach

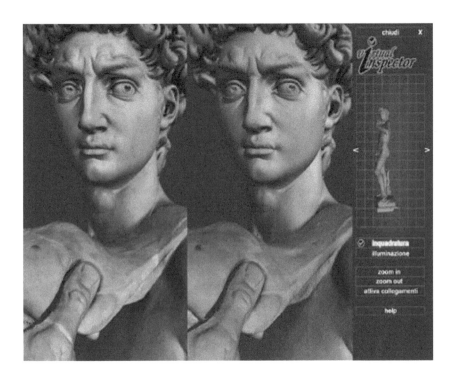

FIGURE 3.9 (SEE COLOR INSERT)
The David model is shown with color mapping; on the left is the pre-restoration status (61 images mapped), while the post-restoration status is shown on the right (another set of 68 images). The two colored David models are rendered in real time with the *Virtual Inspector* system [56].

FIGURE 3.10 (SEE COLOR INSERT)
The image presents a result of the automatic image-to-geometry registration: given the image on the right, the proper projective transformation is computed by finding the best matching between the input image and renderings of the mesh with vertices colored according to a combined normal vector and accessibility shading factor [57].

based on *Mutual Information* [57], a statistical measure of non-linear correlation between two data sources often used in medical images registration. The main idea is to use mutual information as a similarity measure between each image to be registered and some renderings of the geometric 3D model, in order to drive the registration in an iterative optimization framework. The problem can be divided into two sub-problems: finding an approximate solution, the *rough alignment* where we find a view of the geometric model that sufficiently matches the photographic image we want to align. For example, a rough alignment can be found with a simple greedy approach, by generating a large number of suitable initial views and checking which one best matches the input image. Then, this view is refined to find the optimal one, the *fine alignment*, by an iterative process that slightly modifies the view over the 3D model, produces a rendering, and checks the similarity of the rendering with the input image. We demonstrate that some illumination-related geometric properties, such as surface normals, ambient occlusion, and reflection directions can be efficiently used for this purpose, improving the convergence speed of the search. After a comprehensive analysis of such properties we proposed a way to combine these sources of information in order to improve the performance of our automatic registration algorithm [57]. The proposed approach can robustly cover a wide range of real cases and can be easily extended.

3.3.3 Mapping Complex Photographic Detail on 3D Models

Another issue concerning the high-quality mapping of color information over a 3D model is how to manage the very high resolution and redundant color sampling. Just using plain standard or consumer digital cameras one can easily obtain a very high resolution color sampling. Currently, the resolution of consumer digital cameras allows one to obtain a sampling that is an order of magnitude higher than the one of triangulation-based scanning devices; therefore, we might have at least 10 color/reflectance samples falling in the vicinity of each geometric sampled point. A medium size object (e.g., a statue) can be easily sampled with 50-100 images leading to hundred (or even thousand) of million pixels. The need to manage such texturing resolutions in real time opens several issues concerning the selection of the best texture mapping representation, how to extend multiresolution representation to hold both geometry and texture data, and the impact on the rendering speed. Beside the issues concerning the size of the raw data, the problem of mapping complex photographic detail on 3D models raises at least two main problems:

- *Efficient storing of color.* The possible higher density of the color data with respect to geometric data can make unsuitable direct approaches that store the color information over the primitives of the mesh (e.g., color-per-vertex). Texture mapping techniques would be more appropriate, but scanned meshes are usually highly complex and topologically unclean and these conditions make difficult the task of building a good texture parameterization.

- *Coherently mixing the photographic information.* Creating a high-quality texture content is a critical step. Merging all the photos by a resampling process is a valid option, but it should not decrease the quality of the color information (e.g., resampling from multiple images often introduces blurring and ghost effects).

Texture Parameterization

For the specific task of building a mesh parameterization (taking in input the 3D mesh and the mapped photos) one approach could be to determine, given a set of images, an optimal partition of the mesh such that each surface patch could be mapped with (a portion of) one single input image. This approach has been used in a number of early papers (see for example [58, 59]). The main disadvantage is that the number of pieces in which the mesh is partitioned grows up rapidly with the number of involved photos [48].

Another possibility is to adopt classical mesh parameterization techniques that are able to build a good-quality parameterization of the mesh over a texture space. Then, we can resample a texture from the input set of images, at the proper resolution required by the specific application. These approaches (see [60] for a survey on this subject) have often the drawback that, to avoid great distortion in the parameterization, the result is composed by many small texture pieces whose discontinuities can give rise to many visualization issues. A different approach has been recently proposed [61] that tries to overcome these limitations by computing the resulting parameterization in a continuous abstract texture domain that has the same topology of the input mesh and presents no discontinuity.

Resampling Color Data

Given a 3D model and the cameras associated to each input photo, we need a robust algorithm for color resampling from several input photographs, which can produce in output a coherently mixed blending of all the input images either as a texture map or as resampled per-vertex colors. A multivariate blending approach has been recently proposed [62]. This blending framework allows one: to use many different heuristics to characterize the image content (pixel by pixel) in terms of intrinsic quality of each sample; to produce, for each vertex on the surface or pixel on the output texture space, an optimal weighted average of all the available pixel samples which maps on the requested geometric location/parcel.

3.3.4   Advanced Color Data Acquisition: Sampling Surface Reflection Properties

Obviously, we envision future systems able to encode not only the reflected radiation (apparent color), but also able to sample the reflection properties of the surface. The goal is to move from apparent color acquisition towards BRDF sampling or at least the acquisition of approximations of the surface BRDF. To make those solutions practical we need to improve current methods in terms of ease of operation, reduced dependency from highly controlled lighting environment, and reduced processing complexity.

An approximate approach has been recently proposed, with the aim of adopting a very easy to deploy lighting setup that should allow one to remove lighting artifacts from the acquired photos. Flash light is a very easy way to illuminate an object or an environment, but it is rarely considered in most of the computer graphics and computer vision literature concerning color acquisition. This is due to the large amount of artifacts introduced by this specific light source, and to the difficulty in modeling its behavior in space. A simple method of using flash light in the context of color acquisition and mapping on digital 3D models has been recently proposed [63], based on a technique that allows one to characterize the emission in the 3D space of the flash light (see Figure 3.11). The calibration process allows one to acquire accurate information on how the light produced by a given flash is distributed in the 3D space sampled by the camera and how the color values sampled in that region are modified by the flash light. Moreover, given a calibrated flash and the extrinsic parameters for

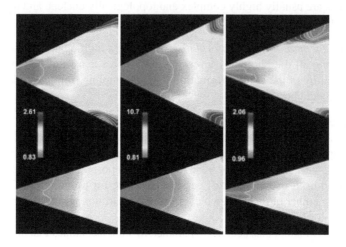

FIGURE 3.11
Some examples of the flash characterization obtained on three different cameras. The images show
plots of an orizontal plane (top) and a vertical plane (bottom) intersecting the working camera space,
for a Nikon reflex camera(Left), a Canon reflex camera (Center) and a compact Casio camera (Right).

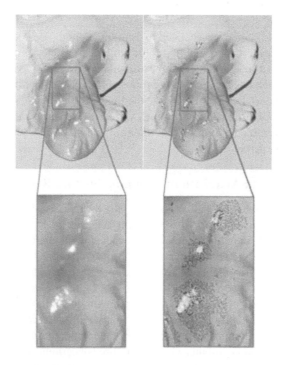

FIGURE 3.12 (SEE COLOR INSERT)
An example of highlights detection. Upper row: an input flash image and the same image after
detection of highlights (blue is the highlight border, cyan is the internal highlight region). Lower row:
two detail views, where highlights candidates selected by taking into account just geometric criteria
(surface normals) are rendered in green and the ones more robustly detected are rendered in cyan.

FIGURE 3.13
An example of shadows detection: left, the original image; right, the shadow detection map.

each single photo (location and direction of view of the camera), we are able to accurately estimate the flash position with respect to the framed object. We are thus able to apply automatic methods to detect and remove artifacts (specular highlights and shadows, see Figures 3.12 and 3.13) from a set of images which are registered to a 3D model. These methods are integrated in the context of a color mapping framework and allow one to obtain improved colored 3D models. These results fit very well the context of cultural heritage, where the acquisition of color has often to be performed on site, with very few constraints on the lighting setup. This approach can be considered as a first result to fill the gap between basic reflected color versus complex BRDF acquisition methods.

Other promising approaches support sophisticated sampling of the reflection properties by adopting *image-based rendering* approaches: instead of producing accurate 3D encoding with poor-quality surface reflection data, we could aim at acquiring and using the classical 2D media, but enhanced with sophisticated reflection encoding. We have a spectrum of possibilities, from the one-view and multiple-lighting approach of Polynomial Texture Maps (PTM) [64], to the more demanding multi-view approach granted by Light Field rendering [65]. In all these cases (see [66] for an overview), we get rid off the 3D model and adopt approaches which sample the artifact with a multitude of images. To produce any view requested at visualization time we process/interpolate the knowledge granted by the sampled image set. In most cases, this implies a huge quantity of image data to be acquired (thus long acquisition times are usually needed), stored, and accessed in real time, for example making web-based visualization not easy.

3.4 MeshLab: An Open Source Tool for Processing 3D Scanned Data

As we have described in previous sections, processing 3D sampled data is not an easy task. We need several algorithms and the pipeline is neither fixed nor easy to be managed, particularly in the case the operator is not an ICT expert. Even if a number of commercial solutions target the market of processing 3D sampled data, we found that the community is also searching for complete software solutions, possibly much cheaper than current SW tools. The introduction of new 3D scanners sold at the cost of a good digital camera makes the request of cheap SW solutions even more pressing. Moreover, most of the commercial packages in this field target the reverse engineering application domain, which has different needs and priorities with respect to the requirements of the CH domain,

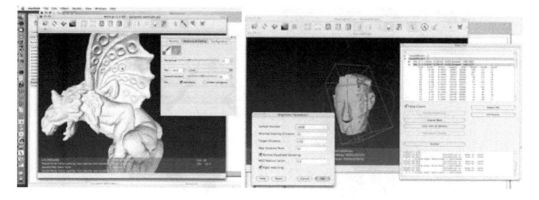

FIGURE 3.14

Snapshots of MeshLab in action: the mesh under processing is interactively displayed and the user can work on it by means of a large set of unattended parametric filters, or by means of interactive tools, like the one shown in the image on the left, where the user is smoothing out some features of the object with some simple mouse strokes; on the right we show a range maps alignment session.

so there is a demand for a set of tools that are more tailored to the processing of sampled 3D historical artifacts.

This objective was at the base of the ISTI-CNR effort in designing and implementing MeshLab as an open source and extendable mesh processing tool [19, 41]. MeshLab was designed as a general 3D mesh processing system tool with the following primary objectives in mind:

- **Mesh processing oriented.** The system should try to stay focused on mesh processing tasks instead of aspiring to contribute to mesh editing and mesh design, where a number of fierce competitors already crowd the software arena (notably Blender, 3D Max, Maya, and many others).

- **Ease of use**. The tool should be designed so that users without high 3D modeling skills could use it (at least for the most basic functionalities).

- **Depth of use**. The tool should be designed so that advanced users can tweak and extend it by adding functionality and or by modifying all the involved parameters.

- **Efficiency.** 3D scanning meshes easily reach several million primitives, so the tool should be able to manage efficiently very large meshes.

As a result, MeshLab presents itself as an intuitive *mesh viewer* application, where digital 3D models, stored in a variety of formats, can be loaded and interactively inspected in an easy way, by simply dragging and clicking on the mesh itself. MeshLab supports an ever growing variety of 3D formats (all the most common formats are supported) to accommodate the broadest set of users. Once having loaded a mesh, the user can work on it by means of a large set of direct parametric filters that perform unattended automatic tasks like surface reconstruction, smoothing, re-meshing or simplification, or by mean of interactive tools (for example, range maps registration).

The system has proved a success over any initial prediction. MeshLab is usually downloaded more than $6,000 - 8,000$ times every month; MeshLab's user community is thousands of users from all over the world. Users come from hundred of universities and renowned commercial companies that have found MeshLab useful in many different contexts, widening the original CH domain we were focusing on.

One of the interesting characteristic of MeshLab is the presence of all the basic resources for processing 3D scanning data. MeshLab already provides tools for: cleaning sampled range maps, performing range maps alignment, merging/reconstruction, simplification, transferral of color information from range maps to meshes, measuring differences between objects, and providing texture parameterization algorithms. MeshLab is an evolving system, being one of the technologies under further development in the framework of the EC IP "3D-COFORM" project (2009-2012) [21]. Some of the color management algorithms described in the previous section are among the new functionalities that we are including in MeshLab.

3.5   Efficient Visualization and Management of Sampled 3D Data

Some issues arise from the very dense sampling resolution granted by modern scanning devices. Being able to sample in the order of ten points per squared millimeter or more (in the case of triangulation-based systems) is of paramount value in many applications which need a very accurate and dense digital description. On the other hand, this information is not easy to process, render, and transfer. Therefore, excessive data density becomes a problem for many applications. In effect, the availability of 3D scanned data was a driving force for intense research on more efficient geometric data management and rendering solutions. Some issues arising from the impressive increase in data complexity (and richness) provided by the evolution of 3D scanning technology are as follows: how to manage/visualize those data on commodity computers; how to improve the ease of use of the visualization tools (as potential users are often not expert with interactive graphics); finally, due to the accuracy of the 3D medium, we can think to use it as the main representation media, able both to represent an artifact but also to integrate other multimedia information.

3.5.1   Simplification and Multiresolution Management of Huge Models

Data complexity can be managed by adopting a *data simplification* approach, reducing the data resolution at the expenses of a (controlled) loss of geometric accuracy. Many solutions have been proposed for the accurate simplification of 3D triangulated surfaces, usually based on the iterative elimination of selected vertices or faces, driven by an error-minimization cost function. This approach allows the construction of any level of resolution we need, usually with a rather expensive computation (from a few seconds to a few hours, depending on the solution used and the complexity of the initial surface) which has to be executed only once. Simplification is very handy to produce models which fit the specific application requirements (e.g., a simple small model for a web presentation which should be downloadable in a given short time, or a model to be used for an accurate rapid reproduction by a specific 3D printer).

Another approach is the so-called *multiresolution encoding* (i.e., storing not just the final simplified model but all the intermediate results reached during the iterative simplification). All these intermediate results have to be encoded in an efficient data structure (the specific *multiresolution representation scheme*) that should allow the interactive application to extract different resolutions on the fly, supporting data feeding rates to the GPU compatible with real-time applications (visualization is an example of those). A view-dependent variable resolution representation can be produced for each frame from these schemes, according to the current view specification (e.g., higher resolution for the portions in foreground, progressively lower resolution for data in the background) and the requested visualization accuracy.

Recent research on *multiresolution schemes* has produced a number of solutions based on higher granularity than previous methods (i.e., instead of focusing on single triangles, patches of triangles become the elementary entity). These solutions allow one to manage huge 3D scanned models at interactive frame rate on consumer PC's [67, 68]. *Virtual Inspector* [56] is an example of a

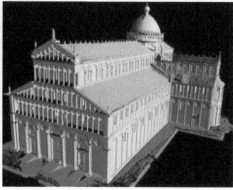

FIGURE 3.15 (SEE COLOR INSERT)
An example of two different visualization modes: on the left we render just the sampled points, on the right we present a rendering where shading of the surface element plays an important role for an improved insight over the details of the represented architecture (from a scanned model of the Cathedral of Pisa, Italy).

visualization system which adopts this approach to support inspection of large complex 3D models in real time (see Figure 3.9).

After around fifteen years of intense research on simplification and multiresolution technologies, we can nowadays consider those technologies sufficiently mature. But there is still some lack of information in the 3D scanning community (at the application level) on the potential of those methods. One of the more common negative concerns raised by practitioners against 3D scanning is the size of the models obtained, which according to this common believe makes them unusable in real applications. This is not true, since the availability of simplification and multiresolution technologies can successfully cope with the data complexity of 3D scanned data. In this context, multiresolution encoding is of paramount value, supporting transparent real-time selection of the level of accuracy and producing at interactive rates a data density that best fits the current application requirements.

3.5.2 Mesh-Based versus Point-Based Encoding and Rendering

We have presented briefly the *triangle-based* simplification and multiresolution methodologies (i.e., data optimization/encoding strategies based on geometric processing applied to a triangle mesh). *Point-based* representations have also been used a lot in the 3D scanning domain, especially to present data acquired with TOF scanning devices. It is important to distinguish between naive point-based and the more sophisticated point-based representation and rendering approaches available. With the term *naive point-based* we mean the use of simple 3D points to render sampled data. This is the usual approach of many scanning devices, which usually render the samples as colored points, using the color channel to map the perceived reflectivity of the sampled surface or the estimated accuracy of the sample.

Naive point rendering is very easy to implement, but it is also a major cause of the very poor acceptance of 3D scanning in several applicative domains. Just splatting points on the screen gives very poor images, where it is not easy at all to understand the relations between different elements, perception of depth is lacking, it is very hard to spot errors or inaccuracies of the data (see Figure 3.15). This does not mean that point-based rendering is a wrong solution, but that choosing a naive implementation can be a very bad decision.

First, we should remember that even if we endorse a point-based approach, all the processing phases presented in Section 3.2 are valid and have to be performed in an accurate and correct way.

The several point clouds (or range maps) acquired have to be aligned precisely one to the other and globally (note that the alignment results can be checked and assessed only when we have the possibility to render the points as a surface, adding shading). A merging phase should be applied even if we want to endorse a point-based representation: it is not sufficient to simply join the several point clouds in a bigger set, because sampled data are redundant and contain noise; we have already seen that a merging phase (performing an intelligent interpolation among sampled data) can improve the data and reduce unnecessary redundancy. Moreover, we need to adopt multiresolution to make large dataset tractable (as an example, the Pisa Dome shown in Figure 3.15 is around 200 mega samples after merging). Now that we have established that we have to process the dataset even when a point-based approach is endorsed, let us focus on visualization constraints. Even when we use points, we should be able to render color (this is very easy, since we can assign a single color to each point) but also to render the surface slope associated to each point. This means that the representation should store a normal vector for each sample, and this vector has to be computed by integrating a small region around the given sample. Many different advanced point-based representations and rendering methodologies have been presented in the last few years. Presenting an overview of the extensive research results on point-based graphics is well beyond the focus of this chapter; this specific domain is the topic of a series of successful symposia, the *EG Point-Based Graphics* series; interested readers can consult those symposia proceedings or a recent tutorial on point-based techniques [69]. Point-based representation as well can be managed in an efficient manner by endorsing simplification or multiresolution methods. The latter are usually based on hierarchical structures that allow one to encode the dataset at different levels of detail and to extract view-dependent representation very easily in real time [70–72].

3.5.3 Usability of Virtual Heritage Worlds

Ease of use of visual CH tools oriented to ordinary people (that are still not very competent with 3D graphics and computer games, especially if not part of the young generation) is an important factor for their success. One of the most complicated actions to perform nowadays with 3D tools is to drive navigation in the virtual space, especially when we have a clear objective (e.g., I want to reach a specific location in the scene and see my focus of interest from a specific view). Therefore, free navigation should be requested only in those cases where this action really adds something to the learning experience. The risk is to have the visitor losing orientation (e.g., discovering himself lost in void space, maybe just because he turned his back to the scene and is erroneously looking at the sky), losing faith that he can drive the inspection or navigation session and quitting the system.

Other important characteristics of a visualization system are its flexibility, completeness, and configurability. To fulfill this objective developers could be induced to design complicated systems characterized by a very complete set of functionalities and involute user interfaces (for an example, consider the commercial 3D modeling systems). Conversely, while designing our *Virtual Inspector* tool [56] as a system oriented to non-expert users (e.g., museum visitors), our approach was to define a restricted set of functionalities and to provide the tool with an easy installation interface for the selection of the subset of these functionalities that the designer of the specific installation wants to endorse (e.g., to build up a new museum kiosk).

Another important factor of success nowadays is web availability of those resources, to allow a much wider community of users to experiment, navigate and enjoy the reconstructed scenes. Implementing a Virtual Heritage reconstruction to make it deployable on the web introduces a number of design constraints, but the technology is almost ready to support that type of distribution channel.

3.5.4   Not Just 3D Data: Adding Other Knowledge

Visualization tools usually focus on just the visual presentation of the shape characteristics of the artifact. This is not enough if we want to provide a comprehensive presentation of the global knowledge available on the artifact. On the other hand, the 3D shape can become some sort of visual 3D map which allows us to integrate, links and present all the available information in an intuitive and visually pleasing way. The goal is therefore to transform standard 3D browsers into tools able to connect the shape (or region of the latter) with all the available multimedia (MM) data that is linked with the artifact. This opens the wider topic on how to support and implement data annotation on 3D models.

Hot spots can be a very handy resource to associate multimedia data to any point or region of a 3D model. This allows one to design interactive presentations where the 3D model becomes a natural visual index to historical/artistic information, for example presented using standard HTML format and browsers. But hot spots are only a first step, which implements a point-wise association metaphor. We should devise more flexible and powerful instruments to associate information to 3D meshes (e.g., supporting links between sub-regions of the surface to any possible MM document). Therefore, tools should provide easy-to-use features for the segmentation of the shape model in components. Segmentation can be driven by pure shape-oriented criteria (usually, working on surface curvature is one of the most diffuse approaches) but it should also include the user in the loop, by giving him the lead in driving the segmentation process according to the metadata that characterize the specific artifact and the message he wants to convey with the specific segmentation.

Data presentation of a multitude of information tokens can become an issue (many different types of point-wise links, associated to different types of information; several different segmentations available for the same object, each one focusing on a different interpretation or context). Intellingent approaches to data hiding and visualization will have to be endorsed also in this type of applications, as it has already been the case of other scientific visualization or data analytics contexts. Moreover, visualization instruments should be extended to provide tools supporting the easy integration and update of the data associated to the 3D digital model, supporting a democratic and cooperative approach such as the one at the base of the Wikipedia effort.

3.5.5   Presenting 3D Data on the Web

In the modern world, we cannot avoid considering the issues related to the web-based access and distribution of 3D data. The peculiar aspect of scanned data with respect to other 3D models is the average size of the models. To make these data usable on the web we should either deploy efficient geometric compression technology [73], or adopt remote rendering approaches [74]. A web-based application which could boost considerably the usage of 3D scanning is GoogleEarth (or similar systems), since those applications could allow very nice opportunities for announcing on the web the availability of 3D models, showing to the users their availability while navigating the geographical space. Another very promising platform is the WebGL specifications for the inclusion of 3D graphics among the data directly managed by common web browsers [75]. The new generation of web browsers (now in beta version) will therefore support natively digital 3D content, without the need to install specific plug-ins. Some examples of how 3D content could be represented and processed using WebGL are presented on [76]

3.6   3D Digitization: How to Improve Current Procedures and Make It More Practical and Successful

While 3D scanning hardware and software technologies have considerably evolved in the last years and their use in the cultural heritage field has gained consent and acceptance, these technologies are far from being easy to use and error-free. Many problems can arise during a scanning campaign; most of these issues are related to various deficiencies in the technologies and the improper background/skills of the operators. Some of these problems have already been briefly mentioned through the chapter; we discuss them in detail in this section.

3.6.1   Technology - Limitations Perceived by Practitioners

First of all, 3D scanning technologies are not able to capture many kinds of materials that may occur in the CH field: transparent or semi-transparent surfaces (like glass, jewels, and some stones); mirroring, polished and very shiny objects (like metals); fluff and fuzzy substances like feathers, furs, or some tissues. All those materials are quite difficult to sample with the current off-the-shelf optical technologies. For some of the previous cases, experimental research projects have shown the feasibility of technological solutions that overcome these limitations (in some cases, by adopting enhanced 2D image media, rather than a pure 3D encoding); in several cases, their practical applicability to real projects has still to be assessed.

The working space of the devices is another issue. The CH domain is a very good example of an application where the objects of interest can span the entire interval: from tiny objects (few millimeters) to an entire building or even an archaeological site or a city. With such a wide domain, an important feature of a scanning device would be the flexibility of the working space supported. Conversely, the devices often offer only a very restricted working space (or require a time-consuming calibration to select a slightly different working range). This forces one to select different devices for the different acquisition contexts, increasing the technical skills required to master technology and the overall cost. In this sense, current acquisition methodologies based on pure images (multi stereo matching and similar approaches) present a big advantage with respect to the classical scanning devices, due to the much wider working space of digital cameras (changing focal lenses is much easier than switching to a different scanning device).

Moreover, as already noted before, most of the current scanning technologies focus only on shape, ignoring the issues related to the acquisition of the optical appearance of the scanned surface. Even when the adopted technologies are able to capture the "color" (for example by mapping photos over the surface), in almost all cases this leads only to the acquisition of the apparent reflected color, without considering the reflectance properties of the surface (e.g., its shininess). The acquisition of a correct representation of the reflectance properties of a surface together with its 3D shape is still the domain of research experiments and it is not an off-the-shelf technology. Another sensible issue of 3D scanning is the cost of the hardware equipment: usually it is very high and it can become prohibitive for many low-budgeted cultural heritage institutions. On the other hand, the field is growing and very cheap solutions are appearing on the market. On the positive side we have to note that the research in this field is very active, so we can easily hope that in the future the capabilities of 3D scanning hardware will improve, gradually covering the shortcomings listed above.

3.6.2   Misuse of Technology

The people involved in a scanning campaign can be roughly divided in two sets: *technical staff*, who actually perform the scanning task with a not-so-strong CH background and sensibility; *CH operators/experts*, who know very well the specific field and the digitization objectives, but often lack a deep knowledge of the technological details. Obviously there are some notable exceptions, typically in most of the successful 3D scanning projects run so far. But, especially if the field increases in terms of number of devices sold and application testbeds run, the situation will turn more and more towards the case of users with a limited technological background. Designers of software tools should keep this in mind while designing the tools.

Selecting the Wrong Device

As we have sketched in the previous sections, there are many possible technologies available, each one with its own pros and cons. We should underline that scanning system producers are usually very poor in illustrating the conditions which make a given device not fit for the planned task. External constraints (like for example the availability of a given device or of consolidated esperience with a specific technology) might affect the choice of the preferred hardware, leading to the selection of a non-optimal device.

Wrong Data Post-Processing

The wish to provide clean and nice results to the purchaser of a scanning service often causes the undocumented (and often not required) editing of the acquired data. 3D scanning technologies are often unable to completely recover the whole surface, leaving holes and unsampled regions in the final model. Smoothing can reduce the sampling noise in the final model, but it can also cancel important high-frequency details (e.g., traces of deterioration of the surface, traces of the tools used to produce the artwork). Any editing action over the originally acquired data should be documented in the most evident way in order to make it possible to distinguish between *ground truth* sampled data and the parts that have been added or heavily edited by the subsequent geometric processing phases. Excessive data smoothing or inaccurate simplification are clear examples of actions which could destroy data accuracy and trustability.

3.6.3   Better Management of 3D Data

On the other hand, lack or incomplete technical knowledge of 3D scanning can cause various inconveniences in the overall process of creating trusted digital reproductions of works of art.

Evaluating and Assessing Data Quality

Being able to evaluate in an objective way the final quality of a 3D scanned model is a basic resource for a wide adoption of this technology. In contrast to a more traditional media such as photography, where established protocols and procedures exist and the quality can be determined by an analysis of the (digital) photo and its provenance data, quality assessment of the final digital 3D model is much more complex. Quality depends on a large number of factors: scanning HW characteristics, modality of use of the scanning device, pipeline and parameters of the SW post-processing which transforms the raw data in the final 3D model. Therefore the whole pipeline, including HW and SW used to produce the digital object, has to be considered while assessing the quality of the digital model. Provenance data are therefore more complex than the ones characterizing the 2D media (photographs). While common practices for the use of general 3D visualization in the research and communication of cultural heritage have been already discussed in the London Charter [77], there is still an absence of clear protocols for the management of 3D scanning technologies and data. Some attempts should be made to define the concept of provenance data for 3D models, in a sufficiently general and standardized way. This is one of the major focuses of the EC research initiative IST IP "3D-COFORM" [21]. 3D-COFORM technlogies are steered towards a comprehensive management of provenance data, by providing integrated management at the level of both the modeling tools and the repository for 3D assets and metadata [78].

The availability of free and standard tools supporting the quality assessment (ideally, by considering both geometry and reflectance data) would be of paramount value, especially for the usually not-technical people who order a scanning campaign. This is an area that desperately needs research and practical results made available to the community. Even a well informed consumer at this point does not have the tools and standards available to make a purchasing decision and a good assessment of the results that are produced by a (usually expensive) project commissioned to an external service company.

Preserving the Data for the Future

An interesting point of discussion is which could be the most fruitful use of the acquired data; a given artifact can be scanned now for a specific application, but once we have the digital model we can envision many different applications of the same 3D model in the future. Therefore, the results of scanning efforts should be made available for future uses, even by different entities. This can be also a way to recover from the cost of the digitization: CH institutions should develop policies to play a role in the market of digital 3D content, finding on the market the financial resources to perform 3D digitization.

Another consideration, that is often neglected, is the *long-term preservation* of these data. While for traditional media, like written material, photography, and films, the issues about the preservation, classification, and future access to the acquired data is a rather well studied subject, there are no established practices for the preservation of digital 3D content. Without going into details, let us just mention here the issues related to the format of the delivered 3D data: often the results of a 3D scanning campaign are delivered only as files encoded in a closed format, accessible only through proprietary software whose longevity is not guaranteed at all. We are seriously risking that the data produced in several important initiatives are buried in a hidden format without the possibility of being utilized in the near future.

3.7 Conclusions

The chapter presented the software technologies available for processing and visualization of sampled 3D data. Due to the vast amount of material published in the current literature on this subject, we were forced to present here just a selection of the approaches proposed. Therefore, the bibliography cites only a very small subset of the available literature (otherwise the bibliography could have easily become excessively long). One of the focuses of the presentation has been to give more emphasis to the open problems or to subjects which are currently under study, rather than presenting in detail the algorithms mentioned. Therefore, the current chapter should be read by consulting also the papers cited in literature, according to the interests of the reader. Some of those papers are very good resources for gathering a state of the art of the specific problem and for getting a more complete list of citations to the other works in the selected sub-domain. Finally, as usually happens with review papers on a subject where the authors have dedicated time and efforts, we have probably given an emphasis on our own results and papers; we apologize to the reader for that but, as we explicitly mentioned in the text, some results are presented as a representative of several other research efforts which produced comparable outputs.

Acknowledgments

We acknowledge the financial support of the EC IST IP project "3D-COFORM" (IST-2008-231809) and the project "START" (funded by the Tuscany Region, 2008-2011).

Bibliography

[1] F. Blais, "A review of 20 years of range sensor development," in *Videometrics VII, Proceedings of SPIE-IS&T Electronic Imaging, SPIE Vol. 5013*, pp. 62–76, 2003.

[2] F. Remondino and S. El-Hakim, "Image-based 3D modelling: a review," *The Photogrammetric Record*, vol. 21, no. 115, pp. 269–291, 2006.

[3] P. Cignoni and R. Scopigno, "Sampled 3D models for CH applications: a viable and enabling new medium or just a technological exercise?," *ACM Journal on Computing and Cultural Heritage*, vol. 1, no. 1, pp. 1–23, 2008.

[4] B. Curless and S. Seitz, "3D Photography," in *ACM SIGGRAPH Course Notes, Course No. 19*, 2000.

[5] G. Sansoni, M. Trebeschi, and F. Docchio, "State-of-the-art and applications of 3D imaging sensors in industry, cultural heritage, medicine, and criminal investigation," *Sensors*, vol. 9, no. 1, pp. 568–601, 2009.

[6] R. Lange, P. Seitz, A. Biber, and R. Schwarte, "Time-of-flight range imaging with a custom solid-state image sensor," in *Proceedings of the SPIE Vol. 3823*, pp. 180–191, 1999.

[7] M. Goesele, N. Snavely, B. Curless, H. Hoppe, and S. M. Seitz, "Multi-view stereo for community photo collections," in *Proceedings of ICCV 2007*, pp. 1–8, 2007.

[8] H. Bay, A. Ess, T. Tuytelaars, and L. van Gool, "Speeded-up robust features (surf)," *Computer Vision and Image Understanding (CVIU)*, vol. 110, no. 3, pp. 346–359, 2008.

[9] NextEngine, Inc., "The NextEngine laser scanning system." More info on: https://www.nextengine.com/, 2010.

[10] M. Vergauwen and L. Van Gool, "Web-based 3D reconstruction service," *Machine Vision Applications*, vol. 17, no. 6, pp. 411–426, 2006.

[11] M. Vergauwen and L. Van Gool, "The Arc3D web-based reconstruction tool." More info on: http://www.arc3d.be/, 2010.

[12] F. Blais, J. Taylor, L. Cournoyer, M. Picard, L. Borgeat, L.-G. Dicaire, M. Rioux, J.-A. Beraldin, G. Godin, C. Lahanier, and G. Aitken, "Ultra-high resolution imaging at 50m using a portable XYZ-RGB color laser scanner," in *International Workshop on Recording, Modeling and Visualization of Cultural Heritage (Ascona, Switzerland, May 22-27)*, 2005.

[13] F. Bernardini and H. E. Rushmeier, "The 3D model acquisition pipeline," *Computer Graphics Forum*, vol. 21, pp. 149–172, March 2002.

[14] INUS Technology, "Rapidform™- 3D Scanning and Inspection Software from INUS Technology, Inc. and Rapidform, Inc." More info on: http://www.rapidform.com/, 2010.

[15] InnovMetrics, "PolyWorks™inspection and modeling solutions." More info on: http://www.innovmetric.com/, 2010.

[16] Raindrop Geomagic, "Geomagic™studio: Reverse Engineering and Custom Design Software." More info on: http://www.geomagic.com/, 2010.

[17] B. Curless, "Vrippack users guide," Tech. Rep. TR-UUCS-04-019, University of Washington (http://www-graphics.stanford.edu/software/vrip/), November 25, 2006.

[18] M. Callieri, P. Cignoni, F. Ganovelli, C. Montani, P. Pingi, and R. Scopigno, "VCLab's tools for 3D range data processing," in *VAST 2003* (A. C. D. Arnold and F. Niccolucci, eds.), (Brighton, UK), pp. 13–22, Eurographics, Nov. 5-7, 2003.

[19] P. Cignoni, M. Callieri, M. Corsini, M. Dellepiane, F. Ganovelli, and G. Ranzuglia, "Meshlab: an open-source mesh processing tool," in *Sixth Eurographics Italian Chapter Conference*, pp. 129–136, 2008.

[20] "EPOCH: The European Network of Excellence on ICT Applications to Cultural Heritage (IST-2002-507382)." More info on: http://www.epoch-net.org/, 2008.

[21] "EU IP 3DCOFORM: Tools and expertise for 3D collection formation (FP7/2007-2013 grant 231809)." More info on: http://www.3d-coform.eu/, 2010.

[22] P. J. Besl and N. D. McKay, "A method for registration of 3-D shapes," *IEEE Transactions on Pattern Analysis and Machine Intelligence*, vol. 14, pp. 239–258, Feb. 1992.

[23] Y. Chen and G. Medioni, "Object modelling by registration of multiple range images," *International Journal of Image and Vision Computing*, vol. 10, pp. 145–155, Apr. 1992.

[24] K. Pulli, "Multiview registration for large datasets," in *Proceedings of the 2nd International Conference on 3D Digital Imaging and Modeling*, pp. 160–168, IEEE, 1999.

[25] N. Gelfand, N. J. Mitra, L. J. Guibas, and H. Pottmann, "Robust global registration," in M. Desbrun and H. Pottmann, editors, *Eurographics Association, ISBN 3-905673-24-X.*, pp. 197–206, 2005.

[26] G. Guidi, J.-A. Beraldin, and C. Atzeni, "High-accuracy 3D modeling of cultural heritage: the digitizing of Donatello's "Maddalena," *IEEE Transactions on Image Processing*, vol. 13, no. 3, pp. 370–380, 2004.

[27] P. Cignoni, C. Montani, C. Rocchini, and R. Scopigno, "External memory management and simplification of huge meshes," *IEEE Transactions on Visualization and Computer Graphics*, vol. 9, no. 4, pp. 525–537, 2003.

[28] M. Garland and P. Heckbert, "Surface simplification using quadric error metrics," in *SIGGRAPH 97 Conference Proceedings*, Annual Conference Series, pp. 209–216, Addison Wesley, Aug. 1997.

[29] H. Hoppe, "New quadric metric for simplifying meshes with appearance attributes," in *Proceedings of IEEE Conference on Visualization (VIS99)*, New York, pp. 59–66, ACM Press, Oct. 25–28, 1999.

[30] H. P. A. Lensch, J. Kautz, M. Goesele, W. Heidrich, and H.-P. Seidel, "Image-based reconstruction of spatial appearance and geometric detail," *ACM Transaction on Graphics*, vol. 22, pp. 234–257, Apr. 2003.

[31] M. Levoy, K. Pulli, B. Curless, S. Rusinkiewicz, D. Koller, L. Pereira, M. Ginzton, S. Anderson, J. Davis, J. Ginsberg, J. Shade, and D. Fulk, "The Digital Michelangelo Project: 3D scanning of large statues," in *SIGGRAPH 2000, Computer Graphics Proceedings*, Annual Conference Series, pp. 131–144, Addison Wesley, July 24-28, 2000.

[32] R. J. Campbell and P. J. Flynn, "A survey of free-form object representation and recognition techniques," *Computer Vision and Image Understanding*, vol. 81, no. 2, pp. 166–210, 2001.

[33] A. Fasano, P. Pingi, P. Cignoni, C. Montani, and R. Scopigno, "Automatic registration of range maps," *Computer Graphics Forum (Proceedings of Eurographics '05)*, vol. 24, no. 3, pp. 517–526, 2005.

[34] U. Castellani, M. Cristani, S. Fantoni, and V. Murino, "Sparse points matching by combining 3D mesh saliency with statistical descriptors," *Computer Graphics Forum*, vol. 27, no. 2, pp. 643–652, 2008.

[35] D. Aiger, N. J. Mitra, and D. Cohen-Or, "4-points congruent sets for robust pairwise surface registration," in *SIGGRAPH '08: ACM SIGGRAPH 2008 papers* New York, pp. 1–10, ACM, 2008.

[36] G. H. Bendels, P. Degener, R. Wahl, M. Körtgen, and R. Klein, "Image-based registration of 3D-range data using feature surface elements," in *5th International Symposium on Virtual Reality, Archaeology and Cultural Heritage (VAST 2004)* (Y. Chrysanthou, K. Cain, N. Silberman, and F. Niccolucci, eds.), pp. 115–124, Eurographics, December 2004.

[37] J. Davis, S. Marshner, M. Garr, and M. Levoy, "Filling holes in complex surfaces using volumetric diffusion," in *First International Symposium on 3D Data Processing, Visualization and Transmission (3DPVT'02)*, pp. 428–438, IEEE Computer Society, 2002.

[38] M. Kazhdan, M. Bolitho, and H. Hoppe, "Poisson surface reconstruction," in *SGP '06: Proceedings of the fourth Eurographics symposium on Geometry processing* (Aire-la-Ville, Switzerland), pp. 61–70, Eurographics Association, 2006.

[39] T. Ju, "Robust repair of polygonal models," *ACM Transactions on Graphics*, vol. 23, no. 3, pp. 888–895, 2004.

[40] J. Podolak and S. Rusinkiewicz, "Atomic volumes for mesh completion," in *EG/ACM Symposium on Geometry Processing (SGP)*, pp. 33–41, Eurographics, 2005.

[41] P. Cignoni, "MeshLab: an open source, portable, and extensible system for the processing and editing of unstructured 3D triangular meshes." More info on: http://meshlab.sourceforge.net/, 2010.

[42] P. Liepa, "Filling holes in meshes," in *EG/ACM Symposium on Geometry Processing (SGP)*, pp. 200–206, Eurographics, June 2003.

[43] L. S. Tekumalla and E. Cohen, "A hole filling algorithm for triangular meshes," Technical Report TR-UUCS-04-019, University of Utah, 2004.

[44] A. Sharf, M. Alexa, and D. Cohen-Or, "Context-based surface completion," *ACM Transactions on Graphics*, vol. 23, no. 3, pp. 878–887, 2004.

[45] G. H. Bendels, R. Schnabel, and R. Klein, "Detail-preserving surface inpainting," in *The 6th International Symposium on Virtual Reality, Archaeology and Cultural Heritage* (M. Mudge, N. Ryan, and R. Scopigno, eds.) (Pisa, Italy), pp. 41–48, Eurographics Association, 2005.

[46] C. Donner, T. Weyrich, E. d'Eon, R. Ramamoorthi, and S. Rusinkiewicz, "A layered, heterogeneous reflectance model for acquiring and rendering human skin," *ACM Transactions Graphics*, vol. 27, no. 5, pp. 1–12, 2008.

[47] A. Ghosh, T. Hawkins, P. Peers, S. Frederiksen, and P. Debevec, "Practical modeling and acquisition of layered facial reflectance," *ACM Transactions Graphics*, vol. 27, no. 5, pp. 1–10, 2008.

[48] M. Callieri, P. Cignoni, and R. Scopigno, "Reconstructing textured meshes from multiple range RGB maps," in *7th International Fall Workshop on Vision, Modeling, and Visualization 2002*, Erlangen, pp. 419–426, IOS Press, Nov. 20–22, 2002.

[49] K. Umeda, M. Shinozaki, G. Godin, and M. Rioux, "Correction of color information of a 3D model using a range intensity image," in *Fifth International Conference on 3-D Digital Imaging and Modeling (3DIM 2005)*, pp. 229-236, June 13-16, 2005.

[50] A. Troccoli and P. K. Allen, "Relighting acquired models of outdoor scenes," in *Fifth International Conference on 3-D Digital Imaging and Modeling (3DIM 2005)*, pp. 245-252, June 13-16, 2005.

[51] H. Rushmeier and F. Bernardini, "Computing consistent normals and colors from photometric data," in *Proceedings of the Second International Conference on 3D Digital Imaging and Modeling* (Ottawa, Canada), pp. 99–108, 1999.

[52] P. Debevec, C. Tchou, A. Gardner, T. Hawkins, C. Poullis, J. Stumpfel, A. Jones, N. Yun, P. Einarsson, T. Lundgren, M. Fajardo, and P. Martinez, "Digitizing the Parthenon: estimating surface reflectance properties of a complex scene under captured natural illumination," in *VMV04 Proceedings*, p. 99, 2004.

[53] R. Tsai, "A versatile camera calibration technique for high accuracy 3D machine vision metrology using off-the-shelf TV cameras and lenses," *IEEE Journal of Robotics and Automation*, vol. RA-3, Aug. 1987.

[54] T. Franken, M. Dellepiane, F. Ganovelli, P. Cignoni, C. Montani, and R. Scopigno, "Minimizing user intervention in registering 2D images to 3D models," *The Visual Computer*, vol. 21, pp. 619–628, Sept. 2005.

[55] M. Dellepiane, M. Callieri, F. Ponchio, and R. Scopigno, "Mapping highly detailed color information on extremely dense 3D models: the case of David restoration," *Computer Graphics Forum*, vol. 27, no. 8, pp. 2178–2187, 2008.

[56] M. Callieri, F. Ponchio, P. Cignoni, and R. Scopigno, "Virtual inspector: a flexible visualizer for dense 3D scanned models," *IEEE Computer Graphics and Applications*, vol. 28, no. 1, pp. 44–55, 2008.

[57] M. Corsini, M. Dellepiane, F. Ponchio, and R. Scopigno, "Image-to-geometry registration: a mutual information method exploiting illumination-related geometric properties," *Computer Graphics Forum*, vol. 28, no. 7, pp. 1755–1764, 2009.

[58] C. Rocchini, P. Cignoni, C. Montani, and R. Scopigno, "Multiple textures stitching and blending on 3D objects," in *Rendering Techniques '99* (D. Lischinsky and G. Ward, eds.), pp. 119–130, Springer-Verlag Wien, 1999.

[59] C. Rocchini, P. Cignoni, C. Montani, and R. Scopigno, "Acquiring, stitching and blending diffuse appearance attributes on 3D models," *The Visual Computer*, vol. 18, no. 3, pp. 186–204, 2002.

[60] M. S. Floater and K. Hormann, "Surface parameterization: a tutorial and survey," in *Advances in Multiresolution for Geometric Modeling* (N. A. Dodgson, M. S. Floater, and M. A. Sabin, eds.), Mathemathics and Vision, pp. 157–186, Springer, 2005.

[61] N. Pietroni, M. Tarini, and P. Cignoni, "Almost isometric mesh parameterization through abstract domains," *IEEE Transactions on Visualization and Computer Graphics*, vol. 16, no. 4, pp. 621–635, 2010.

[62] M. Callieri, P. Cignoni, M. Corsini, and R. Scopigno, "Masked photo blending: mapping dense photographic dataset on high-resolution sampled 3D models," *Computers and Graphics*, vol. 32, no. 4, pp. 464–473, 2008.

[63] M. Dellepiane, M. Callieri, M. Corsini, P. Cignoni, and R. Scopigno, "Improved color acquisition and mapping on 3D models via flash-based photography," *ACM Journal on Computers and Cultural Heritage*, vol. 2, pp. 1–20, Feb. 2010.

[64] T. Malzbender, D. Gelb, and H. Wolters, "Polynomial texture maps," in *SIGGRAPH '01: Proceedings of the 28th annual conference on computer graphics and interactive techniques* (New York), pp. 519–528, ACM Press, 2001.

[65] M. Levoy and P. Hanrahan, "Light Field Rendering," in *Computer Graphics Proceedings, Annual Conference Series, 1996 (ACM SIGGRAPH '96 Proceedings)*, pp. 31–42, 1996.

[66] M. Levoy, "Light fields and computational imaging," *IEEE Computer*, vol. 39, no. 8, pp. 46–55, 2006.

[67] P. Cignoni, F. Ganovelli, E. Gobbetti, F. Marton, F. Ponchio, and R. Scopigno, "Adaptive tetrapuzzles: efficient out-of-core construction and visualization of gigantic multiresolution polygonal models," *ACM Transactions on Graphics (SIGGRAPH 2004)*, vol. 23, no. 3, pp. 796–803, 2004.

[68] P. Cignoni, F. Ganovelli, E. Gobbetti, F. Marton, F. Ponchio, and R. Scopigno, "Batched multi triangulation," in *IEEE Visualization 2005*, pp. 27–35, 2005.

[69] L. Kobbelt and M. Botsch, "A survey of point-based techniques in computer graphics," *Computers & Graphics*, vol. 28, no. 6, pp. 801–814, 2004.

[70] S. Rusinkiewicz and M. Levoy, "QSplat: A multiresolution point rendering system for large meshes," in *Computer Graphics Proceedings, Annual Conference Series (SIGGRAPH 00)*, pp. 343–352, ACM Press, July 24-28, 2000.

[71] E. Gobbetti and F. Marton, "Layered point clouds – a simple and efficient multiresolution structure for distributing and rendering gigantic point-sampled models," *Computers & Graphics*, vol. 28, December 2004.

[72] E. Gobbetti and F. Marton, "Far Voxels - a multiresolution framework for interactive rendering of huge complex 3D models on commodity graphics platforms," *ACM Transactions on Graphics*, vol. 24, pp. 878–885, August 2005. Proc. SIGGRAPH 2005.

[73] J. Rossignac, "Surface simplification and 3D geometry compression," in *Handbook of Discrete and Computational Geometry*, (J. E. Goodman and J. O'Rourke, eds.) CRC Press, 1997, 2004, vol. 2, 2004.

[74] D. Koller, M. Turitzin, M. Levoy, M. Tarini, G. Croccia, P. Cignoni, and R. Scopigno, "Protected interactive 3D graphics via remote rendering," *ACM Trans. Graph*, vol. 23, no. 3, pp. 695–703, 2004.

[75] Khronos Group, "WebGL - OpenGL ES 2.0 for the Web." More info on: http://www.khronos.org/webgl/, 2010.

[76] Di Benedetto, Marco, "SpiderGL - 3D Graphics for Next-Generation WWW (a JavaScript 3D Graphics library which relies on WebGL for realtime rendering)." More info on: http://www.spidergl.org/, 2010.

[77] The London Charter, "For the use of 3D visualization in the research and communication of cultural heritage." More info on: http://www.londoncharter.org/, 2010.

[78] M. Theodoridou, Y. Tzitzikas, M. Doerr, Y. Marketakis, and V. Melessanakis, "Modeling and querying provenance by extending CIDOC CRM," *Distributed and Parallel Databases*, vol. 27, no. 2, pp. 169–210, 2010.

4

ARC3D: A Public Web Service That Turns Photos into
3D Models

David Tingdahl

Katholieke Universiteit Leuven
`Email: tingdahl@esat.kuleuven.be`

Maarten Vergauwen

GeoAutomation
`Email: maarten.vergauwen@geoautomation.com`

Luc Van Gool

Katholieke Universiteit Leuven
`Email: Luc.VanGool@esat.kuleuven.be`

CONTENTS

4.1   Introduction

Increasingly, cultural heritage professionals make use of 3D capture technologies, both for study and for documentation. There is a wide range of such technologies to choose from. One problem is that there does not seem to be a single one that is optimal for all objects or scenes. Thus, also taking into account cost and resources, there often is a need to choose different tools for different aspects of a scanning campaign. The overall scene layout may be captured with one, and then the detailed shapes of smaller pieces with another, for instance. An excellent example is the thorough work by Remondino and colleagues [1], who have combined the complementary strengths of LIDAR scanning, interactive photogrammetric reconstruction, and automatic, uncalibrated structure-from-motion. Here we present a tool for the latter, the first one that was available and which is still free for non-commercial use: ARC3D. But, as said, one must not expect this tool to supply all 3D models one desires. Nonetheless, we believe it can be of great use for museums, excavations, conservation, etc. As shown in Strecha et al. [2], under appropriate conditions for such methods to work, they are quite competitive with LIDAR scanning (i.e., time-of-flight based laser scanning), but require a smaller investment, by far, of time and money.

Before we describe the ARC3D tool in more detail, it may be necessary to explain what uncalibrated structure-from-motion actually is. It is a technique that extracts a 3D model from photographs of an object. The nice thing is that the user only needs to load the images into the computer, which then does the rest. In contradistinction with traditional photogrammetric techniques, the process is fully automated. One does not have to know about the camera settings, nor about the viewpoints from where the different images were taken. With the proliferation of digital photography, taking lots of pictures has become common practice anyway and the creation of 3D models may therefore require only very little overhead from the user's point of view. This said, it helps a lot if some basic guidelines for taking the photos are followed when the goal is to also produce such 3D models. More about all this is to come. It is also important to note that automatically generated 3D data are hardly ever immediately satisfactory for their practical use. Often, some additional filtering or simplification is needed. The goal of such post-processing can be to enhance the quality, but also to reduce the size of the models. In order to support such post-processing, ARC3D is easy to use in combination with MeshLab, a state-of-the-art toolkit for 3D shape processing. MeshLab has been created by the CNR-ISTI group at Pisa and is still regularly updated. It is freely available via sourceforge [3]. See Chapter 3 for a complete introduction to the Meshlab toolkit.

The structure of the paper is as follows. Section 4.2 gives an overview of the ARC3D + MeshLab tools. Section 4.3 describes the principles behind the ARC3D design. Section 4.4 gives guidelines on how images should be taken in order to maximize the chance of successful 3D data extraction. Section 4.5 takes a practical case – one of the Mogao caves in China – as a way to illustrate the steps in the process. Section 4.6 gives some further examples of 3D models produced with ARC3D. Section 4.7 concludes the chapter.

4.2   System Overview

ARC3D (ARC stands for Automated Reconstruction Conduit) is a web service to which users can upload images. The web service returns 3D data extracted from these images. A digital photo-camera and an Internet connection suffice to generate 3D. This web service is meant to bring 3D capture at the fingertips of virtually all cultural heritage professionals. Apart from taking care that good quality pictures are taken, the user does not need any particular skills in terms of computer programming or geometry.

Basically, there are three steps to distinguish in the overall process of turning the images into the final 3D model:

- Images of the object or the scene to be reconstructed are first checked by the user for quality, and then uploaded to the web service [4] which computes dense depth maps. One depth map is created for each image. Such a map shows the distances to the points of the object that are visible for that camera, as the map intensity in each point. This is the basic ARC3D web service.

- A dedicated plug-in puts the dense depth maps into a format that can be read and handled by MeshLab. But this plug-in does actually more than just that. In combination with the confidence values given out by ARC3D for the points in the different depth maps, filtering operations are selected and run automatically, yielding higher quality depth maps as a starting point for further operations on the basis of MeshLab. Chapter 3 explains this plug-in in more detail.

- Then the user can apply several tools in MeshLab, for a variety of purposes, including the interactive registration of the depth maps into complete surfaces, as well as mesh refinement and simplification.

The remainder of this chapter focuses on the first step (i.e., ARC3D and how it extracts 3D from the images). This web service runs on computers at the University of Leuven. As we want to make the explanation of what it does accessible for various groups of users, we avoid mathematical explanations. The math behind ARC3D can be found in [5] and readers who want to read even more background material can have a look at [6] and [7].

4.2.1   System Components

Figure 4.1 shows a schematic overview of the client-server setup of the 3D web service. The client part (user) is found at the top of the figure. The server side is at the bottom. On his PC, the user runs two programs, the *upload tool* (A) and the *modelviewer tool* (B). Details about these tools follow in Sections 4.2.2 and 4.2.3. Both of them can be downloaded from the ARC3D website [4].

In the upload tool, digital images can be imported. Upon registering with the ARC3D web service, the user can transfer these images (indicated C) over the Internet to the Leuven server. There, fully automatically a set of parallel processes are launched, to compute dense 3D information from the uploaded images. The parallel processes are run on a cluster of Linux PC's (D). When the server is ready with the processing, an email is sent to the user, to inform him that the results can be downloaded from the server through FTP. They consist of a dense depth maps for every image and the calibration parameters of the different cameras (E).

The modelviewer tool shows the results to the user. Every depth map in the image set can be rendered as a 3D surface mesh, where unwanted areas like background or low-quality parts of the scene can be masked out. This filtering is aided by the confidence maps, one coming with each depth map. By interactively setting thresholds on the minimum confidence the user wants each depth value in the map to have, lots of noise can be removed with such simple interaction. The meshes can be saved in a variety of formats.

4.2.2   Upload Tool

The first tool a user of the 3D web service encounters is the upload tool. This is the program that uploads images to the server via the Internet. Figure 4.2 shows a screenshot of the GUI of this tool. First, the user selects the images which he wants to upload. Thumbnails of these images are shown on the left and a large version of the image is shown in the main panel when a thumbnail is clicked.

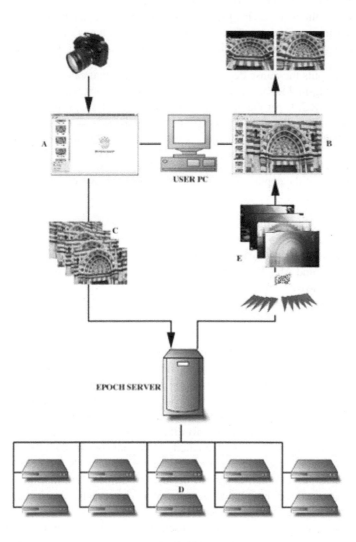

FIGURE 4.1
The ARC3D client-server setup. Images (C) are uploaded to the server, with the upload tool installed
on the PC of the user (A). Connected to the server, a PC cluster (D) extracts the camera parameters
and 3D information. The results (E) can be downloaded via FTP and visualized on the user PC with
the modelviewer tool (B).

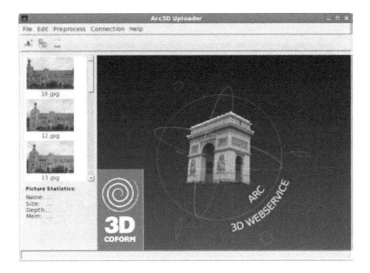

FIGURE 4.2
Upload tool. Thumbnails of the images to be uploaded are shown as well as some statistics. When satisfied, the user uploads the images to the reconstruction server.

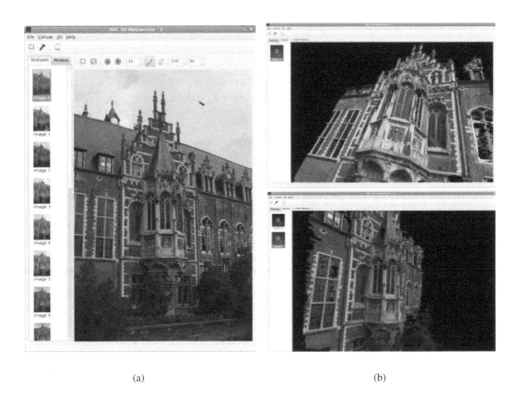

(a)   (b)

FIGURE 4.3
Modelviewer tool. The layout of the tool deliberately resembles that of the upload tool, with thumbnails on the left and a large image in the main panel. The tool allows the user to create textured wireframe representations of the depth maps. (a) The user selects the depth image to be visualized. (b) Two 3D representations of the depthmaps.

FIGURE 4.4
Modelviewer tool. The operator has automatically masked the blue sky with a simple click inside the region.

Images can be removed from the list or extra images can be added. On the bottom left, statistics on the image such as the number of pixels, the size on disk, and the format are shown. When the user is satisfied with the selected images, he can upload them to the reconstruction server. In order to do so, the user first needs to authenticate himself with his user name and password. The upload is started and a progress dialog shows the speed of the upload. In order to limit the upload- and processing time, the user can decide to downscale the images by a certain percentage.

4.2.3   Modelviewer Tool

Upon completion of the 3D reconstruction (i.e., of the per image dense depth maps) the user is notified by email that the results are ready. They are stored in a zip file on the ESAT FTP server in a directory with a random name whose parent is not listable. This makes sure the user's reconstructions are safe from prying eyes. The zip file contains the original images, the calibration of the cameras, the dense depth maps, and quality maps for every image. The results can be inspected by means of the modelviewer tool, a screenshot of which is shown in Figure 4.3(a). The layout of the modelviewer tool purposely resembles that of the upload tool. A thumbnail is shown for every image in the set. If clicked, a large version of the image is shown in the main window. A triangulated 3D model can be created for the current image, using the depth map and camera of this image. Every pixel of the depth map can be put in 3D and a triangulation in the image creates a 3D mesh, using the texture of the current image. Every depth map has its corresponding confidence map, indicating the quality or certainty of the 3D coordinates of every pixel. The user can select a quality threshold. High thresholds yield models with fewer but more certain points. Lower thresholds yield more complete models including areas with a lower reconstruction quality. If required, certain areas of the image can be masked or grayed out by drawing with a black brush in the image. Areas that are masked will not be part of the 3D model. Homogeneous regions like the sky yield bad reconstruction results and can automatically be discarded by the user. A simple click inside such a region starts a mask-growing algorithm that covers the homogeneous region, as can be seen in Figure 4.4 where the sky above Arenberg Castle in Leuven is automatically covered by a mask. The resulting 3D model is displayed in an interactive 3D widget. Two viewpoints of the Castle example are shown in Figure 4.3(b). The model can be exported in different 3D file formats like VRML2, Open Inventor, OBJ, or OpenSG's native file format.

4.3   Automatic Reconstruction Pipeline

After uploading the images to the ARC3D server, no further user interaction is needed, but also not possible. Hence, important prerequisites are high fidelity and robustness on the server part. In order to avoid frustration and multiple uploads at the client side, the server should maximize the chance of obtaining a good 3D result, even from sequences of images that are not perfectly suited for the purpose. This requirement has lead us to the development of a hierarchical, highly-parallelized, and opportunistic 3D reconstruction scheme.

The following sections explain the different steps in the images-to-3D process. Non-technical readers may want to skip these sections and go immediately to Section 4.4. Readers who want to know even more about the technical aspects are referred to [5].

4.3.1   Pipeline Overview

A schematic flowchart of the reconstruction pipeline is shown in Figure 4.5. As usual, rectangles represent procedures or actions. Parallelograms represent data structures. The input of the pipeline is found on the top-left and consists of the set of images the user has uploaded to the server using the upload client. At the bottom right the result can be seen consisting of the dense 3D depth maps (and the corresponding confidence maps, as well as the camera settings, positions, and orientations, which are all not shown in the figure). The ARC3D reconstruction pipeline can be seen to consist of roughly five steps:

1. **Global image comparison.** The first step computes a set of image pairs that can be used for matching, including the Subsampling and Global Image Comparison steps. In this step, the images are first subsampled (hence the hierarchical nature of the pipeline). Since images can be uploaded in non-sequential order, we have to figure out which images can be matched. This is the task of the Global Image Comparison algorithm, which yields a set of image pairs that are candidates for pairwise matching. Typically, such images will have been taken from nearby viewpoints.

2. **Matching.** In this step, feature points are extracted on the subsampled images. All possible matching candidates of step 1 are now tried. Based on the resulting pairwise matches, all image triplets are selected that stand a chance to together already yield a successful 3D reconstruction. This step corresponds to the Pairwise and Projective Triplet Matching boxes in Figure 4.5.

3. **Self-calibration.** The self-calibration algortithm finds the intrinsic parameters of the camera. Image triplets are used because they are the smallest sets of images from which the cameras can be calibrated (assuming extra conditions, e.g., like them having the same internal parameters, including the same focal lengths).

4. **Sparse reconstruction**. Using the self-calibration results, a sparse reconstruction is computed by triangulating the matching feature points between views. The result subsequently upscaled to full resolution.

5. **Dense matching**. This step is responsible for the dense matching, yielding a dense depth map for every image.

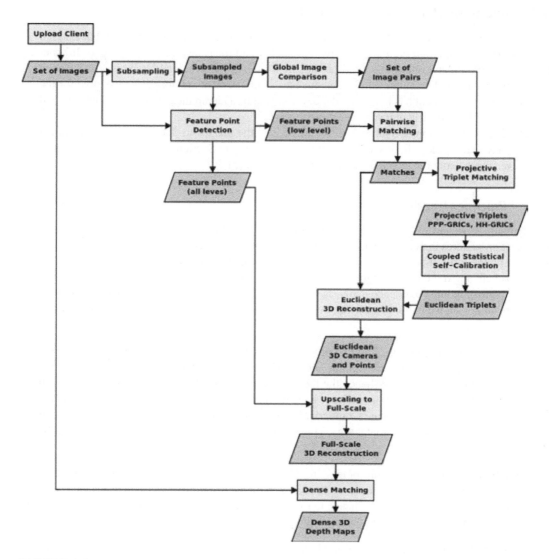

FIGURE 4.5
Global flowchart of the reconstruction pipeline.

Next, we describe the global characteristics of the ARC3D pipeline, like being hierarchical, opportunistic, and parallelized.

4.3.2 Opportunistic Pipeline

Classic 3D reconstruction pipelines make use of the fact that the set of input images is taken in a sequential manner. This helps the reconstruction process tremendously because only consecutive pairs of images must be matched for 3D reconstruction. Unfortunately the 3D web service described in this chapter can not rely on this assumption. Users can upload images in non-sequential order. This has an impact on the matching step, the reconstruction step, and the dense matching step.

4.3.3 Hierarchical Pipeline

In general the quality and accuracy of the resulting depth maps is proportional to the size of the input images. However, computing feature-points and matches on large-scale images is very time-consuming and not so stable a process. That is why all incoming images are first subsampled a number of times until they reach a typical size in the order of $1,000 \times 1,000$ pixels. As can be seen in Figure 4.5 most of the processing is performed on these subsampled images. It is only in the upscaling step (at the bottom right) that the result is upgraded from the low-resolution to the higher resolution of the input image. This hierarchical approach combines a number of advantages. First it is more stable and has a higher chance of success than direct processing of the high-resolution images. Indeed, it is easier to find matching features between small images because the search range is smaller and therefore fewer false matches are extracted. The upscaling step receives a very good initialization from the lower levels. This means that only a small search and fast optimization need to be performed.

4.3.4 Parallel Pipeline

Several operations in the reconstruction pipeline have to be performed many times and independently of each other. Image comparison, feature extraction, pairwise or triplet matching, dense matching, etc., are all examples of such operations. The pipeline is implemented as a Python script which is automatically triggered by the SQL database when a new job arrives. The script has to go through several steps and every step can only be started when the previous one has finished. Inside one step, however, the processing is parallelized. The server on which the script runs has access to a queuing system, as shown in Figure 4.6. When an operation must be performed, the server sends the job to the queuing system which returns a job-id. The queuing system has access to a PC cluster and continuously checks the memory and CPU load of each of the nodes. When a node is free, the job on top of the queue is sent to that node. When the job has finished, the server is automatically notified by the creation of a file, the name of which contains the job-id.

4.4 Practical Guidelines for Shooting Images

4.4.1 Introduction

The ARC3D reconstruction process is completely automatic and aims to support all standard consumer cameras on the market. In order to facilitate as many configurations as possible, there are a number of guidelines to follow when shooting images for ARC3D. To assist in the appreciation of these rules, a short and informal explanation of the basic principles behind ARC3D follows.

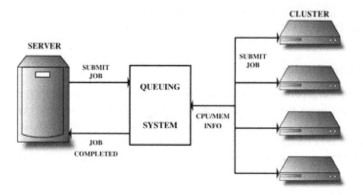

FIGURE 4.6

Queuing system. The server sends a job to the system. The queue has access to a PC cluster and sends the job to any of the nodes that becomes free.

Humans perceive depth mainly by the means of stereo vision (i.e., our eyes see the same object from slightly different viewpoints). The brain can determine the depth from the disparity (difference in position) of the object between the two views. The brain is aware of the optical configuration of our eyes and is able to provide us with an image with correct depth information. In simple words, the same principle is employed in ARC3D where the different positions of the same scene points in the different images are computed. However, ARC3D is unaware of the optics of the camera used to take the image. These parameters have to be determined in a process known as self-calibration. In summary, ARC3D needs to *find and match correspondences* between *enough views* provided the scene contains enough information for *self-calibration*.

All of this comes down to a number of guidelines to follow when selecting a scene and shooting images. Some of them are stringent and absolutely necessary for a successful reconstruction while others may affect the quality of the resulting output.

4.4.2   Image Shooting

The following guidelines should be observed:

- Shoot multiple pictures of the same scene, but viewed from slightly different directions. Walk with the camera in an arc around the scene, while keeping the scene in frame at all times. Figure 4.7 shows a schematic view of a good sequence of views of a scene.

- (Critical) Keep the same zoom setting for all images in the sequence. The auto calibration assumes that the same setting was used throughout the sequence. If the auto calibration fails, the reconstruction fails. In a future version of ARC3D this condition may be relaxed, due to the specification of the focal length in the EXIF data of modern cameras.

- (Critical) Do not pan from the same location. It is not possible to determine enough 3D information from such a sequence, schematics of which are given in Figure 4.8.

- Do not walk straight towards the scene while shooting images.

- Shoot many pictures to ensure a broad selection, but they must not be panned images. Rule: it is better to shoot too many pictures than too few. As a guide, a minimum of five or six images are required for a good reconstruction. Less than four and the reconstruction will fail. One should keep in mind that only points in the scene that are visible in at least two images can be reconstructed in 3D.

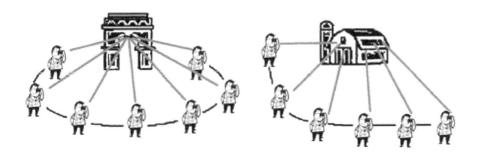

FIGURE 4.7
A diagram of camera positions for a good sequence.

(a) Panning from multiple locations.   (b) Panning from single location.

FIGURE 4.8
Diagrams of two bad sequences. Do not pan the camera without translating.

- If one has to stand too far away from the scene to make it fit in the field of view of the camera, the resolution (the detail of the scene) may become too low. In that case one can also start with a part of the scene and very gradually, while in the meantime changing position, include in field of view new parts of the scene. However, one should keep enough overlap between the different pictures (e.g., for every next picture the newly added information is only about 5% with respect to the previous picture).

- Center the camera to one point in the scene. Try to take pictures with the same point in their center, while walking around the scene.

4.4.3   Scene Selection

ARC3D is able to reconstruct a wide variety of scenes, ranging from small artifacts such as pottery and statues, to caves, cathedrals, and natural scenes such as mountain ranges. However, there are a number of issues to keep in mind while choosing a scene:

- (Critical) Avoid purely planar scenes (i.e., scenes that only consist of a single, planar surface). The auto-calibration step of ARC3D requires a scene with enough 3D information and a scene only containing a flat wall will definitely cause the reconstruction to fail. The case study in Section 4.5.2 describes a technique to avoid this.

- Scenery with a lot of intense texture is much more suitable for this 3D technique than scenery with un-textured areas. Examples of areas with low texture are: white walls, sky, human skin, etc. Examples of high texture: paintings, brick wall, irregular surfaces, etc.

- ARC3D is not able to reconstruct moving objects and aims to automatically disregard them. If possible, they should be avoided since they might degrade the result.

- Reflective surfaces can not be reconstructed (e.g., windows, mirrors).

4.5   Case Study: Reconstruction of the Mogao Caves of Dunhuang

The Mogao Caves complex is a UNESCO world heritage site and one of the most astonishing Chinese ancient cultural sites [8]. The caves are located in the Gobi Desert, close to the town of Dunhuang, in northwestern China. Situated at an important crossroads of the ancient Silk Road, Dunhuang prospered from wealthy caravans transporting goods between China and western Asia and India. In the 4th century AD, Buddhist monks began to carve out shrines in the rock at Mogao. The Mogao Caves soon became a site of pilgrimage and, promoted by rich merchants exploiting the Silk Road, turned into one of the greatest collections of Buddhist art in the world. An outside view of the Mogao Caves can be seen in Figure 4.9(a). The caves were carved during a time span of more than one thousand years and show great diversity, both in structural and artistic aspects. A typical early cave from the Northern Wei Dynasty (386-581 AD) contains a central sculptured column, around which monks performed walking meditations. In later caves, the central column was often omitted and more intricately detailed sculptures were added. The Mogao Caves reached their peak during the Tang Dynasty (618-906 AD), with caves number 96 and 130 as the most outstanding examples. They contain 30 meter tall sitting Buddhas. Also the murals have changed their characteristics; from the wild and crude brush strokes of the early caves to the finer details of the later ones. Today, 492 caves have been preserved, containing a total of 45,000 square meters of murals. We were invited to the

(a)   (b)

FIGURE 4.9

Mogao caves in Dunhuang, China. (a) A view of the mountain wall in which the Mogao Caves are carved. In the front is cave 96, containing one of the enormous sitting Buddha statues. Entrances to more regular caves (cave 322 among them) can also be seen. (b) Mogao cave 322, facing the west wall.

Mogao Caves to demonstrate the 3D reconstruction capabilities of ARC3D. In more than one way, these caves pose enormous challenges to 3D capturing. As has already been described, some of these caves are very high and many parts are therefore difficult to reach. Moreover, the caves with a central Buddha often leave a small space between the statue and the cave walls. Add to that the importance of capturing all the omnipresent murals and sculptures of about 500 caves, and the need for a very easy to use and flexible method becomes evident. The local archaeological team is therefore on the lookout for techniques that may work for their site.

As a first test case, we modeled the largest part of one cave. What follows is a description of the work flow, from image acquisition to post-processing, resulting in a mostly complete model of the cave.

4.5.1   3D Reconstruction of Mogao Cave 322

Mogao cave 322 can be seen in Figure 4.9(b). This is a typical Tang Dynasty cave with a square floor of roughly 4x4 meters and a tapered roof. The cave is supposed to reflect vitality, mirroring the rising power and prosperity of the empire. The west wall is what appears in front upon entering the cave. A doubly recessed niche contains a central Buddha in Lotus position flanked by his Disciples, Bodhisattvas and Devarijas. These seven statues are mostly original. The south and north walls (Figure 4.10(a) and Figure 4.10(b)) are both flat and contain exquisite murals. The south wall depicts a preaching scene while the north features the Amitabha Sutra. The thousand Buddhas motifs covering the slanted roof (Figure 4.10(d)) as well as parts of the walls are painted in a two-dimensional, periodic pattern.

This cave lends itself well to 3D reconstruction. Since the core part of ARC3D consists of automatically finding and matching local features between images, the strong texture covering virtually all parts is very helpful. The murals are completely covering the walls and the roof. Although the thousand Buddhas motif might seem repetitive, each of them is painted separately and contains enough individual characteristics to be robustly matched. In addition, the sandstone and clay materials reflect diffusely, thereby sparing the 3D reconstruction troublesome specular reflections. The main challenge lies in the structural complexity of the statues. There are plenty of fine details and many self-occlusions.

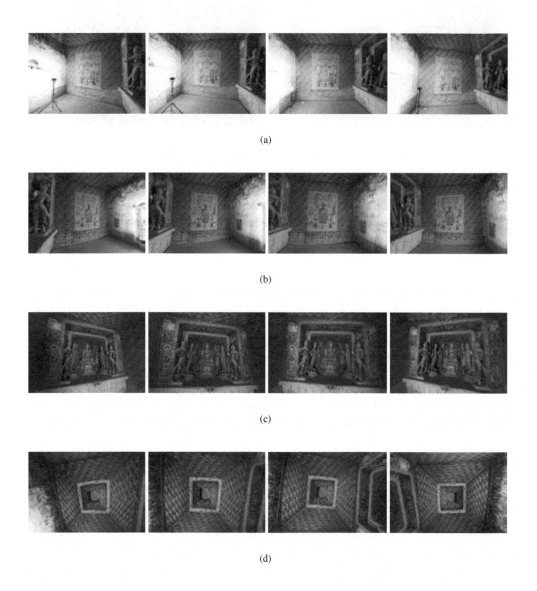

(a)

(b)

(c)

(d)

FIGURE 4.10

Input images used for reconstructing Mogao cave 322. For every sequence we show four out of the ten images used as input to ARC3D. (a) The south wall. The light setup can also be seen to the left. (b) The north wall with a mural depicting the Amitabha Sutra. (c) The west wall. (d) The tapered roof.

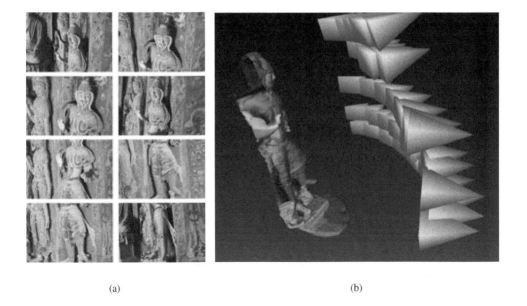

(a)   (b)

FIGURE 4.11
Reconstructing the Devarija statue. (a) Eight of the images used as input. (b) Camera positions, as computed by ARC3D. Each pyramid represents the viewing direction of the camera. The pyramid basis represents the image plane and the apex is the center of projection.

4.5.2   Image Capturing

A Canon EOS-1Ds Mark III digital camera with a tripod and a 14 mm lens was used to capture the images. Although the camera produces images with 20 mega pixels (MP), only half the resolution was used as input to ARC3D. The small quality gain from 10 to 20 MP does not justify the huge increase in computation time. A wide angle lens (like the 14 mm lens we used) is generally recommended for ARC3D, as long as it does not distort the images too much. The larger field of view allows for more features in the image and may improve the matching process. It also gives a larger depth of field and thus lowers the risk of out-of-focus blurring. Finally, it allows more light to reach the sensor, thus reducing image noise. The cave contains no light source other than the sparse daylight emerging from the doorway. To increase the amount of light and to control the shadows, an artificial light source together with a planar diffuser were placed at the entrance (east) wall, as shown in Figure 4.10(a) The cave was reconstructed in five separate sequences: The south, west, and east walls, the ceiling, and a close-up sequence of one of the statues. The west wall was shot by starting close to the south wall and moving the camera parallel to the wall between images, until the other end was reached. Ten images were taken and the whole niche was kept in the field of view for all of them. Some of the images can be seen in Figure 4.10(c). The planarity of the north and south walls could potentially cause the self-calibration stage of ARC3D to break down. To prevent this, parts of the surrounding walls were kept in the field of view for each picture (see Figure 4.10(a) and Figure 4.10(b)). This provided enough out-of-plane information to successfully perform the self-calibration. Also here, ten images per wall were taken. The Devarija statue to the right of the niche was shot by moving the camera in arcs. The tripod was successively lowered between each completed arc as show in Figure 4.11(b). Care had to be taken to ensure sufficient overlap (about 50%) between the images from two successive arcs. For this sequence, 38 images were used of which some can be seen in Figure 4.11(a).

FIGURE 4.12 (SEE COLOR INSERT)
Views of the reconstruction of Mogao cave. The texture is removed in some of the images to better visualize the underlying structure.

FIGURE 4.13 (SEE COLOR INSERT)
Close-up views of the reconstruction of Mogao cave 322.

FIGURE 4.14
Eight of the input images used to reconstruct the Arc de Triomphe.

4.5.3 Result

After submitting the sequences to ARC3D and obtaining the results, the resulting depth maps were imported into MeshLab for post processing. The Poisson Reconstruction [9] tool was used to remove noise and to close holes. ARC3D reconstructs all meshes in their own reference frame, since no information about global position and orientation (aka geo-reference) can be obtained from the images only. Before merging two sequences, the meshes thus had to be transformed into an arbitrary, but common, reference frame. This can be done manually by picking one of the meshes as reference, to which frame the other meshes are scaled, rotated, and translated. Manually transforming the meshes is a time-consuming task, and in this case we used a newly developed tool which performs this automatically. This function is intended to be available to the public in future versions of ARC3D. A video displaying the complete model was rendered using Blender [10], an open-source software package for 3D modeling. Figure 4.12 and Figure 4.13 show some of its snapshots. For a more complete experience, we encourage the reader to watch the video present in the 3D gallery section of the ARC3D website [4].

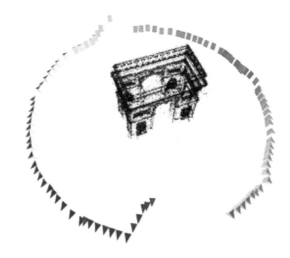

FIGURE 4.15
Image viewpoints for Arc de Triomphe sequence. Note that it was impossible to capture images from the middle of the boulevards radiating from the Place de l'Étoile.

4.6   Examples

The previous section showed how ARC3D can be employed for the 3D reconstruction of an enclosed environment such as a cave. In computer vision parlance, this is an inside-out type of problem. In this section, a number of other scenarios are demonstrated to show ARC3D's capability of dealing with a wide variety of scenes.

4.6.1   A Complete Building: Arc de Triomphe

As a first, complementary example, we take an "outside-in" type of scene, where the camera is moved around the object of interest. The Arc de Triomphe, one of the best known buildings of Paris, was reconstructed using 106 images. Some of those can be seen in Figure 4.14. The open space surrounding the landmark permits the capturing of images from all possible directions. This makes it possible to reconstruct the Arc de Triomph from one long sequence of images. Shooting was made along a mostly circular trajectory around the landmark, as displayed in Figure 4.15, based on ARC3D output about the camera positions. The result is a complete 3D model, which can be seen in Figures 4.16 and 4.17.

Another similar example, but of lower scale, is the reconstruction of a stelae. These stone pillars are typically surrounded by open space and may also be reconstructed in one shooting. Figure 4.18 shows a reconstruction of a stelae from the Calakmul archaeological site in Mexico. Interesting to see is that the inscriptions on the stelae, which are only vaguely visible to the human eye, can clearly be perceived when the texture is removed.

4.6.2   Environment Scene

ARC3D is not only able to reconstruct individual objects; complete models of natural environments can also be created. Figure 4.19 shows the reconstruction of a mountain scene in Alberta, Canada.

FIGURE 4.16
3D reconstruction of Arc de Triomphe.

FIGURE 4.17
3D reconstruction of Arc de Triomphe.

FIGURE 4.18
Reconstruction of a stelae in Calakmul, Mexico. When the texture is removed, the legibility of the carvings is greatly increased.

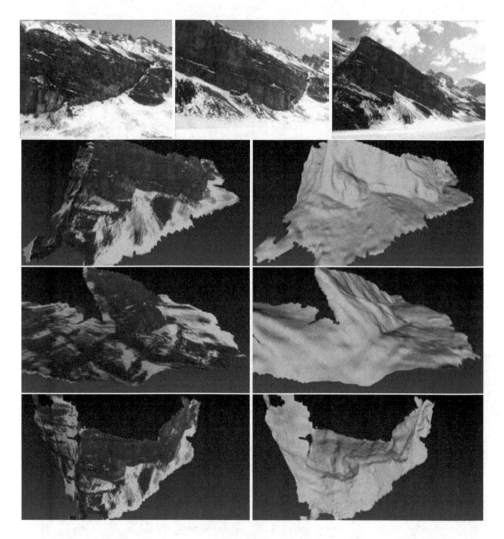

FIGURE 4.19
Three of the 30 input images (top row) and the resulting reconstruction of a mountain range.

FIGURE 4.20 (SEE COLOR INSERT)
Duomo, Prato, Italy. Some of the uploaded images are shown to the left and reconstruction to the right.

FIGURE 4.21
Christ statue at Park Abbey, Leuven, Belgium. The overall scene is well reconstructed but additional close-up images of the statue could have provided finer details.

4.6.3 Further Examples

Two other examples created by the ARC3D user community are displayed in Figure 4.20 and Figure 4.21.

4.7 Conclusions

For a broad range of objects and scenes, self-calibrating 3D reconstruction methods based solely on images are a viable option. ARC3D is such a system, which is freely available as a web service for non-commercial use.

In the future, we plan to extend this web service in several ways. On the agenda are the creation of more automated tools for putting the depth maps into registration, for the extraction of more consistent textures, and for the use of omni-directional images (rather than the perspective images used exclusively now).

Acknowledgments

The authors gratefully acknowledge support by the EC ICT project 3D-COFORM.

Bibliography

[1] F. Remondino, S. Girardi, A. Rizzi, and L. Gonzo, "3D modeling of complex and detailed cultural heritage using multi-resolution data," *JOCCH*, vol. 2, no. 1, 2009.

[2] C. Strecha, W. von Hansen, L. Van Gool, P. Fua, and U. Thoennessen, "On benchmarking camera calibration and multi-view stereo for high resolution imagery," in *CVPR08*, pp. 1–8, 2008.

[3] "MeshLab, open source mesh processing tool." http://meshlab.sourceforge.net.

[4] "Arc3d website." http://www.arc3d.be.

[5] M. Vergauwen and L. Van Gool, "Web-based 3D reconstruction service," *Mach. Vision Appl.*, vol. 17, no. 6, pp. 411–426, 2006.

[6] R. I. Hartley and A. Zisserman, *Multiple View Geometry in Computer Vision.* Cambridge University Press, ISBN: 0521540518, second ed., 2004.

[7] T. Moons, L. Van Gool, and M. Vergauwen, "3D reconstruction from multiple images: Part 1 principles," *Foundations and Trends in Computer Graphics and Vision*, vol. 4, 2010.

[8] "The Dunhuang Academy." http://enweb.dha.ac.cn.

[9] M. Kazhdan, M. Bolitho, and H. Hoppe, "Poisson surface reconstruction," in *SGP '06: Proceedings of the fourth Eurographics symposium on Geometry processing* (Aire-la-Ville, Switzerland), pp. 61–70, Eurographics Association, 2006.

[10] "Blender, open source 3D content creation suite." http://www.blender.org.

5

Accurate and Detailed Image-Based 3D Documentation of Large Sites and Complex Objects

Fabio Remondino

B. Kessler Foundation (FBK)

`Email: remondino@fbk.eu`

CONTENTS

5.1   Introduction

The environment and its heritage sites or objects suffer from wars, natural disasters, weather changes, and human negligence. According to UNESCO [1], "a heritage can be seen as an arch between what we inherit and what we leave behind ... Heritage is our legacy from the past, what we live with today and what we pass on to future generations." Thus the importance of cultural heritage documentation is well recognized and there is an increasing pressure to document and preserve them also digitally. Therefore 3D data become a critical component to permanently record the shapes of important objects so that they might be passed down to future generations. The continuous development of new sensors, data capture methodologies, and multi-resolution 3D representations and the improvement of existing ones are contributing significantly to the documentation, conservation, and presentation of heritage information and to the growth of research in the cultural heritage field. This is also driven by the increasing requests and needs for the digital documentation of heritage sites at different scales and resolutions, for the general mapping and digital preservation of our environment but also for studies, analysis, interpretation, etc.

In the last few years, great efforts have been focused on what we inherit as cultural heritage and on their documentation, in particular for visual man-made or natural heritages, which received a lot of attention and benefits from sensor and imaging advances. This has produced firstly a large number of projects, mainly led by research groups, which have realized very good quality and complete digital models [2–9], and secondly has spurred the creation of guidelines describing standards for correct and complete documentations. The actual technologies and methodologies for cultural heritage documentation [10] allow the generation of very realistic 3D results (in terms of geometry and texture). Photogrammetry [11, 12] and Laser Scanning [13, 14] are the most used techniques for applications like accurate surveying, archaeological documentation, digital conservation or restoration, VR/CG, GIS, 3D repositories and catalogs, web geographic systems, visualization and animation, etc. But despite all the possible applications and the constant pressure of international organizations, a systematic and well-judged use of 3D models in the cultural heritage field is still not yet employed as a default approach for several reasons: (1) the "high cost" of 3D; (2) the difficulties in achieving good 3D models by everyone; (3) the consideration that it is an optional process of interpretation (an additional "aesthetic" factor) and documentation (2D is enough); (4) the difficulty of integrating 3D worlds with other more standard 2D material. But the availability and use of 3D computer models of heritages opens a wide spectrum of further applications and permits new analysis, studies, interpretations, and conservation policies as well as digital preservation and restoration. Thus virtual heritages should be more and more frequently used due to the great advantages that the digital technologies are giving to the heritage world and to recognize the documentation needs stated in the numerous charters and resolutions.

This chapter reviews some important 3D documentation requirements and specifications, and the actual surveying and 3D modeling methodologies, and then will provide an in-depth look at the photogrammetric approach for accurate and detail reconstruction of objects and sites. The chapter is organized as follows: the next section reviews the issues related to reality-based methods for large and complex site surveying and modeling, with the actual digital techniques and sensors; the photogrammetric approach will be described in Section 3, with all the steps of the 3D documentation pipeline and the open research issues in image-based modeling. Finally, some concluding remarks will end the chapter.

5.2   Reality-Based 3D Modeling

Metric 3D documentation and modeling of scenes or objects should be intended as the entire procedure that starts with the surveying and data acquisition, data processing and 3D information generation, visualization and preservation for further uses like conservation, education, restoration, GIS applications, etc. A technique is intended as a scientific procedure (e.g., image processing) to accomplish a specific task while a methodology is a group or combination of techniques and activities combined to achieve a particular task.

Reality-based techniques (e.g., photogrammetry, laser scanning, etc.) [15] employ hardware and software to survey the reality as it is, documenting the actual visible situation of the site. Non-real approaches are instead based on computer graphics software (3D Studio, Maya, Sketchup, etc.) or procedural modeling [16] and they allow the generation of 3D data without any particular survey or knowledge of the site. Nowadays the digital documentation and 3D modeling of cultural heritage is mostly based on reality-based techniques and methodologies and it generally consists of [128]:

- metric recording and processing of a large amount of three (possibly four) dimensional multi-source, multi-resolution, and multi-content information;

- management and conservation of the achieved 3D (4D) models for further applications;

- visualization and presentation of the results to visually present the information to other users;

- digital inventories and sharing for data retrieval, education, research, conservation, entertainment, walkthrough, or tourism purposes.

In most of the practical applications, the generation of digital 3D models of heritage objects or sites for documentation and conservation purposes requires a technique or a methodology with the following properties:

- *accuracy*: precision and reliability are two important factors of the surveying work and final 3D results, unless the work is done for simple and quick visualization;

- *portability*: the technique should be as portable as possible due to accessibility problems, absence of electricity, location constraints, etc;

- *low cost*: most surveying missions have limited budgets and they cannot affort expensive documentation instruments;

- *fast acquisition*: most sites or excavation areas allow a limited time for the 3D recording and documentation not to disturb restoration works or the visitors;

- *flexibility*: due to the great variety and dimensions of sites and objects, the technique should allow different scales and it should be applicable in any possible condition.

All these properties are often not applicable to a specific technique, therefore most of the surveying projects related to large and complex sites integrate and combine multiple sensors and techniques. Nevertheless, the image-based approach has different advantages compared to the range-based one and it is often selected although it sometimes requires the user's interaction and some experience in both data acquisition and processing.

5.2.1 Techniques and Methodologies

The generation of reality-based 3D models of large and complex sites or structures is nowadays performed using methodologies based on passive sensors and image data [17], optical active sensors and range data [14, 18, 19], classical surveying (e.g., total stations or GNSS), or an integration of the aforementioned techniques [20–24]. For architecture engineering and in the construction community, geometric 3D models can also be generated from existing 2D drawings or maps, interactively and using extrusion functions [25]. The choice or integration depends on the required accuracy, site and object dimensions, location constraints, instrument's portability and usability, surface characteristics, working team experience, project's budget, final goal, etc.

Optical active sensors like laser scanners (pulsed, phase-shift, or triangulation-based instruments) and stripe projection systems have received in the last years a great attention for 3D documentation and modeling applications. They deliver directly ranges (e.g., distances thus 3D information in the form of unstructured point clouds) and are getting quite common in the heritage field, despite their high costs, weight, and the usual lack of good texture. For the surveying, the instrument should be placed in different locations or the object is moved in a way that the instrument can see it from different viewpoints. Successively, the raw data needs errors and outliers removal, noise reduction, and sometimes holes filling before the alignment or registration in a unique reference system to produce a unique point cloud of the surveyed scene or object. The registration is generally done in two steps: (1) manual or automatic raw alignment using targets or the data itself and (2) final global alignment based on iterative closest points [26] or least squares method procedures [27]. After the global alignment, redundant points should be removed before a surface model is produced and textured.

According to [28], the scanning results are a function of:

- intrinsic characteristics of the instrument (calibration, measurement principle, etc.);

- characteristics of the scanned material in terms of reflection, light diffusion, and absorption (amplitude response);

- characteristics of the working environment;

- coherence of the backscattered light (phase randomization);

- dependence from the chromatic content of the scanned material (frequency response).

Terrestrial range sensors work from very short ranges (a few centimeters) up to a few kilometers, in accordance with surface proprieties and environment characteristics, delivering 3D data with positioning accuracy from some hundred of microns up to some millimeters. Range sensors, coupled with GNSS/INS sensors, can also be used on aerial platform [29, 30] generally for DTM/DSM generation, city modeling, and archaeological applications. Although aware of the potentialities of the image-based approach and its recent developments in automated and dense image matching [31–35], the easy usability and reliability of optical active sensors in acquiring unstructured and dense point clouds is generally much higher. This fact has made active sensors like laser scanners a very common 3D recording solution, in particular for non-expert users. Active sensors generally also include radar instruments, which are generally not considered as optical sensors, nevertheless they are often used in the cultural heritage 3D documentation or environmental mapping.

On the other hand, image data require a mathematical formulation to transform the two-dimensional image measurements into three-dimensional coordinates. Image-based modeling techniques (mainly photogrammetry and computer vision) are generally preferred in cases of lost objects, monuments, or architectures with regular geometric shapes, low-budget projects, good experience of the working team, or time or location constraints for the data acquisition and processing. Moreover, images contain all the information useful to derive the 3D shape of the surveyed scene as well as its graphical appearance (texture). Image-based 3D modeling generally requires some user interaction in the different steps of the modeling pipeline to derive accurate results, limiting its use only to experts, while for quick 3D results, useful mainly for simple visualization fully automated approaches are available [36]. Generally at least two images are necessary to derive 3D information of corresponding image points but there are also some methods to derive 3D data from a single image [37, 38].

Many works reported how the photogrammetric approach generally allows surveys at different levels and in all possible combinations of object complexities, with high quality requirements, easy usage and manipulation of the final products, few time restrictions, good flexibility, and low costs [13, 39, 40]. Different comparisons between photogrammetry and laser scanning were also presented in the literature [41–43] although it can not be stated a priori which is the best technique. In case of large sites surveying and 3D modeling, the best solution is the integration of multiple sensors and techniques.

5.2.2   Multi-Sensor and Multi-Source Data Integration

Nowadays the state-of-the-art approach for the 3D documentation and modeling of large and complex sites uses and integrates multiple sensors and technologies (photogrammetry, laser scanning, topographic surveying, etc.) to (1) exploit the intrinsic potentials and advantages of each technique, (2) compensate for the individual weaknesses of each method alone, (3) derive different geometric levels of detail (LOD) of the scene under investigation, and (4) achieve more accurate and complete geometric surveying for modeling, interpretation, representation, and digital conservation issues. 3D modeling based on multi-scale data and multi-sensors integration is indeed providing the best 3D results in terms of appearance and geometric detail. Each LOD is showing only the necessary information while each technique is used where best suited.

Since the nineties, multiple data sources have been integrated for industrial, military, and mobile mapping applications. Sensor and data fusion were then applied also in the cultural heritage domain, mainly at terrestrial level [4, 20], but in some cases also with satellite, aerial, and ground information for a more complete and multi-resolution 3D survey [5, 23, 44].

The multi-sensor and multi-resolution concept (Figure 5.1) should be distinguished between

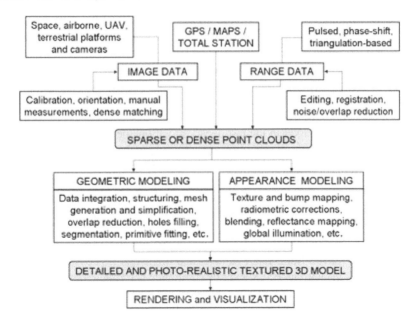

FIGURE 5.1

The multi-sensor and multi-resolution 3D modeling pipeline based on passive and optical active sensors for the generation of point clouds and textured 3D models.

(1) geometric modeling (3D shape acquisition, registration, and further processing) where multiple resolutions and sensors are seamlessly combined to model features with the most adequate sampling step and derive different geometric levels of detail (LOD) of the scene under investigation and (2) appearance modeling (texturing, blending, simplification, and rendering) where photo-realistic representations are sought, taking into consideration variations in lighting, surface specularity, seamless blending of the textures, user's viewpoint, simplification, and LOD.

Beside images acquired in the visible part of the light spectrum, it is often necessary to acquire extra information provided by other sensors working in different spectral bands (e.g., IR, UV) in order to study the object more closely. Thermal infrared information is useful to analyze historical buildings, their state of conservation, reveal padding, older layers, and back structure of frescoes, while near IR is used to study paintings, revealing pentimenti, and preparatory drawings. On the other hand, the UV (ultraviolet) radiations are very useful in heritage studies to identify different varnishes and over-paintings, in particular with induced visible fluorescence imaging systems [45]. All of this multi-modal information needs to be aligned and often overlapped to the geometric data for information fusion, multi-spectral analysis, or other diagnostic applications [46].

5.2.3 Standards in Digital 3D Documentation

"It is essential that the principles guiding the preservation and restoration of ancient buildings should be agreed and be laid down on an international basis, with each country being responsible for applying the plan within the framework of its own culture and traditions" (The Venice Charter, i.e. The International Charter for the Conservation and Restoration of Monuments and Sites, 1964). Even if this was stated more than 40 years ago, the need for a clear, rational, standardized terminology and methodology, as well as an accepted professional principles and technique for interpretation, presentation, digital documentation, and presentation is still evident. Furthermore "...Preservation of the digital heritage requires sustained efforts on the part of governments, creators, publishers, relevant industries and heritage institutions. In the face of the current digital divide, it is necessary to reinforce international cooperation and solidarity to enable all countries to ensure creation, dissemination, preservation and continued accessibility of their digital heritage" (The UNESCO Charter on the Preservation of the Digital Heritage, 2003). Therefore, beside computer recording and modeling, our heritages require more international collaborations and information sharing to digitally preserve them and make them accessible in all the possible forms and to all the possible users and clients.

From a more technical point of view, many image-based modeling packages as well as range-based systems came out on the market in the last decades to allow the digital documentation and 3D modeling of objects or scenes. Many new users are approaching these methodologies and those who are not really familiar with them need clear statements and information to know if a package or system satisfies certain requirements before investing. Therefore technical standards for the 3D imaging field must be created, like those available for the traditional surveying or CMM. Apart from standards, comparative data and best practices are also needed, to show not only advantages but also limitations of systems and software. In these respects, the German VDI/VDE 2634 contains acceptance testing and monitoring procedures for evaluating the accuracy of close-range optical 3D vision systems (particularly for full-frame range cameras and single scan). The American Society for Testing and Materials (ASTM) with its E57 standards committee is trying to develop standards for 3D imaging systems for applications like surveying, preservation, construction, etc. The International Association for Pattern Recognition (IAPR) created the Technical Committee 19 - computer vision for cultural heritage Applications - with the goal of promoting computer vision Applications in cultural heritage and their integration in all aspects of IAPR activities. TC19 aims at stimulating the development of components (both hardware and software) that can be used by researchers in cultural heritage like archaeologists, art historians, curators, and institutions like universities, museums, and research organizations. As far as the presentation and visualization of the achieved 3D models is concerned, the London Charter (http://www.londoncharter.org/) is seeking to define the basic objectives and principles for the use of 3D visualization methods in relation to intellectual integrity, reliability, transparency, documentation, standards, sustainability, and access of cultural heritage.

The Open Geospatial Consortium (OGC) developed the GML3, an extensible international standard for spatial data exchange. GML3 and other OGC standards (mainly the OpenGIS Web Feature Service (WFS) Specification) provide a framework for exchanging simple and complex 3D models. Based on the GML3, the CityGML standard was created, an open data model and XML-based format for storing, exchanging, and representing 3D urban objects and in particular virtual city models.

5.3   Photogrammetry

Photogrammetry [11, 47, 48] is the science of obtaining accurate, metric, and semantic information from photographs (images). Photogrammetry turns 2D image data into 3D data like digital models, rigorously establishing the geometric relationship between the image and the object as it existed at the time of the imaging event. Once the relationship, described with the collinearity principle (Figure 5.2), is correctly recovered, accurate 3D information about a scene can be strictly derived from its imagery. Photogrammetry can be done using underwater, terrestrial, aerial, or satellite imaging sensors. Generally the term Remote Sensing [49] is more associated to satellite imagery and their use for land classification and analysis or changes detection.

The photogrammetric method generally employs a minimum of two images of the same static scene or object acquired from different points of view. Similar to human vision, if an object is seen in at least two images, the different relative positions of the object in the images (the so-called parallaxes) allow a stereoscopic view and the derivation of 3D information of the scene seen in the

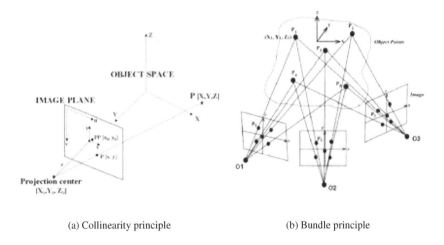

(a) Collinearity principle   (b) Bundle principle

FIGURE 5.2
Collinearity principle (a): camera projection center, image point, and object point P are lying on a straight line. The departure from the theoretical collinearity principle is modeled with the camera calibration procedure. Photogrammetric image triangulation principle or bundle adjustment (b): the image coordinates of homologue points measured in different images are simultaneously used to calculate their 3D object coordinates by intersecting all the collinearity rays.

overlapping area of the images. Photogrammetry can therefore be applied using a single image (e.g., monoplotting, ortho-rectification, etc.), or two (stereo) or multiple images (block adjustment).

Photogrammetry is used in many fields, from traditional mapping and 3D city modeling to the video games industry, from industrial inspections to movie production, from heritage documentation to the medical field. Traditionally, photogrammetry was always considered a manual and time-consuming procedure but in the last decade many developments have led to great improvements in the performances of the technique and now many semi- or fully automated commercial procedures are available (just to mention some packages: Australis, ImageModeler, iWitness, PhotoModeler, ShapeCapture for terrestrial photogrammetry; PCI Geomatica, ERDAS LPS, Bae System Socet-Set, Z-I ImageStation mainly for satellite and aerial photogrammetry). If the goal is the recovery of a complete, detailed, precise, and reliable model, some user interaction in the photogrammetric modeling pipeline does not matter. If the purpose is just the recovery of a 3D model usable for simple visualization or virtual reality applications, fully automated procedures can also be adopted. In the case of heritage sites and objects, the advantages of photogrammetry become readily evident: (1) images contain all the information required for 3D modeling and accurate documentation (geometry and texture); (2) taking images of an object is usually faster and easier; (3) image measurements help avoid potential damage caused by contact surveyings; (4) an object can be reconstructed even if it has disappeared or considerably changed using available or archived images [50,51]; (5) photogrammetric instruments (cameras and software) are generally cheap, very portable, easy to use, and with very high accuracy potentials; (6) the recent developments in automated surface measurement from images revealed results very similar to those achieved with range sensors [33]. Nevertheless, the integration of the photogrammetric approach with other measurement techniques (like laser scanner) should not be neglected and their combination is leading so far to quite good documentation results as it allows the use of the inherent strength of both approaches [4,9,20,44]. Compared to other techniques which also use images to derive 3D information (like computer vision, shape from shading, shape from texture, etc.), photogrammetry does not aim at the full automation of the procedures but it has always

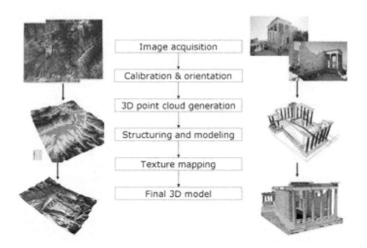

FIGURE 5.3

Photogrammetric 3D modeling pipeline: from the image data acquisition to the generation of the final textured 3D model. According to the application and type of scene or object, sparse or dense point clouds are generated using interactive or automated procedures.

as its first goal the recovery of metric and accurate results. The most well known sister technique of photogrammetry is computer vision, which aims at deriving 3D information from images in a fully automated way using a projective geometry approach. The typical vision pipeline for scene modeling is named *Structure from Motion* [32, 36, 52–54] and it is getting quite common in the 3D heritage community, mainly for visualization and VR applications as it is not yet fully accurate for the daily documentation and recording of large and complex sites.

The entire photogrammetric workflow (Figure 5.3) used to derive metric and accurate 3D information of a scene from a set of images consists of:

- camera calibration and image orientation,

- 3D measurements,

- structuring and modeling,

- texture mapping and visualization.

Compared to the range sensors (e.g., laser scanner) workflow, the main difference stays in the 3D point cloud derivation: indeed range sensors deliver directly the 3D data while photogrammetry requires the mathematical processing of the image data (through the calibration, orientation, and further measurement procedures) to derive the required sparse or dense point clouds useful to digitally reconstruct a scene or an object.

In some cases, 3D information could also be derived from a single image using object constraints [38,55,56] or estimating surface normals instead of image correspondences (shape from shading [37], shape from texture [57], shape from specularity [58], shape from contour [59], shape from 2D edge gradients [60]).

5.3.1 Image Data Acquisition

Nowadays a large amount of image products is available in terms of geometrical resolution (footprint or Ground Sample Distance), spectral resolution (number of spectral channels), and costs. Used for mapping, documentation and visualization purposes, the available images can be acquired from a wide variety of acquisition tools: middle- and high-resolution satellite sensors (GeoEye, WorldView, IKONOS, Quickbird, SPOT-5, Orbview, IRS, ASTER, Landsat), large format and linear array digital aerial cameras (ADS40, DMC, ULTRACAM), space and aerial radar platforms (AirSAR, ERS, Radarsat), model helicopters and low altitude platforms (UAV) with consumer digital cameras or small laser scanner on-board, small format digital aerial cameras (DAS, DIMAC, DSS, ICI), linear array panoramic cameras, still video cameras, camcorders, or even mobile phones. Moreover GNSS and INS/IMU systems allow precise localization and navigation and are beginning to be integrated also in terrestrial cameras. Terrestrial cameras can be traditionally distinguished as (1) amateur/consumer and panoramic cameras, employed for VR reconstructions, quick documentation, and forensic applications, with a relative accuracy potential around 1:25,000; (2) professional SRL cameras, used in industrial, archaeological, architectural, or medical applications, which allow up to 1:100,000 accuracy; (3) photogrammetric cameras, used mainly for industrial applications and with an accuracy potential of 1:200,000. Nowadays consumer digital cameras are coming with more than 8-10 mega pixel sensors, therefore highly precise measurements could be achieved also with these sensors, even if the major problems in these types of cameras are given by the lens/objective, unstable electronics, and some cheap components. Consumer digital cameras contain frame CCD (Charge Coupled Device) or CMOS (Complementary Metal-Oxide Semiconductor) sensors while industrial and panoramic cameras can have linear array sensors.

Between the available aerial platforms, particular attention is now given to the UAVs (Unmanned Aerial Vehicles), i.e., remotely operated aircrafts able to fly without a human crew. They can be controlled from a remote location or fly autonomously based on pre-programmed flight plans. There is a wide variety of UAVs shapes, dimensions, configurations, and characteristics. The most used UAVs in geomatics applications are model helicopters which use integrated GNSS/INS, stabilizer platform, and digital cameras on-board. They are generally employed to get images from otherwise hardly accessible areas and derive DSM/DTM. Traditionally UAVs were employed in the military and surveillance domain, but nowadays they are getting quite common also in the heritage field, as they are able to provide unique top-down or oblique views that can be extremely useful for site studies, inspection, and documentations. Typical UAVs span from a few centimeters up to 2-3 metres and are able to carry up to 50 kilogram instruments (cameras or laser scanners) on-board. In some cases, kites, balloons, or zeppelins are also meant as UAVs, even if they are more unstable and therefore complicated to control and maneuver on the field.

5.3.2 Camera Calibration and Image Orientation

Camera calibration and image orientation are procedures of fundamental importance, in particular for all those geomatics applications which rely on the extraction of precise 3D geometric information from imagery. The early theories and formulations of orientation procedures were developed more than 50 years ago and today there is a great number of procedures and algorithms available [61]. A fundamental criterion for grouping these procedures is the used camera model (i.e., the projective or the perspective camera model). Camera models based on perspective approaches require stable optics, not many image correspondences, and have high stability. On the other hand, projective methods can deal with variable focal length, but are quite unstable and need more parameters and image correspondences to derive 3D information.

In photogrammetry, the calibration and orientation procedures, performed respectively to retrieve the camera interior and exterior parameters, use the collinearity principle and a multi-image approach (Figure 5.2). The *interior orientation* (IO) parameters, recovered with a calibration procedure, consist of the camera constant (or focal length f), position of the principal point (x_0, y_0), and some additional parameters (APs) to model systematic errors due for example to lens distortion. The calibration procedure is defined as the determination of geometric deviations of the physical reality from a geometrically ideal imaging system: the pinhole camera. The calibration can be performed for a

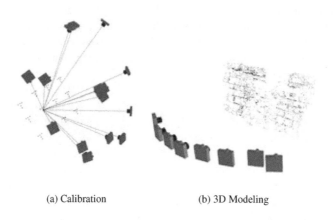

(a) Calibration   (b) 3D Modeling

FIGURE 5.4
(a) Most favorable image network configuration for accurate camera calibration. (b) Typical configuration for object 3D modeling.

single image or using a set of images all acquired with the same set of interior parameters. A camera is considered calibrated if the focal length, principal point offset, and its APs (strictly dependent on the used camera settings) are known. In many applications only the focal length is recovered while for precise photogrammetric measurements all the calibration parameters are generally employed. On the other hand, the *exterior orientation* (EO) parameters consist of the three positions of the perspective center (X_0, Y_0, Z_0) and the rotations (ω, ϕ, κ) around the three axes of the coordinate system. These parameters are generally retrieved using two images (relative orientation) or simultaneously with a set of images (bundle solution).

In all photogrammetric applications, camera calibration and image orientation are recovered separately using a photogrammetric bundle adjustment solution [62–64]. Camera calibration and image orientation can also be performed simultaneously, leading to a self-calibrating bundle-adjustment [65, 66], i.e., a bundle adjustment extended with some APs to compensate for systematic errors due to the uncalibrated camera. But in practical cases, rather than simultaneously recovering the camera internal parameters and reconstructing the object, it is always better first to calibrate the camera at a given setting using the most appropriate network and afterwords recover the camera poses and object geometry using the calibration parameters at the same camera setting. Indeed the image network most suitable for camera calibration is different from the typical image configuration suitable for object reconstruction (Figure 5.4).

In the computer vision community, calibration and orientation are often solved simultaneously leading to the *Structure from Motion* approaches where an uncalibrated image sequence, generally with very short baselines, is used to recover the camera parameters and reconstruct 3D shapes [52, 53, 67, 68]. But the results reported an accuracy of 1:400 in the best case [32], limiting their use in applications requiring only nice-looking 3D models.

The most common set of APs employed to calibrate CCD / CMOS cameras and compensate for systematic errors is the 8-term "physical" model originally formulated by Brown in 1971 [69]:

$$\Delta x = -\Delta x_0 \qquad +\bar{x} \cdot \frac{\Delta f}{f} + \bar{x} \cdot (K_1 r^2 + K_2 r^4 + K_3 r^6)$$
$$+P_1 \cdot (r^2 + 2\bar{x}^2) + 2 \cdot P_2 \cdot \bar{x}\bar{y} + A \cdot \bar{x} + S_x \cdot \bar{y}$$
$$\Delta y = -\Delta y_0 \qquad +\bar{y} \cdot \frac{\Delta f}{f} + \bar{y} \cdot (K_1 r^2 + K_2 r^4 + K_3 r^6)$$
$$+2 \cdot P_1 \cdot \bar{x}\bar{y} + P_2 \cdot \left(r^2 + 2\bar{y}^2\right) \tag{5.1}$$

where: $\bar{x}$ = x - x_0, $\bar{y}$ = y - y_0 and $r^2 = \bar{x}^2 + \bar{y}^2$

Besides the main interior parameters x_0, y_0, and f, Equation 5.1 contains three terms K_i for the radial distortion, two terms P_i for the decentering distortion, one term A for the affinity factor, and one term S_x to correct for the shear of the pixel elements in the image plane. The last two terms are rarely if ever significant in modern digital cameras and in particular for heritage applications but very useful in very high accuracy applications. Numerous investigations on different sets of APs have been performed over the past years, yet this model, despite being more than 30 years old, still holds up as the optimal formulation for complete and accurate digital camera calibration. The three APs K_i used to model the radial distortion Δr (also called barrel or pincushion distortion) are generally expressed via the odd-order polynomial $\Delta r = K_1 r^3 + K_2 r^5 + K_3 r^7$, where r is the radial distance from the image center. A typical Gaussian radial distortion profile Δr is shown in Figure 5.5(a), which illustrates how radial distortion varies with the focal length of the camera. The coefficients K_i are usually highly correlated, with most of the error signal generally being accounted for by the cubic term $K_1 r^3$. The K_2 and K_3 terms are typically included for wide-angle lenses and in high-accuracy machine vision and metrology applications. Recent research has demonstrated the feasibility of empirically modeling radial distortion throughout the magnification range of a zoom lens as a function of the focal length written to the image EXIF header [70]. Decentering distortion is instead due to a lack of centering of lens elements along the optical axis. The decentering distortion parameters P_1 and P_2 are invariably strongly projectively coupled with x_p and y_p. Decentering distortion is usually an order of magnitude or more less than radial distortion and it also varies with focus, but to a much less extent, as indicated by the decentering distortion profiles shown in Figure 5.5. The projective coupling between P_1 and P_2 and the principal point offsets increase with the increasing of the focal length and can be problematic for long focal length lenses. The extent of coupling can be diminished through the use of a 3D object point array, the adoption of higher convergence image, as well as 90-degree-rotated images.

The basic principle of photogrammetry is the collinearity model which states that a point in object space, its corresponding point in an image, and the projective center of the camera lay on a straight line (Figure 5.2). For each point j measured in the image i, the collinearity model, extended to account also for systematic errors, states:

$$x_{ji} - x_{0,i} = -f_i \cdot \frac{r_{11,i}\left(X_j - X_{0,i}\right) + r_{12,i}\left(Y_j - Y_{0,i}\right) + r_{13,i}\left(Z_j - Z_{0,i}\right)}{r_{31,i}\left(X_j - X_{0,i}\right) + r_{32,i}\left(Y_j - Y_{0,i}\right) + r_{33,i}\left(Z_j - Z_{0,i}\right)} + \Delta x_i$$
$$y_{ji} - y_{0,i} = -f_i \cdot \frac{r_{21,i}\left(X_j - X_{0,i}\right) + r_{22,i}\left(Y_j - Y_{0,i}\right) + r_{23,i}\left(Z_j - Z_{0,i}\right)}{r_{31,i}\left(X_j - X_{0,i}\right) + r_{32,i}\left(Y_j - Y_{0,i}\right) + r_{33,i}\left(Z_j - Z_{0,i}\right)} + \Delta y_i \tag{5.2}$$

where:

- x_{ji}, y_{ji} are the image coordinates of the measured tie point j in the image i;

- f_i, $x_{0,i}$, $y_{0,i}$ are the focal length and principal point of the image i;

- $X_{0,i}$, $Y_{0,i}$, $Z_{0,i}$ are the position of the camera perspective center for the image i;

- $r_{11,i}$, $r_{12,i}$,...,$r_{33,i}$ are the nine elements of the rotation matrix R_i created with the three rotation angles (ω, ϕ, κ);

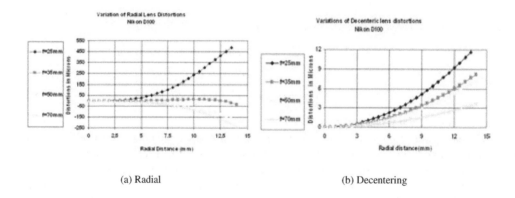

(a) Radial   (b) Decentering

FIGURE 5.5
Typical behavior of the radial (a) and decentering (b) distortion according to the variation of the image radius and focal length.

- X_j, Y_j, Z_j are the 3D object coordinates of the measured point j;

- $\Delta x_i, \Delta y_i$ are the two terms (Additional Parameters) used to extend the basic collinearity model to meet the physical reality and model some systematic errors during the calibration procedure (as stated in Eq. 5.1).

All measurements performed on digital images refer to a pixel coordinate system (u, v) while collinearity equations refer to the metric image coordinate system (x, y). The conversion from pixel to image coordinates is performed with an affine transformation knowing the sensor dimensions and pixel size. For each measured image point (that should be visible in multiple images) a collinearity equation is written. All the equations form a system of equations and the solution of the bundle adjustment is generally obtained with an iterative least squares method (Gauss-Markov model), thus requiring some good initial approximations of the unknown parameters. The bundle adjustment provides a simultaneous determination of all system parameters along with estimates of the precision and reliability of the unknown parameters. Furthermore, correlations between the IO and EO parameters and the object point coordinates, along with their determinability, can be quantified. To enable the correct scaling of the recorded object information, a known distance between two points visible in the images should be used while for absolute positioning (geo-referencing) at least seven external informations (for example three points with known 3D coordinates, generally named *ground control points*) need to be imported in the bundle procedure. Depending on the parameters which are considered either known a priori or treated as unknowns, Equation 5.2 may result in the following cases:

- *self-calibrating bundle*: all parameters on the right-hand side are unknown (IO, EO, object point coordinates, and additional parameters);

- *general bundle method*: the IO is known and fixed, the image coordinates are used to determine the EO and the object point coordinates;

- *spatial resection*: the IO and object point coordinates are known, the EO needs to be determined or the object point coordinates are available and the IO and EO have to be determined;

- *spatial intersection*: knowing the IO and EO, the object point coordinates have to be determined.

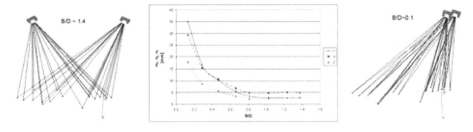

FIGURE 5.6
The behavior of the computed object coordinates accuracy with the variation of the B/D ratio between a pair of images.

Camera calibration and image orientation can be performed fully automatically by means of coded targets which are automatically recognized in the images and used as tie points in the successive bundle adjustment. Otherwise, if no targets are placed in the scene, the tie points are generally measured manually. Indeed automated markerless image orientation is a quite complex task, in particular for convergent terrestrial images. The Automatic Aerial Triangulation (AAT) has reached a significant level of maturity and reliability demonstrated by the numerous commercial softwares available on the market. On the other hand, in terrestrial photogrammetry, commercial solutions for the automated orientation of markerless sets of images are still pending and the topic is still an open research issue [71]. Few commercial packages are available to automatically orient video sequences which are generally of very low geometric resolution and so not really useful for detailed 3D modeling projects.

As previously mentioned, the photogrammetric reconstruction method relies on a minimum of two images of the same object acquired from different viewpoints. Defining B the baseline between the two images and D the average camera-object distance, a reasonable B/D (base-to-depth) ratio between the images should ensure a strong geometric configuration and reconstruction that is less sensitive to noise and measurement errors. A typical value of the B/D ratio in terrestrial photogrammetry should be around then 0.5 - 0.75, even if in practical situations it is often very difficult to fulfill this requirement. Generally, the larger the baseline, the better the accuracy of the computed object coordinates, although large baselines lead to problems in finding automatically the same correspondences in the images, due to strong perspective effects. The influence of increasing the baseline between the images is shown in Figure 5.6, which reports the computed theoretical precision of the object coordinates with the variation of the B/D ratio. It is clearly visible how small baselines have a negative influence on the accuracy of the computed 3D coordinates while larger baselines help to derive more accurate 3D information. In general the accuracy of the computed object coordinates (σ_{XYZ}) depends on the image measurement precision (σ_{xy}), image scale, and geometry (e.g., the scale number S computed and the mean object distance/camera focal length), an empirical factor q, and the number of exposures k [72]:

$$\sigma_{XYZ} = qS\sigma_{xy}/k^{1/2} \qquad (5.3)$$

Summarizing the calibration and orientation phase, as also stated in different studies related in particular to terrestrial applications [66, 72–74], we can affirm that:

- the accuracy of an image network increases with the increase of the base-to-depth (B/D) ratio and using convergent images rather than images with parallel optical axes;

- the accuracy improves significantly with the number of images where a point appears. But measuring the point in more than four images gives less significant improvement;

- the accuracy increases with the number of measured points per image but the increase is not significant if the geometric configuration is strong and the measured points are well defined (like targets) and well distributed in the image;

- the image resolution (number of pixels) influences the accuracy of the computed object coordinates: on natural features, the accuracy improves significantly with the image resolution, while the improvement is less significant on well-defined large resolved targets;

- self-calibration (with or without known control points) is reliable only when the geometric configuration is favorable, mainly highly convergent images of a large number of (3D) targets well-distributed spatially and throughout the image format;

- a flat (2D) testfield could be employed for camera calibration if the images are acquired at many different distances from the testfield, to allow the recovery of the correct focal length;

- at least 2-3 images should be rotated 90 degrees to allow the recovery of the principal point, i.e., to break any projective coupling between the principal point offset and the camera station coordinates, and to provide a different variation of scale within the image;

- a complete camera calibration should be performed, in particular for the lens distortions. In most cases, particularly in modern digital cameras and for unedited images, the camera focal length can be found, albeit with less accuracy, in the header of the digital images. This can be used on uncalibrated cameras if self-calibration is not possible or unreliable.

- if the image network is geometrically weak, correlations may lead to instabilities in the least-squares estimation. The use of inappropriate APs can also weaken the bundle adjustment solution, leading to over-parameterization, in particular in the case of minimally constrained adjustments.

The described mathematical model and collinearity principle are both valid for frame cameras while for linear array sensors a different mathematical model should be employed [75]. Empirical models based on simple affine, projective or DLT transformation were also proposed, finding their main application in the processing of high-resolution satellite imagery [76].

5.3.3   3D Measurements

Once the calibration parameters are known and the images are oriented, the scene measurements can be performed with manual, semi-automated, or automated "matching" procedures. The measured 2D image primitives and correspondences (points or edges) are converted into unique 3D object coordinates (3D point cloud) using the collinearity principle and the known exterior and interior parameters previously recovered.

Manual or semi-automated measurements are performed in monocular or stereoscopic vision when few points are necessary to determine the 3D geometry of an object, e.g., for architectures or for 3D city modeling, where the main corners and edges need to be identified to reconstruct the 3D shape (Figure 5.7(a,b,c)) [77, 78]. A relative accuracy in the range 1:5,000-15,000 is generally expected for such kinds of 3D models.

Automated procedures are instead employed when dense surface measurements and reconstructions are required, for example to derive a digital surface or terrain model (DSM or DTM) or to document detailed and complex objects like reliefs, statues, excavations areas, etc. (Figure 5.7(d,e,f)).

In certain situations, like architectural and city modeling, a mix of automated and semi-automated methods could be necessary in order to derive all the necessary information and geometric details [79, 80].

Automated Measurement Procedures

The latest developments in automated image matching are demonstrating the great potentiality of

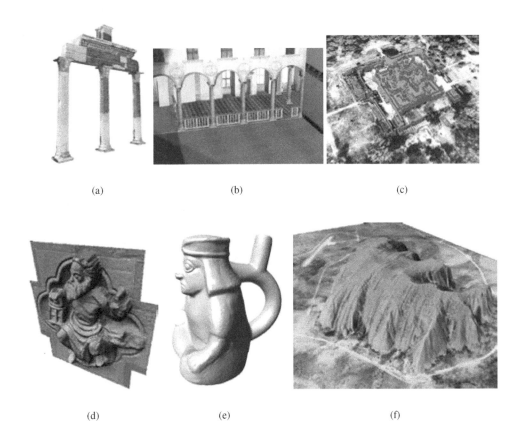

(a) (b) (c)

(d) (e) (f)

FIGURE 5.7 (SEE COLOR INSERT)
3D models derived from aerial or terrestrial images and achieved with interactive measurements (a, b, c) or automated procedures (d, e, f).

the 3D reconstruction method at different scales of work, comparable to point clouds derived using range sensors and with a reasonable level of automation. Overviews on stereo image matching can be found in [31, 81], while [82] compared multi-view matching techniques. In the vision community, the surface measurement is generally performed using stereo-pair depth maps [53, 83] with turntable, visual-hull, controlled environment and uniform backgrounds [84, 85] or multi-view methods aiming at reconstructing a surface which minimize a global photometric discrepancy functional, regularized by explicit smoothness constraints [34, 86, 87]. Experiments are generally performed with indoor low-resolution or Internet images but the results are in any case promising. Some free tools are also available online although reconstruction accuracy in the range 1:50-1:400 images were reported [32, 53, 82]. On the other hand, automated photogrammetric matching algorithms [33, 88, 89] are usually area-based techniques; they rely on cross-correlation or Least Squares Matching algorithms applied on stereo or multiple images providing higher accuracy results in the range of 1:1,000-1:5,000.

Automated procedures involve the automatic establishment of correspondences between primitives extracted from two or more images. The problem of image correlation has been studied for more than 30 years but still many problems exist: complete automation, occlusions, poor or untextured areas, repetitive structures, moving objects (including shadows), radiometric artifacts, capability of retrieving all the details, transparent objects, applicability to diverse camera configuration, etc. In the aerial and satellite photogrammetry the problems are limited and almost solved (a proof is the large amount of commercial software for automated image triangulation and DTM/DSM generation), whereas in terrestrial applications the major problems are related to the three-dimensional characteristics of the surveyed object and the often convergent or wide-baseline images. These configurations are generally hardly accepted by modules developed for DTM/DSM generation from aerial or satellite images.

The state of the art in image matching is the multi-image approach, where multiple images are matched simultaneously and not only pairwise. In [82] the different stereo and multi-view image matching algorithms are classified according to six fundamental properties: the scene representation, photo-consistency measure, visibility model, shape prior, reconstruction algorithm, and initialization requirements. According to [90], multi-view image matching and 3D reconstruction algorithms can be classified in (1) voxel-based approaches which require the knowledge of a bounding box containing the scene and that present an accuracy limited by the resolution of the voxel grid [87, 91, 92]; (2) algorithms based on deformable polygonal meshes which demand a good starting point (like a visual hull model) to initialize the corresponding optimization process, therefore limiting their applicability [84, 85]; (3) multiple depth maps approaches which are more flexible, but require the fusion of individual depth maps into a single 3D model [86, 93], and (4) patch-based methods which represent scene surfaces by collections of small patches (or surfels) [94]. On the other hand, in [33] the image matching algorithms are classified in area-based or feature-based procedures, i.e., using the two main classes of matching primitives: image intensity patterns (windows composed of gray values around a point of interest) and features (edges and regions). This leads respectively to area-based (e.g., cross-correlation or Least Squares Matching (LSM) [95]) and feature-based (e.g., relational, structural, etc.) matching algorithms.

Once the primitives are extracted from the images, they are converted in 3D information using the known camera parameters. Area-Based Matching (ABM), especially the LSM method and its extended concept called Multi-Photo Geometrical Constraints LSM [96, 97], has a very high accuracy potential (up to 1/25 pixel on well defined targets) if well textured image patches are used. ABM generally uses small areas around interest points extracted, e.g., with interest operators [98–100]. Disadvantages of ABM are the need for small searching range for successful matching, the possible smooth results if too large image patches are used and, in case of LSM, the requirement of good initial values for the unknown parameters (although this is not the case for other techniques such as graph-cut). Furthermore, matching problems (i.e., blunders) might occur in areas with occlusions, lack of or repetitive texture or if the surface does not correspond to the assumed model (e.g., planarity of the matched local surface patch). On the other hand, Feature-Based Matching (FBM) involves the extraction of features like points, edges, or regions which are then associated with some attributes (descriptors) useful to characterize and match them [101, 102]. A typical strategy to match characterized features is the computation of similarity measures from the associated attributes. Compared to ABM, FBM techniques are more flexible with respect to surface discontinuities, less sensitive to image noise, and require less approximate values. The accuracy of the feature-based matching is limited by the accuracy of the feature extraction process. Moreover, due to the sparse and irregularly distributed nature of the extracted features, FBM techniques generally deliver point clouds less dense compared to ABM approaches. For all these reasons, the combination of ABM and FBM techniques is the best choice for a powerful surface measurement approach [33, 103].

5.3.4 Structuring and Modeling

Once the scene's main features and details are manually or automatically extracted from the images and converted into 3D information, the produced 3D point cloud should be structured in the most flexible and precise way to accurately represent the 3D measurement results and provide an optimal scene's description and digital representation. Indeed unstructured 3D data are generally not very useful except for quick visualization and most of the applications require structured results, generally in terms of mesh or TIN (Triangular Irregular Network).

In some applications, like building reconstruction, where the object is mainly described with planar or cylindrical patches, a small number of 3D points are sufficient for the 3D modeling, and the structuring and surface generation are achieved with few triangular patches or fitting particular surfaces to the data (Figure 5.8(a, b, c)) [78, 104]. Indeed automated reconstruction methods are still not performing properly with such kinds of objects in all the possible situations and too many assumptions limit the usability of the proposed approaches in the daily 3D documentation [105–107]. In other applications, like terrain modeling, statues, or complex object reconstruction, the surface generation requires very dense point clouds and smart algorithms to produce correct a mesh or TIN and successively photo-realistic results (Figure 5.8(d, e, f)).

The goal of the surface generation can be stated as follows: given a set of sample points P_i assumed to lie on or near an unknown surface S, create a surface model S′ approximating S. A surface reconstruction procedure (also called surface triangulation) cannot exactly guarantee the recovery of S, since we have information about S only through a finite set of sample points P_i. Generally, with the increasing of the sampling density, the output result S′ is more likely to be topologically correct and convergent to the original surface S. A good sample should be dense in detailed and curved areas while sparse in flat parts. Usually if the input data does not satisfy certain properties required by the triangulation algorithm (good point distribution, high density, little noise, etc.), current commercial software tools will produce incorrect results. Sometimes to improve the quality of the reconstruction, additional topological information about the surface (for example breaklines) is added to allow a more correct digital reconstruction of the scene's geometry.

The conversion of the measured 3D point cloud into a consistent polygonal surface is generally based on four steps:

- pre-processing: erroneous points are eliminated, noise is smoothed out, and points are added to fill gaps. The data may also be resampled to generate a model of efficient size;

- determination of the global topology of the object's surface: the neighborhood relations between adjacent parts of the data are determined, using some global sorting step and possible constraints (such as breaklines), mainly to preserve special features such as edges;

- generation of the polygonal surface: triangular (or tetrahedral) networks are created, satisfying certain quality requirements, e.g., a limit on the network element size or no intersection of breaklines;

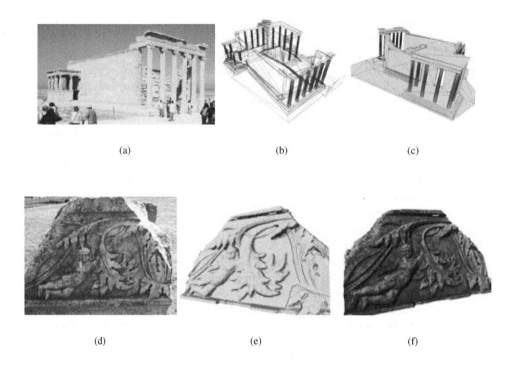

(a)   (b)   (c)

(d)   (e)   (f)

FIGURE 5.8
An architectural object (a) which can be digitally reconstructed with sparse point cloud and few triangular patches (b,c) [8]. A detailed bass-relief (d) which needs an automated surface measurement procedure to derive a dense point cloud and a textured 3D model (e, f).

- post-processing: editing operations (edge corrections, triangle insertion, polygon editing, holes filling, non-manifold parts fixing, etc.) are commonly applied to refine and correct the generated polygonal surface [108, 109]. Those errors are visually unpleasant, might cause lighting blemishes due to the incorrect normals, and the computer model will also be unsuitable for reverse engineering or physical replicas. Moreover, over-sampled areas should be simplified while under-sampled regions should be subdivided [110].

In commercial packages, the global topology of the surface and the mesh generation are generally performed automatically while pre- and post-processing operations still need manual intervention.

5.3.5 Texturing and Visualization

A 3D model can be visualized in wireframe, shaded, or textured mode (Figure 5.9). A textured 3D geometric model is probably the most desirable 3D object documentation since it gives, at the same time, a full geometric and appearance representation and allows unrestricted interactive visualization and manipulation at a variety of lighting conditions. The photo-realistic representation is achieved mapping a color image (generally an orthophoto in the case of aerial or satellite applications) onto the 3D geometric data. The 3D data can be in the form of points or triangles (mesh), according to the applications and requirements. The texturing of 3D point clouds (point-based rendering techniques [111]) allows a faster visualization but for detailed and complex 3D models it is not an appropriate method. On the other hand, in case of meshed or TIN data, the texture is automatically mapped if the camera parameters are known (e.g., if it is a photogrammetric model), otherwise homologue points between the 3D mesh and the 2D image to-be-mapped should be identified (e.g., if the model has been generated using range sensors). This is the bottleneck of the texturing phase in terrestrial applications as it is still an interactive procedure and no automated commercial solutions are available, although some automated approaches were proposed [46, 112, 113]. Indeed the identification of homologue points between 2D and 3D data is a hard task, much more complex than image to image or geometry to geometry registration. Furthermore, in applications involving infrared or multi-spectral images, it is generally quite challenging to identify common features between 2D and 3D data. In practical cases, the 2D-3D registration is done with the well known DLT approach [114] (often referred to as Tsai method [115]), where homologue points between the 3D geometry and a 2D image to-be-mapped are used to retrieve the interior and exterior unknown camera parameters. The color information is then projected (or assigned) to the surface polygons using a color-vertex encoding or a mesh parameterization.

In computer graphics applications, the texturing can also be performed with techniques able to graphically modify the derived 3D geometry (displacement mapping) or simulating the surface irregularities without touching the geometry (bump mapping, normal mapping, parallax mapping).

The texture mapping phase goes much further than simply projecting one or more static images over the 3D geometry. Problems can rise firstly from the time-consuming image-to-geometry registration and then due to variations in lighting, surface specularity, and camera settings. Often the images are exposed with the illumination at imaging time but it may need to be replaced by illumination consistent with the rendering point of view and the reflectance properties (BRDF) of the object [116]. High dynamic range (HDR) images might also be acquired to recover all scene details [117] while color discontinuities and aliasing effects must be removed [118–120].

The photo-realistic product needs then to be visualized, e.g., for communication and presentation purposes. In case of large and complex models the point-based rendering technique does not give satisfactory results and does not provide realistic visualization. Moreover the visualization of a 3D model is often the only product of interest for the external world, remaining the only possible contact with the 3D data. Therefore a realistic and accurate visualization is often required. Furthermore the ability to easily interact with a huge 3D model is a continuing and increasing problem. Indeed model sizes (both in geometry and texture) are increasing at a faster rate than computer hardware advances

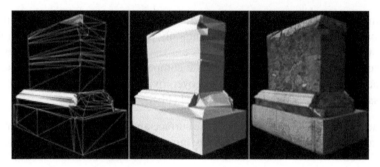

FIGURE 5.9
Image-based 3D model of an archaeological find obtained with interactive procedures and displayed in wireframe, shaded, and textured mode.

and this limits the possibilities of interactive and real-time visualization of the 3D results. Due to the generally large amount of data and its complexity, the rendering of large 3D models is done with multi-resolution approaches displaying the large meshes with different levels of detail (LOD), simplification, and optimization approaches [121–123].

5.3.6 Main Applications and Actual Problems

Photogrammetry can currently deliver mainly three types of 3D models which can be used in geo-related applications:

- digital terrain and surface models (DTM, DSM);

- 3D city models;

- cultural and natural heritage models.

Digital terrain and city models are the basis for any GIS or geo-related study and application. Terrain models are produced, mainly automatically, using aerial or satellite stereo images for the generation of topographic maps, analysis of changes, environmental models, etc. Digital surface models are realized for special applications like environmental studies, vegetation analysis, water management, deformation analysis, etc.

3D city models are generated with semi-automated procedures as more reliable and productive than fully automated methods which require many post-editing corrections. 3D city models are generally required for city planning (Figure 5.10(a)), telecommunication, disaster management, location-based services (LBS), real estate, media and advertising, games and simulation, transportation analysis, navigation, energy providers and heating dispersion studies, and archaeological documentation (Figure 5.10(b)).

Digital heritages are required at different scales and level of detail, from large site [5, 44, 51, 124] to single structure and buildings [125], statue (Figure 5.10(c)), artifacts, and small findings. Beside management, conservation, restoration, and visualization issues, digital models of heritages are very useful for advanced databases where archaeological, geographical, architectural, and semantical information can be linked to the recovered 3D geometry and consulted, e.g., online [126].

The actual problems and main challenges in the 3D metric surveying of large and complex sites or objects arise in every phase of the photogrammetric pipeline, from the data acquisition to the visualization of the achieved 3D results. Selecting the correct platform and sensor, the appropriate measurement and modeling procedures, designing the production workflow, assuring that the final result is in accordance with all the given technical specifications and being able to fluently display

(a)   (b)

(c)   (d)

FIGURE 5.10
3D city model generated for urban planning (a) and archaeological 3D documentation (b) [79]. 3D model of the Great Buddha of Bamiyan (Afghanistan) (c) and its current empty niche after the destruction in 2001 [50] (d).

and interact with the achieved 3D model are the greatest challenges. In particular, key issues and challenges arise from:

- **New sensors and platforms**: new digital sensors and technologies are frequently coming on the market but the software to process the acquired data are generally coming much later. The development and use of new sensors requires the study and test of innovative sensor models, and the investigation of the related network structures and accuracy performance. Of particular interest here are high-resolution satellite and aerial linear array cameras, terrestrial panoramic cameras, and laser scanners.

- **Automation**: the automation of photogrammetric processing is one of the most important issues when it comes to efficiency and costs of data processing. Many researchers and commercial solutions have turned towards semi-automated approaches, where the human capacity in image content interpretation is paired with the speed and precision of computer-driven measurement algorithms. But the success of automation in image analysis, interpretation, and understanding depends on many factors and is still a hot topic of research. The progress is slow and the acceptance of results depends on the quality specifications of the user and final goal of the 3D model. An automated procedure should be judged in terms of the datasets that it can handle but nowadays we can observe that (1) sensor calibration and image orientation can be done automatically [71, 127], (2) scaling and geo-referencing still needs interaction while seeking the control points in the images, (3) DSM generation can be done automatically [32, 33, 35, 91] but may need substantial post-editing if, in aerial- and satellite-based applications, a DTM is required, (4) orthoimage generation is a fully automatic process, (5) object extraction and modeling is mainly done in a semi-automated mode to achieve reliable and precise 3D results.

- **Integration of sensors and data**: the combination of different data sources allows one to derive different geometric levels of detail and exploit the intrinsic advantages of each sensor [8, 44]. The integration so far is mainly done at model-level (for example at the end of the modeling pipeline) while it should also be possible at data-level to overcome the weakness of each data source. In a more general 3D modeling view, an important issue is the increasing use of hybrid sensors, in order to collect as many different cues as possible.

- **On-line and real-time processing**: in some applications there is a need for very fast processing thus requiring new algorithmic implementation, sequential estimation, and multi-core processing. The internet is also a great help in this sector and web-based processing tools [54] for image analysis and 3D model generation [36] are available although limited to specific tasks and not ideal to collect CAD information and accurate 3D models.

5.4   Conclusions

The chapter has presented on overview of the existing techniques and methodologies to derive reality-based 3D models, with a deeper description of the photogrammetric method. Photogrammetry is the art of turning 2D images into accurate 3D models. Images are acquired from terrestrial, aerial, or satellite sensors or can be found in archives. An emerging platform of particular interest in the cultural heritage field is Unmanned Aerial Vehicles (UAVs) like model helicopters which deliver high-resolution vertical or oblique views that can be extremely useful for site studies, inspection, and documentations. Photogrammetry relies on the use of at least two images combined with a stable and consolidated mathematical formulation to derive 3D information with estimates of precision and reliability of the unknown parameters from measured correspondences (tie points) in the images. The correspondences can be extracted automatically or semi-automatically according to the object and project requirements. Photogrammetry is employed in different applications like 3D documentation, conservation, restoration, reverse engineering, mapping, monitoring, visualization, animation, urban planning, deformation analysis, etc. In the case of archaeological and cultural heritage sites or objects, photogrammetry provided for accurate 3D reconstructions at different scales and for hybrid 3D models (e.g., terrain model + archaeological structures as shown in Figure 5.7(c) and Figure 5.10(b)). Nowadays 3D scanners are also becoming a standard source for input data in many application areas, but image-based modeling still remains the most complete, cheap, portable, flexible, and widely used approach, although large experience in data acquisition and processing is highly required. Furthermore, for large sites' 3D documentation, the integration with range sensors is generally the best solution.

Despite the fact that 3D documentation is not yet the state of the art in the heritage field, the reported examples show the potentialities of the photogrammetric method to digitally document and preserve our heritages as well as share and manage the collected digital information. The image-based approach, together with active sensors, Spatial Information Systems, 3D modeling, and visualization and animation software are still in a dynamic state of development, with even better application prospects for the near future.

Acknowledgments

The author would like to thank Prof. A. Gruen (ETH Zurich, Switzerland), Prof. G. Guidi (Politecnico of Milan, Italy), Dr. S. El-Hakim and A.J. Beraldin (NRC Canada), S. Girardi and A. Rizzi (FBK Trento, Italy), S. Benedetti (Benedetti Graphics), Dr. A. M. Manferdini and Prof. M. Gaiani (Univ. of Bologna, Italy), co-authors in some publications, research, and field works for the reported projects and examples.

Bibliography

[1] UNESCO. http://www.unesco.org.

[2] M. Levoy, K. Pulli, B. Curless, S. Rusinkiewicz, D. Koller, L. Pereira, M. Ginzton, S. Anderson, J. Davis, J. Ginsberg, J. Shade, and D. Fulk. The digital Michelangelo project: 3D scanning of large statues. *SIGGRAPH Computer Graphics Proceedings*, pages 131–144, 2000.

[3] J.-A. Beraldin, M. Picard, S. F. El-Hakim, G. Godin, V. Valzano, A. Bandiera, and D. Latouche. Virtualizing a byzantine crypt by combining high resolution textures with laser scanner 3D data. *Proceedings of VSMM 2002*, pages 3–14, 2002.

[4] J. Stumpfel, C. Tchou, N. Yun, P. Martinez, T. Hawkins, A. Jones, Emerson B, and P. Debevec. Digital reunification of the Parthenon and its sculptures. In *Proceedings of Virtual Reality, Archaeology and Cultural Heritage (VAST 2003)*, pages 41–50, 2003.

[5] A. Gruen, F. Remondino, and L. Zhang. The Bamiyan project: multi-resolution image-based modeling. In *Recording, Modeling and Visualization of Cultural Heritage*. E. Baltsavias, A. Gruen, L. Van Gool, M. Pateraki (Eds), Taylor & Francis/Balkema, 2005. ISBN 0-415-39208-X.

[6] G. Guidi, B. Frischer, M. Russo, A. Spinetti, L. Carosso, and L. L. Micoli. Three-dimensional acquisition of large and detailed cultural heritage objects. *Machine Vision Applications*, 17:349–360, 2006.

[7] K. Ikeuchi, T. Oishi, and J. Takamatsu. Digital bayon temple - e-monumentalization of large-scale cultural-heritage objects. In *Proceedings ASIAGRAPH*, 1(2):99–106, 2007.

[8] S. El-Hakim, J.A. Beraldin, F. Remondino, M. Picard, L. Cournoyer, and E. Baltsavias. Using terrestrial laser scanning and digital images for the 3D modelling of the Erechteion, Acropolis of Athens. *Proceedings of DMACH Conference on Digital Media and its Applications in Cultural Heritage*, pages 3–16, 2008. Amman, Jordan.

[9] F. Remondino, S. El-Hakim, S. Girardi, A. Rizzi, S. Benedetti, and L. Gonzo. 3D virtual reconstruction and visualization of complex architectures - the 3D-arch project. *International Archives of the Photogrammetry, Remote Sensing and Spatial Information Sciences*, 38(5/W10), 2009. on CD-ROM.

[10] K. Ikeuchi and D. Miyazaki. *Digitally Archiving Cultural Objects*. Springer, 2008. p. 503, ISBN: 978-0-387-75806-0.

[11] E. M. Mikhail, J. S. Bethel, and J. C. McGlone. *Introduction to Modern Photogrammetry*. Wiles and Sons, 2002. 496 p., ISBN: 0-471-30924-9.

[12] K. Kraus. *Photogrammetry: Geometry from Images and Laser Scans*. Walter de Gruyter, 2007. 459 p., ISBN 978-3110190076.

[13] English Heritage. *Metric Survey Specifications for English Heritage*. English Heritage, 2005. Available at www.english-heritage.org.uk.

[14] G. Vosselman. *Laser Scanning*, volume Encyclopedia of Geographic Information Systems, pages 298–372. Springer-Verlag, 2008. Vol. 1883, ISBN: 978-0-387-30858-6.

[15] A. Gruen. Reality-based generation of virtual environments for digital earth. *International Journal of Digital Earth*, 1(1), 2008.

[16] P. Mueller, P. Wonka, S. Haegler, A. Ulmer, and L. Van Gool. Procedural modeling of buildings. In *Proceedings of ACM SIGGRAPH 2006 / ACM Transactions on Graphics*, 25(3):614–623, 2006.

[17] F. Remondino and S. El-Hakim. Image-based 3D modelling: a review. *The Photogrammetric Record*, 21(115):269–291, 2006.

[18] F. Blais. A review of 20 years of range sensors development. *Journal of Electronic Imaging*, 13(1):231–240, 2004.

[19] P. Cignoni and R. Scopigno. Sampled 3D models for CH applications: a viable and enabling new medium or just a technological exercise. *ACM Journal on Computing and Cultural Heritage*, 1(1), 2008.

[20] S. El-Hakim, J. A. Beraldin, M. Picard, and G. Godin. Detailed 3D reconstruction of large-scale heritage sites with integrated techniques. *IEEE Computer Graphics and Application*, 24(3):21–29, 2004.

[21] G. Guidi, J-A. Beraldin, and C. Atzeni. High accuracy 3D modeling of cultural heritage: the digitizing of Donatello. *IEEE Transactions on Image Processing*, 13(3):370–380, 2004.

[22] L. De Luca, P. Veron, and M. Florenzano. Reverse engineering of architectural buildings based on a hybrid modeling approach. *Computer and Graphics*, 30:160–176, 2006.

[23] P. Rônnholm, E. Honkavaara, P. Litkey, H. Hyypp, and J. Hyypp. Integration of laser scanning and photogrammetry. *International Archives of Photogrammetry, Remote Sensing and Spatial Information Sciences*, 36(2/W52):355–362, 2007.

[24] I. Stamos, L. Liu, C. Chen, G. Woldberg, G. Yu, and S. Zokai. Integrating automated range registration with multiview geometry for photorealistic modelling of large-scale scenes. *International Journal of Computer Vision*, 78(2-3):237–260, 2008.

[25] X. Yin, P. Wonka, and A. Razdan. Generating 3D building models from architectural drawings: A survey. *IEEE Computer Graphics and Applications*, 29(1):20–30, 2009.

[26] J. Salvi, C. Matabosch, D. Fofi, and J. Forest. A review of recent range image registration methods with accuracy evaluation. *Image and Vision Computing*, 25(5):578–596, 2007.

[27] A. Gruen and D. Akca. Least squares 3D surface and curve matching. *ISPRS Journal of Photogrammetry and Remote Sensing*, 59(3):151–174, 2005.

[28] A.J. Beraldin, M. Rioux, L. Cournoyer, F. Blais, M. Picard, and J. Pekelsky. Traceable 3D imaging metrology. In *Proceedings Videometrics IX - SPIE Electronic Imaging Proceedings*, Vol. 6491, pages B.1–B–11, 2007.

[29] J. Shan and C. Toth. *Topographic Laser Ranging and Scanning: Principles and Processing*. CRC Press, 2008. p. 590, ISBN: 978-1420051421.

[30] G. Vosselman and H. G. Maas. *Airborne and Terrestrial Laser Scanning*, page 318. CRC Press, Whittles Publishing, 2010. ISBN: 978-1904445-87-6.

[31] M. Z. Brown, D. Burschka, and G. D. Hager. Advance in computational stereo. *IEEE Transactions on Pattern Analysis and Machine Intelligence*, 25(8):993–1008, 2003.

[32] M. Goesele, N. Snavely, B. Curless, H. Hoppe, and S. M. Seitz. Multi-view stereo for community photo collections. In *Proceedings ICCV 2007*, Rio de Janeiro, Brazil, 2007.

[33] F. Remondino, S. El-Hakim, A. Gruen, and L. Zhang. Development and performance analysis of image matching for detailed surface reconstruction of heritage objects. *IEEE Signal Processing Magazine*, 25(4):55–65, 2008.

[34] V. H. Hiep, R. Keriven, P. Labatut, and J. P. Pons. Towards high-resolution large-scale multi-view stereo. In *Proceddings CVPR2009*, 2009. Kyoto, Japan.

[35] H. Hirschmueller. Stereo processing by semi-global matching and mutual information. *IEEE Transactions on Pattern Analysis and Machine Intelligence*, 30(2):328–341, 2008.

[36] M. Vergauwen and L. Van Gool. Web-based 3D reconstruction service. *Machine Vision and Applications*, 17(6):411–426, 2006.

[37] B. K. P. Horn and M. J. Brooks. *Shape from Shading*. MIT Press, Cambridge, 1989. p. 586, ISBN: 978-0-262-08183-2.

[38] F.A. Van den Heuvel. 3D reconstruction from a single image using geometric constraints. *ISPRS Journal for Photogrammetry and Remote Sensing*, 53(6):354–368, 1998.

[39] G. Pomaska. Image acquisition for digital photogrammetry using off the shelf and metric cameras. *CIPA International Symposium Potsdam*, 2001.

[40] D. DÁyala and P. Smars. Minimum requirement for metric use of non-metric photographic documentation. *University of Bath Report*, 2003.

[41] Bôhler. Comparison of 3D scanning and other 3D measurement techniques. *In Recording, Modeling and Visualization of Cultural Heritage*. E. Baltsavias and A. Gruen and L. Van Gool and M. Pateraki (Eds), Taylor and Francis / Balkema, 2005. ISBN 0-415-39208-X.

[42] F. Remondino, A. Guarnieri, and A. Vettore. 3D modeling of close-range objects: photogrammetry or laser scanning? In *Proceedings of Videometrics VIII, SPIE-IS&T Electronic Imaging*, Vol. 5665, pages 216–225, 2005.

[43] P. Grussenmeyer, T. Landes, T. Voegtle, and K. Ringle. Comparison methods of terrestrial laser scanning, photogrammetry and tacheometry data for recording of cultural heritage buildings. *International Archives of the Photogrammetry, Remote Sensing and Spatial Information Sciences*, 37(5):213–218, 2008.

[44] G. Guidi, F. Remondino, M. Russo, F. Menna, A. Rizzi, and S. Ercoli. A multi-resolution methodology for the 3D modeling of large and complex archaeological areas. *International Journal of Architectural Computing*, 7(1):40–55, 2009.

[45] A. Pelagotti, L. Pezzati, A. Piva, and A. Del Mastio. Multispectral UV fluorescence analysis of painted surfaces. In *Proceedings of 14th European Signal Processing Conference (EUSIPCO)*, Firenze, Italy, 2006.

[46] F. Remondino, A. Pelagotti, A. Del Mastio, and F. Uccheddu. Novel data registration techniques for art diagnostics and 3D heritage visualization. *IX Conference on Optical 3D Measurement Techniques*, 1:1–10, 2009.

[47] T. Luhmann, S. Robson, S. Kyle, and I. Harley. *Close range photogrammetry: Principles, methods and applications*. Whittles, 2002. p. 528, ISBN: 978-0-470-10633-4.

[48] W. Linder. *Digital Photogrammetry: A Practical Course*. Springer, 2009. p. 220, ISBN 978-3540927242.

[49] T. Lillesand, R. W. Kiefer, and J. W. Chipman. *Remote Sensing and Image Interpretation*. Wiley and Sons, 2007. p. 1128, ISBN 978-0470052457.

[50] A. Gruen, F. Remondino, and L. Zhang. Photogrammetric reconstruction of the Great Buddha of Bamiyan. *The Photogrammetric Record*, 19(107):177–199, 2004.

[51] M. Sauerbier, M. Kunz, M. Fluehler, and F. Remondino. Photogrammetric reconstruction of adobe architecture at Tucume, Peru. *International Archives of Photogrammetry, Remote Sensing and Spatial Information Sciences*, 36(5/W1), 2004. International Workshop on Processing and Visualization using High Resolution Imagery, Pitsanulok, Thailand (on CD-ROM).

[52] D. Nister. Automatic passive recovery of 3D from images and video. In *IEEE Proceedings of the 2nd International Symp. on 3D Data Processing, Visualization, and Transmission (3DPVT 2004)*, pages 438–445, 2004.

[53] M. Pollefeys, L. Van Gool, M. Vergauwen, F. Vergbiest, K. Cornelis, J. Tops, and R. Kock. Visual modelling with a hand-held camera. *International Journal of Computer Vision*, 59(3):207–232, 2004.

[54] N. Snavely, S.M. Seitz, and R. Szeliski. Modelling the world from Internet photo collections. *International Journal of Computer Vision*, 80(2):189–210, 2008.

[55] A. Criminisi, I. Reid, and A. Zisserman. Single view metrology. *Proceedings International Conference on Computer Vision*, pages 434–442, 1999.

[56] S. El-Hakim. A practical approach to creating precise and detailed 3D models from single and multiple views. *International Archives of Photogrammetry and Remote Sensing*, 33(B5):122–129, 2000. Amsterdam, The Netherlands.

[57] J. R. Kender. Shape from texture. In *Proceedings DARPA IU Workshop*, 1978.

[58] G. Healey and T. O. Binford. Local shape from specularity. *Proceedings ICCV 1987*, 1987. London, UK.

[59] D. Meyers, S. Skinner, and K. Sloan. Surfaces from contours. *ACM Transactions on Graphics*, 11(3):228–258, 1992.

[60] S. Winkelbach and F. M. Wahl. *Shape from 2D Edge Gradient*, volume Pattern Recognition. Springer-Verlag, 2001. Vol. 2191.

[61] B. Wrobel. Minimum solutions for orientation, *Calibration and Orientation of Cameras in Computer Vision*, pages 7–56. Springer, 2001. Vol. 34, ISBN: 978-3-540-65283-0.

[62] D. Brown. The bundle-adjustment - progress and prospects. In *International Archives of Photogrammetry*, Vol.21(3), 1976.

[63] S. I. Granshaw. Bundle adjustment methods in engineering photogrammetry. *Photogrammetric Record*, 10(56):181–207, 1980.

[64] B. Triggs, P. F. McLauchlan, R. Hartley, and A. Fitzgibbon. Bundle adjustment - A modern synthesis, *Proceedings of the International Workshop on Vision Algorithms: Theory and Practice*, pages 298–372. Springer-Verlag, 1999. Vol. 1883, ISBN: 3-540-67973-1.

[65] C. S. Fraser. Digital camera self-calibration. *ISPRS Journal of Photogrammetry and Remote Sensing*, 52(4):149–159, 1998.

[66] A. Gruen and H. Beyer. System calibration through self-calibration, *Calibration and Orientation of Cameras in Computer Vision*, pages 163–193. Springer, 2001. Vol. 34, ISBN: 978-3-540-65283-0.

[67] T. Rodriguez, P. Sturm, P. Gargallo, N. Guilbert, A. Heyden, J. M. Menendez, and J. I. Ronda. Photorealistic 3D reconstruction from handheld cameras. *Machine Vision and Applications*, 16(4):246–257, 2005.

[68] S. Agarwal, N. Snavely, I. Simon, S. M. Seitz, and R. Szelinski. Building Rome in a day. In *Proceedings ICCV2009*, 2009. Kyoto, Japan.

[69] D. Brown. Close-range camera calibration. *Photogrammetric Engineering*, 37(8):855–866, 1971.

[70] C.S. Fraser and S. Al Ajlouni. Zoom-dependent camera calibration in digital close-range photogrammetry. *Photogrammetric Engineering and Remote Sensing*, 72(9):1017–1026, 2006.

[71] L. Barazzetti, F. Remondino, and M. Scaioni. Combined use of photogrammetric and computer vision techniques for fully automated and accurate 3D modeling of terrestrial objects. *Videometrics, Range Imaging and Applications X, Proceedings of SPIE Optics+Photonics*, 7447, 2009. San Diego, CA, USA.

[72] C. S. Fraser. Network design, *Close-range Photogrammetry and Machine Vision*, pages 256–282. Whittles Publishing, 2001.

[73] T. A. Clarke, J. G. Fryer, and X. Wang. The principal point and CCD cameras. *Photogrammetric Record*, 16(92):293–312, 1998.

[74] S. El-Hakim, J. A. Beraldin, and F. Blais. Critical factors and configurations for practical 3D image-based modeling. *6th Conference on 3D Measurement Techniques*, 2:159–167, 2003. Zurich, Switzerland.

[75] D. Poli. *Modelling of Spaceborne Linear Array Sensors*. PhD thesis, IGP, ETH Zurich, 2005. Mitteilungen Nr.85, ISSN 0252-9335 ISBN 3-906467-50-3.

[76] T. Toutin. Geometric processing of remote sensing images: models, algorithms and methods. *International Journal of Remote Sensing*, 25(10):1893–1924, 2004.

[77] A. Gruen and X. Wang. CC-modeler: a topology generator for 3D city models. *ISPRS Journal of Photogrammetry and Remote Sensing*, 53:286–295, 1998.

[78] S. El-Hakim. Semi-automatic 3D reconstruction of occluded and unmarked surfaces from widely separated views. *International Archives of Photogrammetry, Remote Sensing and Spatial Information Sciences*, 34(B5):143–148, 2002. Corfu, Greece.

[79] A. Gruen and X. Wang. Integration of landscape and city modeling: The pre-hispanic site Xochicalco. *International Archives of the Photogrammetry, Remote Sensing*, 34(5/W3), 2008, on CD-ROM.

[80] S. El-Hakim, F. Remondino, F. Voltolini, and L. Gonzo. Effective high resolution 3D geometric reconstruction of heritage and archaeological sites from images. *Proceedings of 35th CAA Conference*, pages 43–50, 2007. Berlin, Germany.

[81] D. Scharstein and Szeliski R. A taxonomy and evaluation of dense two-frame stereo correspondence algorithms. *International Journal of Computer Vision*, 47(1-3):7–42, 2002.

[82] S.M. Seitz, B. Curless, J. Diebel, D. Scharstein, and R. Szeliski. A comparison and evaluation of multi-view stereo reconstruction algorithm. In *Proceedings CVPR*, pages 519–528, 2006.

[83] C. Strecha, T. Tuytelaars, and L. Van Gool. Dense matching of multiple wide-baseline views. In *Proceedings ICCV 2003*, pages 1194–1201, 2003.

[84] C. Hernandez and F. Schmitt. Silhouette and stereo fusion for 3D object modeling. *Computer Vision and Image Understanding*, 96(3):367–392, 2004.

[85] Y. Furukawa and J. Ponce. Carved visual hulls for image-based modeling. *International Journal of Computer Vision*, 81(1):53–67, 2009.

[86] M. Goesele, B. Curless, and S. M. Seitz. Multi-view stereo revisited. In *Proceedings CVPR 2006*, pages 2402–2409, 2006.

[87] G. Vogiatzis, Hernandez E.C., P.H.S. Torr, and R. Cipolla. Multiview stereo via volumetric graph-cuts and occlusion robust photo-consistency. *IEEE Transactions on Pattern Analysis and Machine Intelligence*, 29(12):2241–2246, 2007.

[88] N. D'Apuzzo. *Surface measurement and tracking of human body parts from multi station video sequences*. PhD thesis, ETH Zurich, Switzerland, 2003. Mitteilungen Nr 81, Institute of Geodesy and Photogrammetry, p. 147.

[89] T. Ohdake and H. Chijatsu. 3D modelling of high relief sculpture using image-based integrated measurement system. *International Archives of the Photogrammetry, Remote Sensing and Spatial Information Sciences*, Vol.36(5/W17), 2005, on CD-ROM.

[90] Y. Furukawa and J. Ponce. Accurate, dense, and robust multi-view stereopsis. *IEEE Transactions on Pattern Analysis and Machine Intelligence*, 32(8):1362–1376, 2009.

[91] J.-P. Pons, R. Keriven, and O.D. Faugeras. Multi-view stereo reconstruction and scene flow estimation with a global image-based matching score. *International Journal of Computer Vision*, 72(2):179–193, 2007.

[92] S. Sinha, P. Mordohai, and M. Pollefeys. Multi-view stereo via graph cuts on the dual of an adaptive tetrahedral mesh. *Proceedings ICCV 2007*, 2007.

[93] C. Strecha, R. Fransens, and L. Van Gool. Combined depth and outlier estimation in multi-view stereo. In *Proceedings CVPR 2006*, page 2394–2401, 2006.

[94] M. Lhuillier and L. Quan. A quasi-dense approach to surface reconstruction from uncalibrated images. *IEEE Transactions on Pattern Analysis and Machine Intelligence*, 27(3):418–433, 2005.

[95] A. Gruen. Adaptive least squares correlations: a powerful matching technique. *South African Journal of Photogrammetry, Remote Sensing and Cartography*, 14(3):175–187, 1985.

[96] A. Gruen and E. Baltsavias. Geometrically constrained multiphoto matching. *Photogrammetric Engineering and Remote Sensing*, 54(5):633–641, 1988.

[97] E. Baltsavias. *Multiphoto geometrically constrained matching*. PhD thesis, ETH Zurich, Switzerland, 1991. Mitteilungen Nr 49, Institute of Geodesy and Photogrammetry, pp. 221.

[98] W. Foerstner and E. Guelch. A fast operator for detection and precise location of distinct points, corners and center of circular features. *ISPRS Conference on Fast Processing of Photogrammetric Data*, pages 281–305, 1987, Interlaken, Switzerland.

[99] C. Harris and M. Stephens. A combined edge and corner detector. In *Proceedings of Alvey Vision Conference*, pages 147–151, 1988.

[100] E. Rosten and T. Drummond. Machine learning for high-speed corner detection. *Proceedings ECCV2006*, pages 430–443, 2006, Graz, Austria.

[101] D. Lowe. Distinctive image features from scale-invariant keypoints. *International Journal of Computer Vision*, 60(2):91–110, 2004.

[102] H. Bay, A. Ess, T. Tuytelaars, and L. Van Gool. Surf: Speeded up robust features. *Computer Vision and Image Understanding*, 110(3):346–359, 2004.

[103] L. Zhang. *Automatic Digital Surface Model (DSM) generation from linear array images*. PhD thesis, IGP, ETH Zurich, 2005. Mitteilungen Nr. 90.

[104] P.E. Debevec, C.J. Taylor, and J. Malik. Modelling and rendering architecture from photographs: A hybrid geometry and image-based approach. *ACM SIGGRAPH'96*, 1996.

[105] T. Werner and A. Zisserman. New technique for automated architectural reconstruction from photographs. *Proceedings ECCV2002*, 2:514–555, 2002.

[106] S. Cornou, M. Dhome, and P. Sayd. Architectural reconstruction with multiple views and geometric constraints. *Proceedings BMV Conference*, 2003.

[107] M. Farenzena and A. Fusiello. 3D surface models by geometric constraints propagation. *Proceedings CVPR 2008*, pages 1–8, 2008. Anchorage (AK), USA.

[108] P. Liepa. Filling holes in meshes. *EG/ACM Proceedings of Symposium on Geometry Processing*, pages 200–206, 2003.

[109] T. Weyrich, M. Pauly, R. Keiser, S. Heinzle, S. Scandella, , and M. Gross. Post-processing of scanned 3D surface data. *Eurographics Symposium on Point-Based Graphics*, pages 85–94, 2004.

[110] T. K. Dey, J. Giesen, S. Goswami, J. Hudson, R. Wenger, and W. Zhao. Undersampling and oversampling in sample based shape modelling. *Proceedings of IEEE Visualization*, pages 83–90, 2001.

[111] L. Kobbelt and M. Botsch. A survey of point-based techniques in computer graphics. *Computers and Graphics*, 28(6):801–814, 2004.

[112] H. Lensch, W. Heidrich, and H. Seidel. Automated texture registration and stitching for real world models. *Proceedings 8th Pacific Graphics Conference on Computer Graphics and Application*, pages 317–327, 2000.

[113] M. Corsini, M. Dellepiane, F. Ponchio, and R. Scopigno. Image-to-geometry registration: a mutual information method exploiting illumination-related geometric properties. *Computer Graphics Forum*, 28(7):1755–1764, 2009.

[114] Y. I. Abdel-Aziz and H. M. Karara. Direct linear trans-formation from comparator coordinates into object space coordinates in close-range photogrammetry. In *Proceedings of the Symposium on Close-Range Photogrammetry*, pages 1–18, 1971. Falls Church (VA) USA.

[115] R. Y. Tsai. An efficient and accurate camera calibration technique for 3D machine vision. In *Proceedings CVPR 1986*, pages 364–374, 1986.

[116] H. P. A. Lensch, J. Kautz, M. Goesele, W. Heidrich, and H.-P. Seidel. Image-based reconstruction of spatial appearance and geometric detail. *ACM Transaction on Graphics*, 22(2):234–257, 2003.

[117] E. Reinhard, G. Ward, S. Pattanaik, and P. Debevec. *High dynamic range imaging: acquisition, display and image-based lighting.* Morgan Kaufmann Publishers, 2005.

[118] P. Debevec, C. Tchou, A. Gardner, T. Hawkins, C. Poullis, J. Stumpfel, A. Jones, N. Yun, P. Einarsson, T. Lundgren, M. Fajardo, and P. Martinez. Estimating surface reflectance properties of a complex scene under captured natural illumination. *USC ICT Technical Report ICT-TR-06*, 2004.

[119] K. Umeda, M. Shinozaki, G. Godin, and M. Rioux. Correction of color information of a 3D model using a range intensity image. In *Proceedings of 5th International Conference on 3-D Digital Imaging and Modeling (3DIM 2005)*, page 229–236, 2005.

[120] M. Callieri, P. Cignoni, M. Corsini, and R. Scopigno. Masked photo blending: Mapping dense photographic dataset on high-resolution sampled 3D models. *Computer and Graphics*, 32(4):464–473, 2008.

[121] D. Luebke, M. Reddy, J. Cohne, A. Varshney, B. Watson, and R. Huebner. *Level of detail for 3D graphics.* Morgan Kaufmann Publishers, 2002. p. 432, ISBN: 1-55860-838-9.

[122] P. Cignoni, F. Ganovelli, E. Gobbetti, F. Marton, F. Ponchio, and R Scopigno. Batched multi triangulation. *Proceedings IEEE Visualization*, pages 207–214, 2005.

[123] A. Dietrich, E. Gobbetti, and S.-E. Yoon. Massive-model rendering techniques: a tutorial. *Computer Graphics and Applications*, 27(6):20–34, 2007.

[124] A. Gruen and S. Murai. High-resolution 3D modeling and visualization of Mount Everest. *ISPRS Journal of Photogrammetry and Remote Sensing*, 57:102–113, 2002.

[125] F. Remondino, S. El-Hakim, E. Baltsavias, M. Picard, and L. Grammatikopoulos. Image-based 3D modeling of the Erechteion, acropolis of Athens. *International Archives of Photogrammetry, Remote Sensing and Spatial Information Sciences*, 37(B5-2):1083–1091, 2008.

[126] A. M. Manferdini, F. Remondino, S. Baldissini, M. Gaiani, and B. Benedetti. 3D modeling and semantic classification of archaeological finds for management and visualization in 3D archaeological databases. In *Proceedings of 14th International Conference on Virtual Systems and MultiMedia (VSMM 2008)*, pages 221–228, 2008.

[127] S. Cronk, C. Fraser, and H. Hanley. Automatic metric calibration of colour digital cameras. *Photogrammetric Record*, 21(116):355–372, 2006.

[128] P. Patias. Cultural heritage documentation. *Applications of 3D Measurement from Images*. Whittles Publishing, pp. 225-250, 2008.

6

Digitizing the Parthenon: Estimating Surface Reflectance under Measured Natural Illumination

Paul Debevec, Chris Tchou, Andrew Gardner, Tim Hawkins, Charis Poullis, Jessi Stumpfel, Andrew Jones, Nathaniel Yun, Per Einarsson, Therese Lundgren, Marcos Fajardo

University of Southern California
Email: debevec@ict.usc.edu, tchouster@gmail.com,
androo.gardner@gmail.com, tim@lightstage.com,
charalambos@poullis.org, jessi@cs.caltech.edu,
jones@ict.usc.edu, corellian_knight@hotmail.com,
per.einarsson@dice.se, therese.lundgren@gmail.com,
marcosss@gmail.com

Philippe Martinez

Ecole Normale Superieure
Email: pmartine@ens.fr

CONTENTS

6.1 Introduction

Digitizing objects and environments from the real world has become an important part of creating realistic computer graphics. Capturing geometric models has become a common process through the use of structured lighting, laser triangulation, and laser time-of-flight measurements. Recent projects such as [1–3] have shown that accurate and detailed geometric models can be acquired of real-world objects using these techniques.

To produce renderings of an object under changing lighting as well as viewpoint, it is necessary to digitize not only the object's geometry but also its reflectance properties: how each point of the object reflects light. Digitizing reflectance properties has proven to be a complex problem, since these properties can vary across the surface of an object, and since the reflectance properties of even a single surface point can be complicated to express and measure. Some of the best results that have been obtained [2, 4, 5] capture digital photographs of objects from a variety of viewing and illumination directions, and from these measurements estimate reflectance model parameters for each surface point.

Digitizing the reflectance properties of outdoor scenes can be more complicated than for objects since it is more difficult to control the illumination and viewpoints of the surfaces. Surfaces are most easily photographed from ground level rather than from a full range of angles. During the daytime the illumination conditions in an environment change continuously. Finally, outdoor scenes generally exhibit significant mutual illumination between their surfaces, which must be accounted for in the reflectance estimation process. Two recent pieces of work have made important inroads into this problem. In Yu at al. [6] estimated spatially varying reflectance properties of an outdoor building based on fitting observations of the incident illumination to a sky model, and [7] estimated reflectance properties of a room interior based on known light source positions and using a finite element radiosity technique to take surface interreflections into account.

In this paper, we describe a process that synthesizes previous results for digitizing geometry and reflectance and extends them to the context of digitizing a complex real-world scene observed under arbitrary natural illumination. The data we acquire includes a geometric model of the scene obtained through laser scanning, a set of photographs of the scene under various natural illumination conditions, a corresponding set of measurements of the incident illumination for each photograph, and finally, a small set of Bi-directional Reflectance Distribution Function (BRDF) measurements of representative surfaces within the scene. To estimate the scene's reflectance properties, we use a global illumination algorithm to render the model from each of the photographed viewpoints as illuminated by the corresponding incident illumination measurements. We compare these renderings to the photographs, and then iteratively update the surface reflectance properties to best correspond to the scene's appearance in the photographs. Full BRDFs for the scene's surfaces are inferred from the measured BRDF samples. The result is a set of estimated reflectance properties for each point in the scene that most closely generates the scene's appearance under all input illumination conditions.

While the process we describe leverages existing techniques, our work includes several novel contributions. These include our incident illumination measurement process, which can measure the full dynamic range of both sunlit and cloudy natural illumination conditions, a hand-held BRDF measurement process suitable for use in the field, and an iterative multiresolution inverse global illumination process capable of estimating surface reflectance properties from multiple images for scenes with complex geometry seen under complex incident illumination.

The scene we digitize is the Parthenon in Athens, Greece, digitally laser scanned and photographed in April 2003 in collaboration with the ongoing Acropolis Restoration project. Scaffolding and equipment around the structure prevented the application of the process to the middle section of the temple, but we were able to derive models and reflectance parameters for both the East and West facades. We validated the accuracy of our results by comparing our reflectance measurements to ground truth measurements of specific surfaces around the site, and we generate renderings of the model under novel lighting that are consistent with real photographs of the site. At the end of the paper we discuss avenues for future work to increase the generality of these techniques. The work in this chapter was first described as a Technical Report in [8].

6.2 Background and Related Work

The process we present leverages previous results in 3D scanning, reflectance modeling, lighting recovery, and reflectometry of objects and environments. Techniques for building 3D models from multiple range scans generally involve first aligning the scans to each other [9,10], and then combining the scans into a consistent geometric model by either "zippering" the overlapping meshes [11] or using volumetric merging [12] to create a new geometric mesh that optimizes its proximity to all of the available scans. In its simplest form, a point's reflectance properties can be expressed in terms of its Lambertian surface color - usually an RGB triplet expressing the point's red, green, and blue reflectance properties. More complex reflectance models can include parametric models of specular and retroflective components; some commonly used models are [13–15]. More generally, a point's reflectance can be characterized in terms of its Bi-directional Reflectance Distribution Function (BRDF) [16], which is a 4D function that characterizes for each incident illumination direction the complete distribution of reflected illumination. Marschner et al. [17] proposed an efficient method for measuring a material's BRDFs if a convex homogeneous sample is available. Recent work has proposed models which also consider scattering of illumination within translucent materials [18]. To estimate a scene's reflectance properties, we use an incident illumination measurement process. Marschner et al. [19] recovered low-resolution incident illumination conditions by observing an object with known geometry and reflectance properties. Sato et al. [20] estimated incident illumination conditions by observing the shadows cast from objects with known geometry. Debevec in [21] acquired high-resolution lighting environments by taking high dynamic range images [22] of a mirrored sphere, but did not recover natural illumination environments where the sun was directly visible. We combine ideas from [19, 21] to record high-resolution incident illumination conditions in cloudy, partly cloudy, and sunlit environments. Considerable recent work has presented techniques to measure spatially-varying reflectance properties of objects. Marschner in [4] uses photographs of a 3D scanned object taken under point-light source illumination to estimate its spatially varying diffuse albedo. This work used a texture atlas system to store the surface colors of arbitrarily complex geometry, which we also perform in our work. The work assumed that the object was Lambertian, and only considered local reflections of the illumination. Sato et al. [23] use a similar sort of dataset to compute a spatially-varying diffuse component and a sparsely sampled specular component of an object. Rushmeier et al. [2] use a photometric stereo technique [24, 25] to estimate spatially varying Lambertian color as well as improved surface normals for the geometry. Rocchini et al. [26] use this technique to compute diffuse texture maps for 3D scanned objects from multiple images. Debevec et al. [27] use a dense set of illumination directions to estimate spatially-varying diffuse and specular parameters and surface normals. Lensch et al. [5] presents an advanced technique for recovering spatially-varying BRDFs of real-world objects, performing principal component analysis of relatively sparse lighting and viewing directions to cluster the object's surfaces into patches of similar reflectance. In this way, many reflectance observations of the object as a whole are used to estimate spatially-varying BRDF models for surfaces seen from limited viewing and lighting directions. Our reflectance modeling technique is less general, but adapts ideas from this work to estimate spatially-varying non-Lambertian reflectance properties of outdoor scenes observed under natural illumination conditions, and we also account for mutual illumination. Capturing the reflectance properties of surfaces in large-scale environments can be more complex, since it can be harder to control the lighting conditions on the surfaces and the viewpoints from which they are photographed. Yu et al. [6] solve for the reflectance properties of a polygonal model of an outdoor scene modeled with photogrammetry. The technique used photographs taken under clear sky conditions, fitting a small number of radiance measurements to a parameterized sky model. The process estimated spatially varying diffuse and piecewise constant specular parameters, but did not consider retroreflective components. The process derived two *pseudo-BRDFs* for each surface, one according to its reflectance of light from the sun and one according to its reflectance of light from the sky and environment. This allowed more general spectral modeling but required every surface to be observed under direct sunlight in at least one photograph, which we do not require. Using room interiors [7, 28, 29] estimate spatially varying diffuse and piecewise constant specular parameters using inverse global illumination. The techniques used knowledge of the position and intensity of the scene's light sources, using global illumination to account for the mutual illumination between the scene's surfaces. Our work combines and extends aspects of each of these techniques: we use pictures of our scene under natural illumination conditions, but we image the illumination directly in order to use photographs taken in sunny, partly sunny, or cloudy conditions. We infer non-Lambertian reflectance from sampled surface BRDFs. We do not consider full-spectral reflectance, but have found RGB imaging to be sufficiently accurate for the natural illumination and reflectance properties recorded in this work. We provide comparisons to ground truth reflectance for several surfaces within the scene. Finally, we use a more general Monte-Carlo global illumination algorithm to perform our inverse rendering, and we employ a multiresolution geometry technique to efficiently process a complex laser-scanned model.

6.3 Data Acquisiton and Calibration

6.3.1 Camera Calibration

In this work we used a Canon EOS D30 and a Canon EOS 1Ds digital camera, which were calibrated geometrically and radiometrically. For geometric calibration, we used the Camera Calibration Toolbox for Matlab [30] which uses techniques from [31]. Since changing the focus of a lens usually changes its focal length, we calibrated our lenses at chosen fixed focal lengths. The main lens used for photographing the environment was a 24mm lens focused at infinity. Since a small calibration object held near this lens would be out of focus, we built a larger calibration object 1.2m $\times$ 2.1m from an aluminum honeycomb panel with a 5cm square checkerboard pattern applied (Figure 6.1(a)). Though nearly all images were acquired at $f/8$ aperture, we verified that the camera intrinsic parameters varied insignificantly (less than 0.05%) with changes of $f/stop$ from $f/2.8$ to $f/22$. In this work we wished to obtain radiometrically linear pixel values that would be consistent for images taken with different cameras, lenses, shutter speeds, and $f/stop$. We verified that the "RAW" 12-bit data from the cameras was linear using three methods: we photographed a gray scale calibration chart, we used the radiometric self-calibration technique of [22], and we verified that pixel values were proportional to exposure times for a static scene. From this we found that the RAW pixel values exhibited linear response to within 0.1% for values up to 3,000 out of 4,095, after which saturation appeared to reduce pixel sensitivity. We ignored values outside of this linear range, and we used multiple exposures to increase the effective dynamic range of the camera when necessary.

Most lenses exhibit a radial intensity falloff, producing dimmer pixel values at the periphery of the image. We mounted each camera on a Kaidan nodal rotation head and photographed a diffuse disk light source at an array of positions for each lens at each $f/stop$ used for data capture (Figure 6.1(b)). From these intensities recorded at different image points, we fit a radially symmetric 6th order even polynomial to model the falloff curve and produce a flat-field response function, normalized to unit response at the image center.

Each digital camera used had minor variations in sensitivity and color response. We calibrated these variations by photographing a MacBeth color checker chart under natural illumination with each camera, lens, and $f/stop$ combination, and solved for the best 3×3 color matrix to convert each image into the same color space. Finally we used a utility for converting RAW images to floating-point images using the EXIF metadata for camera model, lens, ISO, f/stop, and shutter speed

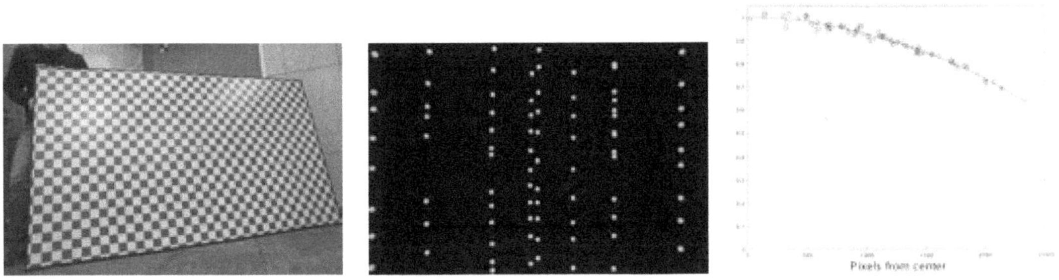

FIGURE 6.1
(a) 1.2m × 2.1m geometric calibration object; (b) Lens falloff measurements for 24mm lens at $f/8$;
(c) Lens falloff curve for (b).

to apply the appropriate radiometric scaling factors and matrices. These images were organized in a PostGreSQL database for convenient access.

6.3.2 BRDF Measurement and Modeling

In this work we measure BRDFs of a set of representative surface samples, which we use to form the most plausible BRDFs for the rest of the scene. Our relatively simple technique is motivated by the principal component analyses of reflectance properties used in [5, 32], except that we choose our basis BRDFs manually. Choosing the principal BRDFs in this way meant that BRDF data collected under controlled illumination could be taken for a small area of the site, while the large-scale scene could be photographed under a limited set of natural illumination conditions.

Data Collection and Registration

The site used in this work is composed entirely of marble, but its surfaces have been subject to different discoloration processes yielding significant reflectance variations. We located an accessible 30cm × 30cm surface that exhibited a range of coloration properties representative of the site. Since measuring the reflectance properties of this surface required controlled illumination conditions, we performed these measurements during our limited nighttime access to the site and used a BRDF measurement technique that could be executed quickly.

The BRDF measurement setup (Figure 6.2), includes a hand-held light source and camera, and uses a frame placed around the sample area that allows the lighting and viewing directions to be estimated from the images taken with the camera. The frame contains fiducial markers at each corner of the frame's aperture from which the camera's position can be estimated, and two glossy black plastic spheres used to determine the 3D position of the light source. Finally, the device includes a diffuse white reflectance standard parallel to the sample area for determining the intensity of the light source.

The light source chosen was a 1,000W halogen source mounted in a small diffuser box, held approximately 3m from the surface. Our capture assumed that the surfaces exhibited isotropic reflection, requiring the light source to be moved only within a single plane of incidence. We placed the light source in four consecutive positions of $0°, 30°, 50°, 75°$, and for each took hand-held photographs at a distance of approximately 2m from twenty directions distributed on the incident hemisphere, taking care to sample the specular and retroreflective directions with a greater number of observations. Dark clothing was worn to reduce stray illumination on the sample. The full capture process involving 83 photographs required forty minutes.

Data Analysis and Reflectance Model Fitting

To calculate the viewing and lighting directions, we first determined the position of the camera from the known 3D positions of the four fiducial markers using photogrammetry. With the camera

FIGURE 6.2
BRDF samples are measured from a 30cm square region exhibiting a representative set of surface reflectance properties. The technique used a hand-held light source and camera and a calibration frame to acquire the BRDF data quickly.

positions known, we computed the positions of the two spheres by tracing rays from the camera centers through the sphere centers for several photographs, and calculated the intersection points of these rays. With the sphere positions known, we determined each light position by shooting rays toward the center of the light's reflection in the spheres. Reflecting the rays off the spheres, we find the center of the light source position where the two rays most nearly intersect. Similar techniques to derive light source positions have been used in [5, 33].

From the diffuse white reflectance standard, the incoming light source intensity for each image could be determined. By dividing the overall image intensity by the color of the reflectance standard, all images were normalized by the incoming light source intensity. We then chose three different areas within the sampling region best corresponding to the different reflectance properties of the large-scale scene. These properties included a light tan area that is the dominant color of the site's surfaces, a brown color corresponding to encrusted biological material, and a black color representative of soot deposits. To track each of these sampling areas across the dataset, we applied a homography to each image to map them to a consistent orthographic viewpoint. For each sampling area, we then obtained a BRDF sample by selecting a 30×30 pixel region and computing the average RGB value. Had there been a greater variety of reflectance properties in the sample, a PCA analysis of the entire sample area as in [5] could have been used.

Looking at Figure 6.3, the data show largely diffuse reflectance but with noticeable retroreflective and broad specular components. To extrapolate the BRDF samples to a complete BRDF, we fit the BRDF to the Lafortune cosine lobe model (Eq. 6.1) in its isotropic form with three lobes for the Lambertian, specular, and retroreflective components:

$$f(\overrightarrow{u}, \overrightarrow{v}) = \rho_d + \sum_i [C_{xy,i}(u_x v_x + u_y v_y) + C_{z,i} u_z v_z]^{N_i} \qquad (6.1)$$

As suggested in [15], we then use a non-linear Levenberg-Marquardt optimization algorithm to

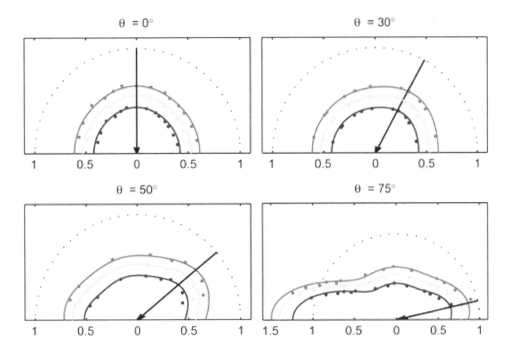

FIGURE 6.3
BRDF data and fitted reflectance lobes are shown for the RGB colors of the tan material sample for the four incident illumination directions. Only measurements within $15°$ of in-plane are plotted.

determine the parameters of the model from our measured data. We first estimate the Lambertian component ρ_d, and then fit a retroreflective and a specular lobe separately before optimizing all the parameters in a single system. The resulting BRDFs (Figure 6.4(b), back row) show mostly Lambertian reflectance with noticeable retroreflection and rough specular components at glancing angles. The brown area exhibited the greatest specular reflection, while the black area was the most retroreflective.

BRDF Inference
We wish to be able to make maximal use of the BRDF information obtained from our material samples in estimating the reflectance properties of the rest of the scene. The approach we take is informed by the BRDF basis construction technique from [5], the data-driven reflectance model presented in [32], and spatially-varying BRDF construction technique used in [34]. Because the surfaces of the rest of the scene will often be seen in relatively few photographs under relatively diffuse illumination, the most reliable observation of a surface's reflectance is its Lambertian color. Thus, we form our problem as one of inferring the most plausible BRDF for a surface point given its Lambertian color and the BRDF samples available.

We first perform a principal component analysis of the Lambertian colors of the BRDF samples available. For RGB images, the number of significant eigenvalues will be at most three, and for our samples the first eigenvalue dominates, corresponding to a color vector of (0.688, 0.573, 0.445). We project the Lambertian color of each of our sample BRDFs onto the 1D subspace S (Figure 6.4(a)) formed by this eigenvector. To construct a plausible BRDF f for a surface having a Lambertian color ρ_d, we project ρ_d onto S to obtain the projected color ρ'_d. We then determine the two BRDF samples whose Lambertian components project most closely to ρ'_d. We form a new BRDF f' by linearly interpolating the Lafortune parameters (C_{xy}, C_z, N) of the specular and retroreflective lobes of these two nearest BRDFs f_0 and f_1 based on distance. Finally, since the retroflective color of a surface usually corresponds closely to its Lambertian color, we adjust the color of the retroflective lobe to correspond to the actual Lambertian color ρ_d rather than the projected color ρ_d'. We do this by dividing the retroreflective parameters C_{xy} and C_z by $(\rho_d')^{1/N}$ and then multiplying by $(\rho_d)^{1/N}$ for each color channel, which effectively scales the retroreflective lobe to best correspond to the Lambertian color ρ_d. Figure 6.4(b) shows a rendering with several BRDFs inferred from new Lambertian colors with this process.

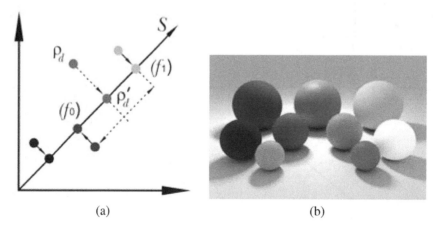

(a)   (b)

FIGURE 6.4
(a) Inferring a BRDF based on its Lambertian component ρ_d; (b) Rendered spheres with measured and inferred BRDFs. Back row: the measured black, brown, and tan surfaces. Middle row: intermediate BRDFs along the subspace S. Front row: inferred BRDFs for materials with Lambertian colors not on S.

6.3.3   Natural Illumination Capture

Each time a photograph of the site was taken, we used a device to record the corresponding incident illumination within the environment. The lighting capture device was a digital camera aimed at three spheres: one mirrored, one shiny black, and one diffuse gray. We placed the device in a nearby accessible location far enough from the principal structure to obtain an unshadowed view of the sky, and close enough to ensure that the captured lighting would be sufficiently similar to that incident upon the structure. Measuring the incident illumination directly and quickly enabled us to make use of photographs taken under a wide range of weather including sunny, cloudy, and partially cloudy conditions, and also in changing conditions.

Apparatus Design

The lighting capture device is designed to measure the color and intensity of each direction in the upper hemisphere. A challenge in capturing such data for a natural illumination environment is that the sun's intensity can exceed that of the sky by over five orders of magnitude, which is significantly beyond the range of most digital image sensors. This dynamic range surpassing 17 stops also exceeds that which can conveniently be captured using high dynamic range capture techniques. Our solution was to take a limited dynamic range photograph and use the mirrored sphere to image the sky and clouds, the shiny black sphere to indicate the position of the sun (if visible), and the diffuse grey sphere to indirectly measure the intensity of the sun. We placed all three spheres on a board so that they could be photographed simultaneously (Figure 6.5). We painted the majority of the board gray to allow a correct exposure of the device to be derived from the camera's auto-exposure function, but surrounded the diffuse sphere by black paint to minimize the indirect light it received. We also included a sundial near the top of the board to validate the lighting directions estimated from the

(a)   (b)

FIGURE 6.5
(a) The incident illumination measurement device at its chosen location on the site; (b) An incident illumination dataset.

black sphere. Finally, we placed four fiducial markers on the board to estimate the camera's relative position to the device.

We used a Canon D30 camera with a resolution of $2{,}174 \times 1{,}446$ pixels to capture images of the device. Since the site photography took place up to 300m from the incident illumination measurement station, we used a radio transmitter to trigger the device at the appropriate times. Though the technique we describe can work with a single image of the device, we set the camera's internal auto-exposure bracketing function to take three exposures for each shutter release at -2, +0, and +2 stops. This allowed somewhat higher dynamic range to better image brighter clouds near the sun, and to guard against any problems with the camera's automatic light metering.

Sphere Reflectance Calibration

To achieve accurate results, we calibrated the reflectance properties of the spheres. The diffuse sphere was painted with flat gray primer paint, which we measured as having a reflectivity of $(0.30, 0.31, 0.32)$ in the red, green, and blue color channels. We further verified it to be nearly spectrally flat using a spectroradiometer. We also exposed the paint to several days of sunlight to verify its color stability. In the above calculations, we divide all pixel values by the sphere's reflectance, producing values that would result from a perfectly reflective white sphere.

We also measured the reflectivity of the mirrored sphere, which was made of polished steel. We measured this reflectance by using a robotic arm to rotate a rectangular light source in a circle around the sphere and taking a long-exposure photograph of the resulting reflection (Figure 6.6(a)). We found that the sphere was 52% reflective at normal incidence, becoming more reflective toward grazing angles due to Fresnel reflection (Figure 6.6(b)). From the measured reflectance data we used a nonlinear optimization to fit a Fresnel curve to the data, arriving at a complex index of refraction of $(2.40 + 2.98i, 2.40 + 3.02i, 2.40 + 3.02i)$ for the red, green, and blue channels of the sphere.

Light from a clear sky can be significantly polarized, particularly in directions perpendicular to the direction of the sun. In our work we assume that the surfaces in our scene are not highly specular, which makes it reasonable for us to disregard the polarization of the incident illumination in our reflectometry process. However, since Fresnel reflection is affected by the polarization of the incoming light, the clear sky may reflect either more or less brightly toward the grazing angles of the mirrored sphere than it should if it were photographed directly. To quantify this potential error, we photographed several clear skies reflected in the mirrored sphere and at the same time took hemispherical panoramas with a 24mm lens. Comparing the two, we found an RMS error of 5% in sky intensity between the sky photographed directly and the sky photographed as reflected in the

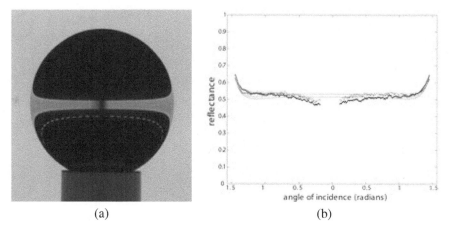

(a)   (b)

FIGURE 6.6
(a) Mirrored sphere photographed under an even ring of light, showing an increase in brightness at extreme grazing angles (the dark gap in the center is due to light source occluding the camera). (b) Fitted Fresnel reflectance curves.

mirrored sphere (Figure 6.7). In most situations, however, unpolarized light from the sun, clouds, and neighboring surfaces dominates the incident illumination on surfaces, which minimizes the effect of this error. In Section 6.6, we suggest techniques for eliminating this error through improved optics.

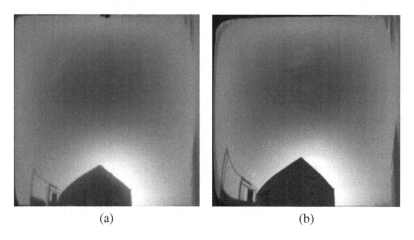

(a)   (b)

FIGURE 6.7
(a) Sky photographed as reflected in a mirrored sphere; (b) Stitched sky panorama from 16 to 24mm photographs, showing slightly different reflected illumination due to sky polarization.

Image Processing and Deriving Sun Intensity

To process these images, we assemble each set of three bracketed images into a single higher dynamic range image, and derive the relative camera position from the fiducial markers. The fiducial markers are indicated manually in the first image of each day and then tracked automatically through the rest of the day, compensating for small motions due to wind. Then, the reflections in both the mirrored and shiny black spheres are transformed to 512×512 images of the upper hemisphere. This is done by forward-tracing rays from the camera to the spheres (whose positions are known) and reflecting the rays into the sky, noting for each sky point the corresponding location on the sphere. The image of the diffuse sphere is also mapped to the sky's upper hemisphere, but based on the sphere's normals rather the reflection vectors. In the process, we also adjust for the reflectance properties of the spheres as described in Section 6.3.3, creating the images that would have been produced by spheres with unit albedo. Examples of these unwarped images are shown in Figure 6.8.

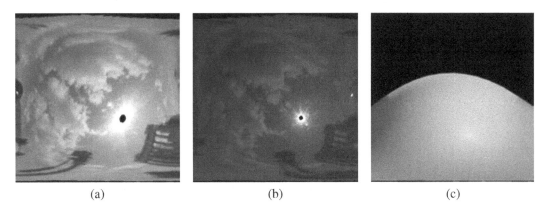

(a) (b) (c)

FIGURE 6.8
Sphere images unwarped to the upper hemisphere for the (a) Mirrored sphere; (b) Shiny black sphere; (c) Diffuse sphere D. Saturated pixels are shown in black.

If the sun is below the horizon or occluded by clouds, no pixels in the mirrored sphere image will be saturated and it can be used directly as the image of the incident illumination. We can validate the accuracy of this incident illumination map by rendering a synthetic diffuse image D' with this lighting and checking that it is consistent with the appearance of the actual diffuse sphere image D. As described in [35], this lighting operation can be performed using a diffuse convolution filter on the incident lighting environment. For our data, the root mean square illumination error for our diffuse sphere images agreed to within 2% percent for a variety of environments.

When the sun is visible, it usually saturates a small region of pixels in the mirrored sphere image. Since the sun's bright intensity is not properly recorded in this region, performing a diffuse convolution of the mirrored sphere image will produce a darker image than actual appearance of the diffuse sphere (Compare D' to D in Figure 6.9). In this case, we reconstruct the illumination from the sun as follows. We first measure the direction of the sun as the center of the brightest spot reflected in the shiny black sphere (with its darker reflection, the black sphere exhibits the most sharply defined image of the sun). We then render an image of a diffuse sphere $D^\star$ lit from this direction of illumination, using a unit-radiance infinite light source 0.53 degrees in diameter to match the subtended angle of the real sun. Such a rendering can be seen in the center of Figure 6.9.

We can then write that the appearance of the real diffuse sphere D should equal the sphere lit by the light captured in the mirrored sphere D' plus an unknown factor α times the sphere illuminated by the unit sun $D^\star$, i.e.,

$$D' + \alpha D^\star = D \tag{6.2}$$

Since there are many pixels in the sphere images, this system is overdetermined, and we compute the red, green, and blue components of α using least squares as $\alpha D^\star \approx D - D'$. Since $D^\star$ was rendered using a unit radiance sun, α indicates the radiance of the sun disk for each channel. For efficiency, we keep the solar illumination modeled as the directional disk light source, rather than updating the mirrored sphere image M to include this illumination. As a result, when we create renderings with the measured illumination, the solar component is more efficiently simulated as a direct light source.

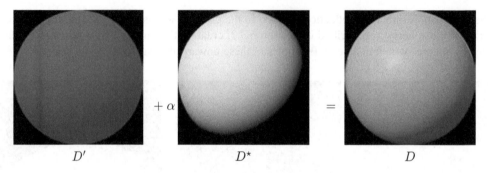

FIGURE 6.9
Solving for sun intensity α based on the appearance of the diffuse sphere D and the convolved mirrored sphere D'.

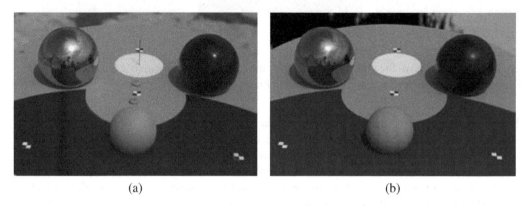

(a)   (b)

FIGURE 6.10
(a) Real photograph of the lighting capture device; (b) Synthetic rendering of a 3D model of the lighting capture device to validate the lighting measurements.

We note that this process does not reconstruct correct values for the remaining saturated pixels near the sun; the missing illumination from these regions is effectively added to the sun's intensity. Also, if the sun is partially obscured by a cloud, the center of the saturated region might not correspond precisely to the center of the sun. However, for our data the saturated region has been sufficiently small that this error has not been significant. Figure 6.10 shows a lighting capture dataset and a comparison rendering of a model of the capture apparatus, showing consistent captured illumination.

6.3.4 3D Scanning

To obtain 3D geometry for the scene, we used a time-of-flight panoramic range scanner manufactured by Quantapoint, Inc., which uses a 950nm infrared laser measurement component [36]. In high-resolution mode, the scanner acquires scans of 18,000 by 3,000 3D points in 8 minutes, with a maximum scanning range of 40m and a field of view of 360 degrees horizontal by 74.5 degrees vertical. Some scans from within the structure were scanned in low-resolution, acquiring one-quarter the number of points. The data returned is an array of (x,y,z) points as well as a 16-bit monochrome image of the infrared intensity returned to the sensor for each measurement. Depending on the strength of the return, the depth accuracy varied between 0.5cm and 3cm. Over five days, 120 scans were acquired in and around the site, of which 53 were used to produce the model in this chapter (Figure 6.11).

FIGURE 6.11
Range measurements, shaded according to depth (top), and infrared intensity return (bottom) for one of 53 panoramic laser scans used to create the model. A fiducial marker appears at right.

Scan Processing

Our scan processing followed the traditional process of alignment, merging, and decimation. Scans from outside the structure were initially aligned during the scanning process through the use of checkerboard fiducial markers placed within the scene. After the site survey, the scans were further aligned using an iterative closest point (ICP) algorithm [9, 10] implemented in the CNR-Pisa 3D scanning toolkit [37] (see Chapter 3). To speed the alignment process, three or more subsections of each scan corresponding to particular scene areas were cropped out and used to determine the alignment for the entire scan.

For merging, the principal structure of the site was partitioned into an $8 \times 17 \times 5$ lattice of voxels 4.3 meters on a side. For convenience, the grid was chosen to align with the principal architectural features of the site. The scan data within each voxel was merged by a volumetric merging algorithm [12] also from the CNR-Pisa toolkit using a volumetric resolution of 1.2cm. Finally, the geometry of a $200m \times 200m$ area of surrounding terrain was merged as a single mesh with a resolution of 40cm.

Several of the merged voxels contained holes due to occlusions or poor laser return from dark surfaces. Since such geometric inconsistencies would affect the reflectometry process, they were filled using semi-automatic tools with Geometry Systems, Inc. GSI Studio software (Figure 6.12).

Our reflectometry technique determines surface reflectance properties which are stored in texture maps. We used a texture atlas generator [38] based on techniques in [39] to generate a 512×512 texture map for each voxel. Then, a low-resolution version of each voxel was created using the Qslim software [40] based on techniques in [41]. This algorithm was chosen since it preserves edge polygons, allowing low-resolution and high-resolution voxels to connect without seams, and since it preserves the texture mapping space, allowing the same texture map to be used for either the high- or low-resolution geometry.

The complete high-resolution model of the main structure used 89 million polygons in 442 non-empty voxels (Figure 6.13). The lowest-resolution model contained 1.8 million polygons, and the surrounding environment used 366K polygons.

6.3.5 Photograph Acquisition and Alignment

Images were taken of the scene from a variety of viewpoints and lighting conditions using the Canon 1Ds camera. We used a semi-automatic process to align the photographs to the 3D scan data. We began by marking approximately 15 point correspondences between each photo and the infrared

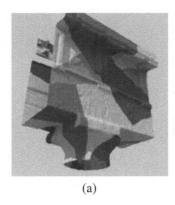

(a)   (b)

FIGURE 6.12 (SEE COLOR INSERT)
(a) Geometry for a voxel colored according to texture atlas regions; (b) The corresponding texture
atlas.

intensity return image of one or more 3D scans, forming a set of 2D to 3D correspondences. From
this we estimated the camera pose using Intel's OpenCV library, achieving a mean alignment error of
between 1 and 3 pixels at $4,080 \times 2,716$ pixel resolution. For photographs with higher alignment
error, we use an automatic technique to refine the alignment based on comparing the structure's
silhouette in the photograph to the model's silhouette seen through the recovered camera as in [42],
using a combination of gradient-descent and simulated annealing.

6.4 Reflectometry

In this section we describe the central reflectometry algorithm used in this work. The basic goal
is to determine surface reflectance properties for the scene such that renderings of the scene under
captured illumination match photographs of the scene taken under that illumination. We adopt an
inverse rendering framework as in [21, 29] in which we iteratively update our reflectance parameters
until our renderings best match the appearance of the photographs. We begin by describing the basic
algorithm and continue by describing how we have adapted it for use with a large dataset.

6.4.1 General Algorithm

The basic algorithm we use proceeds as follows:

1. Assume initial reflectance properties for all surfaces

2. For each photograph:

    (a) Render the surfaces of the scene using the photograph's viewpoint and lighting

    (b) Determine a reflectance update map by comparing radiance values in the photograph to
        radiance values in the rendering

    (c) Compute weights for the reflectance update map

3. Update the reflectance estimates using the weightings from all photographs

4. Return to step 2 until convergence

FIGURE 6.13
Complete model assembled from the 3D scanning data, including low-resolution geometry for the surrounding terrain. High- and medium-resolution voxels used for the multiresolution reflectance recovery are indicated in white and blue.

For a pixel's Lambertian component, the most natural update for a pixel's Lambertian color is to multiply it by the ratio of its color in the photograph to its color in the corresponding rendering. This way, the surface will be adjusted to reflect the correct proportion of the light. However, the indirect illumination on the surface may change in the next iteration since other surfaces in the scene may also have new reflectance properties, requiring further iterations.

Since each photograph will suggest somewhat different reflectance updates, we weight the influence a photograph has on a surface's reflectance by a confidence measure. For one weight, we use the cosine of the angle at which the photograph views the surface. Thus, photographs which view surfaces more directly will have a greater influence on the estimated reflectance properties. As in traditional image-based rendering (e.g., [43]), we also downweight a photograph's influence near occlusion boundaries. Finally, we also downweight an image's influence near large irradiance gradients in the photographs since these typically indicate shadow boundaries, where small misalignments in lighting could significantly affect the reflectometry.

In this work, we use the inferred Lafortune BRDF models described in Sec. 6.3.2 to create the renderings, which we have found to also converge accurately using updates computed in this manner. This convergence occurs for our data since the BRDF colors of the Lambertian and retroreflective lobes both follow the Lambertian color, and since for all surfaces most of the photographs do not observe a specular reflection. If the surfaces were significantly more specular, performing the updates according to the Lambertian component alone would not necessarily converge to accurate reflectance estimates. We discuss potential techniques to address this problem in the future work section.

6.4.2 Multiresolution Reflectance Solving

The high-resolution model for our scene is too large to fit in memory, so we use a multiresolution approach to computing the reflectance properties. Since our scene is partitioned into voxels, we can compute reflectance property updates one voxel at a time. However, we must still model the effect of shadowing and indirect illumination for the rest of the scene. Fortunately, lower-resolution geometry can work well for this purpose. In our work, we use full-resolution geometry (approx. 800K triangles) for the voxel being computed, medium-resolution geometry (approx. 160K triangles) for the immediately neighboring voxels, and low-resolution geometry (approx. 40K triangles) for the remaining voxels in the scene. The surrounding terrain is kept at a low resolution of 370K triangles. The multiresolution approach results in over a 90% data reduction in scene complexity during the reflectometry of any given voxel.

Our global illumination rendering system was originally designed to produce 2D images of a scene for a given camera viewpoint using path tracing [44]. We modified the system to include a new function for computing surface radiance for any point in the scene radiating toward any viewing position. This allows the process of computing reflectance properties for a voxel to be done by iterating over the texture map space for that voxel. For efficiency, for each pixel in the voxel's texture space, we cache the position and surface normal of the model corresponding to that texture coordinate, storing these results in two additional floating-point texture maps.

1. Assume initial reflectance properties for all surfaces.

2. For each voxel V:

   - Load V at high resolution, V's neighbors at medium resolution, and the rest of the model at low resolution.

   - For each pixel p in V's texture space:

     – For each photograph I:

       * Determine if p's surface is visible to I's camera. If not, break. If so, determine the weight for this image based on the visibility angle, and note pixel q in I corresponding to p's projection into I.

       * Compute the radiance l of p's surface in the direction of I's camera under I's illumination.

       * Determine an updated surface reflectance by comparing the radiance in the image at q to the rendered radiance l.

     – Assign the new surface reflectance for p as the weighted average of the updated reflectances from each I.

3. Return to step 2 until convergence

Figure 6.14 shows this process of computing reflectance properties for a voxel. Figure 6.14(a) shows the 3D model with the assumed initial reflectance properties illuminated by a captured illumination environment. Figure 6.14(b) shows the voxel texture-mapped with radiance values from a photograph taken under the captured illumination in (a). Comparing the two, the algorithm determines updated surface reflectance estimates for the voxel, shown in Figure 6.14(c). The second iteration compares an illuminated rendering of the model with the first iteration's inferred BRDF properties to the photograph, producing new updated reflectance properties shown in Fig. 6.14(d). For this voxel, the second iteration produces a darker Lambertian color for the underside of the ledge, which results from the fact that the *black* BRDF sample measured in Section 6.3.2 has a higher proportion of retroreflection than the average reflectance. The second iteration is computed with a greater number of samples per ray, producing images with fewer noise artifacts. Reflectance estimates for three voxels of a column on the East facade are shown in texture atlas form in Figure 6.15. Reflectance properties for all voxels of the two facades are shown in Figures 6.16(b) and 6.19(d). For our model, the third iteration produces negligible change from the second, indicating convergence.

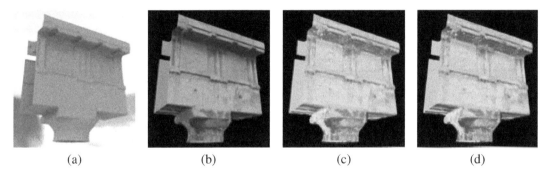

FIGURE 6.14 (SEE COLOR INSERT)
Computing reflectance properties for a voxel (a) Iteration 0: 3D model illuminated by captured illumination, with assumed reflectance properties; (b) Photograph taken under the captured illumination projected onto the geometry; (c) Iteration 1: New reflectance properties computed by comparing (a) to (b). (d) Iteration 2: New reflectance properties computed by comparing a rendering of (c) to (b).

FIGURE 6.15
Estimated surface reflectance properties for an East facade column in texture atlas form.

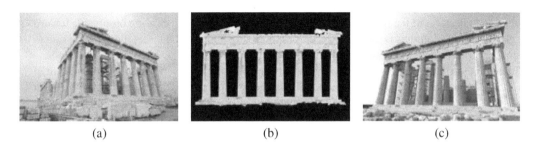

FIGURE 6.16
(a) One of eight input photographs; (b) Estimated reflectance properties; (c) Synthetic rendering with novel lighting.

6.5   Results

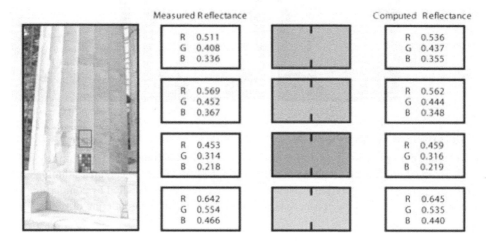

FIGURE 6.17 (SEE COLOR INSERT)
Left: Acquiring a ground truth reflectance measurement. Right: Reflectance comparisons for four locations on the East facade.

We ran our reflectometry algorithm on the 3D scan dataset, computing high-resolution reflectance properties for the two westmost and eastmost rows of voxels. As input to the algorithm, we used eight photographs of the East facade (e.g., Figure 6.16(a)) and three of the West facade, in an assortment of sunny, partly cloudy, and cloudy lighting conditions. Poorly scanned scaffolding which had been removed from the geometry was replaced with approximate polygonal models in order to better simulate the illumination transport within the structure. The reflectance properties of the ground were assigned based on a sparse sampling of ground truth measurements made with a MacBeth chart. We recovered the reflectance properties in two iterations of the reflectometry algorithm. For each iteration of the reflectometry, the illumination was simulated with two indirect bounces using the inferred Lafortune BRDFs. Computing the reflectance for each voxel required an average of ten minutes.

Figures 6.16(b) and 6.19(d) show the computed Lambertian reflectance colors for the East and West facades, respectively. Recovered texture atlas images for three voxels of the East column second from left are shown in Figure 6.15. The images show few shading effects, suggesting that the maps have removed the effect of the illumination in the photographs. The subtle shading observable toward the back sides of the columns is likely the result of incorrectly computed indirect illumination due to the remaining discrepancies in the scaffolding.

Figures 6.19(a) and (b) show a comparison between a real photograph and a synthetic global illumination rendering of the East facade under the lighting captured for the photograph, indicating a consistent appearance. The photograph represents a significant variation in the lighting from all images used in the reflectometry dataset. Figure 6.19(c) shows a rendering of the West facade model under novel illumination and viewpoint. Figure 6.19(e) shows the East facade rendered under novel artificial illumination. Figure 6.19(f) shows the East facade rendered under sunset illumination captured from a different location than the original site. Figure 6.18 shows the West facade rendered using high-resolution lighting environments captured at various times during a single day.

To provide a quantitative validation of the reflectance measurements, we directly measured the reflectance properties of several surfaces around the site using a MacBeth color checker chart. Since

FIGURE 6.18 (SEE COLOR INSERT)
Rendering of a virtual model of the Parthenon with lighting from 7:04am (a), 10:35am (b), 4:11pm (c), and 5:37pm (d). Capturing high-resolution outdoor lighting environments with over 17 stops of dynamic range with time lapse photography allows for realistic lighting.

the measurements were made at normal incidence and in diffuse illumination, we compared the results to the Lambertian lobe directly, as the specular and retroreflective lobes are not pronounced under these conditions. The results tabulated in Figure 6.17 show that the computed reflectance largely agreed with the measured reflectance samples, with a mean error of $(2.0\%, 3.2\%, 4.2\%)$ for the red, green, and blue channels.

6.6 Discussion and Future Work

Our experiences with the process suggest several avenues for future work. Most importantly, it would be of interest to increase the generality of the reflectance properties which can be estimated using the technique. Our scene did not feature surfaces with sharp specularity, but most scenes featuring contemporary architecture do. To handle this larger gamut of reflectance properties, one could imagine adapting the BRDF clustering and basis formation techniques in [5] to photographs taken under natural illumination conditions. Our technique for interpolating and extrapolating our BRDF samples is relatively simplistic; using more samples and a more sophisticated analysis and interpolation as in [32] would be desirable. A challenge in adapting these techniques to natural illumination is that observations of specular behavior are less reliable in natural illumination conditions. Estimating reflectance properties with increased spectral resolution would also be desirable.

In our process the photographs of the site are used only for estimating reflectance, and are not used to help determine the geometry of the scene. Since high-speed laser scan measurements can be noisy, it would be of interest to see if photometric stereo techniques as in [2] could be used in conjunction with natural illumination to refine the surface normals of the geometry. Yu et al. [6] for

FIGURE 6.19 (SEE COLOR INSERT)
(a) A real photograph of the East facade, with recorded illumination; (b) Rendering of the model under the illumination recorded for (a) using inferred Lafortune reflectance properties; (c) A rendering of the West facade from a novel viewpoint under novel illumination. (d) Front view of computed surface reflectance for the West facade (the East is shown in 6.16(b)). A strip of unscanned geometry above the pediment ledge has been filled in and set to the average surface reflectance. (e) Synthetic rendering of the West facade under a novel artificial lighting design. (f) Synthetic rendering of the East facade under natural illumination recorded for another location. In these images, only the front two rows of outer columns are rendered using the recovered reflectance properties; all other surfaces are rendered using the average surface reflectance.

example used photometric stereo from different solar positions to estimate surface normals for a building's environment; it seems possible that such estimates could also be made given three images of general incident illumination with or without the sun.

Our experience calibrating the illumination measurement device showed that its images could be affected by sky polarization. We tested the alternative of using an upward-pointing fisheye lens to image the sky, but found significant polarization sensitivity toward the horizon as well as undesirable lens flare from the sun. More successfully, we used a 91% reflective aluminum-coated hemispherical lens and found it to have less than 5% polarization sensitivity, making it suitable for lighting capture. For future work, it might be of interest to investigate whether sky polarization, explicitly captured, could be leveraged in determining a scene's specular parameters [45].

Finally, it could be of interest to use this framework to investigate the more difficult problem of estimating a scene's reflectance properties under unknown natural illumination conditions. In this case, estimation of the illumination could become part of the optimization process, possibly by fitting to a principal component model of measured incident illumination conditions.

6.7 Conclusion

We have presented a process for estimating spatially-varying surface reflectance properties of an outdoor scene based on scanned 3D geometry, BRDF measurements of representative surface samples, and a set of photographs of the scene under measured natural illumination conditions. Applying the process to a real-world archaeological site, we found it able to recover reflectance properties close to ground truth measurements, and able to produce renderings of the scene under novel illumination consistent with real photographs. The encouraging results suggest further work be carried out to capture more general reflectance properties of real-world scenes using natural illumination.

Acknowledgments

The authors would like to acknowledge the following individuals and organizations which supported this project: Richard Lindheim, David Wertheimer, Neil Sullivan, Nikos Toganidis, Katerina Paraschis, Tomas Lochman, Manolis Korres, Angeliki Arvanitis, James Blake, Bri Brownlow, Chris Butler, Elizabeth Cardman, Alan Chalmers, Paolo Cignoni, Yikuong Chen, Jon Cohen, Costis Dallas, Christa Deacy-Quinn, Paul T. Debevec, Naomi Dennis, Apostolos Dimopoulos, George Drettakis, Paul Egri Costa-Gavras, Darin Grant, Rob Groome, Christian Guillon, Craig Halperin, Youda He, Eric Hoffman, Leslie Ikemoto, Peter James, David Jillings, Genichi Kawada, Shivani Khanna, Randal Kleiser, Cathy Kominos, Jim Korris, Marc Levoy, Dell Lunceford, Donat-Pierre Luigi, Mike Macedonia, Brian Manson, Paul Marty, Hiroyuki Matsuguma, David Miraglia, Chris Nichols, Chrysostomos Nikias, Mark Ollila, Yannis Papoutsakis, John Parmentola, Fred Persi, Dimitrios Raptis, Simon Ratcliffe, Mark Sagar, Roberto Scopigno, Alexander Singer, Judy Singer, Diane Suzuki, Laurie Swanson, Bill Swartout, Despoina Theodorou, Mark Timpson, Rippling Tsou, Zach Turner, Esdras Varagnolo, Greg Ward, Karen Williams, Min Yu and The Work Site of the Acropolis, The Louvre, The Basel Skulpturhalle, The British Museum, The Spurlock Museum, The Herodion Hotel, Quantapoint Inc., Geometry Systems Inc., CNR Pisa, Alia, The US Army, TOPPAN Printing Co Ltd., and The University of Southern California.

Bibliography

[1] M. Levoy, K. Pulli, B. Curless, S. Rusinkiewicz, D. Koller, L. Pereira, M. Ginzton, S. Anderson, J. Davis, J. Ginsberg, J. Shade, and D. Fulk, "The Digital Michelangelo Project: 3D Scanning of Large Statues," *Proceedings of SIGGRAPH 2000*, pp. 131–144, July 2000.

[2] H. Rushmeier, F. Bernardini, J. Mittleman, and G. Taubin, "Acquiring Input for Rendering at Appropriate Levels of Detail: Digitizing a Pietà," *Eurographics Rendering Workshop 1998*, pp. 81–92, June 1998.

[3] K. Ikeuchi, "Modeling from Reality," in *Proceedings Third International Conference on 3-D Digital Imaging and Modeling* (Quebec City), pp. 117–124, May 2001.

[4] S. Marschner, *Inverse Rendering for Computer Graphics*. PhD thesis, Cornell University, August 1998.

[5] H. P. A. Lensch, J. Kautz, M. Goesele, W. Heidrich, and H.-P. Seidel, "Image-Based Reconstruction of Spatial Appearance and Geometric Detail," *ACM Transactions on Graphics*, vol. 22, pp. 234–257, Apr. 2003.

[6] Y. Yu and J. Malik, "Recovering Photometric Properties of Architectural Scenes from Photographs," in *Proceedings of SIGGRAPH 98*, Computer Graphics Proceedings, Annual Conference Series, pp. 207–218, July 1998.

[7] Y. Yu, P. Debevec, J. Malik, and T. Hawkins, "Inverse Global Illumination: Recovering Reflectance Models of Real Scenes from Photographs," *Proceedings of SIGGRAPH 99*, pp. 215–224, August 1999.

[8] P. Debevec, C. Tchou, A. Gardner, T. Hawkins, C. Poullis, J. Stumpfel, A. Jones, N. Yun, P. Einarsson, T. Lundgren, M. Fajardo, P. Martinez, "Estimating Surface Reflectance Properties of a Complex Scene under Captured Natural Illumination," Technical report, USC, 2004. ICT-TR-06.2004.

[9] P. Besl and N. McKay, "A Method for Registration of 3-d Shapes," *IEEE Transactions on Pattern Analysis and Machine Intelligence*, vol. 14, pp. 239–256, 1992.

[10] Y. Chen and G. Medioni, "Object Modeling from Multiple Range Images," *Image and Vision Computing*, vol. 10, pp. 145–155, April 1992.

[11] G. Turk and M. Levoy, "Zippered Polygon Meshes from Range Images," in *Proceedings of SIGGRAPH 94*, Computer Graphics Proceedings, Annual Conference Series (Orlando, Florida), pp. 311–318, ACM SIGGRAPH / ACM Press, July 1994. ISBN 0-89791-667-0.

[12] B. Curless and M. Levoy, "A Volumetric Method for Building Complex Models from Range Images," in *Proceedings of SIGGRAPH 96*, Computer Graphics Proceedings, Annual Conference Series (New Orleans, Louisiana), pp. 303–312, ACM SIGGRAPH / Addison Wesley, August 1996.

[13] G. J. W. Larson, "Measuring and Modeling Anisotropic Reflection," in *Computer Graphics (Proceedings of SIGGRAPH 92)*, vol. 26 (Chicago, Illinois), pp. 265–272, July 1992.

[14] M. Oren and S. K. Nayar, "Generalization of Lambert's Reflectance Model," *Proceedings of SIGGRAPH 94*, pp. 239–246, July 1994.

[15] E. P. F. Lafortune, S.-C. Foo, K. E. Torrance, and D. P. Greenberg, "Non-Linear Approximation of Reflectance Functions," *Proceedings of SIGGRAPH 97*, pp. 117–126, 1997.

[16] F. E. Nicodemus, J. C. Richmond, J. J. Hsia, I. W. Ginsberg, and T. Limperis, "Geometric Considerations and Nomenclature for Reflectance," *National Bureau of Standards Monograph 160*, October 1977.

[17] S. R. Marschner, S. H. Westin, E. P. F. Lafortune, K. E. Torrance, and D. P. Greenberg, "Image-Based BRDF Measurement Including Human Skin," *Eurographics Rendering Workshop 1999*, June 1999.

[18] H. W. Jensen, S. R. Marschner, M. Levoy, and P. Hanrahan, "A Practical Model for Subsurface Light Transport," in *Proceedings of SIGGRAPH 2001*, Computer Graphics Proceedings, Annual Conference Series, pp. 511–518, ACM Press/ACM SIGGRAPH, August 2001. ISBN 1-58113-292-1.

[19] S. R. Marschner and D. P. Greenberg, "Inverse Lighting for Photography," in *Proceedings of the IS&T/SID Fifth Color Imaging Conference*, November 1997.

[20] I. Sato, Y. Sato, and K. Ikeuchi, "Illumination Distribution from Shadows," in *Proceedings of IEEE Conference on Computer Vision and Pattern Recognition (CVPR'99)*, pp. 306–312, June 1999.

[21] P. Debevec, "Rendering Synthetic Objects into Real Scenes: Bridging Traditional and Image-Based Graphics with Global Illumination and High Dynamic Range Photography," in *Proceedings of SIGGRAPH 98*, Computer Graphics Proceedings, Annual Conference Series, pp. 189–198, July 1998.

[22] P. E. Debevec and J. Malik, "Recovering High Dynamic Range Radiance Maps from Photographs," in *Proceedings of SIGGRAPH 97*, Computer Graphics Proceedings, Annual Conference Series, pp. 369–378, Aug. 1997.

[23] Y. Sato, M. D. Wheeler, and K. Ikeuchi, "Object Shape and Reflectance Modeling from Observation," in *SIGGRAPH 97*, pp. 379–387, 1997.

[24] K. Ikeuchi and B. Horn, "An Application of the Photometric Stereo Method," in *6th International Joint Conference on Artificial Intelligence* (Tokyo, Japan), pp. 413–415, August 1979.

[25] S. K. Nayar, K. Ikeuchi, and T. Kanade, "Determining Shape and Reflectance of Hybrid Surfaces by Photometric Sampling," *IEEE Transactions on Robotics and Automation*, vol. 6, pp. 418–431, August 1994.

[26] C. Rocchini, P. Cignoni, C. Montani, and R. Scopigno, "Acquiring, Stitching and Blending Diffuse Appearance Attributes on 3D Models," *The Visual Computer*, vol. 18, no. 3, pp. 186–204, 2002.

[27] P. Debevec, T. Hawkins, C. Tchou, H.-P. Duiker, W. Sarokin, and M. Sagar, "Acquiring the Reflectance Field of a Human Face," *Proceedings of SIGGRAPH 2000*, pp. 145–156, July 2000.

[28] C. Loscos, M.-C. Frasson, G. Drettakis, B. Walter, X. Granier, and P. Poulin, "Interactive Virtual Relighting and Remodeling of Real Scenes," in *Eurographics Rendering Workshop 1999*, June 1999.

[29] S. Boivin and A. Gagalowicz, "Inverse Rendering from a Single Image," in *Proceedings of IS&T Color in Graphics, Imaging, and Vision*, 2002.

[30] J.-Y. Bouguet, "Camera Calibration Toolbox for Matlab," 2002. http://www.vision.caltech.edu/bouguetj/calib_doc/.

[31] Z. Zhang, "A Flexible New Technique for Camera Calibration," *IEEE Transactions on Pattern Analysis and Machine Intelligence*, vol. 22, no. 11, pp. 1330–1334, 2000.

[32] W. Matusik, H. Pfister, M. Brand, and L. McMillan, "A Data-Driven Reflectance Model," *ACM Transactions on Graphics*, vol. 22, pp. 759–769, July 2003.

[33] V. Masselus, P. Dutré, and F. Anrys, "The Free-Form Light Stage," in *Rendering Techniques 2002: 13th Eurographics Workshop on Rendering*, pp. 247–256, June 2002.

[34] S. Marschner, B. Guenter, and S. Raghupathy, "Modeling and Rendering for Realistic Facial Animation," in *Rendering Techniques 2000: 11th Eurographics Workshop on Rendering*, pp. 231–242, June 2000.

[35] G. S. Miller and C. R. Hoffman, "Illumination and Reflection Maps: Simulated Objects in Simulated and Real Environments," in *SIGGRAPH 84 Course Notes for Advanced Computer Graphics Animation*, July 1984.

[36] J. Hancock, D. Langer, M. Hebert, R. Sullivan, D. Ingimarson, E. Hoffman, M. Mettenleiter, and C. Froehlich, "Active Laser Radar for High-Performance Measurements," in *Proceedings IEEE International Conference on Robotics and Automation* (Leuven, Belgium), May 1998.

[37] M. Callieri, P. Cignoni, F. Ganovelli, C. Montani, P. Pingi, and R. Scopigno, "VCLAB's Tools for 3D Range Data Processing," in *VAST 2003 and Eurographics Symposium on Graphics and Cultural Heritage*, 2003.

[38] Graphite, 2003. http://www.loria.fr/ levy/Graphite/index.html.

[39] B. Lévy, S. Petitjean, N. Ray, and J. Maillot, "Least Squares Conformal Maps for Automatic Texture Atlas Generation," *ACM Transactions on Graphics*, vol. 21, pp. 362–371, July 2002.

[40] QSlim, 1999. http://graphics.cs.uiuc.edu/ garland/software/qslim.html.

[41] M. Garland and P. S. Heckbert, "Simplifying Surfaces with Color and Texture Using Quadric Error Metrics," in *IEEE Visualization '98*, pp. 263–270, Oct. 1998.

[42] H. P. A. Lensch, W. Heidrich, and H.-P. Seidel, "A Silhouette-Based Algorithm for Texture Registration and Stitching," *Graphical Models*, vol. 63, pp. 245–262, Apr. 2001.

[43] C. Buehler, M. Bosse, L. McMillan, S. J. Gortler, and M. F. Cohen, "Unstructured Lumigraph Rendering," in *Proceedings of ACM SIGGRAPH 2001*, Computer Graphics Proceedings, Annual Conference Series, pp. 425–432, Aug. 2001.

[44] J. Kajiya, "The Rendering Equation," in *SIGGRAPH 86*, pp. 143–150, 1986.

[45] S. Nayar, X. Fang, and T. Boult, "Separation of Reflection Components Using Color and Polarization," *International Journal of Computer Vision*, vol. 21, pp. 163–186, February 1997.

7

Applications of Spectral Imaging and Reproduction to Cultural Heritage

Simone Bianco, Alessandro Colombo, Francesca Gasparini, Raimondo Schettini

University of Milano Bicocca
Email: simone.bianco@disco.unimib.it, colomboal@disco.unimib.it,
gasparini@disco.unimib.it, schettini@disco.unimib.it

Silvia Zuffi

Isituto per le Tecnologie della Costruzione - Consiglio Nazionale delle Ricerche
Email: zuffi@cs.brown.edu

CONTENTS

7.1 Introduction

Color images code color information according to three channels, corresponding to the red, green, and blue components of the image acquisition device. In recent years, the field of digital imaging has extended the traditional trichromatic RGB paradigm to more than three dimensions, introducing what is called spectral or multispectral imaging. The aim of multispectral imaging is to acquire, to process, to reproduce, and to store images with a higher color quality; indeed it is oriented towards those application sectors that request high-quality treatment of color information [1, 2]. Many experiences exist on the successful application of spectral imaging to cultural heritage, this being a field where the acquisition and reproduction of accurate color information are two fundamental processes.

Traditionally, artifacts are captured with three-channel devices, and the resulting RGB images are processed within the framework of colorimetry in order to accomplish faithful reproduction across different devices or media. Indeed, the colorimetric approach satisfies - with its inherent limitations - cross-device color communication, but it is far from providing a consistent color reproduction for different viewing conditions. The attempts made at exploiting multispectral imaging for the acquisition of cultural heritage artifacts have revealed the several advantages of this approach [3]–[14]. Different hardware configurations have been used in the state of the art. In [3] a cooled monochrome digital camera with a liquid-crystal tunable filter is used. In [4] and [5], instead, a monochrome digital camera and a filter wheel with seven broadband gaussian filters is considered. Cupitt et al. [6] adopted a combination of micro- and macro-scanning, using a CCD area sensor; the sensor is equipped with a color mosaic mask with filter characteristics closely matched to a linear combination of the CIE–1931 XYZ spectral response. Ribès et al. [7] settled out a linear CCD array camera equipped with a built-in half-barrel mechanism that automatically positions a set of thirteen interference filters, ten filters covering the visible spectrum and the other three covering the near infrared. In [8] a monochrome CCD camera is used together with a multispectral lighting system composed of a slides projector with six color filters. In [10] a combination of a high-resolution photographic image and a low resolution multispectral image is used: the multispectral image is captured using a trichromatic digital camera system with two color filters, for a total of nine color channels. In [12] a monochromatic CCD camera with three to six color filters is used, while in [14] a cooled CCD digital camera with a fast-tunable liquid-crystal filter is adopted.

Multispectral imaging not only extends the traditional trichromatic imaging to a higher dimension, but it aims at providing a description of the reflective or transmissive properties of the surface. A more precise color analysis makes multispectral images suitable for monitoring and restoration of artworks, and for any research activity that requires high-quality color information. These images can be also rendered for specific viewing conditions, device, and reproduction media, in order to be disseminated through different communication channels. A generic framework for the application of multispectral imaging in cultural heritage is illustrated in Figure 7.1. Usually multispectral images are reproduced on colorimetric devices, but several works investigate how to exploit the multispectral information in the reproduction phase, in order to achieve faithful reproduction across different viewing conditions [15, 16].

7.2 Colorimetric and Multispectral Color Imaging

Multispectral imaging can advocate many advantages over colorimetric imaging [4, 17]. First of all, it is quite straightforward to produce a correspondent colorimetric version of a multispectral image once viewing conditions are assigned, but it is theoretically impossible to reverse this process. In RGB images a pixel is a triplet of integers that code the amount of R, G, and B digital counts of an RGB device. During the acquisition process, these data are a measure of radiance integrated over the wavelength domain of red, green, and blue device's sensors. The original spectral information is thus critically under-sampled with conventional color acquisition devices and the reproduced colors suffer from metamerism. In a multispectral image a pixel is a vector of real numbers that represents a physical property defined on a per-wavelengths basis. In the case of reflective media, like a painting or photograph, each pixel stores the reflectance spectrum at the point on the artwork surface. Typically thirty-one samples are considered, corresponding to a sampling of the visible spectrum from 400 to 700 nm with steps of 10 nm, as recommended by the Commission Internationale de l'Eclairage (CIE) [18]. The number of samples is not standardized, although several works have investigated the minimum number of sensors to use [19, 19–24, 24]. In [20] the use of five to seven filters is suggested; in [22] the behavior of a multispectral system having two to six filters is studied. In [23, 24] the use

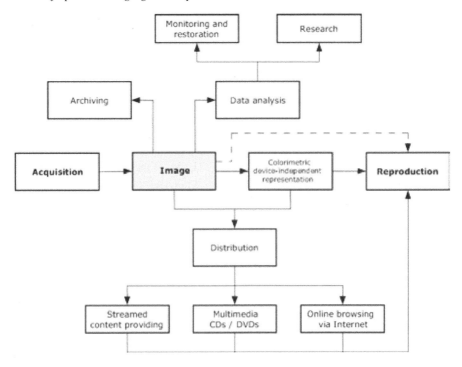

FIGURE 7.1
A general framework for the use of multispectral imaging in the cultural heritage field.

of eight and nine filters is respectively suggested. The multispectral image constitutes a fundamental physical description of the artifact, independent from the environment and observer, which can be targeted to any desired description specific for a given observer and viewing conditions. Being device independent, a multispectral image is invariant across different acquisition devices, allowing comparison of artifacts whose images are taken from different devices. Note however that, being in general the reflective properties of a surface dependent from the geometry of the illumination, it is assumed that the illumination and acquisition geometry are controlled, as it would happen if the artifact surface was measured by a spectrophotometer. Color information captured with RGB devices cannot generate a fully accurate colorimetric representation due to the fact that the sensitivities of the sensor employed do not correspond to those of the standard colorimetric observer [25]. If a multispectral image is available, precise colorimetric coordinates can be computed for each pixel in the image. In addition, multispectral images can also show details in the artifacts that are hard to see, if not impossible to detect, in RGB images.

7.3   Capturing a Multispectral Image

Two different approaches exist for multispectral imaging, called respectively narrow-band and wide-band image capture [26]. They differentiate in the way they sample the wavelengths of the visible spectrum. In the narrow-band approach the acquisition of radiance information is obtained by a set of narrow-band filters, centered in principle one for each wavelength sample. Various technologies are available to produce spectrally narrow filters. One possibility is to realize a filter wheel with narrow bandpass glass filters in front of a camera. This system requires usually costly custom-made filters.

Moreover, a filter wheel is an electro-mechanical tool with several inherent drawbacks: slow band switching, small number of filters, sequential access to color bands, cumbersome design, and limited versatility [27].

More convenient is to realize narrow-band systems using a tunable filter. Spectral transmission of this device can be electronically controlled through the application of voltage or acoustic signal. Tunable filters can provide finer spectral sampling, and rapid and random switching between color bands because there are no moving parts and no discontinuity in the spectral transmission range [27]. In particular a solid state Liquid Crystal Tunable Filter (LCTF) has been widely used [15,28–30]. The peak wavelengths of the LCTF can be controlled, permitting a fine spectral sampling and producing usually thirty-one peaks in the range from 400 to 720 nm [31]. One of the most important advantages of this system is its robustness to arbitrary spectral shapes. In fact a sampling rate of 10 nm (in the ideal case of infinitesimally wide band filters) permits one to reconstruct spectral features which are at least as wide as twice the sampling rate [26]. The LCTF has the advantage of being solid state and reliably repeatable, and can be easily controlled by a computer for an efficient, automated, and relatively fast imaging. On the other hand, a large storage space is required for each acquired target, and registration of the thirty-one images is a serious issue. Moreover this system has severe drawbacks in terms of size, costs, and unwieldiness. The so called wide-band approach has been developed to realize easier multispectral systems. In these systems, the visible spectrum is sampled at a wide step and each adopted sensor is sensitive to light energy in a sufficiently large wavelength interval. Several works have demonstrated that five to eight basis vectors seem to be sufficient for accurate spectral reconstruction [20, 32–36].

Thus it is possible to significantly reduce the number of filters (from the thirty-one adopted in the narrow-band approach), still recovering accurate target reflectances. Wide-band systems have the advantage that they can be assembled from "off the shelf" hardware components typical of scientific research and professional photography. With respect to narrow band acquisition systems, they are much more easily deployed, manageable, flexible in their use, and comparatively cheaper. However, such systems do not perform a direct measure of reflectance, but rather produce data that must be further processed to achieve the true multispectral image [37].

A multispectral acquisition system is composed by a multispectral camera, a processing module to derive reflectance from the acquired radiance images, and a transformation module for the conversion into a colorimetric space, suitable for the colorimetric reproduction on common output devices (Figure 7.2). No true multispectral reproduction devices exist other than prototypes, while multispectral characterization of colorimetric devices is still in its early stages [37]. Several multispectral acquisition systems have been developed and tested. These systems differentiate themselves fundamentally for the number of sensors employed. Usually multispectral cameras rely on a standard B/W digital camera and a set of colored filters. A typical wide-band system uses optical filters to simulate sensors of different sensitivity. Either traditional filters like those used in standard photography or a tunable filter can be adopted. Burns and Berns [34], used a monochrome digital camera installed with seven interference filters while Imai et al. [38] combined a monochrome camera with a filter wheel containing six absorption filters. Methods have also been proposed that use commonly available optical filters and trichromatic digital still cameras [39]. Imai et al. [40] adopted a conventional trichromatic digital camera combined with a series of absorption filters. Seven filter combinations were placed in front of the digital RGB color CCD camera. One was no filter at all, while the remaining six were combinations of Wratten filters. A system based on a commercial color-filter array (CFA) digital camera coupled with a two-position filter slider containing absorption filters has been adopted to facilitate multispectral imaging for imaging cultural heritage [13,41–43]. For each target, two RGB images are taken through each filter, so there are in total six channels for this camera system. Figure 7.3 illustrates typical processes involved in the acquisition of a multispectral image using a wide-band system [37]. First, as the acquired images are usually affected by some form of hardware noise, such noise must be estimated and modeled, so that a noise correction procedure can be established and applied. If the illumination is not evenly distributed across the scene, its

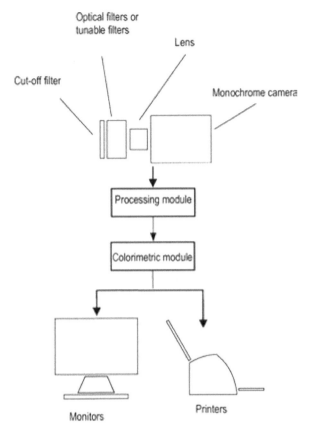

FIGURE 7.2
A system for the acquisition and reproduction of multispectral images.

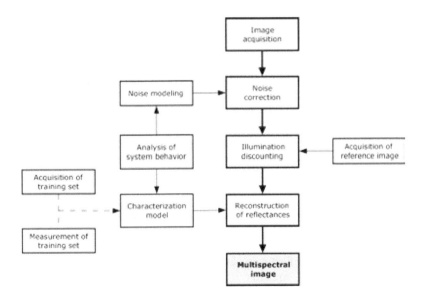

FIGURE 7.3
The acquisition process for a typical wide-band multispectral acquisition system.

uneven effect must then be discounted; this can be done by acquiring a reference image to estimate the effect and correct it. After this pre-processing is done, the acquired images can be fed to the characterization model, which reconstructs the true reflectances of the points in the scene. This model can be built based on an analysis of the system behavior, or empirically derived from the acquisition and measurement of the reflectances of a suitable training set. The need of acquiring large or high-resolution images may force the operator to acquire different parts ("tessels") of the scene separately, and then resort to mosaicking to obtain the whole image [44].

7.4   Imaging and Signal Processing Techniques

In general, the acquisition performed using a given i-th sensor at a single 2-D point $\mathbf{x}$ will return a value $a_i(\mathbf{x})$ in the form [45]:

$$a_i(\mathbf{x}) = \int_{\lambda_1}^{\lambda_2} E(\lambda, \mathbf{x}) R(\lambda, \mathbf{x}) S_i(\lambda) \, d\lambda. \tag{7.1}$$

This value integrates contributions from the energy E that reaches the physical sample observed, the spectral reflectance R of the sample, and the "sensitivity" S_i of the i-th sensor. The integration with respect to the wavelength λ is performed in the range λ_1 to λ_2 of the sensor's sensitivity. If this range exceeds that of the visible light spectrum, then appropriate steps must be taken to cut unwanted radiation off. As introduced before, two different approaches are currently used in multispectral imaging to obtain the reflectance estimate: narrow-band and wide-band.

7.4.1   "Narrow-Band" Multispectral Imaging

In the case of narrow-band systems, the device's sensors are sensitive to a very narrow wavelength interval or the light sources employed show a very narrow emission spectrum. In both cases, assuming that the selective property can be modeled as a delta function, the value $a_i(\mathbf{x})$ obtained from an acquisition at a single point $\mathbf{x}$ can be interpreted as the value of function $E(\lambda, \mathbf{x})R(\lambda, \mathbf{x})S(\lambda)$ at the specific wavelength λ_i, so that, by changing sensors or light sources, different values of this function can be estimated on the whole visible light spectrum. For a given wavelength λ_i, Equation 7.1 then becomes:

$$a_i(\mathbf{x}) = E(\lambda_i, \mathbf{x}) R(\lambda_i, \mathbf{x}) S_i(\lambda_i), \tag{7.2}$$

and if the properties of the illuminant and sensor(s) are known or can be measured, then the values $E(\lambda_i)$ and $S(\lambda_i, \mathbf{x})$ are known and $R(\lambda_i, \mathbf{x})$ can be computed. As an alternative, the output values $a_i(\mathbf{x})$ can be compared with the corresponding values previously obtained from the acquisition of a reference physical sample whose reflectance is known [44]. If the result of this previous acquisition is indicated with $a_i'(\mathbf{x})$, then it is:

$$\frac{a_i(\mathbf{x})}{a_i'(\mathbf{x})} = \frac{E(\lambda_i, \mathbf{x}) R(\lambda_i, \mathbf{x}) S_i(\lambda_i)}{E(\lambda_i, \mathbf{x}) R'(\lambda_i, \mathbf{x}) S_i(\lambda_i)} = \frac{R(\lambda_i, \mathbf{x})}{R'(\lambda_i, \mathbf{x})}, \tag{7.3}$$

where $R'(\lambda_i, \mathbf{x})$ is the (known) reflectance of the reference sample. The value of $R(\lambda_i, \mathbf{x})$ can then be computed using the following equation:

$$R(\lambda_i, \mathbf{x}) = \frac{a_i(\mathbf{x})}{a_i'(\mathbf{x})} R'(\lambda_i, \mathbf{x}). \tag{7.4}$$

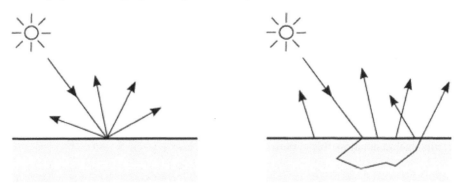

FIGURE 7.4
Model of a pure reflective material (left) and of a translucent material in which subsurface light scattering occurs (right).

If necessary, the model can be further improved to cope with such phenomena as subsurface light scattering and photoluminescence effects such as fluorescence and phosphorescence. Subsurface light scattering is a mechanism of light transport in which light penetrates the surface of a translucent object, is scattered by interacting with the material, and exits the surface at a different point. The light will generally penetrate the surface and be reflected a number of times at irregular angles inside the material, before passing back out of the material at an angle other than the angle it would have if it had been reflected directly off the surface (see Figure 7.4).

7.4.2 "Wide-Band" Multispectral Imaging

The second approach to multispectral acquisition is based on wide-band sensors. In this case, each sensor is sensitive to light energy in a sufficiently large wavelength interval, and the emission of the light source considered has a sufficiently broad spectrum, so that the values $a_i(\mathbf{x})$ obtained from the acquisition cannot be associated to specific wavelengths and do not permit a direct measure of reflectance [44]. Wide-band approaches require a correlation method learned from a suitable training set to relate the output from the multispectral camera at some pixel with the reflectance spectrum of the corresponding surface point in the scene. The output of a generic multispectral camera may be denoted as:

$$\mathbf{a}(\mathbf{x}) = [a_i(\mathbf{x})], \qquad (7.5)$$

where i is an index that varies with the filter used (or the spectral band examined), and $\mathbf{x}$ is a two-dimensional vector identifying the point considered within the acquired scene. If M filters are used, then $\mathbf{a}(\mathbf{x})$ is an M-dimensional vector. The reflectance of the object at point $\mathbf{x}$ is a function of the wavelength λ, and can be denoted as $\mathbf{r}(\lambda, \mathbf{x})$; however, since in practice it is not easy (or even always possible) to give an analytical form to $\mathbf{r}$, a sampling of its value is customarily considered instead. The light spectrum is then sampled at a discrete number of values of λ, and the reflectance is expressed as:

$$\mathbf{r}(\lambda, \mathbf{x}) = [R(\lambda_j, \mathbf{x})], \qquad (7.6)$$

where j is an index that varies with the sample wavelengths. If N sample values of λ are considered, then $\mathbf{r}(\lambda, \mathbf{x})$ is an N-dimensional vector. To establish a correlation between the system output and the corresponding reflectance, the system characterization function:

$$\mathbf{a}(\mathbf{x}) \mapsto \mathbf{r}(\lambda, \mathbf{x}), \qquad (7.7)$$

must be described or estimated in some way. If the value of $R(\lambda_j, \mathbf{x})$ at N different wavelengths values λ_j is wanted, then the discrete form of Equation 7.1 will be written for the i-th sensor (filter) as:

$$a_i(\mathbf{x}) = \sum_{j=1}^{N} E(\lambda_j, \mathbf{x}) R(\lambda_j, \mathbf{x}) S_i(\lambda_j) \Delta\lambda_j, \tag{7.8}$$

M such equations can be written to form a linear system, where M is the number of sensors (filters) used. In algebra notation, this system can be written as:

$$\mathbf{a} = \mathbf{Dr}, \tag{7.9}$$

with:

$$\begin{aligned}
\mathbf{a} &= [a_i(\mathbf{x})], \\
\mathbf{D} &= [d_{ij}(\lambda, \mathbf{x})] = E(\lambda_j, \mathbf{x}) S_i(\lambda_j), \\
\mathbf{r} &= [R(\lambda_j, \mathbf{x})],
\end{aligned} \tag{7.10}$$

and if matrix $\mathbf{D}$ were known, then Equation 7.9 could be solved with respect to $\mathbf{r}$ by means of some system inversion technique.

Methods that perform this inversion belong to what is called direct reconstruction approach [46]. The simplest but the most inaccurate method simply inverts Equation 7.9 by using a pseudoinverse approach or ordinary least squares regression. This method, adopted by Tominaga [47] to recover the spectral distribution of illumination from a six-channel imaging system, is not well applied in practice because this solution is very sensitive to noise [48]. Herzog et al. [49] have proposed a weighted least squares regression based on a weighting matrix to invert the system characteristics under a smoothness constraint. Hardeberg [48] has proposed a method based on a priori knowledge of a spectral reflectance database, but does not consider camera noise.

However, direct reconstruction is not widely used as it requires spectral characterization of the whole image system. Estimation of the illuminant E and of the sensitivity S_i is not straightforward, and a complex illumination geometry (such as multiple and possibly different light sources used together from different angles) would require costly computations as well. The analysis in the frequency domain of reflectance spectra and color signals motivates the recourse to dimensionality reduction techniques involving, in the most common approach, empirical linear models. In general, when representative data are available, linear models are defined on the basis of statistical information, applying Principal Component Analysis (PCA) [50] and Independent Component Analysis (ICA) [51] to estimate the relationship between the acquisition output $\mathbf{a}$ and the sampled reflectance function $\mathbf{r}$. If P samples are available, and their corresponding $\mathbf{a}_k$ and $\mathbf{r}_k$ vectors (with k ranging from 1 to P) are considered, then it is:

$$\mathbf{a}_k = \mathbf{Dr}_k, \tag{7.11}$$

and therefore

$$\mathbf{A}_S = \mathbf{DR}_S, \tag{7.12}$$

with:

$$\mathbf{R}_S = [\mathbf{r}_1|...|\mathbf{r}_P] \text{ and } \mathbf{A}_S = [\mathbf{a}_1|...|\mathbf{a}_P]. \tag{7.13}$$

The (pseudo-)inverse D^- of matrix D can then be computed by inverting Equation 7.12 with some chosen technique, and the reflectance $\mathbf{r}$ for a generic acquisition output $\mathbf{a}$ can thus be computed using the relationship:

$$\mathbf{r} = D^- \mathbf{a}. \tag{7.14}$$

Learning-based reconstruction is the most popular approach for spectral reflectance recovering. Several methods have been implemented to solve Equation 7.14 based on learning processes (see for instance [34] and [52]– [54]). Zhao and Berns [42] have developed a method named "Matrix R Method" to reconstruct spectral reflectance accurately while simultaneously achieving high colorimetric performance for a given illuminant and observer. All these methods do not require the knowledge of the spectral characteristics of the imaging system. However, as they are learning-based techniques their performance is greatly affected by the choice of a calibration target.

Some studies that outline the theoretical bases for the choice of the "training set," which is the set of the colors used to build the empirical model, were recently published [55] – [58]. This set must be "sufficiently representative" of the whole range of possible colors, which intuitively means that the resulting model can actually be extended to any other color. This is not a clear-cut notion, but specific targets that include a varied selection of sample colors, such as the Macbeth ColorChecker and ColorChecker DC [59] are available.

Moreover, as the system data are obtained using a finite number of different optical filters that sample the wavelength of the visible spectra, an important issue relies on the filter selection in terms of shape and number. Assuming that any system noise in the acquired data has been properly corrected, it is reasonable to expect that the best quality will be obtained using all the available filters. It is also reasonable that choosing to employ different subsets of the available filters will give different results, and not all possible subsets will lead to acceptable results when reflectances are reconstructed from the acquired data. However, minimizing the number of filters used is important to reduce operational costs and acquisition time, as well as the amount of data needed to store the acquired spectral images. Furthermore, it is generally not guaranteed that all noise can be corrected, so a greater number of filters may yield a greater amount of noise and actually lead to biased results.

7.4.3 Training Set Selection

Despite the possibility of estimating reflectance on the basis of smoothing constraints, multispectral imaging systems often rely on training sets. The quality of the reflectance estimation depends on the correlation method and training set selected: it improves if a "good" training set is available while a "bad" training set may negatively affect the resulting estimation and a certain correlation method may be rejected because it appears to yield poor results, when these are really due to the training set. There are also cases in which the context dictates the use of a specific correlation method: in this event the quality of the approximation is influenced by the training set alone. In real acquisitions large training sets are unwieldy and changing training sets often for different applications is inefficient, so the importance of having a few small training sets of broad applicability is more likely to be stressed.

Due to all this, the literature to date numbers several methods to address the problem of choosing a "good" training set. Hardeberg et al. [55] have employed an algebraic method to select a training set of low numerosity. They start from a small number of reflectance samples chosen accordingly to their spectral variance. Additional samples are added in order to maximize the volume defined by the vector space spanned by these spectra. The method selects colors based on their reflectances, known usually through measurements. The color whose reflectance vector has maximum norm among all the target colors (the most reflective color in the target) is chosen as the first color of the training set. At step k, the color r not already chosen that maximizes the ratio of the smallest to the largest singular value of the matrix $[r_1 | \ldots | r_{k-1} | r]$, whereby $r_1, \ldots, r_{k-1}$ indicate those colors already selected, becomes the k-th color in the training set.

Pellegri et al. in [56] presented two different approaches to selecting a "good" training set from an initial array of available colors: the *Hue Analysis Method*, based on colorimetric considerations, and the *Camera Output Analysis Method*, mainly based on algebraic and geometrical considerations.

The training sets selected with both the two approaches have low numerosity and broad applicability, and are particularly suitable to be employed in real acquisitions. With the Hue Analysis Method the selection of the training set is based on hue. Assuming that the reflectances of the colors in the target are known, the corresponding LHC coordinates are computed, and the colors are projected onto an HC plane. This plane is then divided into n sectors of equal angular width, and the n colors are chosen as widely spaced as possible covering the whole plane. In the *Camera Output Analysis Method* they developed three strategies to choose colors of the training set well apart from one another. In the first version of the method, principal component analysis is applied to all available colors, so that the principal eigenvectors are identified. Then, for each eigenvector in order of relevance, the color not already chosen "nearest" to that eigenvector is included in the training set, for a total of M vectors. This strategy tries to maximize the orthogonality of the colors chosen for the training set, so that the impact of measurement errors on the geometry of the space is kept to a minimum. The second version of this method maximizes the relative linear distance of training set colors while still taking into account their anisotropic spatial disposition (as described by the eigenvectors). For each eigenvector the color not already chosen with the greatest absolute coordinate on that eigenvector is included in the training set, for a total of M vectors. In the third version of the Camera Output Analysis Method, for each eigenvector, both the color with the greatest coordinate and the color with the smallest coordinate are selected and included in the training set, for a total of $2M$ colors.

A further method mainly based on algebraic and geometrical considerations, called Linear Distance Maximization Method (LDMM), was introduced by Pellegri et al. in [57]. The LDMM selects colors based on their corresponding system output vectors as obtained from an acquisition performed in the chosen operational conditions. The first color of the training set is the color whose associated system output vector has maximum norm among all the target colors (the brightest color in the target for the chosen acquisition conditions). The second selected color is the remaining color that has maximum distance from the first color. At subsequent steps, for each remaining color, the minimum distance from it to the colors already selected is computed, and the remaining color for which such distance is maximum is added to the training set. The distance used is the infinity norm of the difference of the system output vectors corresponding to the colors considered. Mohammadi et al. [58] selected a set of calibration targets among a series of samples using agglomerative hierarchical cluster analysis. Reflectance vectors representative of each different cluster were selected and stored as new calibration targets. They have shown that beyond a threshold number of samples in the calibration target, the performance of reconstruction became independent of the number of samples used in the calculation. Compared to a principal components method, cluster analysis generates directly a selected set in spectral space that is more intuitive than projection of spectral space into a space of eigenvectors.

7.4.4 Filters Selection

One of the most important components in a multispectral acquisition system is the set of optical filters that allows acquisition in different bands of the visible light spectrum. Typically, either traditional optical filters (sometimes mounted on a rotating wheel-frame) or tunable filters are employed, and in both cases there is a certain degree of freedom in the choice of which filters (or, in the case of tunable filters, filter configurations) should be used. The number of basis functions (optical filters) necessary to accurately represent the reflectance spectra depends on the characteristics of the data that one is modeling, and on the characteristics of the functions used in the linear model. There are many studies on the dimensionality of such linear models based on PCA, which is used to seek the set of basis functions that minimizes the correlation among dimensions and identifies those dimensions that are most descriptive of the data set [60]. In general, for natural reflectance spectra, a number of basis functions between six and nine is considered adequate, as emerged from studies from Cohen [61] on a subset of the Munsell surface spectral reflectances collected by Kelley et al. [62], from Maloney [20] on spectral reflectances of natural formations collected by Kirnov [63], and from Jaaskelainen [35] on samples derived from several different kinds of plants.

Performing acquisitions in a fixed environment is a common practice in the field of multispectral imaging. This includes using the same illuminant source and illumination geometry. In this case, information retrieved from the acquisition of a representative target may be extended to other acquisitions performed under the same conditions. In particular, by studying the acquisition of such a target one can determine which filters among those available yield the most significant information in the chosen environmental setup, and use only those filters for all subsequent acquisitions, at least as long as the error observed in the reconstruction of the target colors reflectances is acceptable. By selecting subsets of filters of decreasing numerosity, one can also establish the minimum number of filters required to obtain sufficiently small errors.

Novati et al. [64] proposed the Filter Vectors Analysis Method (FVAM) for choosing subsets of filters among those available to use with a multispectral acquisition system. The FVAM chooses an optimal subset of filters based on a statistical analysis of the acquisition of a representative target performed using all the available filters and tries to identify those filters that yield the most information in the given environmental conditions. They compared this method with the Evenly Spaced Filters method (ESF) that chooses filters so that their transmittance peak wavelengths are as evenly spaced as possible within the considered spectrum. The results of their experiments suggest that the FVAM can not bring substantial improvements over an evenly spaced selection strategy. Usually, a training set is fixed when evaluating filter selection strategies and on the other hand the filter set is considered chosen when evaluating training set selection strategies. With the assumption that a suitable characterization method has been chosen in advance Pellegri et al. [65] investigated the joint effect of training set and filter selection on the accuracy of the spectra reconstruction. They assumed that candidate colors are searched within a large dataset and the filters could be selected among a large set of available filters (or tunable filter configurations) that span the visible light spectrum. In particular, they investigated how the quality of reflectance estimation varies when filter sets of decreasing numerosities are chosen, and to which extent this variation is affected by choosing training sets that differ in their composition and in the number of colors included. Although it could be expected that a larger set of properly chosen filters would lead to better results, minimizing the number of filters used is important to reduce operational costs and acquisition time, as well as the amount of data needed to store the acquired spectral images. The numerosities of both filter set and training set also have an impact on the processing time required to obtain reflectance estimates from system output data.

They considered two methods for the choice of the filters and two methods for the choice of the training set, testing all the possible combinations and selecting filter sets and training sets of decreasing numerosity in each case. For training set selection, they considered a method by Hardeberg et al. [55] and their Linear Distance Maximization Method (LDMM) [57]. Both methods adopt an iterative approach in which the training set colors are chosen one by one; the first color is chosen for its characteristics, while each subsequent color is chosen among the remaining target colors because it "best suits" (in some sense) the set of the colors already chosen. For filter selection, they considered the Evenly Spaced Filters (ESF) method and the Filter Vectors Analysis Method (FVAM) [64]. Both methods identify each available filter by the wavelength at which it shows its transmittance peak. The results of these experiments seem to indicate that increasing the number of filters used does not necessarily lead to improved estimates for reflectance, while decreasing the number of sample colors in the training set generally results in worse estimates. However, the estimation errors observed do not show completely consistent trends; in particular, no clear and decisive indication could be obtained in order to conclude that a specific method or combination of methods performs better than the others.

Connah et al. [24] have investigated the minimum number of sensors to use, while also minimizing reconstruction error. They derived different numbers of optimized sensors, constructed by transforming the characteristic vectors of the data, and simulate reflectance recovery with these sensors in the presence of noise. They find an upper limit of nine optimized sensors above which the noise prevents decreases in error. They also have demonstrated that this level is both noise and dataset dependent, by providing results for different magnitudes of noise and different reflectance datasets. Chatzis et al. [66] proposed a filter selection method based on the Principal Feature Analysis (PFA) [67]. The filter selection problem is defined as a dimensionality reduction process of the complete set of filters and is obtained by selecting a subset that contains most of the essential information. The filters are selected to maximize their degree of orthogonality after projection into the vector space spanned by the most significant eigenvectors.

7.5 Recovery Multispectral Information from RGB Images

According to the definition in [68], multispectral imaging refers to imaging techniques that use more than three filters in order to estimate spectral color information. It is clear from the previous discussion about wide-band systems that a number of sensors equal to the number of spectral samples needed to describe the reflectance spectrum is not necessary, and that many techniques exist for reflectance estimation from sensors' data. Given the ease of use of trichromatic systems, one could however pose the question if an RGB camera could be used to estimate spectral information. In general, the methods designed to be used with wide-band systems could be applied to three-sensor devices, in particular those methods based on matrix inversions, of interest for their computational efficiency. Recovering multispectral information from three-sensor devices is an interesting issue, as practical systems that produce low-cost spectral images can be obtained using two acquisitions of the same scene by a traditional RGB camera, [41]. Berns et al. [13] have developed a practical spectral imaging system that uses a traditional Color-Filter-Array (CFA) camera coupled with two optimized filters. For each target, one RGB image is taken through each filter, so there are six channels for this camera system. This camera system was used to retrieve spectral reflectance factor of each pixel of a famous painting, using the Matrix R reconstruction method [43]. Also Bianco et al. [69] have investigated how low-cost spectral images can be obtained using two acquisitions of the same scene: the former acquired by a traditional RGB imaging device, the latter coupling the same camera with a suitable chosen absorption filter. The combination of these two acquisitions can be considered as acquired by a 6-band imaging device. The details of the experiments carried out are reported in the next section as a practical example of recovering multispectral information from three-sensor devices.

7.5.1 Spectral Based Color Imaging Using RGB Digital Still Cameras

In [69], the performance of three-channel digital still cameras and of modified (3×2-channels) digital still cameras are evaluated in terms of their capability to estimate spectral reflectance information. The experiments showed that very good results can be achieved by simply using the RGB camera as a spectral based imaging device. These results can be further improved combining two different shots of the same scene acquired using the RGB camera with and without a properly chosen absorption filter.

RGB Imaging

The RAW digital data $\mathbf{d}(\mathbf{x}) = [d_i(\mathbf{x})]$, $i = 1, 2, 3$ (corresponding to the RGB device dependent values) of a typical digital still camera can be mathematically modeled for each pixel $\mathbf{x}$ as follows:

$$d_i(\mathbf{x}) = n_i(\mathbf{x}) + t \sum_{j=1}^{N} E(\lambda_j, \mathbf{x}) R(\lambda_j, \mathbf{x}) S_i(\lambda_j) \Delta \lambda_j, \tag{7.15}$$

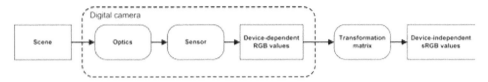

FIGURE 7.5
A typical digital still camera pipeline.

where $\mathbf{r} = [R(\lambda_j, \mathbf{x})]$ is the scene reflectance at the given pixel position $\mathbf{x}$, $\mathbf{e} = [E(\lambda_j, \mathbf{x})]$ is the spectral power distribution of the incident light, $\mathbf{s}_i = [S_i(\lambda_j)]$, $i = 1, 2, 3$ are the spectral sensitivities of the camera filters, t is the exposure time, and $n_i(\mathbf{x})$ is an additive noise term. Equation 7.15 computes the device dependent RGB values for a given pixel; to obtain a device independent representation of it, we have to determine the best 3-by-3 matrix transformation $\mathbf{M}$ (called matrixing) that transforms the RAW RGB values $\mathbf{d}$ into the corresponding sRGB device independent values $\hat{\mathbf{d}} = \left[\hat{d}_i(\mathbf{x})\right]$, $i = 1, 2, 3$ (Figure 7.5):

$$\hat{\mathbf{d}} = \mathbf{dM}. \tag{7.16}$$

There are several methods in literature to compute such a matrix $\mathbf{M}$ (Least Squares, White Point Preserving Least Squares, Polynomial Regression, Non Maximum Ignorance [70]– [72]). The simplest is to compute the best least squares 3×3 matrix for a given data set of known colors.

Spectral Imaging Using a Single RGB Camera Acquisition

The RAW values of the digital camera are used to estimate the generalized pseudo inverse matrix $\mathbf{M}_R$, to reconstruct the spectra of the colors of the training set, $\mathbf{r} = [R(\lambda_j, \mathbf{x})]$, where λ ranges between 400 and 700 nm and is sampled with a step of 10 nm:

$$\mathbf{r} = \mathbf{dM}_R. \tag{7.17}$$

The matrix $\mathbf{M}_R$ is no more a 3×3, but a 3×31. As for the previous imaging method, there are several ways to calculate the matrix $\mathbf{M}_R$, such as Least Squares with Toeplitz matrix, Smoothing Inverse and Linear Models [7,8]. The Generalized Pseudo Inverse is adopted here to estimate $\mathbf{M}_R$:

$$\mathbf{d}^T \mathbf{r} = \mathbf{d}^T \mathbf{dM}_R, \tag{7.18}$$

$$\left(\mathbf{d}^T \mathbf{d}\right)^{-1} \mathbf{d}^T \mathbf{r} = \mathbf{M}_R. \tag{7.19}$$

Spectral Imaging Using Two RGB Camera Acquisitions

Two acquisitions of the same scene are taken: the former is acquired by a traditional RGB camera, the latter coupling the same camera with a suitable chosen absorption filter (Figure 7.6). This time we have three more cues to reconstruct the spectra $\mathbf{r}$: $\mathbf{d}_{AF}(\mathbf{x}) = [d_{AF,i}(\mathbf{x})]$, $i = 1, 2, 3$, mathematically modeled as follows:

$$d_{AF,i}(\mathbf{x}) = n_i(\mathbf{x}) + t \sum_{j=1}^{N} E(\lambda_j, \mathbf{x}) R(\lambda_j, \mathbf{x}) S_i(\lambda_j) F(\lambda_j) \Delta\lambda_j, \tag{7.20}$$

where $\mathbf{f} = [F(\lambda_j)]$ is the spectral transmittance of the absorption filter selected, and the other quantities are the same as Equation 7.15.

Thus Equation 7.17 becomes:

$$\mathbf{r} = [\mathbf{d}|\mathbf{d}_{AF}] \mathbf{M}'_R, \tag{7.21}$$

where now $\mathbf{M}'_R$ is a 6×31 matrix.

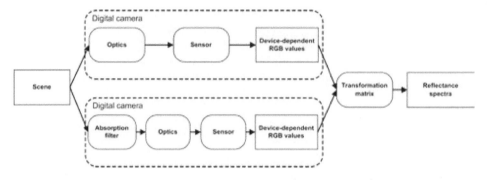

FIGURE 7.6
The six-channel imaging system.

7.6   Storing a Multispectral Image

One major issue in multispectral imaging is given by the large size of multispectral data files. Spectra are customarily sampled at specific wavelength values, and since reflectance is an index that varies in the real range $[0, 1]$, they are basically given as floating point numbers. This, together with the need of sampling at many wavelengths, leads to oversized files, especially when images are concerned [73]. For example, if spectra are sampled from 400 nm to 700 nm at 10 nm intervals, which is a common choice, and samples are recorded as double-precision real numbers, a 21×30 cm image with a resolution of 300 ppi thirty-one bands occupies around 2 GB of storage space, while a 50×80 cm image at 96 ppi requires around 1.4 GB. Even a simple screen-wide image of $1,024 \times 768$ pixels would occupy around 190 MB. Sizes remain large even if samples are recorded as 16-bit integers, which is a common practice. In such a scenario, the transmission of multispectral data over networks (and the Internet in particular), as well as the storage of large archives, are strongly limited, and a clear need for good compression methods arises.

Spectral images have higher degrees of redundancy among their multiple channels than do RGB images. Three types of redundancies can be underlined: spatial, spectral, and precision redundancies. A number of techniques can be applied to obtain a lossless compression in the spatial domain, as it is the same scenario of traditional images [74]–[76]. Moreover, spectra of multispectral images often show regular features that hint to a possible correlation in the spectral domain too, as confirmed by Principal Component Analysis (PCA) [29]. Thus multispectral images can be compressed also in the spectral dimensions by reducing the spectral sampling rate, and/or in the precision dimension decreasing the number of channels and bit depth used in spectral reconstruction. Imai et al. [77] have investigated compression approaches which reduce spatial and spectral sampling rates and precision demands without injuring the reconstruction process. Considering the influence on colorimetric and spectral accuracy of the number of spectral channels, in terms of bit depth and sampling rate, they found promising results in adopting compression techniques which include an arithmetic coding step that takes into account the entropic information of the eigenvector coefficients.

Compression schemes generally use some method to reduce the number of components within each spectrum (compression in the "spectral domain"), and then apply some compression scheme on the resulting image "channels" (compression in the "spatial domain"). Favored methods for the spectral domain are Principal Component Analysis and different sorts of clustering strategies, while JPEG-like compression schemes (based on the Discrete Cosine Transform as well as on Wavelets) and various types of Vector Quantization are usually proposed for the spatial domain [78]–[86]. All these methods are usually applied in their lossy versions (or are intrinsically lossy in nature), and can achieve good compression rates. Parkkinen et al. [87] for instance, have developed three methods for the lossy compression of multispectral images. The first method uses clustering in the compression of the spatial dimensions, and the Wavelet Transform (CL-W) or the Principal Component Analysis (CL-P) in the compression of the spectral dimension. The second method is based on the Wavelet Transform in all the dimensions simultaneously. The third compression method has three variants. The first variant (SP-P) applies the PCA to the spectral dimension and the high-quality, wavelet-based SPIHT-algorithm [88] to the spatial dimensions. The second (JP-P) and the third variant (JP2000) applied respectively the DCT-JPEG and JPEG-2000 to the spatial dimensions.

Murakami et al. [89] proposed a Weighted Karhunen-Loeve transform (WKLT) as spectral transform for multispectral image compression. To reduce the colorimetric error, a weighting matrix that accounts for the color matching functions of human observer as well as for the viewing illuminant is incorporated to the KLT. However, the improvements in colorimetric accuracy are obtained on the cost of poor spectral accuracy in certain wavelengths. Yu et al. [90] have shown that the preservation of spectral accuracy and the reduction of colorimetric error is a tradeoff, which can be controlled by adding a diagonal matrix to the weighting matrix of KLT.

Pellegri et al. [73] have underlined two main drawbacks of lossy compression schemes in the field of multispectral images. First, as one main advantage of multispectral data is their independence from the observer, using psycho-perceptual models with these data may be conceptually inconsistent. Second, as multispectral data are often proposed and collected for use in high-precision contexts, the issue of whether a lossy compression is acceptable, which degree of precision should be preserved, and whether and how this can be guaranteed arises: the application of two (or sometimes more) compression schemes one on top of the other usually worsens this problem. Starting from these considerations they have presented a lossless encoding for multispectral data that takes into account regular features in the spectral domain. They have approximated the spectra with the traditional Differential Code-Pulse Modulation (DCPM). Components at fixed and variable positions within the spectrum wavelength range have been taken as a base for approximation, following some criterion dictated by the spectrum characteristics.

The compression ratios in lossless compression rarely exceeds 2:1, since lossless coding exploits only the statistical redundancy of the image, but in lossy compression they can be tenfold. Lossy methods described in the literature are usually tested on images from airbone devices, mainly Landsat TM [79] and Aviris [87] collections. This makes any direct comparison of compression rates very difficult as these images are rather different from the photographic multispectral images. Moreover they usually adopt generic perception-based scores, or objective scores that do not have precise meaning in terms of color, like the Root Mean Squared Error (RMSE) or the peak-signal-to-noise ratio (PSNR). Yu et al. [90] used PSNR for the evaluation of spectral accuracy, and the $CIEL^*a^*b^*$ ΔE color difference for the evaluation of color reproducibility. They have shown their compression results only on three multispectral images captured by a multispectral camera with sixteen narrow-band color filters. They have compared their method to the WKLT, reporting the compression ratio with respect to both PSNR and the adopted ΔE color difference, showing that it is possible to find a tradeoff between PSNR and color accuracy. Pellegri in his doctoral thesis [91] has evaluated the performance of his lossless approach on several multispectral images available from scholarly research, and showed that compression ratios higher than 1.8:1 and up to 2.6:1 are possible. As a reference, he has considered the established LOCO-I [92] method, which is capable of compression ratios consistently higher than 2.5:1 (40% rate) on 8-bit images. This method has achieved an average compression ratio of roughly 1.25:1 (80% rate) on the considered multispectral images. A better comparison of the different compression methods presented in the literature should be performed on the same set of mutispectral images, and the same measures for spectral accuracy and color reproduction fidelity should be adopted, but it is out of the scope of this review.

7.7 Evaluating System Performance

One open issue in multispectral imaging is to find a spectral metric with a clear meaning and known quantitative features. Such a metric should capture the geometry of the space of spectra, and allow to quantitatively compare spectra while still having high confidence in the corresponding colorimetric results for given illuminant and observer. The definition of a quality metric is difficult as there are neither a completely clear semantics for spectra nor an accepted representation model with an exhaustive physical significance. Nonetheless, it would be beneficial to have at least a clear reference and expectation about how differences given by a spectral metric relate to the corresponding colorimetric differences.

As several spectral metrics have been proposed over the years [93]–[95], it may be useful to attempt a classification. Three main categories have emerged: pure spectral metrics, which take as only input data the two spectra to be compared; mixed metrics, which also take some form of data concerning the illuminant or observer; and metamerism indices, which also use some colorimetric quantity (like, for instance, the difference ΔL^* in $CIEL^*a^*b^*$ space). It is reasonable to expect that the more one goes towards colorimetry, the likelier it is that a good match betweeen the chosen spectral and colorimetric metrics can be found, but also the less significant the result. In fact, it may be better to directly compute the colorimetric distance, as the resulting spectral metric could not even synthetize results under different illuminants. On the other hand, the purer the spectral metric, the more variability can be expected in the results.

Basically, a spectral metric must quantify the diversity of two spectra, which can be done in a number of ways (RMS being the most widely spread). The way diversity is quantified is also responsible for the ability or disability of the metric to cope with dissimilar metamers. Any metric that does not allow differences on different components of two sampled spectra to cancel each other out will not be able to cope with dissimilar metamers. Anyway, it would be useful if this operation could be independent from the number of components in the spectra considered, as this allows comparison of results obtained for spectra sets sampled differently. However, the actual difficulties in establishing a relationship between spectral and colorimetric metrics depend on the observer-illuminant pair considered. Both observer and illuminant act by enhancing the impact of the differences associated to certain wavelengths over the others: the observer also generally enhances differences on dark colors compared to brighter ones, resulting in a sort of non-linear correction. It is then conceivable that a pure spectral metric can only and at most cope with the observer non-linearity, while mixed metrics can also account for wavelength biasing. Also, mixed metrics can act in two steps, the first being adding the observer effect on wavelengths, which is valid for all illuminants, and the second being adding the illuminant effect too. Finally, any pure spectral metric can likely become a mixed metric with the proper changes.

Based on the remarks above, it can be expected that a spectral metric can be built using different elements, each trying to cope with one issue, including quantifying spectral difference, correcting for observer non-linearity, and accounting for illuminant and observer wavelength biasing. See [96] for an example of a mixed metric.

7.8 Multispectral Image Reproduction

7.8.1 Colorimetric Reproduction

The reproduction of multispectral imaging using current colorimetric devices follows an established framework, based on colorimetry as well as on convenient international standards (Figure 7.7).

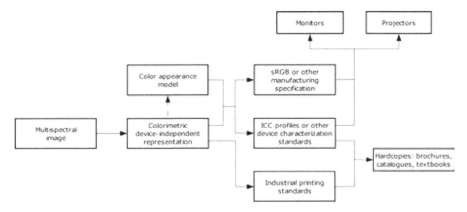

FIGURE 7.7
General scenario for the reproduction of multispectral images on current devices.

Multispectral images are converted to colorimetric representations, given a chosen illuminant and observer pair, and are then transformed to the RGB (or equivalent) representations of the devices involved using some characterization methods, possibly based on some standards like ICC profiles [97]. Alternately, a manufacturing specification like sRGB [98] can be leveraged, if the devices comply with it. In any case, a color appearance model may be applied to account for any specific intended viewing conditions. If, on the other hand, offset or other industrial printing is the target, colorimetric representations will be matched to any suitable industry standard to drive the printing process.

7.8.2   Spectral Characterization

Output devices share the technical problems of input devices with regard to characterization profiles (high-dimensionality color spaces are involved) and profile format standards: however, the situation is different with respect to the nature of devices. In fact, from a color reproduction perspective, there are no spectral devices, but only spectrally characterized devices. In other words, any colorimetric (i.e., current) output device may theoretically be spectrally characterized, in the sense that the spectra (not the colorimetric coordinates) of the reproduced colors may be considered in the characterization profile. Also, adding more inks (in the case of printers), phosphors (monitors), or lights/filters (projectors) may help to improve a spectral characterization, but it is not a necessary step to attempt such a characterization. On the other hand, the physical nature of devices may make it more or less easy to obtain a spectral (or even colorimetric) characterization. This is especially true for printers, whose behavior usually proves highly resistant to analytical modeling, and difficult to capture empirically as well: spectral characterization is likely to be affected by this problem too. However, the worst case for spectral models is that of CRT monitors, as their spectral emissions show spikes and irregular features, which do not match well with reflectance spectra: many LCD monitors also share this problem, in this case due to the spectral distribution of the back-light.

Derhak and Rosen introduced the so-called LabPQR space and a low-dimensional transform to use in spectral color management [99]. This space has explicit colorimetric portions and explicit spectral reconstruction portions. The use of this space makes it possible to visualize spectral gamuts and to begin thinking about the spectral gamut mapping problem.

Finally, an interesting question is whether it will be possible (and feasible) to implement spectral profiles with different rendering intents, similar to what the ICC standard allows today for traditional imaging. Printing and, to a more limited extent, monitor-based visualization, have benefited from the possibility of distinguishing between different kinds of documents, and the underlying principles (such as saturation enhancement, or exploitation of the so-called memory colors, and so on) are not necessarily tied to the colorimetric framework.

Printer Characterization Techniques for spectral-based printer characterization commonly employ analytical models, formulated on the basis of the physics behind the printing process. In the modeling of binary printers, most of the methods are based on the color-mixing model of Neugebauer [100]– [102]. The model, in its original formulation, predicts the outcome of a print with poor accuracy, and several strategies for its upgrading have been proposed. The Yule-Nielsen coefficient [103], introduced to take into account the effects of light scattering in the substrate, increases the model's performance, but the resulting Yule-Nielsen Spectral Neugebauer model still needs solutions to deal with interactions among inks and of inks with paper. According to the Yule-Nielsen Spectral Neugebauer (YNSN) equation, the spectrum of an N-inks halftone print is the weighted sum of 2^N different colors, called Neugebauer primaries, given by all the possible overprints of inks. The weight of each Neugebauer primary is the area that the corresponding combination of inks covers in the halftone cell. The YNSN model for a four-ink halftone print is:

$$R_{print}(\lambda) = \left[\sum_{p=0}^{15} A_p R_p(\lambda)^{\frac{1}{n(\lambda)}} \right]^{n(\lambda)} \tag{7.22}$$

where $R_{print}(\lambda)$ is the reflectance of the printed color and $R_p(\lambda)$ is the reflectance of the p-th Neugebauer primary. The wavelength dependent exponent $n(\lambda)$ is the Yule-Nielsen coefficient. In its original formulation, the Yule-Nielsen coefficient is a constant value with a physical meaning: in absence of scattering its value is 1, in the case of full scattering its value is 2. In recent works, n is considered merely an optimization parameter, and the predictions of the models can be improved by making it a function of the wavelength [104, 105]. In Equation 7.22, A_p is the Neugebauer primary area coverage, that is the percentage of the halftone cell covered by the p-th Neugebauer primary. A model commonly used to compute area coverage is Demichel's, which assumes that drops of ink are placed at random and statistically independent positions inside a unit cell. This model is considered valid for random or rotated halftone screens [106], while it fails for singular screen superposition, although the color deviation observed is not excessively large [107]. For dot-on-dot printing a different formulation must be considered [108].

In literature, many strategies have been proposed to improve the accuracy of Neugebauer models, not always with the intent of better modeling of inks dot gain and interactions. Approaches exist that assume that the Neugebauer primaries can be optimized, or increased in number to face the problem with cellular methods, that treat the Neugebauer equation as an interpolation model [109, 110]. This last strategy has been adopted in spectral-based characterization applications [108].

More complex methods have been introduced to describe optical dot gain, among which the convolution with a point spread function (PSF) [111], or probability models [110]. Alternative approaches describe the spreading of the ink by enlarging the drop impact on the basis of the configuration of its neighbors and the state of the surface [112, 113], or by modeling the physical dot gain with a transmission function defined on a blurred version of the halftone image [111]. Other methods for the improvement of model accuracy employ cellular approaches [114] or ascribe partial uncertainty to the measurements of reflectance. Examples of methods that take this circumstance into account are [115, 116]. A method based on neural networks can be found in [117] while genetic algorithms have been adopted in [118].

7.8.3   Spectral Reproduction

In recent years, multispectral reproduction is attracting increasing attention triggered by its appealing feature of significantly reducing undesirable metamerism effects with respect to traditional colorimetric approaches [2]. In multispectral reproduction the aim is to produce, in print, a color having reflectance equal to that of the input color, specified throughout a reflectance spectrum.

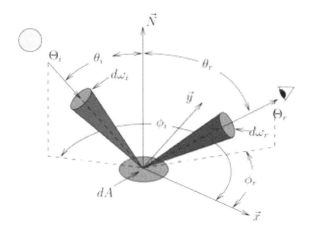

FIGURE 7.8
The geometry of the BRDF.

Since output devices' rendering capabilities have limited gamut, out-of gamut requests must be handled in order to produce a response. Spectral gamut mapping has to be approached in two stages. First, colorimetric error is reduced as low as possible. Once the colorimetric aspect of a process is complete, spectral fidelity can be addressed.

7.8.4 Viewpoint and Lighting Position Invariant Reproduction

The appearance of an object is the result of a complex interaction between the light, the observer, and the material the object is composed of. Digital archives are traditionally composed of acquisitions taken from a single viewpoint and with a fixed-position light source. The acquired image is the result of the photographer's decisions about the positions of the three elements in play. Multispectral imaging allows illuminant invariant digitalization; however, in order to achieve a totally-invariant and full digitalization of an object it is also necessary to record its shape (using, for instance, a 3D scanner) and its reflectance properties taken from multiple viewpoints and using multiple light positions. The law that rules the object optical properties is known as the Bidirectional Reflectance Distribution Function (BRDF) [119]. Once the BRDF of an object is captured, computer graphics models can be applied to generate artificial images from any viewpoint and with any lighting condition. The BRDF (denoted as f_r) is defined as the ratio of the directional reflected radiance to the directionally incident irradiance:

$$f_r(\theta_i, \phi_i; \theta_r, \phi_r) = \frac{dL_r(\theta_i, \phi_i; \theta_r, \phi_r; E_i)}{dE_i(\theta_i, \phi_i)}, \qquad (7.23)$$

where the subscript r is associated to reflected quantities while i is associated to incident quantities; L_r denotes the reflected radiance; and E_i denotes the incident irradiance. The geometry of the BRDF is depicted in Figure 7.8.

As can be seen, a light source illuminates a surface element dA through the element of solid angle $d\omega_i$ from the incident direction expressed in polar coordinates (θ_i, ϕ_i). The reflected flux is in the direction (θ_r, ϕ_r) centered within the solid angle $d\omega_r$. The BRDF is a wavelenght dependent quantity; here, for the sake of simplicity, the wavelength λ has been omitted from notation.

Since real world surfaces are not composed of a single material, the reflection properties vary on a two-dimensional domain. For describing these cases, Dana et al. [120] introduced the Bidirectional Texture Function (BTF) as a spatially modulated BRDF.

In practice, the BRDF or the BTF are usually approximated with a limited set of acquisitions in order to fit a reflection model taken from the Computer Graphics domain.

The device used to physically measure a BRDF is called a gonioreflectometer. Since BRDFs are a function of four dimensions, physically measuring them is very challenging. Gonioreflectometers must be able to sample the four degrees of freedom; this implies that a camera, a light source, and the captured object must rotate and move. There are often problems with consistency of the light source, camera stability, variations in surface geometry, and interreflections. See [121]–[129] for examples of various types of gonioreflectometers.

In computer graphics, various light reflection models have been proposed during the past decades. The first type of reflection models are based on theory of optics and physics. Torrance and Sparrow [130] proposed a model where rough surfaces are approximated through small, mirror-like, and randomly oriented facets. Blinn [131] adapted the Torrance-Sparrow in order to optimize the performance for computer graphics applications. Blinn substituted the Gaussian microfacet distributed function with an ellipsoid of revolution modeled function. Cook and Torrance [132] developed a more general reflection model which is able to approximate the color shift that occurs when the reflectance changes with the incident angle. He and Torrance [133] extended the Cook and Torrance model with polarization, directional Fresnel effects, and the subsurface scattering effect. Shirley et al. [134] presented an approach for polished materials, i.e., those materials having smooth surfaces and a significant subsurface scattering. Finally, the Beard-Maxwell model [135] approximated the material surface using a three-dimensional terrain of micro-facets. They also included shadowing, obscuration, and polarization effects.

The second class of reflection models are the empirical models. These models have been derived on well-fitting the BRDF without regarding the physical meaning of the parameters involved. The Phong reflection model [136] is the most popular reflection model in computer graphics. The model uses one parameter for the diffuse reflection component and two parameters for the specular reflections. Based on Phong's model, Lafortune [137] proposed a model based on fitting the BRDF through the sums of cosine lobes. This model is able to represent complex effects such as retro-reflection and off-specular reflection. Finally, Ward [122] proposed an empirical model derived fitting the data measured by his gonioreflectometer.

7.9 Final Remarks

Traditional approaches to digital archiving used conventional photographic techniques. A first step toward the full digitalization of cultural heritage artifacts had been the adoption of multispectral images acquired with fixed viewpoint and fixed lighting position. The main goals of this approach had been the independence from illuminant and the ability to reproduce accurately art paintings on different devices.

The visual arts system for archiving and retrieval of images (VASARI) [5, 138] was a project founded by the European Community in 1989 aimed to the digitalization of art paintings. The proposed system was able to digitize high-resolution multispectral images (six bands) moving a CCD camera and acquiring multiple shots. The final image was obtained composing the acquisitions set through mosaicking. Subsequently, the MARC project (Methodology for Art Reproduction in Color) [6] applied VASARI results for the printing of high-quality art books. The data provided from both the MARC and VASARI scanners have been processed in order to deliver reliable CIELAB coordinates. Miyake et al. [139] developed a multiband camera system based on a single chip CCD and a five-filters rotating wheel. The device was aimed to acquire the reflectance spectra of art paintings. In 2001, the European Community sponsored another project in the field, the CRISATEL [7] (Conservation Restoration Innovation Systems for Image and Digital Archiving to

Enhance Training, Education and Lifelong Learning). The multispectral system consisted of a CCD camera equipped with thirteen interference filters (ten filters with the bandwidth of 40 nm in the visible range and three filters with bandwidth of 100 nm in the near-infrared region). The system was used to acquire the famous Mona Lisa in order to analyze its materials.

During the last decade surface spectral reflectance reconstruction and pigment analysis have been the main focus of many additional works, see for example [14]– [9].

Totally invariant digitalization and realistic digital reproduction of pieces of art require the acquisition of shape and BRDF, as explained in Section 7.8.4. In some works full or partial illumination invariant reproduction is proposed using photometric approaches (see for example [140, 141]).

Tominaga et al. focused their attention on the digitalization of art paintings. In a first experimental setup [142], they proposed a system composed of a camera, a laser range scanner, and a projector with six color filters. The camera position remained fixed and there were only eight different azimuth incident angles; the system was not able to measure absolute BRDF values. Since the range scanner had some problems in capturing colored surfaces, they proposed, in a successive work [8], to reconstruct the surface geometry using an estimation of the surface normals through photometric stereo. Once the spectral reflectances and the surface normals had been estimated, they fitted the parameters of the Cook and Torrance [132] reflection function; renderings had been generated via ray tracing.

A series of works at Chiba University [143]– [145] resulted in a goniospectral imaging system composed of a 3D laser scanner and a robot arm controlling the light position. A fixed camera was equipped with multi-band filters. The system was limited since there was only one degree of freedom; full-BRDF measurements were not possible. The system has been also applied to paper and cloth samples [146].

Mansouri et al. [147] proposed a prototypal multispectral 3D scanner. The system was composed of a CCD camera equipped with a filters wheel and a digital projector. The camera was able to acquire six-band multispectral images; range maps were generated using the projector through structured light techniques. The system was not able to measure BRDF values since the camera and light source were at fixed positions.

With a similar philosophy, Brusco et al. [148] combined a 3D laser scanner with a spectrograph. The system was aimed to the interactive visualization and the restoration of historical buildings featuring frescoes. As in the previous case, no complete BRDF measurements were possible.

A multi-year research program [13] has been undertaken in the Munsell Color Science Laboratory (MCSL) at Rochester Institute of Technology to build a multispectral image acquisition system and to test spectral reconstruction algorithms. The system will be able to capture the full BRDF and surface geometry of art paintings.

The future of multispectral imaging is the realization of a complete spectral processing pipeline, from acquisition to processing and reproduction. The LabPQR space [99] is showing promising results. The colorimetric aspects of the space make it a very useful hybrid, ensuring that both colorimetry and spectrophotometry can be taken into consideration when spectral gamut mapping takes place.

Bibliography

[1] F. König and P. Herzog, "On the limitations of metameric imaging," in *Proc. of the Conference on Image Processing, Image Quality and Image Capture Systems (PICS-99)*, Savannah, Georgia, USA, pp.163-168, 1999.

[2] B. Hill, "Color capture, color management and the problem of metamerism: Does multispectral imaging offer the solution?," in *Proc. IS&T/SPIE Dev. – Independent Color, Color Hardcopy and Graphics Arts V*, pp. 2–14, 2000.

[3] R. Berns, L. Taplin, F. Imai, E. Day, and D. Day, "Spectral imaging of matisse's pot of geraniums: A case study," in *Proc. IS&T/SID 11th Color Imaging Conference, Spotlight Presentations II, Scottsdale*, Arizona, United States, pp. 149–153, 2003.

[4] K. Martinez, J. Cupitt, D. Saunders, and R. Pillay, "Ten years of art imaging research," *Proceedings of the IEEE*, vol. 90, no. 1, pp. 28–41, 2002.

[5] D. Saunders and J. Cupitt, "Image processing at the national gallery: The vasari project," *The National Gallery Technical Bulletin*, vol. 14, no. 1, pp. 72–85, 1993.

[6] J. Cupitt, K. Martinez, and D. Saunders, "Methodology for art reproduction in colour: the marc project," *Computers and the History of Art*, vol. 6, no. 2, pp. 1–20, 1996.

[7] A. Ribés, H. Brettel, F. Schmitt, H. Liang, J. Cupitt, and D. Saunders, "Color and multispectral imaging with the Crisatel multispectral system," in *PICS*, pp. 215–219, 2003.

[8] S. Tominaga and N. Tanaka, "Spectral image acquisition, analysis, and rendering for art paintings," *Journal of Electronic Imaging*, vol. 17, no. 4, p. 043022, 2008.

[9] H. Maitre, F. Schmitt, J.-P. Crettez, Y. Wu, and J. Y. Hardeberg, "Spectrophotometric image analysis of fine art paintings," in *Proc. IS&T/SID 4th Color Imaging Conference: Color Science, Systems and Applications*, pp. 50–53, 1996.

[10] F. Imai and R. Berns, "High-resolution multi-spectral image archives: A hybrid approach," in *Proc. IS&T/SID 6th Color Imaging Conference: Color Science, Systems and Applications*, pp. 224–227, 1998.

[11] J. Farrell, J. Cupitt, D. Saunders, and B. Wandell, "Estimating spectral reflectances of digital artwork," in *International Symposium on Multispectral Imaging and Color Reproduction for Digital Archives*, pp. 58–64, 1999.

[12] H. Haneishi, T. Hasegawa, A. Hosoi, Y. Yokoyama, N. Tsumura, and Y. Miyake, "System design for accurately estimating the spectral reflectance of art paintings," *Applied Optics*, vol. 39, no. 35, pp. 6621–6632, 2000.

[13] R. S. Berns, "Improving artwork reproduction through 3d-spectral capture and computer graphics rendering: Project overview," tech. rep., Rochester Institute of Technology, College of Science, Center for Imaging Science, Munsell Color Science Laboratory, Rochester, New York, United States, 2006.

[14] F. M. P. B. Ferreira, P. Fiadeiro, S. Nascimento, V. de Almeida, M. Pereira, and J. Bernardo, "Spectral characterization of a hyperspectral system for imaging of large art paintings," in *CGIV 2006 – Third European Conference on Color in Graphics, Imaging and Vision.*, pp. 350–354, 2006.

[15] M. Rosen and X. Jiang, "Lippmann2000: A spectral image database under construction," in *International Symposium on Multispectral Imaging and Color Reproduction for Digital Archives*, Chiba University, Chiba, Japan, October 21-22, 1999.

[16] D. Tzeng and R. Berns, "Spectral-based six-color separation minimizing metamerism," in *Proc. IS&T/SID, 8th Color Imaging Conference: Color Science and Engineering, Systems, Technologies and Applications*, Scottsdale, Arizona, USA, pp. 342-347, 2000.

[17] R. Berns, "Color-accurate image archives using spectral imaging," *Scientific Examination of Art: Modern Techniques in Conservation and Analysis, National Academies Press*, vol. 90, no. 1, pp. 105–119, 2005.

[18] "Commission internationale de l'eclairage." http://www.cie.co.at/cie/.

[19] J. Eem, H. D. Shin, and S. Park, "Reconstruction of surface spectral reflectances using characteristic vectors of munsell colors," in *Proc. IS&T/SID 2nd Color Imaging Conference*, Springfield, Virginia, United States, pp. 127-131, 1994.

[20] L. Maloney, "Evaluation of linear models of surface spectral reflectance with a small number of parameters," *Journal of the Optical Society of America A*, vol. 3, no. 10, p. 1673-1683, 1986.

[21] P. Burns, *Analysis of image noise in multispectral color acquisition*. PhD thesis: Center for Imaging Science, Rochester Institute of Technology, 1997.

[22] R. Lenz, M. Österberg, J. Hiltunen, T. Jaaskelainen, and J. Parkkinen, "Unsupervised filtering of color spectra," *Journal of the Optical Society of America A*, vol. 13, no. 7, p. 1315-1324, 1996.

[23] J. Parkkinen, J. Hallikainen, and T. Jaaskelainen, "Characteristic spectra of Munsell colors," *Journal of the Optical Society of America A*, vol. 6, p. 318322, 1989.

[24] D. Connah, A. Alsam, and J. Hardeberg, "Multispectral imaging: How many sensors do we need?," *Journal of Imaging Science and Technology*, vol. 50, p. 4548, January/February 2006.

[25] G. Sharma and H. Trussell, "Figures of merit for color scanners," *IEEE Transactions on Image Processing*, vol. 6, no. 7, pp. 990–1001, 1997.

[26] F. Imai, M. Rosen, and R. Berns, "Comparison of spectrally narrow-band capture versus wide-band with a priori sample analysis for spectral reflectance estimation," in *Proc. IS&T/SID 8th Color Imaging Conference: Color Science and Engineering, Systems, Technologies and Applications*, Scottsdale, Arizona, USA, pp. 234-240, 2000.

[27] S. Poger and E. Angelopoulou, "Selecting components for building multispectral sensors," in *IEEE CVPR Technical Sketches (CVPR Tech Sketches)*, 2001.

[28] S. Tominaga, "Spectral imaging by a multi-channel camera," *Journal of Electronic Imaging*, vol. 8, pp. 332–341, October 1999.

[29] J. Hardeberg, H. Brettel, and F. Schmitt, "Multispectral image capture using a tunable filter," in *Proc. of SPIE 3963, Color imaging : device-independent color, color hardcopy, and graphic arts V, San Jose, California, United States, pp. 77-88*, 2000.

[30] R. Slawson, Z. Ninkov, and E. Horch, "Hyperspectral imaging: Wide-area spectrophotometry using a liquid-crystal tunable filter," *The Publications of the Astronomical Society of the Pacific*, vol. 111, no. 759, pp. 621–626, 1999.

[31] J. Hardeberg, "Multispectral color image capture using a liquid crystal tunable filter," *Optical Engineering*, vol. 41, p. 25322548, October 2002.

[32] H. Haneishi, T. Hasegawa, N. Tsumura, and Y. Miyake, "Design of color filters for recording artworks," in *IS&T s 50th Annual Conference*, Springfield, VA, pp. 369-372, 1997.

[33] Y. Miyake, Y. Yokoyama, N. Tsumura, H. Haneishi, K. Miyata, and J. Hayashi, "Development of multiband color imaging systems for recording of art paintings," in *Proc. IS&T/SPIE Conference on Color Imaging: Device-Independent Color, Color Hardcopy, and Graphic Arts IV SPIE 3648*, Bellingham, WA, pp. 218-225, 1999.

[34] P. Burns and R. Berns, "Analysis of multispectral image capture," in *Proc. IS&T/SID 4th Color Imaging Conference: Color Science, Systems and Applications*, Springfield, Virginia, USA, pp. 19-22, 1996.

[35] T. Jaaskelainen, J. Parkkinen, and S. Toyooka, "Vector-subspace model for color representation," *Journal of the Optical Society of America A*, vol. 7, pp. 725–730, 1990.

[36] M. J. Vrhel, R. Gershon, and L. Iwan, "Measurement and analysis of object reflectance spectra," *Color Research & Application*, vol. 19, p. 49, 1994.

[37] G. Novati, P. Pellegri, and R. Schettini, "Acquisition, mosaicking and faithful rendering of multispectral images of cultural artifacts," in *Proc. of Multimedia.Information@DEsign for Cultural Heritage - MIDECH 2005*, Milano, Italy, pp 161-166, 2005.

[38] F. H. Imai, D. R. Wyble, R. S. Berns, and D. Tzeng, "A feasibility study of spectral color reproduction," *Journal of Imaging Science and Technology*, vol. 47, pp. 543–553, 2003.

[39] M. Parmar, F. Imai, S. H. Park, and F. Joyce, "A database of high dynamic range visible and near-infrared multispectral images," in *Proc. of SPIE 6817, in Digital Photography IV, pp. 68170M.1-68170M.10*, San Jose, California, USA, 2008.

[40] F. Imai, R. Berns, and D. Tzeng, "A comparative analysis of spectral reflectance estimation in various spaces using a trichromatic camera system," *Journal of Imaging Science and Technology*, vol. 44, pp. 280–287, 2000.

[41] R. S. Berns, L. A. Taplin, M. Nezamabadi, M. Mohammadi, and Y. Zhao, "Spectral imaging using a commercial color-filter array digital camera," in *14th Triennal ICOM-CC meeting*, pp. 743–750, ICOM, 2005.

[42] A. Y. Zhao and R. S. Berns, "Image-based spectral reflectance reconstruction using the matrix r method," *Color Research & Application*, vol. 32, no. 5, pp. 343–351, 2007.

[43] Y. Zhao, *Image Segmentation and Pigment Mapping of Cultural Heritage Based on Spectral Imaging*. PhD thesis, Rochester Institute of Technology, College of Science, Center for Imaging Science, Rochester, New York, United States, 2008.

[44] G. Novati, P. Pellegri, and R. Schettini, "An experience in multispectral mosaicking," in *Proc. Color Imaging X: Processing, Hardcopy, and Applications IX, SPIE Vol. 5667*, pp. 84–94, 2005.

[45] G. Novati, P. Pellegri, and R. Schettini, "An affordable multispectral imaging system for the digital museum," *International Journal of Digital Libraries, Special issue on Digital Museum*, vol. 5, no. 3, pp. 167–178, 2005.

[46] A. Ribés, F. Schmitt, R. Pillay, and C. Lahanier, "Calibration and spectral reconstruction for crisatel: An art painting multispectral acquisition system," *Journal of Imaging Science and Technology*, vol. 49, pp. 563–573, 2005.

[47] S. Tominaga, "Multichannel vision system for estimation surface and illumination functions," *Journal of the Optical Society America A*, vol. 13(11), pp. 2163–2173, 1996.

[48] J. Hardeberg, *Acquisition and reproduction of color images: colorimetric and multispectral approaches*. PhD thesis, Ecole Nationale Supérieure des élécommunications, ENST, Paris, France., 1999.

[49] P. Herzog, D. Knip, H. Stiebig, and F. König, "Colorimetric characterization of novel multiple-channel sensors for imaging and metrology," *Journal of Electronic Imaging*, vol. 8, no. 4, pp. 342–353, 1999.

[50] D. Tzeng and R. S. Berns, "A review of principal component analysis and its applications to color technology," *Color Research & Application*, vol. 30, pp. 84–98, April 2005.

[51] A. Hyvärinen and E. Oja, "Independent component analysis: Algorithms and applications," *Neural Network*, vol. 13, pp. 411–430, 2000.

[52] R. Berns, F. Imai, P. Burns, and D. Tzeng, "Multispectral-based color reproduction research at the Munsell Color Science Laboratory," in *International Society for Optical Engineering*, vol. 3409, pp. 14–25, 1998.

[53] F. H. Imai, "Simulation of spectral estimation of an oil-paint target under different illuminants," tech. rep., Munsell Color Science Laboratory Techincal, 2002.

[54] A. Ribés and F. Schmitt, "A fully automatic method for reconstruction of spectral reflectance curves by using mixture density networks," *Pattern Recognition Letters*, vol. 24, no. 11, pp. 1691–1701, 2003.

[55] J. Hardeberg, H. Brettel, and F. Schmitt, "Spectral characterisation of electronic cameras," in *Proceedings SPIE 3490, Color Imaging: Processing, Printing, and Publishing in Color*, pp. 100–109, 1998.

[56] P. Pellegri, G. Novati, and R. Schettini, "Selection of training sets for the characterisation of multispectral imaging systems," in *Proceedings of IS&T PICS 2003: The Conference on Image Processing, Image Quality, Image Capture, and Systems*, pp. 479–484, 2003.

[57] P. Pellegri, G. Novati, and R. Schettini, "Training set selection for multispectral imaging system characterisation," *Journal of Imaging Science and Technology*, vol. 48, no. 3, pp. 203–210, 2004.

[58] M. Mohammadi, M. Nezamabadi, R. Berns, and L. Taplin, "Spectral imaging target development based on hierarchical cluster analysis," in *Proc. IS&T/SID 12th Color Imaging Conference*, pp. 59-64, 2004.

[59] Gretag ColorChecker. http://www.xrite.com.

[60] B. A. Wandell, "The synthesis and analysis of color images," *IEEE Transaction on Pattern Analysis and Machine Intelligence*, vol. 9, no. 1, pp. 2–13, 1987.

[61] J. Cohen, "Dependency of the spectral reflectance curves of the Munsell color chips," *Psyconomic Science*, vol. 1, pp. 369–370, 1964.

[62] K. L. Kelley, K. S. Gibson, and D. Nickerson, "Tristimulus specification of the Munsell book of color from spectrophotometric measurements," *Journal of the Optical Society America A*, vol. 33, pp. 355–376, 1943.

[63] E. L. Kirnov, *Spectral reflectance properties of natural formations*. 1947.

[64] G. Novati, P. Pellegri, and R. Schettini, "Selection of filters for multispectral acquisition using the filter vectors analysis method," in *Proceedings Color Imaging IX: Processing, Hardcopy, and Applications IX*, R. Eschbach, G.G. Marcu eds., SPIE Vol. 5293 pp. 20-26, 2004.

[65] P. Pellegri, G. Novati, and R. Schettini, "Training set and filters selection for the efficient use of multispectral acquisition systems," in *CGIV 2004 – Second European Conference on Color in Graphics, Imaging and Vision*, pp. 393–297, 2004.

[66] I. Chatzis, V. Kappatos, and E. Dermatas, "Filter selection for multi-spectral image acquisition using the feature vector analysis methods," in *Proceedings of IPROMS 2006*, 2006.

[67] I. Cohen, Q. Tian, X. Zhou, and T. Huang, *Feature selection using principal feature analysis*. Urbana-Champaign: University of Illinois, 2002.

[68] H. R. Kang, *Computational Color Technology*, vol. PM159. SPIE Press Book, 2006.

[69] S. Bianco, F. Gasparini, and R. Schettini, "Spectral based color imaging using rgb digital still cameras," *Colore e Colorimetria: contributi multidisciplinari, Vol II, Collana Quaderni di Ottica e Fotonica 14*, pp. 67–78, 2006.

[70] G. D. Finlayson and M. S. Drew, "Constrained least-squares regression in color spaces," *Journal of Electronic Imaging*, vol. 6, pp. 484–493, 1997.

[71] G. A. F. Seber, *Linear regression analysis*. New York: John Wiley & Sons, 1977.

[72] M. J. Vrhel, *Mathematical methods of color correction*. PhD thesis: Carolina State University, Department of Electrical and Computer Engineering, 1993.

[73] P. Pellegri, G. Novati, and R. Schettini, "Multispectral loss-less compression using approximation methods," in *IEEE ICIP2005*, vol. 2, pp. 638–41, 2005.

[74] M. Ryan and J. Arnold, "The lossless compression of aviris images by vector quantization," *IEEE Transactions on Geoscience and Remote Sensing*, vol. 35, no. 3, pp. 546–550, 1997.

[75] S. R. Tate, "Band ordering in lossless compression of multispectral images," *IEEE Transactions on Computers*, vol. 46, no. 4, pp. 477–483, 1997.

[76] J. Wang, K. Zhang, and S. Tang, "Spectral and spatial decorrelation of landsat-tm data for lossless compression," *IEEE Transactions on Geoscience and Remote Sensing*, vol. 33, no. 5, pp. 1277–1285, 1995.

[77] F. H. Imai, M. R. Rosen, R. S. Berns, N. Ohta, and N. Matsushiro, "Preliminary study on spectral image compression," in *Color Forum Japan*, pp. 67–70, 2000.

[78] G. Abousleman, "Compression of hyperspectral imagery using hybrid dpcm/dct and entropy-constrained trellis coded quantization," in *DCC '95: Proceedings of the Conference on Data Compression*, (Washington, DC, USA), p. 322, IEEE Computer Society, 1995.

[79] P. L. Dragotti, G. Poggi, and A. R. Ragozini, "Compression of multispectral images by three-dimensional SPIHT algorithm," *IEEE Transactions on Geoscience and Remote Sensing*, vol. 38, pp. 416–428, 2000.

[80] G. Gelli and G. Poggi, "Compression of multispectral images by spectral classification and transform coding," *IEEE Transactions on Image Processing*, vol. 8, no. 4, 1999.

[81] C. Rittirong, Y. Rangsanseri, and P. Thitimajshima, "Multispectral image compression using median predictive coding and wavelet transform," in *20th Asian Conference on Remote Sensing*, 1999.

[82] S. Mongkolworaphol, Y. Rangsanseri, and P. Thitimajshima, "Multispectral image compression using FCM-based vector quantization," in *21st Asian Conference on Remote Sensing*, 2000.

[83] Y. Tseng, H. Shih, and P. Hsu, "Compression of hyperspectral images using 3D wavelet transformation," *Asian Journal of GeoInformatics*, vol. 3, no. 1, 2002.

[84] G. R. Canta and G. Poggi, "Compression of multispectral images by address-predictive vector quantization," in *Signal Processing: Image Communications*, pp. 147–159, 1997.

[85] J. Vaisey, M. Barlaud, and M. Antonini, "Multispectral image coding using lattice vq and the wavelet transform," in *IEEE International Conference on Image Processing*, pp. 712–716, 1998.

[86] T. Markas and J. H. Reif, "Multispectral image compression algorithms," in *IEEE Data Compression Conference (DCC'93)*, pp. 391–400, 1993.

[87] J. Parkinnen, M. Hauta-Kasari, A. Kaarna, J. Lehtonen, P. Koponen, and T. Jaaskelainen, "Multispectral image compression," in *2nd International Symposium on Multispectral Imaging*, pp. 49–58, 2000.

[88] A. Said and W. Pearlman, "A new fast and efficient image codec based on set partitioning in hierarchical trees," *IEEE Transactions on Circuits and Systems for Video Technology*, vol. 6, no. 3, pp. 243–250, 1996.

[89] Y. Murakami, H. Manabe, T. Obi, M. Yamaguchi, and N. Ohyama, "Multispectral image compression for color reproduction; weighted klt and adaptive quantization based on visual color perception," in *Proceedings of IS&T/SID, Ninth Color Imaging Conference*, Scottsdale, Arizona, USA, pp. 68-72, 2001.

[90] S. Yu, Y. Murakami, T. Obi, M. Yamaguchi, and N. Ohyama, "Improvements for multispectral image compression for color reproducibility with preservation to spectral accuracy," in *IEEE Conference on Image Processing*, vol. 2, pp. II – 710–13, 2005.

[91] P. Pellegri, *Loss-Less and semantic near loss less compression of photographic multispectral images*. 2005.

[92] M. Weinberger, G. Seroussi, and G. Sapiro, "The loco-i lossless image compression algorithm: Principles and standardization into jpeg-ls," *IEEE Transactions on Image Processing*, vol. 9, pp. 1309–13241, August 2000.

[93] J. Viggiano, "Metrics for evaluating spectral matches: A quantitative comparison," in *CGIV 2004 – Second European Conference on Colour in Graphics, Imaging, and Vision*, pp. 286–291, 2004.

[94] F. Imai, M. Rosen, and R. Berns, "Comparative study of metrics for spectral match quality," in *CGIV 2002–First European Conference on Colour in Graphics, Imaging and Vision*, pp. 492–496, 2002.

[95] D. Kalenova, P. Toivanen, and V. Botchko, "Color differences in a spectral space," in *CGIV 2004 – Second European Conference on Colour in Graphics, Imaging and Vision*, pp. 368–371, 2004.

[96] G. Novati, P. Pellegri, and R. Schettini, "Spectral metrics and their ability to predict perceptual color differences: An evaluation," *Colore e Colorimetria: contributi multidisciplinari, Collana Quaderni di Ottica e Fotonica*, vol. 13, pp. 71–78, 2005.

[97] The International Color Consortium. ICC Profile Format Specification ICC.1:2004-10 (Profile version 4.2.0.0), 2006.

[98] M. Stokes, M. Anderson, S. Chandrasekar, and R. Motta, "A standard default color space for the Internet–SRGB." http://www.color.org/contrib/sRGB.html, 1996.

[99] M. W. Derhak and M. R. Rosen, "Spectral colorimetry using LabPQR: An interim connection space," *Journal of Imaging Science and Technology*, vol. 50, no. 1, pp. 53–63, 2006.

[100] R. Balasubramanian, "Optimization of the spectral neugebauer model for printer characterization," *Journal of Electronic Imaging*, vol. 8, no. 2, pp. 156–166, 1999.

[101] D. Yuan Tzeng and R. S. Berns, "Spectral-based ink selection for multiple-ink printing i. colorant estimation of original objects," in *Proc. 6th IS&T/SID Color Imaging Conference: Colorants Estimation of Original Objects*, pp. 182–187, 1988.

[102] K. Iino and R. S. Berns, "A spectral based model of color printing that compensates for optical interactions of multiple inks," in *AIC Color 97, Proc. 8th Congress International Colour Association*, pp. 610–613, 1997.

[103] J. A. C. Yule and W. J. Neilsen, "The penetration of light into paper and its effect on halftone reproduction," in *TAGA - Technical Association of the Graphic Arts Proceedings*, pp. 65–76, 1951.

[104] D. R. Wyble and R. S. Berns, "A critical review of spectral models applied to binary color printing," *Color Research & Application*, vol. 25(1), pp. 4–19, 2000.

[105] K. Iino and R. Berns, "Building color management modules using linear optimization. i. desktop color system," *Journal Imaging Sci. Technol.*, vol. 42, pp. 79–94, 1998.

[106] G. L. Rogers, "Neugebauer revisited: Random dots in halftone screening," *Color Research & Application*, vol. 23(2), pp. 104–113, 1998.

[107] I. Amidror and R. D. Hersch, "Neugebauer and Demichel: dependence and independence in n-screen superpositions for colour printing," *Color Research & Application*, vol. 25, pp. 267–77, 2000.

[108] R. Balasubramanian, "Printer model for dot-on-dot halftone screens," in *Proc. SPIE Vol. 2413, Color Hard Copy and Graphic Arts IV*, pp. 356–364, 2005.

[109] C. Hains, S. Wang, and K. Knox, *Digital Color Imaging Handbook*, ch. Digital Color Halftone, Chapter 6. CRC Press, Boca Raton, FL, 2003.

[110] J. S. Arney, T. Wu, and C. Blehm, "Modeling the yule-nielsen effect on color halftones," *Journal of Imaging Science & Technology*, vol. 42, pp. 335–340, 1998.

[111] S. Gustavson, "Color gamut of halftone reproduction," in *Proc. SPIE Vol. 2949, Imaging Sciences and Display Technologies*, pp. 264–271, 1996.

[112] P. Emmel and R. Hersch, "A unified model for color prediction of halftoned prints," *Journal of Imaging Science and Technology*, vol. 44(4), pp. 351–359, 2000.

[113] P. Emmel and R. Hersch, "Modeling ink spreading for color prediction," *Journal of Imaging Science and Technology*, vol. 46, pp. 237–246, 2002.

[114] A. U. Agar and J. P. Allebach, "An iterative cellular YNSN method for color printer characterization," in *Proc. of the Sixth IS&T/SID Color Imaging Conference*, pp. 197–200, 1998.

[115] M. Xia, E. Saber, G. Sharma, and A. M. Tekalp, "End-to-end color printer calibration by total least squares regression," *IEEE Transactions on Image Processing*, vol. 8, pp. 700–716, 1999.

[116] C. Lana, M. Rotea, and D. Viassolo, "Characterization of color printers using robust parameter estimation," in *Proc. IS&T/SID 11th Color Imaging Conference: Color Science and Engineering Systems, Technologies, and Applications*, pp. 224–231, 2003.

[117] R. Schettini, D. Bianucci, G. Mauri, and S. Zuffi, "An empirical approach for spectral color printers characterization," in *CGIV 2004 – Second European Conference on Colour in Graphics, Imaging and Vision*, pp. 393–397, 2004.

[118] R. Schettini and S. Zuffi, "Accounting for inks interaction in the yule-nielsen spectral neugebauer model," *Journal of Imaging Science and Technology*, vol. 50(1), pp. 35–(10), 2006.

[119] F. E. Nicodemus, J. C. Richmond, J. J. Hsia, I. W. Ginsberg, and T. Limperis, "Geometric considerations and nomenclature for reflectance," *National Bureau of Standards*, 1977.

[120] K. J. Dana, B. van Ginneken, S. K. Nayar, and J. J. Koenderink, "Reflectance and texture of real-world surfaces," *ACM Transactions on Graphics*, vol. 18, no. 1, pp. 1–34, 1999.

[121] J. F. Murray-Coleman and A. M. Smith, "The automated measurement of brdfs and their application to luminaire modeling," *Journal of the Illuminating Engineering Society*, pp. 87–99, Winter 1990.

[122] G. J. Ward, "Measuring and modeling anisotropic reflection," in *SIGGRAPH*, pp. 265–272, 1992.

[123] M. Turner and J. Brown, "The sandmeier field goniometer: A measurement tool for bidirectional reflectance," in *Nasa Commercial Remote Sensing Verification and Validation Symposium*, 1998.

[124] H. Li, S. C. Foo, K. E. Torrance, and S. H. Westin, "Automated three-axis gonioreflectometer for computer graphics applications," in *Optical Engineering*, vol. 45(4), 2006.

[125] M. Levoy, "Stanford spherical gantry." http://graphics.stanford.edu/projects/gantry/, 2002.

[126] W. Matusik, H. Pfister, M. Brand, and L. McMillan, "A data-driven reflectance model," *ACM Transactions on Graphics*, vol. 22, no. 3, pp. 759–769, 2003.

[127] R. Pointer, J. Barnes, J. Clarke, and J. Shaw, "A new goniospectrophotometer for measuring gonio-apparent materials," *Coloration Technology*, vol. 121, no. 2, pp. 96–103, 2005.

[128] F. Leloup, T. D. Waele, J. Versluys, P. Hanselaer, and M. Pointer, "Full 3D BSDF spectroradiometer," in *Proceedings of the ISCC/CIE Expert Symposium '06: 75 Years of the CIE Standard Colorimetric Observer,*, CIE (Commission International d'Eclairage), Dec. 2006.

[129] L. Technology, "RIT Goniometer: Requirements and specifications," tech. rep., Rochester Institute of Technology, Rochester, New York, United States, October 2007.

[130] K. E. Torrance and E. M. Sparrow, "Theory for off-specular reflection from roughened surfaces," *Journal of the Optical Society of America*, vol. 57, no. 9, pp. 1105–1112, 1967.

[131] J. Blinn, "Models of light reflection for computer synthesized pictures," *SIGGRAPH Computer Graphics*, vol. 11, no. 2, pp. 192–198, 1977.

[132] R. L. Cook and K. E. Torrance, "A reflectance model for computer graphics," *ACM Transactions on Graphics*, vol. 1, no. 1, pp. 7–24, 1982.

[133] X. He, K. Torrance, F. Sillion, and D. Greenberg, "A comprehensive physical model for light reflection," in *SIGGRAPH '91: Proceedings of the 18th Annual Conference on Computer Graphics and Interactive Techniques*, (New York, NY, USA), pp. 175–186, ACM, 1991.

[134] P. Shirley, B. Smits, H. Hu, and E. Lafortune, "A practitioners' assessment of light reflection models," *Pacific Conference on Computer Graphics and Applications*, p. 40, 1997.

[135] J. R. Maxwell, J. Beard, S. Weiner, D. Ladd, and S. Ladd, "Bidirectional reflectance model validation and utilization," tech. rep., Defense Technical Report AFAL-TR-73-303 (Infrared and Optics Division, Environmental Research Institute of Michigan), 1973.

[136] B. T. Phong, "Illumination for computer generated pictures," *Communications of the ACM*, vol. 18, no. 6, pp. 311–317, 1975.

[137] E. Lafortune, S. Foo, K. Torrance, and D. Greenberg, "Non-linear approximation of reflectance functions," in *SIGGRAPH '97: Proceedings of the 24th Annual Conference on Computer Graphics and Interactive Techniques*, (New York, NY, USA), pp. 117–126, ACM Press/Addison-Wesley Publishing Co., 1997.

[138] K. Martinez, J. Cupitt, and D. R. Saunders, "High-resolution colorimetric imaging of paintings," in *Cameras, Scanners, and Image Acquisition Systems*, vol. 1901, pp. 25–36, SPIE, 1993.

[139] Y. Miyake, Y. Yokoyama, N. Tsumura, H. Haneishi, K. Miyata, and J. Hayashi, "Development of multiband color imaging systems for recordings of art paintings," in *Society of Photo-Optical Instrumentation Engineers (SPIE) Conference Series* (G. B. Beretta & R. Eschbach, ed.), vol. 3648 of *Society of Photo-Optical Instrumentation Engineers (SPIE) Conference Series*, pp. 218–225, Dec. 1998.

[140] T. Hawkins, J. Cohen, and P. Debevec, "A photometric approach to digitizing cultural artifacts," in *VAST '01: Proceedings of the 2001 Conference on Virtual Reality, Archeology, and Cultural Heritage*, (New York, NY, USA), pp. 333–342, ACM, 2001.

[141] T. Malzbender, D. Gelb, and H. Wolters, "Polynomial texture maps," in *SIGGRAPH '01: Proceedings of the 28th Annual Conference on Computer Graphics and Interactive Techniques*, (New York, NY, USA), pp. 519–528, ACM, 2001.

[142] S. Tominaga, T. Matsumoto, and N. Tanaka, "3d recording and rendering for art paintings," in *Proc. IS&T/SID 9th Color Imaging Conference: Color Science and Engineering: Systems, Technologies, and Applications*, pp. 337–341, 2001.

[143] H. Haneishi, T. Iwanami, N. Tsumura, and Y. Miyake, "Goniospectral imaging of 3D objects," in *Proc. IS&T/SID 6th Color Imaging Conference*, pp. 173–176, 1998.

[144] H. Haneishia, T. Iwanami, T. Honma, N. Tsumura, and Y. Miyake, "Goniospectral imaging of three-dimensional objects," *Journal of Imaging Science and Technology*, vol. 45(5), pp. 451–456, 2001.

[145] T. Nakaguchi, M. Kawanishi, N. Tsumura, and Y. Miyake, "Optimization of camera and illumination direction on goniospectral imaging method," *Journal of The Society of Photographic Science and Technology of Japan*, vol. 68(6), pp. 532–537, 2005.

[146] Y. Akao, N. Tsumura, P. Herzog, Y. Miyake, and B. Hill, "Gonio-spectral imaging of paper and cloth samples under oblique illumination conditions based on image fusion techniques," *The Journal of Imaging Science and Technology*, vol. 48(3), pp. 227–234, 2004.

[147] A. Mansouri, A. Lathuiliere, F. S. Marzani, Y. Voisin, and P. Gouton, "Toward a 3D multi-spectral scanner: An application to multimedia," *IEEE MultiMedia*, vol. 14, no. 1, pp. 40–47, 2007.

[148] N. Brusco, S. Capeleto, M. Fedel, A. Paviotti, L. Poletto, G. Cortelazzo, and G. Tondello, "A system for 3D modeling frescoed historical buildings with multispectral texture information," *Machine Vision and Applications*, vol. 17, no. 6, pp. 373–393, 2006.

8

Did Early Renaissance Painters Trace Optically Projected Images? The Conclusion of Independent Scientists, Art Historians, and Artists

David G. Stork
Ricoh Innovations and Stanford University
Email: artanalyst@gmail.com

Jacob Collins
c/o Adelson Galleries
New York

Marco Duarte
Princeton University
Email: mduarte@princeton.edu

Yasuo Furuichi
Consultant
Kanagawa, Japan

Dave Kale, Ashutosh Kulkarni
Stanford University

M. Dirk Robinson
Ricoh Innovations
Email: dirkr@rii.ricoh.com

Sara J. Schechner
Harvard University
Email: schechn@fas.harvard.edu

Christopher W. Tyler
Smith-Kettlewell Eye Research Institute
Email: cwt@ski.org

Nicholas C. Williams
Towan Headland

CONTENTS

8.1   Introduction

In 2000, the contemporary painter, photographer and set designer David Hockney claimed that some Western artists, as early as 1420, secretly built optical projectors, projected portions of a sunlit scene or subject onto their supports (canvas, panel, ...), traced these images and later applied paint [11]. In the words of Hockney and his collaborator, thin-film physicist Charles Falco,

> Our thesis is that certain elements in certain paintings made as early as c. 1430 were produced as a result of the artist using either concave mirrors or refractive lenses to project the images of objects illuminated by sunlight onto his board/canvas. The artist then traced some portions of the projected images, made sufficient marks to capture only the optical perspective of other portions, and altered or completely ignored yet other portions where the projections did not suit his artistic vision. As a result, these paintings are composites containing elements that are "eyeballed" along with ones that are "optics-based." Further, starting at the same time, the unique look of the projected image began to exert a strong influence on the appearance of other works even where optical projections had not been directly used as an aid.

We refer to central projection claim as the *direct tracing* claim and the claimed that artists *saw* such projected images and they strove to duplicate elements of this new "optical" ideal, even without directly tracing projected images as the *indirect influence* claim. This direct tracing claim, if proven, would have great import for the history of optics and would show that projected images were recorded roughly two centuries earlier than scholars previously thought. Confirmation of the theory would also radically alter our understanding of artists' praxis in the early Renaissance.

The optical projection theory (or tracing theory) has been widely promoted in the popular media, in a BBC documentary, a website, and a number of non-expert-peer-reviewed papers by its two promoters. It has also been examined thoroughly by a wide range of international expert in the relevant domains of computer vision, pattern recognition, conservation, history of optics and art (including a four-day symposium in Ghent in 2003), as well as by a number of realist painters. Our goal here is to bring together the evidence, the arguments and the counter-arguments by independent experts, all in order to render a final judgement about the optical tracing claim, at least for the early

Renaissance. Clearly, this chapter can provide but a summary of the key evidence and arguments; readers should consult the primary cited literature for full details.[1]

In Section 8.2 we describe the optical tracing theory in a bit more detail and explore its scholarly and even philosophical foundations in order to determine what aspects of the claims can—and cannot—be tested, even in principle. That is, we clarify the scope of any scholarly analysis. Then in Section 8.3 we consider the image evidence from key paintings and the arguments concerning the projection claim, both pro and con; in particular we describe alternative non-optical explanations for the "optical" evidence. We turn in Section 8.4 to the important matter of contemporary documentary evidence—or lack of evidence—in support of the projection claim and the proponents' speculation that this lack of evidence is due to artists protecting "trade secrets" or fearing the Inquisition. In Section 8.5 we examine the physical evidence and material culture of the time to see if adequate materials and knowledge about appropriate projections existed. In Section 8.6 we describe other general developments circa 1430—such as the rise in oil paints and worn spectacles—that may better explain the rise in realism in art of that time. In Section 8.7 we question whether even if artists *had* traced images in the early Renaissance whether it would have led to an increase in realism such as is found. In Section 8.8 we summarize the independent scholarly consensus about the tracing theory and we conclude in Section 8.9 with some speculations on the further use of rigorous computer analysis in the study of art.

8.2 The Projection Theory

As mentioned above, artist David Hockney, after surveying the grand sweep of the development of Western painting, claimed to have identified a newfound realism or "optical" style circa 1430 that he and his colleague, Charles Falco, attribute to some artists secretly building optical projectors, tracing projected images of sunlit tableaus or subjects on their supports (canvas, oak panel, ...), and then applying paint [11].

It is unclear which of these two claims—the *direct tracing* claim or the *indirect influence* claim—is the more important or central to Hockney, but there is no doubt that the direct tracing claim is central to the *team* of Hockney and Falco; it is also the claim that has captured the public's imagination, that appears prominently in Hockney's book, in his BBC documentary, in television stories, in public lectures, and in other publications.

8.2.1 Philosophical and Logical Foundations of the Projection Theory

Hockney is an artist and, as such, we do not expect traditional academic rigor in his speculations or claims, but most of the subsequent work and promotion of the projection theory was done with scientist Falco, so we are justified in exploring, with traditional rigor, the intellectual foundations of their claims and evidence.

It pays to examine the philosophical and logical foundations of the claims in order to determine what kind of evidence can and cannot be brought to bear, and if the claims even rise to the status amenable to objective test. We consider first the indirect influence claim. Recall that Hockney states that artists circa 1430 *saw* optically projected images and that some artists took these as a new ideal to duplicate—even if partially, with modification—in their works. What kind of objective scholarly tests could ever prove or disprove such a claim? Consider putative contemporary documentary evidence, for instance early Renaissance writings by artists, patrons, or critics. Surely if an artist wrote that he had seen projected images, and possibly delighted in them, and that he deliberately

---

[1] Many of these papers are available from www.diatrope.com/stork/TechnicalPublications.html.

sought to duplicate some of the visual properties in his paintings, that would be strong and convincing evidence, especially if objective image analysis (somehow) corroborated this claim. As we shall see in Section 8.4, though, there is little or no such evidence for the period in question. Absent such documentary evidence, what visual or material evidence might we then evaluate, such as visual evidence in the paintings themselves? Surely the existence of optical devices (e.g., concave mirrors) of sufficiently high optical quality is a necessary—but not sufficient—pre-requisite. We shall see in Section 8.5, however, that there is little persuasive supporting evidence for such claims.

It is hard to imagine persuasive visual evidence, at least for the period in question, other than if the painting included purely "optical" features arising only in projectors, such as blur spots—such as appear in some paintings much later by Jan Vermeer (1632–1675), including *The Milkmaid*. Rigorous computer vision and image analysis—for instance perspective, lighting, brush stroke, and color analysis—would be of little or no value in testing the influence claim, at least in the early Renaissance, because there are numerous additional confounding influences, such as the introduction of new media (oil paints), social and cultural changes (secular, scientific, and humanistic subjects), and so on. There are non-projection optical influences as well, such as the well-documented rise in the use of spectacles [15], which would enable artists—especially those over 30 or 40 years of age—to see both distant subjects and a nearby canvas in sharp focus. In short, there would seem to be no way to disentangle the many complex, indirect, and interacting influences to prove projected images exerted some indirect influence upon artists of the time.

Moreover, whose scholarly or subjective judgements on which works or passages possess the "optical look" should be favored? How would we objectively decide between the differing and competing impressions of several scholars or artists? Indeed, the award-winning professional realist artists among the current authors (J. Collins and N. Williams) often disagree with Hockney on numerous such matters. How do we objectively test which of them is right? Such debates arise in humanistic studies seeking new interpretations, but these bear little weight in objective tests of scientific or technical matters, such as this debate over the projection theory. In short, there seems to be no satisfactory objective and scholarly answer to such questions.

For these reasons we shall not consider further the indirect influence claim, except to note that, as developed in Section 8.7, the "optical look" of such a projection has the specific character of an extremely narrow depth of field projecting most of the scene as blurred, a look that is not found in a single painting from the 14th and 15th centuries, which seems to constitute strong evidence against the indirect influence hypothesis in its explicit form.

The philosophical and methodological drawbacks attending the *direct tracing* claim are subtle, but no less severe than the ones just described. Note that Hockney and Falco's claim, above, was that an artist such as van Eyck and Campin would secretly build a projector and then trace some portions and freely draw others where projections "did not suit his artistic vision" and, as such, any painting would be "composits" containing elements that are 'eyeballed' along with ones that are 'optics-based'. But how can we know, *a priori*, which are which? Surely we cannot rely on a single modern artist's impressions, including Hockney's, on this matter of "artistic vision." After all, suppose other artists (such as the artists among the present authors)—or indeed art historians or scientists—have different views on that matter. Who is right? Nor can we pick and choose which features are "optical" after they have been "fit" with an optical model—as to do so would risk confirming a pre-conceived "conclusion." Such a danger of creating "just so" stories is evident elsewhere in science, particularly evolutionary theory. Arguing in such a way would make the projection "theory" technically *non-falsifiable* [29], and thus devoid of explanatory power—thus not even an acceptable theory. Ernst Gombrich, too, developed this point in the context of scientific explanations of perception in visual arts [9]. Proponents would be free to alter their claims as evidence or analyses disconfirm theory, always in order to confirm their pre-determined "conclusion." In short, such revisions would expose the fact that the projection claims do not constitute a true scholarly theory. In fact, though, such major revisions have arisen several times in the debate, except to note that, as developed in Section 8.7, the "optical look" of such a projection has the specific character of an extremely narrow depth of field projecting most of the scene as blurred, a look that is not found in a single painting from the 14th and 15th centuries, which seems to constitute strong evidence against the *indirect influence* hypothesis in its explicit form.

It is important, too, to be clear about the logic of the optical hypothesis. One might imagine that the use of optical projection would imply that the perspective of the painting would be perfectly accurate, and that the occurrence of obvious deviations from accurate perspective would constitute evidence against the projection hypothesis. In fact, however, the opposite is the case, and Hockney and Falco consistently argue that the presence of perspective flaws is "proof" of its validity. Why is this the case? It is because they acknowledge that the optics of this period would use spherical lenses, and optical projection with spherical lenses is subject to spherical aberration, so that there would only be a small zone of the optical projection clear enough to be usable to guide the painter. They estimate the size of the usable zone to be 30 cm in diameter, beyond which the lens (or the canvas) would have to be moved to bring the next portion of the painting into focus. Such shifts would disrupt the continuity of the perspective between the two zones, and hence predict the occurrence of breaks in perspective in works painted by the projection method.

However, this analysis does not imply the reverse implication, that any errors in perspective are a convincing argument for the use of optical projection. Precisely such errors in perspective were almost universal in paintings before the proposed date of the introduction of the optical projection technique (1430), simply because the painters were painting by eye and did not understand the logic or the rules of perspective. Thus, finding errors in perspective after this date is most easily explained by the same logic, and carries no implication of any change in technique. The only form of perspective analysis that would constitute plausible evidence for the use of optics is if the perspective was perfect within 30 cm zones of paintings but showed errors between these zones of accuracy [62]. Not only do the theory's proponents not report any example that fits this description, they do not even perform this dual comparison in any of their analysis. They either report errors in perspective without establishing zones of full accuracy, or argue for zones of perfect accuracy without comparing them to adjacent zones. In fact, as we shall see, even in these cases the analysis was flawed, and the perspective is in fact inaccurate in all the cases that they use the argument from perfection.

In the final analysis the main evidence adduced by Hockney and Falco is the composition of the paintings themselves. As an example, consider Jan van Eyck's Arnolfini portrait (Figure 8.1). In their web posting for the 2001 *Art and Optics* symposium at New York University, Hockney and Falco wrote: "van Eyck placed a convex mirror at the center of this [Arnolfini] masterpiece, the very mirror which, turned around, he may well have used to construct this image." Note especially that this quote refers to "this image"; indeed the painting is displayed, prominently, *in full*, on the home page of that website. (The full image also appears in Hockney's book, the cover of the journal bearing their first technical article, and indeed elsewhere.) This quote, in full context, gives no hint that the claim might refer to a teeny portion of the image, or what portion that might be. Shortly after that conference, Stork showed that the full Arnolfini image was in such poor perspective—even within putative "exposures"—that it was extremely unlikely that projections were used throughout the body of the painting. (Stork also showed that the focal length of the convex mirror differed significantly from that of the putative projection mirror, and hence the theory's proponents claim that "the very mirror may well have been used..." was false on another ground, as we review in Section 8.3.3.) Later, Hockney and Falco focused technical attention on the splendid chandelier or *lichtkroon* (Dutch, "light crown"), Hockney asserting in a high-profile television broadcast: "That chandelier is *in perfect perspective*," as it would be were it traced under optical projections. In that broadcast Hockney then demonstrated his claim by tracing the projected image of a similar chandelier—the arms *as well as the decorative structures most distant from the arms*. That is, he demonstrated that the believed van Eyck would have traced the *full* chandelier.

Hockney is not even accurate in his description of the chandelier in this painting. He argues that it was optically projected because it is "seen from head on (not from below as you would expect)." By this assertion, he means as you would expect from the composition of the painting as a whole. As pointed out by Tyler [62] in a thorough analysis of the errors throughout the original Hockney book, this claim is obviously false, since the front arms of the chandelier are clearly much higher in the painting than the rear ones, in every feature from the candles to the lowest ornaments or *crockets*. This arrangement must imply that the chandelier is seen from below. Thus, this argument for the optical projection hypothesis completely collapses on even casual examination.

Moreover, shortly thereafter, Stork and Criminisi showed rigorously that the full chandelier was not in perspective—not even close [3, 5, 31, 32]. More specifically, they showed that if the physical Arnolfini chandelier was fairly symmetric, then its image in the painting was not in perspective. In this way, they refuted Hockney's projection claim. In response, Hockney and Falco then claimed that the decorative structures or crockets were "soldered on" by hand to the base arms and hence would have been haphazardly arrayed—in short, that the *full* chandelier image was in proper perspective, just that the chandelier itself was highly irregular and asymmetric [13].

Then Stork and Criminisi used rigorous photogrammetry of a modern casting of a 15th century chandelier as well as of large, appropriate Renaissance chandeliers and a prayer book holder, and in one case direct physical measurement with a tape measure *in situ*, all to show that all of these chandeliers were far more symmetric than is consistent with the projection claim [5]. Furthermore, experts in *dinanderie*—the decorative metalwork of the early Renaissance—pointed out that the arms on such a chandelier were fashioned from a *single* mold and then arrayed around the central staff, thus making the chandeliers highly symmetric—too symmetric to be consistent with the theory's proponents' claims. In that era, decorative structures were *never* soldered onto arms of dinanderie.

Stork and Criminisi also showed that a talented realist painter could paint highly complex chandeliers in excellent perspective—far better than did van Eyck—without any aids whatsoever, optical or mechanical. In this way they undermined experimentally Hockney's key motivation.

In summary, then, the proponents' claims about the Arnolfini portrait went through the following revisions, in order:

1. The *full image*—"this image"—was executed using direct tracing from the convex mirror depicted in the painting.

2. Just the full (assumed nearly symmetric) *chandelier image* was executed using direct tracing from some alternative and unknown (concave) mirror.

3. The *full chandelier image* was executed using direct tracing from a concave mirror, but the physical chandelier was *asymmetric* (because the crockets were "soldered on" to the arms).

4. Just the image of the nearly symmetric chandelier *base arms* were executed using direct tracing from a concave mirror, the (asymmetric) crockets added "by eye."

Clearly, this successive retraction of claims—a jettisoning of the claim about the depicted convex mirror, and then an overall reduction in painting area to teeny portion of their original claim—is not a principled refinement due to improved measurements or the inclusion of more data, as is typical in scientific research, but instead a wholesale revision of the key aspects of the claims as independent scholars rebut each claim in turn.

Our goal in rehearsing this history of the debate over the Arnolfini portrait is not to merely rebut the tracing claim, but instead to expose these *ad hoc, ex post facto* revisions to the proponents' claims as evidence and independent analysis shows that their previous claims were wrong. In short, if the projection "theory" allows such *ad hoc, ex post facto* alterations and major revisions to the claims under the subjective and debatable impressions concerning "artist's vision," well then the "theory" is devoid of explanatory power and not a true scholarly theory at all.

A closely related issue to the construction of "just so" stories centers on how a theory accommodates and explains truly new evidence—evidence that was not available when a claim was made, but which is surely relevant. Consider for a moment a case touted by Falco: Walter and Luis Alvarez's bold claim that a comet strike caused the extinction of dinosaurs roughly 65 million years ago. This claim received very strong support *later* when the "smoking gun" impact site beneath the shore of the Yucatan Peninsula and associated higher-than-expected concentrations of the element iridium were discovered. There remains some debate over this theory, but the relevance of this corroborating evidence is beyond question. This newly found evidence fit very well into their theory, without requiring *ad hoc* modifications.

In this regard, and by contrast, the optical projection theory fails. Consider the theory's proponents' claim that van Eyck secretly built an epidiascope—a simple projector to form the image of a flat object such as an artwork or document—to copy and enlarge his silverpoint study of Cardinal Niccolò Albergati (1431), which we shall revisit in Section 8.3.4. Stork explored a number of non-optical copying methods too, including the use of a reducing compass or proportional compass, *compasso da reduzione* or *Reductionszirkel*, known from as early as Roman times, and their relation to the fidelity and "relative shifts" in the van Eyck works [5, 30]. Later, Thomas Ketelsen and his team (which included two physicists) published their dramatic discovery of nine tiny pinprick holes along the contours of the silverpoint study. These holes are completely consistent with the use of a reducing compass, whose metal tipped legs would indeed leave such marks in the silverpoint; these pinprick holes play *no role whatsoever* in the optical projection theory. In light of their dramatic discovery, this European team concluded that van Eyck used such a reducing compass—not an epidiascope. (The "relative shifts" and scaling of the ear can also be easily explaining by the reducing compass, as we discuss below). Again, this newfound evidence fit perfectly with the *mechanical* compass explanation, refuting the optical explanation.

Given this very strong new evidence for mechanical (not optical) copying, Falco then claimed that this evidence was irrelevant to the oil copy because such holes could not be carbon dated and one could not know when or why they were placed in the silverpoint [24, quoted in]. (He also claimed, without evidence, that the nine pinprick holes were "too few" to enable an artist to achieve the fidelity found in the oil copy—a claim that is contradicted by the experimental evidence based on the work of several realist artists [5].) Our point here is to highlight the fact that rather than incorporating and explaining this newfound distinctive pinprick evidence in the optical explanation, the promoters of the projection theory were forced to retreat and claim the evidence is irrelevant in an *ad hoc*, *ex post facto* way. Nor, too, did proponents provide evidence of another copy of the silverpoint study that might bolster their additional claim that the pinprick holes were used for another copy.

Hockney's inspiration for his theory came when he marveled at the "photographic" quality in the works of French Neoclassical painter Jean Auguste Dominique Ingres (1780–1867) evident in an exhibition of this artist's works. Hockney concludes, based in large part on the quality of Ingres's lines, that Ingres used some form of optical aid, specifically a camera lucida, a simple device that allows the artist to see the subject optically superposed on his support such as paper [8]. But is optics the only technique that would yield a change in the quality of line stressed by Hockney?

The following is by the artist A. S. Hartrick, who trained in Paris under the academic master Fernand Cormon, a student of Alexandre Cabanel, Eugène Fromentin, and Jean-François Portaels. Hartrick describes a 19th century non-optical method from French that Ingres might have used:

> Fernand Cormon, the master from whom I learnt most in my student days in Paris, had a method which I believe may prove valuable to others... Setting himself at that distance from his model which would give him approximately the same scale as that on which his figure would appear when the whole decoration was viewed at once, and also that at which he could see the whole figure of his model and of his drawing as of about the same size, he would sketch in the main movement and construction of his figure with a few bold lines, and fix the main distribution, as well as the weight of the accents. Over this drawing he then placed a sheet of tracing paper. Moving nearer the model if necessary, he next searched the character of the contour and all details of the features and extremities most thoroughly, working with pencil and modeling all up as far as they could be carried, till finally he had a finished study, usually no more than a foot high.

By this means much of the freshness of a sketch was retained, because the drawing could be completed in a reasonable time before the model or the artist was tired. Afterwards these drawings were carefully squared [copied using a grid] and enlarged to any size he desired [10, pp. 72–73].

Our goal here is not to argue Ingres used the method described by Hartrick. Rather, it is to offer evidence that it is a plausible alternative in order to re-emphasize the need to consider multiple explanations, and especially the need to reject through evidence—or better yet, disprove—competing explanations. It is simply methodologically unsatisfactory to create an explanation, select assumptions consistent with it, and ignore alternative explanations and contradictory evidence,

Consider, too, more closely the key philosophical issue of the burden of proof. The projection theory is clearly a revisionist theory, intended to overturn the standard view that the precision of the paintings was due to the talent of the artist, new media such as oil paints, etc., as the proponents themselves often stress. As such, the burden of proof for the new theory lies foursquare upon the proponents' shoulders. They cannot point to a painting, advance the claim it was executed by tracing optical projections, and then demand others "disprove" their claim, of course. (This would be like claiming that the exceptional works were painted by aliens from another galaxy, then saying it must be so unless others produce evidence against such aliens.) Rather, they must show that it is far more plausible that the work was executed using optics than it was using traditional, non-optical methods, such as rulers, reducing compasses, grid constructions, "eyeballing," and so on. In the absence of such compelling evidence and reasoning, we must reject their claim. Likewise it is not sufficient for proponents to somehow "fit" the visual evidence using an optical model (especially when there are many optical degrees of freedom) or even "re-enact" or "demonstrate" that an optical procedure might *conceivably* have been used. They must also show that one cannot fit the evidence with non-optical explanation. In most of their papers, proponents do not acknowledge even the possibility of specific alternative, non-optical explanations, and in no cases do they rule out such alternative explanations, as we shall see below.

8.3   Image Evidence

We begin with a bit of background, then move to specific paintings, claims, and counter-claims.

8.3.1   Background

As shown in Section 8.4, below, a wide range of experts conclude that there is no persuasive documentary evidence that artists of the early Renaissance saw images projected onto a screen, traced them during the execution of their works, certainly nothing one would expect for a procedure claimed to have fundamentally transformed art and art praxis. Nor is there is a persuasive explanation why artists in guilds or ateliers devoted to developing and sharing technical information, such as van Eyck and Campin, would have kept this important information as a confidential "trade secret." Quite the contrary: such artists freely advertised their discoveries—or hints *about* their discoveries—in order to attract patrons and apprentices [21]. As such, then, the theory's proponents have focused on visual evidence within the paintings themselves. In this section we summarize the image evidence and the arguments for and against the optical projection claims, organized very roughly according to their importance to the theory. This section can be considered an updated version of, and indeed confirmation of, an earlier overview [35]. We give merely a brief summary of the evidence and arguments; interested readers should consult the original papers, as cited, for more details.

An immediate question arises about the timing of the change in style that is Hockney's chief evidence for the use of optical projection, because the putative "sudden transition" has a very fluid boundary variously attributed as occurring within decades as much as 100 years apart, pinned by time boundaries that are mutually contradictory in many cases [64]. Thus, the pictorial evidence for the boundary is given as between 1423 (Fabriano) and 1436 (van Eyck), between 1438 (Pisanello) and 1553 (Moroni), in depictions of armour between 1450 (Pisanello) and 1460 (Mantegna), between 1475 (Melozzo da Forlì) and 1514 (Raphael), between 1514 (Cranach) and 1560 (Moroni) and finally between 1525 and 1595. Incidentally, in making this latter transnational comparison, Hockney somewhat implausibly defines the transition as occurring at the same time in both Northern and Southern Europe. In summary, Hockney makes no comment on the floating discrepancy in the mutually contradictory timings of his evidence. One could suppose that different artists picked up the ideas at different times from each other, but all explicit statements of the hypothesis are that it was a nearly universal transition that occurred suddenly (with subsequent evolution of the optical technologies). Rather than sort through these conflicting dates, we shall focus on the period 1430–1550.

8.3.2 Lorenzo Lotto, *Husband and Wife* (1543)

Hockney and Falco claim their central evidence in the entire debate centers on the carpet pattern in Lorenzo Lotto's *Husband and Wife* (1543) in the Hermitage Museum in St. Petersburg Russia, which they call their "Rosetta Stone," claiming that it is "simply not possible" that the painting was executed without optics, and that this visual evidence "proves" that Lotto used optics. The painting was the subject of the first technical paper on the projection theory [12], and is discussed in several of the proponents' other papers and presentations.

Claim

In brief, these proponents point to perspective anomalies in the pattern of the carpet which they explain by Lotto secretly building a concave-mirror projector and projecting the image of a (nearly symmetric) carpet onto his canvas. They claim he traced the pattern in three sections, refocussing his projector between these "exposures" to overcome its limited depth-of-field [12]. The proponents adjust a number of parameters in their putative projector (mirror focal length, facial area, locations, etc.) to "fit" the image evidence. Proponents also pointed to an "indistinct" or "blurry" passage at the top of the keyhole which they explain with the highly unorthodox claim that Lotto traced this passage "blurry" to reconcile the current sharp projected image there with his memory of the previously out-of-focus image there—a claim they apply to no other painting, and not even to other equivalent passages within this painting.

The proponents also point to a passage in Lotto's *Libro di spese* or personal notebook as documentary support for their claim about Lotto, as we discuss immediately below [15].

Rebuttal

The Hockney and Falco claim rested on their unstated and untestable assumption that the specific physical carpet in Lotto's studio was symmetric, at least to about 2%—roughly their claimed precision of fit. Stork pointed out, however, that such handmade and transported "Lotto carpets" (later named for this artist) were typically asymmetric upon creation and would have become even more asymmetric by the time they arrived in Lotto's studio [23, 34]. After all, such carpets were hand-knotted in what became present-day rural Turkey by uneducated young girls working side-by-side. After the girls tied knots for months, the carpets were then taken down from vertical looms, thus relieving months-long stresses in the weave and hence altering the shapes of the decorative patterns. Then, the carpets were rolled and transported in donkey carts hundreds of miles over dirt roads, loaded onto ships for the rough journey to Venice, then likely unrolled and displayed and moved in shops for extended periods. Indeed, such carpets surviving in museum collections are usually asymmetric—far more than is consistent with proponents' stated precision of tracing. There seems to be no way, even in principle, for anyone to prove that Lotto's *particular* carpet in his studio was symmetric to less than 2%, as would be a necessary first step in supporting the tracing claim for this painting.

If Lotto traced within an "exposure," then the perspective in the corresponding passage should be accurate or coherent to Hockney and Falco's stated precision of three significant figures. Tyler showed conclusively, however, that the perspective in the carpet deviated by more than that precision, even within a single putative "exposure" [62]. He also showed that the carpet had global coherence but local incoherence—precisely the opposite of what we would expect had the carpet been traced under multiple projections.

The Hockney and Falco claim about the origin of the "blurry" region at the top of the keyhole is unorthodox, and used in no other painting in the debate—indeed not even elsewhere in this painting where we might expect it, specifically the transition from the closest putative "exposure" to the middle "exposure." Close inspection of the painting shows that the passage was executed with a somewhat large brush. There is no visual evidence in that passage for marks indicating that Lotto traced an image—no pencil marks, no incisions, no alternative colors, no partially hidden drawings, and so forth.

Robinson and Stork levied an even more serious challenge to the projection claim for this painting. They showed that the setup in the Hockney and Falco concave mirror projector was fundamentally flawed as it did not include Lotto's 116-cm-wide canvas. When the canvas is included into their setup, however, the light from the carpet would be blocked and not reach the mirror; the full optical setup in the putative projector simply cannot work [25, 43]. Using sophisticated ray-tracing software, Robinson and Stork showed that when Lotto's canvas is included into the setup, as required, the light from the carpet is forced to strike the putative concave projection mirror at a large angle, leading to the significant off-axis aberrations of astigmatism and coma—images too blurry to reveal the fine detail in the painting. Most importantly, the blurriness is nearly the same at different distances into the tableau. The ray tracing simulations show that putative images would not have gone in and out of focus, as is central to the proponents' arguments. In short, these aberrations preclude the kinds of depth-of-field phenomena central to the optical claim: the fundamental phenomenon (refocusing to overcome limited depth-of-field) underlying the Hockney and Falco explanation simply would not occur in the setup they presented.

Finally, every technical aspect of the specific documentary passage adduced as support by Hockney and Falco in fact contradicts their image evidence and optical claims. Whereas Lotto's *Libro di spese* (personal notebook) refers to a "big" mirror, the projection proponents infer a mirror *small* indeed (diameter roughly 2.5 cm); whereas the *Libro* refers to a *breakable crystal* or glass mirror, the proponents claim instead an *unbreakable metal* mirror; whereas the *Libro* states the mirror cost an "enormous sum," the proponents' small metal mirror would have been *inexpensive*. Further, there is no textual evidence in the *Libro* to indicate the mirror was concave (or could even produce an image) as needed for a projector, rather than the much more common convex or plane mirrors. Nor is there any description of the complicated tracing procedure. Indeed, had such a mirror had such a remarkable projection capability, there is every reason to believe that Lotto would have made extensive comments about its wondrous capabilities in his private notebook, as did Giambattista della Porta about a century later when he discovered the "magical" image-projection effects of concave mirrors. Surely Lotto was not afraid of revealing trade secrets through his personal notebooks.

For these reasons, these and other independent scholars rejected the optical projection claim for this, the central evidence in the debate—Hockney and Falco's "Rosetta Stone."

8.3.3   Jan van Eyck, *Portrait of Giovanni Arnolfini and His Wife* (1434)

The second-most important work in the debate is van Eyck's *Portrait of Giovanni Arnolfini and His Wife* of 1434 in the National Gallery London, striking for its heightened realism and for appearing near one of the dates Hockney ascribed to the change in realism in Western art (Figure 8.1). This painting figures prominently in Hockney's book [11], appears on the home page of the New York Institute for the Humanities *Art and Optics* website, and on the cover of the issue of *Optics and Photonics News* carrying the first scholarly article by the proponents [12].

FIGURE 8.1 (SEE COLOR INSERT)
Jan van Eyck, *Portrait of Giovanni Arnolfini and His Wife* (1434), 82.2 × 60 cm, oil on oak panel.
National Gallery London.

Claim

As we saw in Section 8.2.1, Hockney and Falco have retreated and altered their claims about this work, first claiming the entire painting—"this image"—was executed by tracing an image projected by the convex mirror depicted within the work ("the very mirror"), then that just the full chandelier was traced, and finally that just the chandelier arms were traced but not the decorative structures. Although Hockney and Falco place error bars around the purported two-dimensional location of the *bobeches* or candle holders, they do not explain how they calculate those locations, what assumptions they needed to make, nor do they cite the rigorous methods of computer vision relevant to that claim [3]. Most importantly, they give no evidence that achieving such purported accuracy demands the use of optics—merely one of many steps needed to rule out the default non-optical claim.

Rebuttal

As mentioned in Section 8.2.1, above, the full Arnolfini room could not have been executed from a projection from the depicted convex mirror, turned around because: (1) the perspective within putative "exposures" is incoherent, (2) the estimated focal length of that mirror is too short, (3) there is not enough light in the room, (4) such a large mirror produces a blur spot too large to reveal the fine detail found in the painting, (5) such a hand-blown distorted mirror would produce an even blurrier image, and (6) such a convex mirror was lined with a rough, unpolished coating of molten lead or other metals, so it could not function as a mirror anyway (cf. Section 8.5).

Stork and Criminisi showed that the image of the full chandelier is not in perspective—not even close [3, 5, 31, 38, 39]. It was the fact that led Hockney and Falco to retreat from the claim "that chandelier is in perfect perspective."

If we jump over the proponents' several intermediate claims to their final, much amended claim— that just the arms of the chandelier were executed from projections—we confront a number of problems, unanswered questions, and even apparently unanswerable questions. Other artists, and of course Hockney himself on CBS's *60 Minutes*, feel van Eyck would have traced the decorative structures, so we must ask: why would the proponents' final claim be that van Eyck had traced the tiny, mostly hidden arms at the back of the chandelier but not the crockets that appear so prominently at the left? By what objective principle or independent evidence can we decide such matters? Further, Hockney and Falco give no evidence that an artist getting those portions in good perspective would require an artist to use projections. In fact, Stork and Criminisi showed that at least one realist artist could execute two complex chandeliers entirely "by eye" in better perspective than van Eyck. While hundreds of millions of modern people have seen photographs and television images in excellent perspective, it seems that artists trained the way we know that artists of the early Renaissance were trained can paint such an image in good perspective by eye. In the Renaissance, such artists were selected in youth according to their talents, then apprenticed to masters, and spent endless years of apprenticeship studying life drawing and copying works (without optical aids).

8.3.4 Jan van Eyck, *Portrait of Niccolò Albergati* (1431 and 1432)

The next works we consider are also by van Eyck: a small study portrait in silverpoint of Cardinal Niccolò Albergati from 1431 in the Kupferstich Kabinet in Dresden Germany and a larger copy in oil from 1432 in the Kunsthistorisches Museum in Vienna. Here the proponents claim that the artist copied the silverpoint study by means of an *epidiascope*.

Claim

The proponents claim that van Eyck secretly used an *epidiascope*, or simple opaque projector, to copy and enlarge the silverpoint study. (The epidiascope was unknown from that era [16].) There are two classes of evidence proponents highlight. First, the fidelity of contours is high: portions of contours, suitably scaled, overlap quite accurately. Second, proponents report that for a given relative shift or offset of the images, only a *portion* of the contours have good correspondence, but if the contours are then shifted with respect to each other, then a different portion of contours overlap. To explain this "relative shift" or "relative offset" evidence, Hockney and Falco claim van Eyck traced part of the projected image and then "accidentally bumped" the projector—thereby shifting one image with respect to the other—and then continued tracing. They explicitly claim van Eyck "made a mistake" in this regard.

Rebuttal

The proponents claim that the fidelity found in the van Eyck works and that the evidence of relative shifts demands an optical aid. However, experiments show that modern professional realist artists can achieve the fidelity found in the van Eyck works without using optics [5]. Moreover, the experimental evidence shows that the "relative shift" evidence can be easily explained as van Eyck merely placing and scaling the ear such as we find in the final work, presumably by eye and for purely artistic reasons. In fact, realist artists and teachers of life drawing point out that novice drawers and even some accomplished artists working rapidly, frequently place the ear "too close" to the front of the face—just as we find in van Eyck's silverpoint [30, 31, 38, 39].

There is, further, an immediate problem with the Hockney and Falco claim that van Eyck made a "mistake" by bumping his putative epidiascope. The artist would *surely* have noticed any resulting mismatch or misalignment of image contours, the projected image and the traced contours already committed to the support as well as the contours that border different traced passages. We built a simple epidiascope of the type proposed by Hockney and Falco and deliberately "bumped" the mirror to shift the alignment of the two images. The mismatch between the contours was extremely conspicuous, especially for the few moments that one of the images was *moving*. It seems inconceivable that van Eyck, working closely on an important commission, would not have noticed such a bump "mistake." Instead, it is all but certain that van Eyck deliberately shifted the position of the ear for artistic or compositional reasons.

As mentioned above, Thomas Ketelsen and his team, which included two physicists, discovered the first truly new evidence in the debate over these works, that is, evidence that was not available to proponents when they created their explanation: tiny pinprick holes along the contours in the silverpoint study [19, 20]. Such distinctive physical evidence is entirely consistent with the use of a reducing compass as the mark of a tip of the compass; such evidence plays no role whatsoever in the optical explanation. Indeed, Ketelsen and his team conclude van Eyck used a reducing compass, not a concave mirror epidiascope. Note too that the reducing compass dates from Roman times whereas the epidiascope was unknown at that time of van Eyck [16].

Falco later tried to explain away this key pinprick evidence, claiming it was irrelevant to the oil copy and stating without evidence that nine holes are "too few" for van Eyck to have achieved the fidelity found in the works [24]. However, experimental evidence from realist painters shows that excellent fidelity can be achieved "by eye," and even better fidelity with a few measurements provided by a reducing compass [5]. Furthermore, the nine holes found by Ketelsen and his team indicate a lower limit to the number of measurements made by van Eyck; it is quite possible that the artist made *many* measurements with a reducing compass by choosing different pairs of the nine holes, or only *lightly* touching the device to the silverpoint study, thus leaving no additional pinprick holes. In short, Ketelsen et al.'s mechanical explanation fits all the visual and contextual evidence better than does the Hockney and Falco epidiascope explanation.

We note in passing another work by van Eyck: his highly realistic or "optical" *Portrait of a Man in a Turban* of 1433 in the National Gallery London, a work widely believed to be a self-portrait. As Hockney himself admits, self portraits cannot be executed by tracing optical projections and thus we can confidently conclude that van Eyck did not need to use a complex, secret optical device to attain the realism that characterizes his œuvre. There are numerous highly realistic self-portraits from the Renaissance and later—from Albrecht Dürer to Carracci to Diego Velàzquez—and we can likely be confident none of them were executed by the secret use of optical projections.

8.3.5   Robert Campin, *The Mérode Altarpiece* (1430)

The Hockney and Falco claim for *The Mérode Altarpiece* is significant in that, if verified, this triptych would be the earliest recording of the image of an illuminated object projected by an optical element such as a concave mirror or converging lens—the first step toward the chemical recording of an image in photography, nearly four centuries later.

Claim

In brief, the projection theory proponents claim that Campin secretly built a concave mirror projector, took St. Joseph's bench and its trellis out into the sunlight, projected its image onto this panel support, and traced the trellis in "exposures," each refocusing to overcome the purported projector's limited depth-of-field. Hockney and Falco initially pointed to a single-break change in perspective between the front of the trellis and the back, which they attribute to Campin repositioning his mirror. Later, they favored a two-break explanation and pointed to tiny "kinks" in the upper-left-to-lower-right (UL-to-LR) slats, and claimed such kinks would not have arisen had Campin used a straightedge.

Rebuttal

In the first place, it seems quite improbable that Campin would have needed to go to the elaborate lengths of an optical projection to draw the very simple trellis consisting of crossing parallel slats. It seems that any artist of the *trecento* would know how to draw frame and connect across the diagonals to make a trellis. Moreover, Stork showed that a very simple geometrical or mechanical construction, specifically that the artist merely traced diagonal slats, could explain the change in vanishing points between the front and back of the trellis [33]. The same form of mechanical construction could explain a "three-exposure" model as well. Hockney and Falco claimed that kinks in some individual slat images were due to abutting exposures under different focus conditions, and that the kinks precluded the use of a straightedge. However, they showed only the visual evidence from the UL-to-LR slats. Kulkarni and Stork reasoned that if Campin refocused, there would likely be kinks in the *other* slats, the lower-left-to-upper-right slats (LL-to-UR), at that same depth; it would be very unlikely that orthogonal slats would just happen to be straight there. However, when Kulkarni and Stork checked those LL-to-UR slats, they found that there were no kinks at that depth of the orthogonal kinks [21]. Kulkarni and Stork also discovered that there was an alignment of kinks in the LL-to-UR slats *into a range of depths* in the scene, rather than at a single putative refocusing depth. Such an alignment is incompatible with Hockney and Falco's projection claim.

Kulkarni and Stork argued, moreover, that it would have been quite difficult for Campin to trace the dim projected slat images *between* the kinks as straight, such as we find them. Finally, they showed that all the kink evidence—angles and separations—could be explained by Campin using a subtly kinked straightedge or mahl stick, both widely used at that time. All the evidence—changes in perspective, teeny kinks—fit naturally into the mechanical explanation, where Campin would have used a slightly kinked ruler or mahl stick.

8.3.6   Georges de la Tour, *Christ in the Carpenter's Studio* (1645)

To support his projection claim about de la Tour (and Caravaggio, Section 8.3.7, below) Hockney relies on evidence of *lighting* rather than of perspective. As such, a new range of computer vision techniques have been brought to bear on the analysis of his claim.

Claim

The de la Tour painting, now in the Louvre Museum in Paris, is so realistic to Hockney that he believes that de la Tour must have secretly traced an optical projection [11, pp. 128–129]. Because the single candle depicted in the work simply cannot produce enough light (as Hockney and Falco themselves admit), Hockney claims that when de la Tour painted Christ, some very bright light source was "in place of" St. Joseph, and when St. Joseph was painted, some very bright light source was "in place of" Christ [29]. Hockney shows two "half paintings," as we could call them, illustrating his claim.

Rebuttal

Beyond the issue of identifying a portable light source available in 1645 bright enough to provide sufficient light for an optical projection (for which Hockney and Falco provide no suggestion), the problem of whether de la Tour traced projected images thus comes down to answering the question: where is the source of illumination in the tableau? If it is a dual source—one each "outside the picture," or "in place of the other figure"—then the evidence would be at least *consistent* with Hockney's claim. If however, it is somewhere else—for instance the candle, the location Hockney explicitly rejects—then Hockney's optical claim would fail for this work.

A number of computer vision methods all yield a plausible answer to this question. The simplest method is cast-shadow analysis. The single, best-defined cast shadow in the entire tableau—the shadow of St. Joseph's right hand cast onto the beam below—clearly shows the the *candle* (not Christ) is the location of the illumination. When the full set of identifiable cast shadows is used, the shadow lines overlap strongly at the position of the candle [36]. Indeed, when the cast-shadow evidence is pooled or integrated by Bayesian statistical methods, the probability is far higher in the location of the candle than in place of either the other figures [6]. Finally, a few cast shadows of woodworking tools beneath St. Joseph point to a source in place of St. Joseph—the precise *opposite* of Hockney's claim. (Such shadows are completely consistent with de la Tour painting the tableau without projections but with an assistant holding a source above those tools.)

Another, independent, computer vision method for inferring the location of the illumination is based on the occluding-contour algorithm, which takes as input the pattern of lightness along the outer or occluding contour of an object, such as Christ's knee, shin, and so on [24]. Stork and Johnson applied this algorithm to the figures in this painting and showed that, as before, the best single location for the light source was at the location of the candle [51].

Yet another method for inferring the position of the illuminant is based on the pattern of lightness on planar surfaces, here the floor. Stork and Kale solved the equations for the appearance model of a planar Lambertian surface (such as the floor in this painting) and used it to estimate the position of a point source consistent with the pattern of lightness found on the floor [15, 50]. Although the results here were not as definitive as in the previous cases, the results were more consistent with the light being in place of the candle than in place of the other figures, thus supporting the rebuttal of Hockney's claim.

There is other lighting evidence in the tableau that cannot be analyzed by the above methods, for instance the pattern of light on solid surfaces such as Christ's chest or St. Joseph's thigh. To exploit such information one must assume a three-dimensional model of these objects [49]. To this end, Stork and Furuichi build a full three-dimensional computer graphics model of the tableau and adjusted the position of a virtual light source "in place of" St. Joseph and also "in place of" the candle. They found that the setup with the light in place of the candle led to a rendering that matched the painting—lightness on surfaces, directions of cast shadows, etc.—far better than if the light was in place of St. Joseph. In this way they corroborated the other conclusions from other methods that the candle was the source of illumination.

In short, the bulk of the visual lighting evidence is inconsistent with Hockney's claim about the location of the source, and thus contradicts his tracing theory as applied to this painting.

8.3.7 Caravaggio, *The Calling of St. Matthew* (1599–1600)

As with the de la Tour painting just described, Hockney's projection claim for this Caravaggio painting in the Contarelli Chapel in San Luigi dei Francesi in Rome, seems to be that Caravaggio traced an optical projection. Again as in the case of the de la Tour, the claim's resolution centers on whether the illumination is direct sunlight (needed for a projector) or instead a local, and hence artificial, source (and insufficient for an optical projector) [17, 20].

Claim

Hockney gives little visual evidence concerning this work, save for his informal impressions. He states: "Instead, there is a single light source, very strong from the right " [11]. By referring to "source," rather than explicitly "sun," Hockney seems to be inferring that the source was local (a conclusion consistent with some technical analyses, below). What Hockney apparently did not realize when he wrote that statement is that such a local source cannot provide enough illumination for a projector and thus Caravaggio would not have traced a projection.

Rebuttal

As in the case of the de la Tour, simple cast-shadow analysis of the shadows on the rear wall suggests that the light source is local in this painting, though this conclusion is based on reasonable (but untestable) assumptions about objects outside the frame of the picture. Model-based analysis of the pattern of light on the rear wall is not quite definitive [50, 51]. The dramatic reduction of luminance on the left of the wall is hard to reconcile with solar illumination. Sophisticated computer graphics modeling suggests a local illuminant for the rear wall, but a more distant source for the individual figures or that each figure was executed individually, making it relatively easy to give them all the same overall brightness.

In summary, the lighting evidence shows that the illumination may have been local and hence not compatible with Hockney's optical claim.

8.3.8   Hans Memling, *Flower Still-Life* (c. 1490)

Given his motivations for the tracing theory as an explanation for the rise in realism, it is a bit unusual that Hockney would claim this Memling work in the Thyssen-Bornemisza Museum in Madrid was executed using optics. After all, the carpet pattern is extremely simple, especially for an artist of Memling's stature and abilities and the painting surely lacks the "optical look" touted by Hockney. Why would an artist—*any* artist—employ a complicated optical system to draw such a simple pattern, one devoid of the subtleties and visual richness that motivated the projection theory?

Claim

Hockney's claims for *Flower Still-Life* follow the arguments for Campin's *Mérode Altarpiece* (cf., Section 8.3.5). That is, he believes Memling built a projector based on a concave mirror or converging lens, projected the image of the front of his table onto this canvas, and traced that image. Because such an optical system would have a limited depth-of-field (range of objects acceptably in focus), the artist might have had to then refocus his projector for the back half of the carpet. In doing so, he might have tipped and moved his mirror, thereby moving the horizon line and vanishing points. In support of this explanation, Hockney shows that the central vanishing point defined by the front half of the carpet are slightly higher than those defined by the back half. He shows no other vanishing points.

Rebuttal

Hockney failed to test the coherence of perspective in the front half of the carpet by drawing perspective lines at an angle to the direction of view, that is, construct additional vanishing points. Stork performed that additional perspective analysis and revealed that both the front half and the back half of the carpet are not in good perspective [37]. In fact, the angular deviations from perfect perspective within each half of the carpet are roughly twice those of the *change* in angle for lines from the front and the back halves defining the central vanishing points. In short, the evidence *against* the claim the carpet was in perspective is twice as salient as the evidence that is—at best—consistent with Hockney's optical claim. As such, we must reject the optical projection claim.

8.3.9   Hans Holbein, *The Ambassadors* (1533)

Hockney uses Holbein's *The Ambassadors* (1533) in the National Gallery London as a compelling example of the interest in optics, at least by the early sixteenth century.

Claim

Hockney and Falco point to two sources of visual evidence in support of their claim that Holbein traced an optical projection in this work. First, they point to perspective anomalies in the books depicted on the lower shelf in the painting, anomalies Hockney claims show that each was traced under a different optical projection. Second, they point to the famous anamorphic skull in this painting and claim Holbein refocussed a projector to overcome its limited depth of field. They find that with careful selection of optical parameters, particularly refocusing positions, they can find a line on the skull's jaw whose shape repeats [13].

Rebuttal

We can easily dismiss the optical explanation for the first source of visual evidence. Perspective anomalies of the sort found in the books on the shelf appear very frequently throughout art of the time and before, including Medieval frescoes (which of course could not be executed using optics). In short, perspective anomalies of this sort prove nothing whatsoever about the possible use of optics.

However, the original anamorphic projection could have been readily realized by viewing a skull in a slanted mirror and reaching out to outline the features of the skull on the mirrors slanted surface. The outlines could then be traced onto transparent paper and transferred to the painting without any understanding of the geometry of the optical projection [62]. The other interesting feature of this painting is that, despite its central feature of a sidetable displaying numerous astronomical and geometrical instruments, not a single mirror, lens or optical device is depicted. This does not give much support to the idea that Holbein was enamoured with the use of optics for the depiction of difficult objects. Surely he would have shown at least a few of the lenses and curved mirrors that Hockney and Falco supposed to have been in such vogue among the artists of this time. Hockney also claims that the globes in this painting are "marvellously accurate in their foreshortening," "perfect" and "precise," providing further evidence for Holbein's use of the optical projection method. Yet it is clear from inspection of longitude lines as they converge towards the handle that the longitude lines of the terrestrial globe are distorted, and the reconstruction of this geometry shows numerous inaccuracies consistent with brilliant painting "by eye" rather than accurate optical projection [65]. Thus the idea that Holbein was demonstrating a newfound infatuation with the use of optics with the anamorphic skull and other features of this painting becomes implausible on examination of the details of the painting.

8.3.10 Hans Holbein, *Georg Gisze* (1532)

Hockney mentions briefly and in passing Memling's portrait of Georg Gisze [11] in the Gemäldegalerie, Staatliche Museen, Berlin. Given the context, however, it appears that he is claiming that Memling traced projections for this painting too.

Claim

Hockney points to a break in the perspective line in the carpet in the work, and suggests that such a break is consistent with Holbein refocusing a mirror projector, much as in the case of Memling's *Flower Still-Life* (Section 8.3.8). Moreover, the coin box on the table is rendered in a different perspective than the front edge of the table, perhaps because it was executed under a projection from a different mirror position.

Rebuttal

Recall again that the essence of the Hockney and Falco theory is that the perspective in paintings based on optical projection should be locally consistent yet globally disrupted. A telling feature of the tapestry carpet in this painting is that, transforming the perspective as if it were viewed directly from above reveals that many of the components of the rug and the objects upon it have distorted local perspective. None of the rosettes have consistent symmetry: one is strongly rhomboidal, one is rectangular rather than square, and one has inconsistent symmetry. This, of course, means that there were in fact no consistent cues by which the transformation could be rigorously performed, but it was done to the best compromise by eye. It should be clear that many of the features have inconsistent distortions, and that the circular bases of the glass and the sand shaker are particularly distorted (again in inconsistent directions). Note that Hockney's entire analysis of the global inconsistency in this painting consists of just two white lines, the upper of which has no relation to any feature of the rug, and especially nothing that would align with the feature identified by the lower line. Thus the idea that it reveals a global inconsistency is not supportable [62]. We are arguing that all the features exhibit local inconsistency, and that the global organization is, in fact, largely consistent (as indicated by the straightness of the border lines in the rug). Thus the pattern of perspective disruptions is exactly the opposite from that expected on the Hockney and Falco hypothesis, and is completely consistent with what would be expected of an artist with an excellent eye (as Holbein undoubtedly was) attempting to approximate the design of a complex object viewed in extreme perspective without the use of any mechanical aids.

Another bizarre feature of the painting is that the table has the shape of a narrow triangle (after perspective correction), rather than having rectangular sides. The possibility of this construction being the actual shape of a real table is excluded by the fact that the book at upper left would fall off the table if the far side did not have a corner.

In attributing the change in fabric depictions to optical projection, Hockney neglects 250 years of intensive development of the artistic culture, comparable to neglecting the difference between Joshua Reynolds and the artists of today. More tellingly, he neglects the classic work of Gentile da Fabriano's *The Adoration of the Magi* (1423), which incorporates fabrics even more complex than those shown on pp. 37, 39 and 41 of *Secret Knowledge*. Fabriano's fabrics are reproduced on p. 70 of the book, where Hockney argues that they remain "essentially flat" and are judged as non-optical. Yet the complexity of their design is just the sort of thing that Hockney is offering as evidence for the use of optical projection. Moreover, close inspection of the cape of the kneeling Magi Melchior in this painting reveals that the texture is indeed strongly folded, though not as heavily shadowed as the van Eyck painting with which it is compared. The Fabriano work thus shows that artists before the supposed "transition" could paint complex fabric patterns without optics, and tends to support the idea of a gradual evolution of the painting style for fabrics from 1300 to 1600, as opposed to the concept of a sudden stylistic change attributable to optics in as early as 1420 [62].

8.4 Documentary Evidence

It has been noted by the projection theory proponents, and widely by experts in history of optics and art, that there is no documentary evidence that any artist saw an image of an illuminated object projected onto a screen and traced one during the execution of any of their works during the early Renaissance. A four-day symposium and accompanying proceedings devoted to examining Hockney's theory, especially the matter of documentary evidence, unanimously rejected the tracing claim, in large part for this reason [7, 40]. As workshop organizer Christof Lüthy summarized, "With respect to the 15th century, the idea that the Flemish Realism could be derived from the use of mirrors was roundly rejected." Likewise, "Taken together, the material, the visual and the textual evidence presented in these articles, makes the Hockney-Falco thesis extremely unlikely as far as its application for the period before the first textual reference to image projection around 1550 is concerned. The material evidence flatly contradicts the Hockney-Falco thesis, and while the textual evidence on its own cannot fully exclude the discovery of image projection, taken together with the material evidence of poor quality mirrors, the painterly use of image projection becomes extremely unlikely" [22].

The earliest scholar cited by Hockney and Falco in support of the tracing theory is the great Arab optical scientist, Ibn al-Haytham often Latinized to Alhacen [1, 31]. A. I. Sabra, who has translated all of Ibn al-Haytham's works, rejects any suggestion that this scientist created such projections:

"And yet, as already noted by M. Nazif, there is no report in [al-Haytham's] *Optics* of a composed picture inside the dark room" [23, in, p. 54]. Surely this optical scientist, who wrote more than 100 books, did not hide some projection experiment to preserve a "trade secret," nor was this Arab fearful of Inquisition, which would nevertheless arise centuries later! Lefèvre summarizes the evidence that artists traced projected images as early as claimed by Hockney and Falco:

> But there's a problem. There is, to date, not a single piece of direct evidence to support this [tracing] suggestion: there is not one example of a camera obscura or even a single part of one that dates from the 17th century, there are no written documents to confirm such devices were employed by artists of this time, no receipts for related materials or other unambiguous hints. [23, p. 5].

Saint Paul referred to the poor quality of Roman metal mirrors when he compared the flawed, dark view people have of this world to the clear knowledge of God that awaits them: "For now we see in a mirror dimly, but then face to face." Sara Schechner, an historian of science specializing in early instruments, has further shown that other cultural and documentary evidence reinforces what we find in analyzing the extant mirrors of antiquity through the early Renaissance—that the images were crude and dim [28].

The earliest documentary evidence to support the possibility of tracing comes well over a century after Hockney and Falco claim the procedure revolutionized art, specifically in the 1558 writings of Giambattista della Porta—the well-funded and highly connected magician and optical experimenter in Italy.

It pays to take a moment to clarify a possibly misleading reference to mirrors and projections from before the time of van Eyck. Falco has pointed to a number of passages in *Le Roman de la Rose* which discuss concave mirrors and the images they form. However, every one of these passages describes an image projected into space between the mirror and the viewer, rather than the far more difficult procedure of projecting an image onto a screen such as a canvas. *Le Roman de la Rose* bears no description of such projection—the type needed by the projection theory.

Hockney and Falco speculate that this lack of evidence was because artists sought to preserve "trade secrets" but Pamela O. Long's study shows, instead, that in the early Renaissance artisans and artists freely announced their discoveries in order to attract patrons and apprentices [21]. In the very rare cases of true secrets, such as those of the Venetian glass makers, knowledge *about* the existence of such secrets was well known. We have no credible documentary evidence *about* mirror projection "trade secrets." In short, the proponents' speculation for the lack of purported tracing procedure is not supported by expert scholarship.

8.5 Material Culture and Re-Enactments

Hockney and Falco assert that master painters of the early 15th century, such as Robert Campin and Jan van Eyck, used glass or metal mirrors to project images onto canvas where they could be easily traced to give lifelike detail. In particular, Hockney has pointed to the convex glass mirror in van Eyck's Arnolfini portrait (1434) and asserted in the *Art and Optics* website, "If you were to reverse the silvering, and then turn it round, this would be all the optical equipment you would need for the meticulous and natural-looking detail in the picture." Elsewhere Hockney claims that van Eyck also used a concave glass mirror to enlarge or reduce drawings and that later artists employed good quality, flat, glass mirrors to reverse images while retaining details. Falco, for his part, has made similar arguments for small concave metal mirrors. However, inspection of surviving mirrors and related objects shows that they were too crude to offer the early Renaissance painter an optical short-cut to a naturalistic image of his subject. Moreover, there was no mention of using mirrors to project an image in medieval optical works, and no material evidence survives that could have performed the task even if artisans or scholars had thought of doing this [28].

Bronze was the most common material for ancient mirrors which were cast into a slightly convex disk and polished by hand. The principal challenge was to prevent air holes and blisters, or the oxidization of impurities or threads of unmixed metal, which would cause pockmarks, cracks, or veins in the surface of the casting. Medieval mirrors of metal were also small, dark, and convex and their reflectivity was limited by the rough casting and being hand-polished. Moreover, such mirrors were extremely rare. Concave metal mirrors were seldom mentioned outside of the context of burning mirrors. Burning mirrors had very short focal lengths and were not figured or shaped accurately enough to project an image even at that range. Note that a deformed concave mirror yields a blurry, useless image—not a deformed sharp one [8].

Progress in the manufacture of glass mirrors was very slow and stymied by the difficulties in preparing the glass, making it transparent, shaping it, and foliating it. Glass made in Europe in the 14th and 15th centuries was tinted dark green or brown and filled with numerous air bubbles. The "broad" technique of forming glass panes produced a thick, almost opaque, uneven sheet of glass. The reflection off its surface was very distorted and mirrors made by backing it with lead were poor. The newer "crown" technique developed around 1330 produced thin disks of glass that had deep furrows and ridges, which could not be foliated with lead to make a mirror. The striations and bubbles in glass panes formed by either technique refracted light in a very irregular manner thus yielding poor images.

Crude spheres were much easier to form than plate glass. Consequently, glass mirrors that date from the 14th and 15th centuries were indeed convex like those seen in the Renaissance paintings that fascinate Hockney. The glass blower gathered molten glass on the end of his blowpipe and blew a bubble. While still on the blowpipe, small, thin spheres of glass were coated inside with molten lead, tin, antimony, or a mixture of these metals. When the metal and glass cooled, the sphere was cut into pieces to form convex mirrors. The reflected image from these convex mirrors was blurry since these were far from perfect spheres. Hockney's assertion that the convex mirrors (like that depicted in the Arnolfini portrait) could be reversed in their frames in order to serve as concave mirrors is false. The metal-coated interior would not be smooth, polished, or shiny, nor could it stand up to polishing. No method existed to coat the outer surface of the sphere. In fact, no concave, converging glass mirrors are known from this period; there was no method to make them. Thus, Hockney's claim that van Eyck and others used concave mirrors to project images onto canvas is moot.

8.5.1 Re-Enactments

There is also a problem with Hockney and Falco's modern re-enactments of purported early Renaissance procedures. One practical mistake that Hockney made was to assume that a modern shaving mirror has optical characteristics similar to mirrors from the past. Our cheapest mirror today is far superior to any mirror from 500 years ago, even accounting for the well-understood effects of rusting, corrosion, and so on. We cannot expect modern qualities of reflectivity or image production from old apparatus. Historical arguments that do not take this into account are prone to error [28].

Falco sets out to make a "suitable concave mirror using only technology that would have been available in the 15th century, with the goal of producing a 'mirror lens' of the specifications we calculated from Lotto's painting." For this he uses modern aluminum and brass stock and five grades of grinding/polishing compound. He writes, "Historians tell us that artisans were grinding glass spectacles by the 14th century, so they certainly had abrasive compounds at that time." While artisans did have abrasives, they were not as pure as modern ones. Moreover, they did not work in aluminum, and 15th century brass was a different alloy than modern brass. So, this purported re-enactment is not using materials comparable to historical ones.

Second, Falco uses a technique of grinding two pieces of metal together to generate a matched pair of concave and convex spherical surfaces. This is a well-known modern technique used by makers of telescope mirrors. However, this technique for grinding spherical surfaces was not introduced until the 17th century when astronomers required better telescope lenses (and later mirrors) than spectacle makers (and metal workers) were producing. Prior to that, lens makers ground their lenses in concave molds that were created by hammering copper into a rough, curved shape. They did not even use a template as a spherical control, much less a file to remove the hammer blows. As for metal mirrors, they were convex and made by casting. Two metal surfaces were not ground together in their creation. This means that lenses and mirrors were aspheric in Lotto's day and not made by Falco's method. In short, Falco's reenactment is anachronistic.

There is, further, a significant logical problem underlying Falco's "re enactments." Even if we grant that 15th century artisans had the "right" raw materials or tools at their disposal, there is no logical reason to conclude that they would have put these together in the same way and for the same purpose that later individuals have thought to put them together. To claim otherwise leads to silly conclusions—e.g., a claim that Aristotle could have discovered electrical current in the fourth century B.C. because he had coins of dissimilar metals, parchment, gold wire, and salt water; in short, all the ingredients of Volta's electrical pile of 1800. The progression from "could have" to "did" is even more logically suspect.

Thus, Falco's conclusion—"It is quite easy to fabricate concave mirrors of suitable focal length, diameter, and resolution for 15th century artists to have used to project images"—is fallaciously made [32].

8.6 Non-Optical Contexts

We mention that it may be no coincidence that the transition to the "optical look" Hockney identifies near 1430 is the same date as the emergence of the use of oil paints. Indeed, Jan van Eyck is sometimes called the "father of modern oil painting," though oil paints were used in a few cases before him. Oil paints afford a wider range of lightness—whiter whites and blacker blacks—richer, more saturated colors, and a number of layering and glazing techniques, which reach the apotheosis in the works of Rembrandt, who would apply as many as 50 layers of oil paint in a given passage. Much of the "optical look" is due to shading, sfumato, chiaroscuro, unrelated to the accuracy of contours related to any putative tracing of projected images. Moreover, this is also the time in the well-documented rise in the use of spectacles [15], [62]. Spectacles would allow an artist, especially one over 35 or so, to see distant subjects and close painting.

We note in passing that sculpture changed dramatically during this period as well and became far more "realistic." Consider, for instance, the evolution in style from the anonymous architectural statuary in Western (Royal) Portal of Chartres Cathedral (c. 1145) to Donatello's *David* (c. 1440) to Michelangelo's *Pietà* (1499). This remarkable rise in sculptural realism and expressiveness was, of course, unrelated to any development of technical tools analogous to those in the optical claim.

8.7 The "Value" in Tracing

Although the above discussion centered on the tracing claim, we must not forget the full extent of Hockney and Falco's claim: that tracing itself helped lead to the heightened realism or "optical look" of the *ars nova*. Tracing surely aids in capturing contours, of course, but it does not help in capturing subtleties in color, shading, and tone. The sight of a full-color projected image might aid an artist, but the contours alone are much like a child's color book. It is extremely difficult to *paint* directly under projections (as Hockney himself admits), and it would impede rather than aid rendering of color.

As Tyler [62] points out, one important issue is the true nature of the "optical look" that supposedly inspired the change in painting style in the 1420s. In the book, Hockney shows examples of the true "optical look" that is obtained by projection of the still-life scene of a bowl of fruit through an optical lens of about 15 cm diameter. With this large lens made to modern standards, the scene has a depth of field of only an inch or so of sharp focus. All the rest of the objects are heavily blurred. Hockney argues explicitly that it is precisely the characteristics of the projection of the optical image, including out-of-focus regions, that should have appeared in paintings at the time that optical projection first came into play.

In contrast, the large-scale van Eyck paintings that supposedly represent the style inspired by the "optical look" are almost preternaturally sharp throughout the scene. This property of ubiquitous clarity had, in fact, been characteristic of paintings since Greek and Roman times. It was nothing new. Conversely, as one can see from the projected image of still life [11, p. 104], the true "optical look" is extremely fuzzy, and would have been more likely to have inspired French Impressionism than the Renaissance precision. In fact, there have been compelling suggestions that the looseness of the late paintings of Impressionists such as Claude Monet was due to the reduced optical quality of their own eyes over time [61]. Thus, the look of the paintings is essentially the opposite of what would be predicted from Hockney's "optical look" hypothesis [62].

8.8 Scholarly Consensus

As far as we know from published scholarly works (rather than websites, blogs, letters to the editor, YouTube videos, and so on), the independent scholarly consensus—indeed unanimous consensus—is to reject the Hockney direct tracing claim. Nearly a dozen scientists or technologists, eight historians of optics and art, and two curators have published scholarly works rejecting the theory, at least for the works in question. Participants in the four-day workshop devoted to testing the tracing theory unanimously, and in no uncertain terms, rejected its claims [40].

Consider, too, the more informal literature of book reviews. A few early reviews of *Secret Knowledge* expressed intrigue with the bold tracing theory, but the vast majority—especially those by experts in art of the period—were strongly critical. We are aware of but a single book review to date (by an English professor) that finds Hockney's evidence and arguments persuasive [60]. Even Hockney's long-time collaborator and broad contributor to Hockney's book, art historian Martin Kemp of Oxford University, recently acknowledged his skepticism about the central claim of the optical projection theory:

> My own view is that Campin and van Eyck may well have been inspired by optically generated images—the camera obscura was well known to mediaeval natural philosophers— but *probably did not actually use them directly at any stage in the making of their pictures.* [18].

In short, Kemp too is skeptical about the central and explicit claim of the Hockney theory: that the "optical look" arose in Western art circa 1430 because some artists traced optically projected images.

8.9 Conclusions

We have examined the claim that some European artists secretly traced optically projected images during the execution of passages in some of their works as early as circa 1430 and that such a procedure was key to the rise of a newfound realistic, photographic, or "optical" look in the *ars nova* or new art of that time. We find that the theory itself, as stated, relies on subjective and ultimately untestable premises about which portions of an image an artist would or would not have traced. The theory's proponents have exploited this lack of a theoretical foundation to alter and retreat from claims in an *ex post facto* and *ad hoc* way. Further, the theory's proponents rarely explore in adequate depth alternative non-optical explanations for the visual evidence in paintings. We analyze the visual evidence in key paintings adduced in support of the tracing theory and find, without exception, that alternative, non-optical explanations are as plausible—and indeed generally far more plausible—than the optical explanations, especially in light of independent physical evidence and constraints. We also review briefly the documentary record for the period in question (c. 1420–1550) and find no persuasive evidence to support the direct artistic use of such projections. We examine, and ultimately reject, the speculation that this lack of documentary evidence was due to artists protecting "trade secrets" or fearing the Inquisition.

It is clear that the overwhelming—and to our knowledge unanimous—conclusion of independent scholars writing on this subject is to reject the Hockney direct tracing theory, at least for the period in question (1430–1550). Moreover it has not been demonstrated that tracing was needed for this rise, or in fact helped it at all. Of course, every scholar should be, and to our knowledge indeed is, open to new evidence that may arise, and no rebutter is so irresponsible as to have claimed to have "disproven" the tracing claim. Instead, we—and rebutters more generally—claim merely to have rebutted every aspect of Hockney and Falco's frequent claims to have "proven" their direct tracing claim, at least in the early Renaissance.

Despite this broad scholarly rejection of the direct tracing claim, we reiterate that we do not take a stand—for or against—Hockney's alternative claim of artistic influence, i.e., that some artists *saw* and were indirectly *influenced* by projected images. None of the technical analysis, such as referenced in Section 8.3, shed much light on the influence claim, though the lack of supporting contemporary documentary evidence in the early Renaissance and the issues of burden of proof for a revisionist theory argue against the indirect influence claim.

Although computer vision, pattern recognition, and image analysis long predate Hockney's speculations, his theory has motivated the development of a number of algorithms in particular, and the general acceptance of computer methods in the study of art. We feel that this, then, may be Hockney's most important legacy in this general domain. A number of scholars have moved past the tracing claims to address a wider range of questions in the history of art, research that is leading to new techniques and shedding new light on art and art praxis [41, 42, 44–48, 50, 52].

Acknowledgments

We gratefully acknowledge insights provided by numerous scientists, conservators, and historians of optics and art, as well as by anonymous reviewers. We would like to thank especially Martin Kemp of Oxford University and Walter Liedtke of the Metropolitan Museum of Art for their insights, but in doing so do not mean to imply they agree with every point presented here. The first author expresses thanks to the Getty Research Institute, Santa Monica, CA, for extended reader's privileges in its superb research library.

Bibliography

[1] Ibn al-Haytham. *The optics of Ibn al-Haytham, Books I-III*. The Warburg Institute, University of London, London, UK, 1989.

[2] Antonio Criminisi. *Accurate visual metrology from single and multiple uncalibrated images*. ACM Distinguished Dissertation Series. Springer-Verlag, London, 2001.

[3] Antonio Criminisi. Machine vision: The answer to the optical debate? *Optical Society of American Annual Meeting*, Rochester, NY (abstract), 2004.

[4] Antonio Criminisi and David G. Stork. Did the great masters use optical projections while painting? Perspective comparison of paintings and photographs of Renaissance chandeliers. In Josef Kittler, Maria Petrou, and Mark S. Nixon, editors, *Proceedings of the 17th International Conference on Pattern Recognition*, volume IV, pages 645–648, 2004.

[5] Marco Duarte and David G. Stork. Image contour fidelity analysis of mechanically aided enlargements of Jan van Eyck's *Portrait of Cardinal Niccolò Albergati*. *Leonardo*, 43(1):43–50, 2010.

[6] Richard O. Duda, Peter E. Hart, and David G. Stork. *Pattern classification*. John Wiley and Sons, New York, NY, Second edition, 2001.

[7] Sven Dupré, editor. *Early Science and Medicine: A Journal for the Study of Science, Technology and Medicine in the Pre-modern Period: Optics, instruments and painting 1420–1720: Reflections on the Hockney-Falco Thesis*, volume X, no. 2. Brill Academic Publishers, Leiden, The Netherlands, 2005.

[8] David Falk, Dieter Brill, and David Stork. *Seeing the Light: Optics in nature, photography, color, vision and holography*. Wiley, New York, NY, 1986.

[9] Ernst H. Gombrich. *Art and illusion: A study in the psychology of pictorial representation*. Phaidon, London, 1960.

[10] A. S. Hartrick. *Drawing from drawing as an educational force to drawing as an expression of the emotions*. Sir Isaac Pitman & Sons, Ltd., New York, NY, 1921.

[11] David Hockney. *Secret knowledge: Rediscovering the lost techniques of the old masters*. Viking Studio, New York, NY, 2001.

[12] David Hockney and Charles M. Falco. Optical insights into Renaissance art. *Optics and Photonics News*, 11(7):52–59, 2000.

[13] David Hockney and Charles M. Falco. Quantitative analysis of qualitative images. In *Electronic Imaging*. SPIE Press, Bellingham, WA, 2005.

[14] Vincent Ilardi. *Renaissance vision from spectacles to telescopes*. American Philosophical Society, Philadelphia, PA, 2007.

[15] David Kale and David G. Stork. Estimating the position of illuminants in paintings under weak model assumptions: An application to the works of two Baroque masters. In Bernice E. Rogowitz and Thrasyvoulos N. Pappas, editors, *Electronic Imaging: Human vision and electronic imaging XIV*, volume 7240, pages 72401M1–12. SPIE/IS&T, Bellingham, WA, 2009.

[16] Martin Kemp. *The science of art: Optical themes in Western art from Brunelleschi to Seurat.* Yale U. Press, New Haven, CT, 1990.

[17] Martin Kemp. Science in culture: Caravaggio's optical naturalism. *Nature*, 420(6914):364, 2002.

[18] Martin Kemp. Imitation, optics and photography: Some gross hypotheses. In Wolfgang Lefévre, editor, *Inside the Camera Obscura: Optics and art under the spell of the projected image*, pages 243–264. Max-Planck-Institute für Wissenschaftsgeschichte, Berlin, Germany, 2007.

[19] Thomas Ketelsen, Olaf Simon, Ina Reiche, Silke Merchel, and David G. Stork. Evidence for mechanical (not optical) copying and enlarging in Jan van Eyck's *Portrait of Niccolò Albergati*. In *Optical Society of American Annual Meeting*, Rochester, NY, 2004. OSA.

[20] Thomas Ketelsen, Olaf Simon, Ina Reiche, and Silke Merchel. New information on Jan van Eyck's portrait drawing in Dresden. *Burlington magazine*, CXLVII(1224):169–175, 2005.

[21] Ashutosh Kulkarni and David G. Stork. Optical or mechanical aids to drawing in the early Renaissance? A geometric analysis of the trellis in Robert Campin's *Mérode Altarpiece*. In Kurt S. Niel and David Fofi, editors, *Electronic Imaging: Machine vision applications II*, volume 7251, pages 72510R1–9. SPIE/IS&T, Bellingham, WA, 2009.

[22] Roberta Lapucci. *Caravaggio e l'ottica/Caravaggio and optics.* Privately published, Florence, IT, 2005.

[23] Wolfgang Lefévre, editor. *Inside the camera obscura– Optics and art under the spell of the projected image.* Max-Planck-Institut für Wissenschaftsgeschichte, Berlin, Germany, 2007.

[24] Eric J. Lerner. Van Eyck's dividers: Simple geometry? *Optics and Photonics News*, 16(7/8):9, 2005.

[25] Pamela O. Long. *Openness, secrecy, authorship: Technical arts and the culture of knowledge from antiquity to the Renaissance.* Johns Hopkins U. Press, Baltimore, MD, 2001.

[26] Christoph Lüthy. Reactions of historians of science and art to the Hockney thesis: Summary of the European Science Foundation's conference of 12–15 November, 2003. *Optical Society of American Annual Meeting*, Rochester, NY (Abstract), 2004.

[27] Rosamond E. Mack. Lotto: A carpet connoisseur. In David Alan Brown, Peter Humfrey, and Mauro Lucco, editors, *Lorenzo Lotto: Rediscovered master of the Renaissance*, pages 58–67. National Gallery, Washington, DC, 1997.

[28] Peter Nillius and Jan-Olof Eklundh. Automatic estimation of the projected light source direction. In *IEEE Conference on Computer Vision and Pattern Recognition (CVPR01)*, volume 1, pages 1076–1083, 2001.

[29] Karl Raimund Popper. *Conjectures and refutations.* Routledge, London, UK, 1962.

[30] M. Dirk Robinson and David G. Stork. Aberration analysis of the putative projector for Lorenzo Lotto's
Husband and wife: Image analysis through computer ray-tracing. In David G. Stork and Jim Coddington, editors, *Computer image analysis in the study of art*, volume 6810, pages 68100H–1–11. SPIE/IS&T, Bellingham, WA, 2008.

[31] Abdelhamid I. Sabra. Alhacen's optics in Europe: Some notes on what it said and what it did not say. In Wolfgang Lefévre, editor, *Inside the camera obscura: Optics and art under the spell of the projected image*, pages 53–57, Berlin, Germany, 2007. Max-Planck-Institute für Wissenschaftsgeschichte.

[32] Sara J. Schechner. Against the Hockney-Falco thesis: Glass and metal mirrors of the 15th century could not project undistorted images. In *Technical Digest of Frontiers in Optics 2004*. Optical Society of America, Washington, DC, 2004.

[33] Sara J. Schechner. Between knowing and doing: Mirrors and their imperfections in the Renaissance. In Sven Dupré, editor, *Early Science and Medicine: A Journal for the Study of Science, Technology and Medicine in the Pre-modern Period: Optics, instruments and painting 1420–1720: Reflections on the Hockney-Falco Thesis*, volume X, no. 2, pages 137–162. Brill Academic Publishers, Leiden, The Netherlands, 2005.

[34] David G. Stork. Color and illumination in the Hockney theory: A critical evaluation. In *Proceedings of the 11th Color Imaging Conference (CIC11)*, volume 11, pages 11–15, Scottsdale, AZ, 2003. IS&T.

[35] David G. Stork. Did Jan van Eyck build the first 'photocopier' in 1432? In Reiner Eschbach and Gabriel G. Marcu, editors, *Electronic imaging: Color imaging IX: Processing, hardcopy and applications*, pages 50–56, Bellingham, WA, 2004. SPIE.

[36] David G. Stork. Optics and realism in Renaissance art. *Scientific American*, 291(6):76–84, 2004.

[37] David G. Stork. Optics and the old masters revisited. *Optics and Photonics News*, 15(3):30–37, 2004.

[38] David G. Stork. Were optical projections used in early Renaissance painting? A geometric vision analysis of Jan van Eyck's *Arnolfini portrait* and Robert Campin's *Mérode Altarpiece*. In Longin J. Latecki, David M. Mount, and Angela Y. Wu, editors, *SPIE electronic imaging: Vision geometry XII*, pages 23–30, Bellingham, WA, 2004. SPIE.

[39] David G. Stork. Asymmetry in 'Lotto carpets' and implications for Hockney's optical projection theory. In Bernice E. Rogowitz, Thrasyvoulos N. Pappas, and Scott J. Daly, editors, *SPIE electronic imaging: Human vision and electronic imaging X*, volume 5666, pages 337–343, Bellingham, WA, 2005. SPIE.

[40] David G. Stork. Did early Renaissance painters trace optical projections? Evidence pro and con. In Longin J. Latecki, David M. Mount, and Angela Y. Wu, editors, *SPIE Electronic Imaging: Vision geometry XIII*, volume 5675, pages 25–31, Bellingham, WA, 2005. SPIE.

[41] David G. Stork. Did Georges de la Tour use optical projections while painting *Christ in the Carpenter's Studio*? In Amir Said and John G. Apolstolopoulos, editors, *SPIE electronic imaging: Image and video communications and processing*, volume 5685, pages 214–219, Bellingham, WA, 2005. SPIE.

[42] David G. Stork. Did Hans Memling employ optical projections when painting *Flower Still-Life*? *Leonardo*, 38(2):57–62, 2005.

[43] David G. Stork. Optique et réalisme dans l'art de la Renaissance. *Revue Pour la Science*, 327:74–86, 2005.

[44] David G. Stork. Spieglein, Spieglein and der Wand. *Spektrum der Wissenschaft: Forschung und Technik in der Renaissance*, Spezial 4/2004(4):58–61, 2005.

[45] David G. Stork. Tracing the history of art: Review of *Early Science and Medicine: Optics, instruments and painting, 1420-1720: Reflections on the Hockney-Falco theory*. *Nature*, 438(7070):916–917, 2005.

[46] David G. Stork. Computer vision, image analysis and master art, Part I. *IEEE Multimedia*, 13(3):16–20, 2006.

[47] David G. Stork. Mathematical foundations for quantifying shape, shading and cast shadows in realist master drawings and paintings. In Gerhard X. Ritter, Mark S. Schmalz, Junior Barrera, and Jaakko T. Astola, editors, *SPIE electronic imaging: Mathematics of data/image pattern recognition, compression and encryption with applications IX*, volume 6314, pages 63150K–1–6, Bellingham, WA, 2006. SPIE.

[48] David G. Stork. Aberration analysis of the putative projector for Lorenzo Lotto's *Husband and wife*. *Optical Society of American Annual Meeting*, San Jose, CA (Abstract), 2007.

[49] David G. Stork. Imaging technology enhances the study of art. *Vision Systems Design*, 12(10):69–71, 2007.

[50] David G. Stork. Locating illumination sources from lighting on planar surfaces in paintings: An application to Georges de la Tour and Caravaggio. In *Optical Society of American Annual Meeting*, Rochester, NY, 2008. Optical Society of America.

[51] David G. Stork. Locating illumination sources from lighting on planar surfaces in paintings: An application to Georges de la Tour and Caravaggio. In *Optical Society of American Annual Meeting*, page (abstract), Rochester, NY, 2008. OSA.

[52] David G. Stork. New insights into Caravaggio's studio methods: Revelations from computer vision and computer graphics modeling. *Renaissance Society of American Annual Meeting*, Los Angeles, CA (abstract):102, 2009.

[53] David G. Stork and Jim Coddington, editors. *Computer image analysis in the study of art*, volume 6810. SPIE/IS&T, Bellingham, WA, 2008.

[54] David G. Stork, Jim Coddington, and Anna Bentkowska-Kafel, editors. *Computer vision and image analysis of art*. SPIE/IS&T, Bellingham, WA, 2010 (forthcoming).

[55] David G. Stork and Marco Duarte. Computer vision, image analysis and master art, Part III. *IEEE Multimedia*, 14(1):14–18, 2007.

[56] David G. Stork and Marco Duarte. Revisiting computer image analysis and art. *IEEE Multimedia*, 14(3):108–109, 2007.

[57] David G. Stork and Yasuo Furuichi. Image analysis of paintings by computer graphics synthesis: An investigation of the illumination in Georges de la Tour's *Christ in the Carpenter's Studio*. In David G. Stork and Jim Coddington, editors, *Computer image analysis in the study of art*, volume 6810, pages 68100J–1–12. SPIE/IS&T, Bellingham, WA, 2008.

[58] David G. Stork and M. Kimo Johnson. Computer vision, image analysis and master art, Part II. *IEEE Multimedia*, 14(3):12–17, 2006.

[59] David G. Stork and M. Kimo Johnson. Estimating the location of illuminants in realist master paintings: Computer image analysis addresses a debate in art history of the Baroque. In *Proceedings of the 18th International Conference on Pattern Recognition*, volume I, pages 255–258, Hong Kong, 2006. IEEE Press.

[60] David L. Sweet. Review of *Secret Knowledge* and *Vermeer's Camera*. *Art in America*, April:39–40, 2002.

[61] Patrick Trevor-Roper. *The world through blunted sight*. Bobbs-Merrill, New York, NY, 1970.

[62] Christopher W. Tyler. 'Rosetta stone?' Hockney, Falco and the sources of 'opticality' in Renaissance art. *Leonardo*, 37(5):397–401, 2004.

[63] Christopher W. Tyler and David G. Stork. Did Lorenzo Lotto use optical projections when painting *Husband and wife*? In *Optical Society of American Annual Meeting*, Rochester, NY, 2004. Optical Society of America.

[64] Christopher W. Tyler. Critical commentary on Hockneys 'Secret Knowledge.' *Leonardo On-line, 37(5), 2004 (doi:10.1162/0024094041955971)*.

[65] Christopher W. Tyler. The Geometry of Globes. *In Webexhibits: Art and Optics. http://www.webexhibits.org/hockneyoptics/post/tyler5.html*

FIGURE 1.9
3D reconstructive model of the VAP House, view from south west.

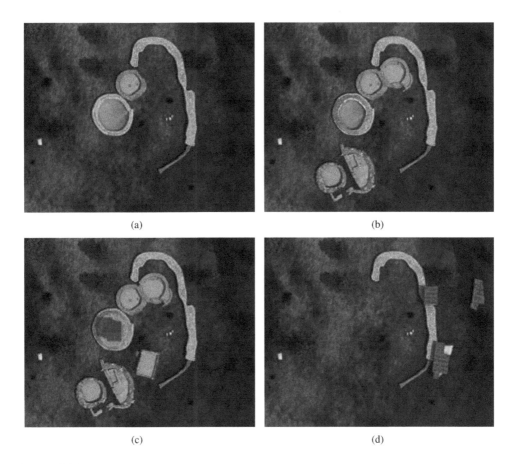

(a)   (b)

(c)   (d)

FIGURE 1.14
Multilayered 3D model of the Polizzello Acropolis: (a) phase I, half of the eighth century BC; (b) phase II, from the half of seventh to the beginning of sixth century BC; (c) phase III, end of sixth century BC; (d) phase IV, between the end of fifth and the beginning of fourth century BC.

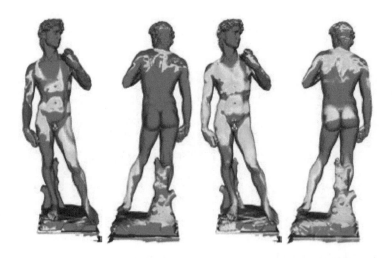

FIGURE 2.19

Exposure of David's surface to dust, mist, or other contaminations. This visualization shows, using a false-color ramp, the different classes of exposition produced by the simulation (red: absence of fall, blue: high density of fall), under a maximal angle of random fall of 5 degrees (on the left) and 15 degrees (on the right).

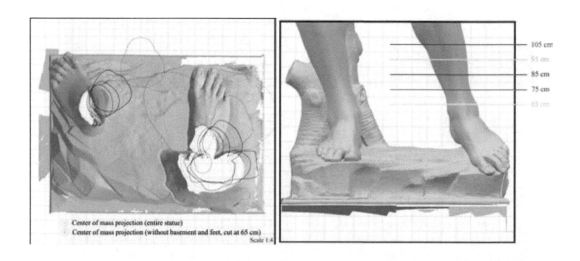

FIGURE 2.21

Visualization of the projection of the center of mass (marked by a yellow circle) and of the profiles of some cut-through sections (ankles, knees, and groin; see the respective height of those cut-through sections in the right-most image).

FIGURE 2.22
The digital model is used as an index to the scientific investigations performed on selected points or on subregions of the statue's surface.

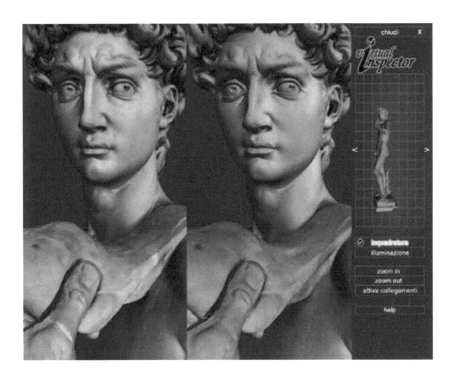

FIGURE 3.9
The David model is shown with color mapping; on the left is the pre-restoration status (61 images mapped), while the post-restoration status is shown on the right (another set of 68 images). The two colored David models are rendered in real time with the *Virtual Inspector* system [56].

FIGURE 3.10

The image presents a result of the automatic image-to-geometry registration: given the image on the right, the proper projective transformation is computed by finding the best matching between the input image and renderings of the mesh with vertices colored according to a combined normal vector and accessibility shading factor [57].

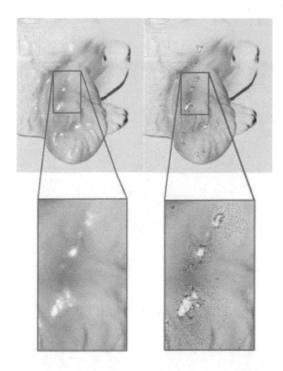

FIGURE 3.12

An example of highlights detection. Upper row: an input flash image and the same image after detection of highlights (blue is the highlight border, cyan is the internal highlight region). Lower row: two detail views, where highlights candidates selected by taking into account just geometric criteria (surface normals) are rendered in green and the ones more robustly detected are rendered in cyan.

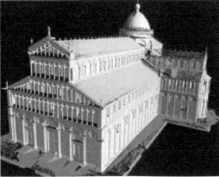

FIGURE 3.15

An example of two different visualization modes: on the left we render just the sampled points, on the right we present a rendering where shading of the surface element plays an important role for an improved insight over the details of the represented architecture (from a scanned model of the Cathedral of Pisa, Italy).

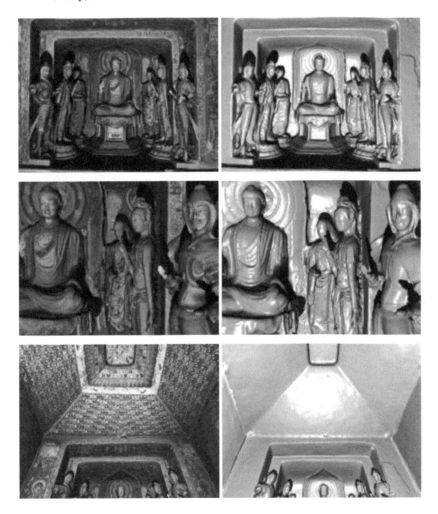

FIGURE 4.12

Views of the reconstruction of Mogao cave. The texture is removed in some of the images to better visualize the underlying structure.

FIGURE 4.13
Close-up views of the reconstruction of Mogao cave 322.

FIGURE 4.20
Duomo, Prato, Italy. Some of the uploaded images are shown to the left and reconstruction to the right.

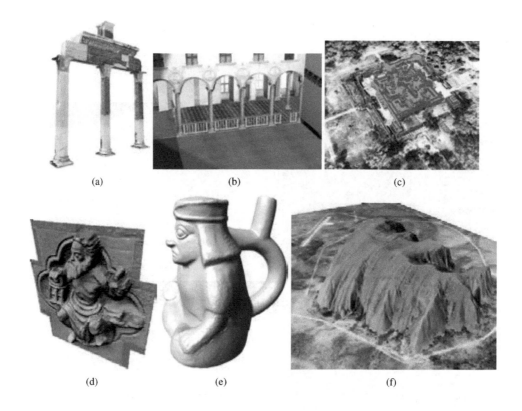

(a) (b) (c)

(d) (e) (f)

FIGURE 5.7
3D models derived from aerial or terrestrial images and achieved with interactive measurements (a, b, c) or automated procedures (d, e, f).

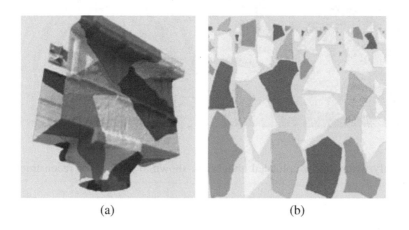

(a) (b)

FIGURE 6.12
(a) Geometry for a voxel colored according to texture atlas regions; (b) The corresponding texture atlas.

(a)   (b)   (c)   (d)

FIGURE 6.14

Computing reflectance properties for a voxel (a) Iteration 0: 3D model illuminated by captured illumination, with assumed reflectance properties; (b) Photograph taken under the captured illumination projected onto the geometry; (c) Iteration 1: New reflectance properties computed by comparing (a) to (b). (d) Iteration 2: New reflectance properties computed by comparing a rendering of (c) to (b).

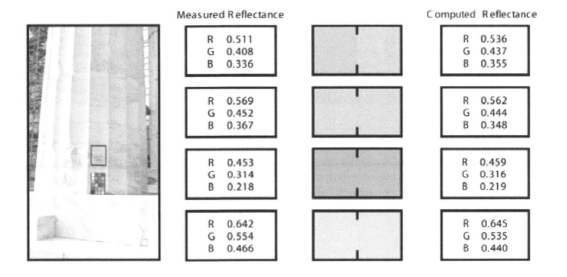

FIGURE 6.17

Left: Acquiring a ground truth reflectance measurement. Right: Reflectance comparisons for four locations on the East facade.

(a)   (b)

(c)   (d)

FIGURE 6.18
Rendering of a virtual model of the Parthenon with lighting from 7:04am (a), 10:35am (b), 4:11pm
(c), and 5:37pm (d). Capturing high-resolution outdoor lighting environments with over 17 stops of
dynamic range with time lapse photography allows for realistic lighting.

FIGURE 6.19

(a) A real photograph of the East facade, with recorded illumination; (b) Rendering of the model under the illumination recorded for (a) using inferred Lafortune reflectance properties; (c) A rendering of the West facade from a novel viewpoint under novel illumination. (d) Front view of computed surface reflectance for the West facade (the East is shown in 6.16(b)). A strip of unscanned geometry above the pediment ledge has been filled in and set to the average surface reflectance. (e) Synthetic rendering of the West facade under a novel artificial lighting design. (f) Synthetic rendering of the East facade under natural illumination recorded for another location. In these images, only the front two rows of outer columns are rendered using the recovered reflectance properties; all other surfaces are rendered using the average surface reflectance.

FIGURE 8.1
Jan van Eyck, *Portrait of Giovanni Arnolfini and His Wife* (1434), 82.2 × 60 cm, oil on oak panel.
National Gallery London.

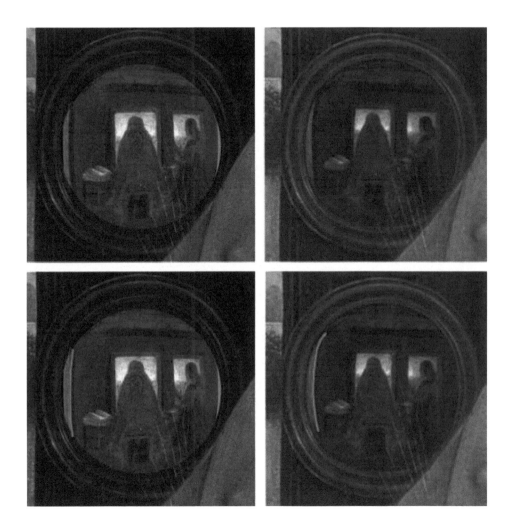

FIGURE 9.14

Left column: Image of Mary reflected in the convex mirror we would expect had Memling accurately portrayed such a tableau. Curvature and scene configuration are those of panel (c) in Figure 9.5. *Right column*: Image of Mary reflected in the convex mirror in Memling's original painting. Visual inspection shows that reflection painted by the artist exhibits a more dramatic bending of left side of the wall (e.g., notice the linear structure highlighted in red in the right-bottom panel). This bend is not compatible with the best mirror curvature predicted by our analysis. Notice the lower curvature of the same linear structure highlighted in red in the left-bottom panel.

(a)   (b)

FIGURE 10.1
(a), (b) Examples of foxing.

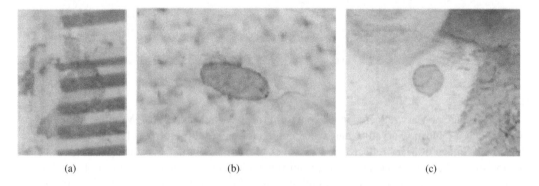

(a)   (b)   (c)

FIGURE 10.2
(a), (b) and (c) Examples of water blotches.

(a)   (b)

FIGURE 10.18

(a) An example of background with a variation of colors. Also a foxing stain is visible in the last line of text. (b) Image in Figure 10.18(a) after restoration.

FIGURE 11.7
Top row: showing four frames from four different sequences exhibiting pathological motion. Second row from top: PM/blotch detection in red/green respectively. Third row: restoration employing standard detection without PM detection. Incorrectly reconstructed areas are highlighted in cyan. Last row: restoration with PM detection embedded in the blotch removal process. Note the better preservation of the PM regions while still rejecting blotches.

(a)   (b)

FIGURE 11.8
Examples of semi-transparent defects. The slightly visible semi-transparent blotch is circled in the right image.

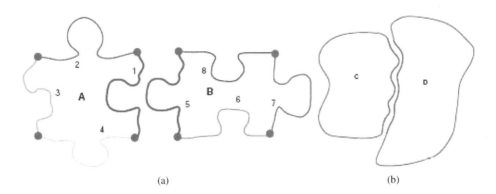

(a)   (b)

FIGURE 12.2

Two figures from [7] demonstrate the differences between reassembling commercial jigsaw puzzles
and reconstructing broken artifacts. As shown in (a), jigsaw pieces have readily identifiable corners
(red dots) allowing programs to easily separate portions of the boundary (shown as different color
curves) that will match with some other unique puzzle piece. Additionally, each boundary segment
is a smooth planar curve having an *isthmus* or neck that is highly indicative for finding that unique
matching puzzle piece. Figure (b) shows two hypothetical fragment boundaries. Note here that the
problem is made much more difficult as corners are not easily identifiable and may not indicate the
beginning or end of a curve that will uniquely match some other fragment. Worse, any portion of a
boundary curve may match to any other fragment and the curve itself may match equally well with
numerous similar boundaries from other fragments.

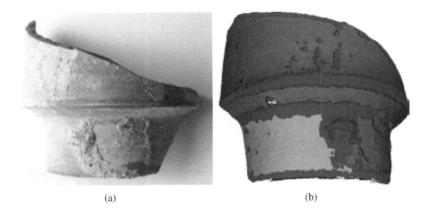

(a)   (b)

FIGURE 12.9

(a) An archaeological pot fragment excavated in Petra, Jordan. (b) A segmentation of the fragment
surface in (a) from [53] obtained by dividing the surface into regions that are well approximated by
a single quadratic surface patch. (b) Four quadratic surface patches indicated in orange, red, green,
and brown. Blue surface points are those points that do not lie close to any of the estimated quadratic
patches and typically include blemishes on the sherd surface created by calcification, chipping, and
weathering. This asymmetric surface data can make automatic estimation of the vessel axis and
profile curve problematic, and exclusion of these regions can greatly improve the accuracy and
robustness of sherd surface estimation approaches.

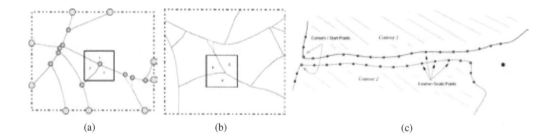

(a)   (b)   (c)

FIGURE 12.12

In these modified figures from [7], we define the geometric features used by various approaches for assembly. In (a), the outer contour of the square ceramic tile, i.e., the *intact boundary,* Ω_{outer}, is shown in red; *fracture boundaries,* Ω_{inner}, are shown in blue; *vertices,* $\Omega_{inner\perp}$, are shown as green points; and *outer vertices,* $\Omega_{outer\perp}$, are shown as yellow points. Note that most automatic assembly algorithms including [8–11,22], only consider the fracture boundaries, Ω_{inner}, and perhaps the vertices, $\Omega_{inner\perp}$. The black boxes in Figures (a,b) show "T" and "Y" junctions respectively that ceramics tend to generate when fractured. These junctions are typically high curvature locations along the fragment fracture boundary and are important for reconstruction as they often denote origin points for fracture boundaries that match to only one other fragment (c).

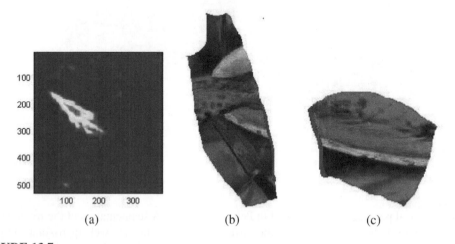

(a)   (b)   (c)

FIGURE 13.7

Similarity matrix (a) for the contour sequences of two fragments (b), (c). The correctly matched contour segments are shown in blue in (b), (c).

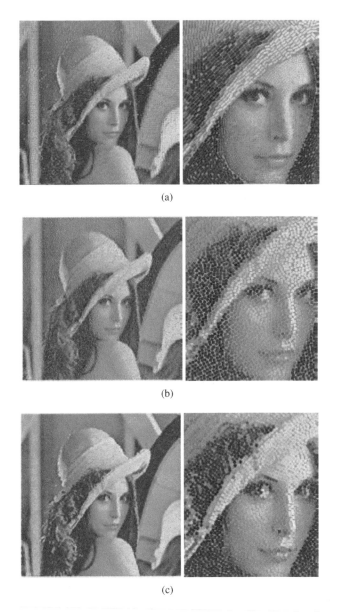

FIGURE 15.6
Examples of mosaics: (a) Schlechtweg et al. [39], (b) Liu et al. [32], (c) Battiato et al. [6].

FIGURE 15.7
Other examples of ancient mosaics: (a) the original image, (b) Battiato et al. [4].

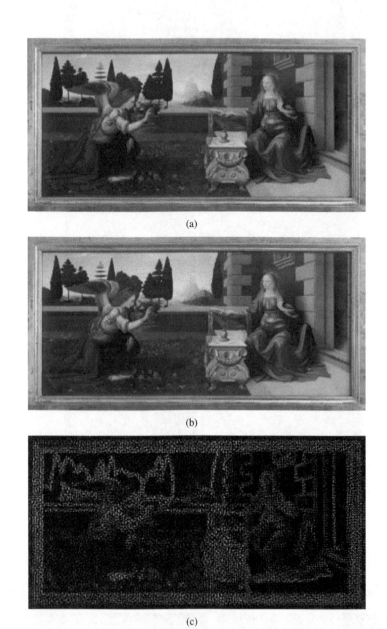

(a)

(b)

(c)

FIGURE 17.7
Leonardo da Vinci, "Annunciazione": (a) the original image, (b) the watermarked image, (c) difference between (a) and (b) multiplied by a factor of 32 (courtesy of Ministero per i Beni e le Attività Culturali, Soprintendenza per il Patrimonio Storico Artistico ed Etnoantropologico e per il Polo Museale della Città di Firenze [1]).

9

A Computer Analysis of the Mirror in Hans Memling's Virgin and Child *and* Maarten van Nieuwenhove

Silvio Savarese

University of Michigan
Email: silvio@umich.edu

David G. Stork

Ricoh Innovations and Stanford University
Email:artanalyst@gmail.com

Andrey Del Pozo

University of Illinois at Urbana-Champaign
Email:delpozo2@uiuc.edu

Ron Spronk

Queen's University,
Email:spronkr@queensu.ca

CONTENTS

9.1   Introduction

Recently, a number of scholars trained in computer vision, pattern recognition, image processing, computer graphics, and art history have developed methods for addressing problems in the history and interpretation of fine art [45, 46]. These methods, when paired with art historical and contextual knowledge, have shown modest but promising successes in authenticating works through statistical analysis of brush strokes [17] or patterns of dripped paint [16, 52], revealing the studio condition and working methods of realist artists [49], testing for artists' use of optical aids, [31, 37] and dewarping of images depicted in convex mirrors to provide new views into artists' studios [4]. These methods do not *replace* the expertise of the trained art scholar but rather are tools to *enhance* such expertise, much as a microscope expands the scholarly power of a biologist. In this chapter we use computer image analysis of depicted convex mirrors to shed light on a number of questions in the art history of the early Renaissance.

Convex mirrors begin to appear prominently in the paintings of the early Northern Renaissance, as symbols of wealth, as metaphors for an all-seeing Deity, as objects of inherent visual interest, and as showpieces for artists' technical mastery. Prominent examples from the north and, somewhat later, Italy include: Jan van Eyck, *Portrait of Giovanni Arnolfini and His Wife* (1434); Robert Campin, *Heinrich von Werl and St. John the Baptist* (1438); Petrus Christus, *Saint Eligius* (1449); Hieronymus Bosch, *The Seven Deadly Sins* (c. 1480); Hans Memling, *Vanity* (c. 1485) and *Maarten van Nieuwenhove Diptych* (1487); Quentin Metsys, *The Money Lender and His Wife* (1514); Parmigianino, *Self-Portrait in a Convex Mirror* (c. 1524); Laux Furtenagel, *Hans Burgkmair and His Wife* (1527); Pieter Bruegel the Elder, *Prudence* (1559); Caravaggio, *Martha and Mary Magdalene* (c. 1598). Plane mirrors have appeared in European painting as well, of course, but there seem to be few if any clear depictions of *concave* mirrors around this time.

Recent computer image analyses have provided tools for addressing a number of unresolved questions in the art history of the Renaissance. Criminisi and his colleagues modeled the optics of convex mirrors to dewarp the mirror images in Jan van Eyck's *Portrait of Giovanni Arnolfini and His Wife* and Robert Campin's *Heinrich von Werl and St. John the Baptist* [4]. They could then, in software, perform the inverse optical transform to correct or "dewarp" the image depicted in the mirror. This inverse transform depends upon the single unknown parameter R, the radius of curvature or "bulginess" of the mirror (Figure 9.4). More accurately speaking, in their case the transform depends upon the unknown *ratio* of the facial radius to the radius of curvature. They adjusted the computer model mirror's radius of curvature such that door jams, windows, and so on in the reconstructed image were as rectilinear as possible. In this way they revealed—for the first time in nearly 600 years—new views into the respective tableau rooms. These authors found, nevertheless, that even after best overall dewarping in *Arnolfini*, there remained slight distortions which could be attributed to the artist changing his viewing position, or that the mirror shape differed from the ideal, or that the artist's hand was not uniformly true while copying the reflected image. They found the reconstructed space in *Heinrich von Werl* conformed sufficiently well to the laws of perspective that they could use rigorous visual metrology to estimate the relative heights of figures in the mirror space [3]. They also used the visual information recovered from the dewarped images and a variety of other clues to attribute again this painting to Robert Campin.

Stork extended such analysis to address David Hockney and Charles Falco's hypothesis that van Eyck used the very mirror he *depicted* within the painting to *execute* the rest of the painting by tracing a projected image [11]. That is, they hypothesized that van Eyck turned the con*vex* mirror around and used it as a con*cave* projection mirror, to project an image of the tableau onto the oak panel support, trace it, and then fill it in with paint. Stork estimated the overall size of the mirror in *Arnolfini*; then together with the relative mirror curvature provided by Criminisi et al., he computed the absolute radius of curvature of this convex mirror, R. He then computed this mirror's focal length, f_{mir}, which in the paraxial ray approximation is simply $f_{mir} = R/2$ [8]. He also created a computer graphics model of the tableau to estimate the location and focal length of a putative optical projector for this painting, f_{proj}. He found these two focal lengths differed significantly, and thus he rejected the suggestion that van Eyck might have used the depicted mirror to build a projector for executing this painting. Stork also estimated the smallest blur spot of such a mirror and showed that this spot was too large—that is, the image too blurry—for an artist to trace the fine detail such is found in this painting [31, 32].

Computer graphics models of tableaus of paintings have been used to answer art historical questions. Johnson et al. [18] built a model of Vermeer's *Girl with a Pearl Earring*, and then adjusted the position of the model of the lighting source until the rendered model matched the painting most closely. In this way, they estimated the direction of the illumination. Their main goal, though, was to integrate estimation from a number of disparate sources such as the cast shadows, lightness along

FIGURE 9.1

Hans Memling, *Virgin and Child and Maarten van Nieuwenhove* (1487), detail, 22 × 44 cm. Notice that the mirror reflects the donor in the same room, silhouetted against the windows unifying the two panels. Image courtesy President and Fellows, Harvard College.

occluding contours, and lightness throughout an approximate facial model [18]. Computer graphics has also answered questions in art unrelated to convex mirrors. Stork and Furuichi created a computer graphics model of Georges de la Tour's *Christ in the Carpenter's Studio*, and compared the rendered images with the light in two locations: in place of the depicted candle, and "in place of the other figure" (i.e., in place of Christ when St. Joseph was painted and vice versa) [49]. In this way they found that for this painting—and for all others in de la Tour's nocturne œuvre they tested—that the best fit position was at the candle. In this way they rebutted David Hockney's optical projection claim, at least for *Christ in the Carpenter's Studio*.

Simple computer image analysis (elementary perspective analysis) has addressed another art historical claim about Memling and mirrors, but in this case *concave* mirrors. David Hockney claimed that Memling executed *Flower Still-Life* (c. 1490) by means of tracing an image projected by a concave mirror [11]. In support of his claim, Hockney pointed to the difference in location between the central vanishing point defined by the front half of the carpet and that defined by the back half of the carpet. He hypothesized that Memling refocussed a projector to overcome its limited depth of field and in doing so Memling tipped the projection mirror and thereby changed the location of these central vanishing points. However, Stork did a full perspective analysis of the front half of the carpet, and of the back half of the carpet, which included tests for secondary vanishing points. This analysis showed that each portion of the carpet was not in proper perspective, though it should be if executed under optical projections [37]. Such evidence thus rebuts Hockney's optical projection claim.

In Section 9.2 we clarify the art historical question we address, one centered on the devotional practice associated with Memling's diptych. As we shall see, its answer relies on an analysis of the image of the mirror depicted in its left panel. In Section 9.3 we describe our computer graphics modeling of the mirror and tableau. In Section 9.3 we introduce the mathematical tools that we employ in this analysis. In Section 9.5 we present our results, and in Section 9.6 we present our conclusions and implications for the understanding of this diptych. We conclude by speculating on other art historical problems that might be addressed by such computer methods. An early version of this manuscript appeared in [26].

9.2 Memling's Diptych

Many devotional half-figured portrait diptychs, such as Memling's *Virgin and Child and Maarten van Nieuwenhove*, apparently did not function while hanging on a wall or column, nor were they fully opened with their wings placed at a straight angle of 180° when used in private devotional practice. Instead, such objects were often used in a standing position with their hinged wings at an obtuse angle and it has been suggested that this was also the case with Hans Memling's famed diptych in Bruges. However, Memling introduced some remarkable pictorial discrepancies between the wings of his diptych which point to the possibility that this object was actually hung on a wall or column. In such a situation, the left wing with the Virgin and Child would have been secured to the wall in a stationary position, while the wing with the portrait of the *Virgin and Child and Maarten van Nieuwenhove* could be opened and closed or otherwise manipulated.

The two hinged wings appear to depict a single, unified space, since the stone parapet in the foreground, decked with a carpet, continues over both panels. The donor's prayer book is placed on a fold of the Virgin's red robe that spills onto the right wing. In this unified space, the donor sits to the Virgin's left and is turned towards her. He appears to be situated in the same plane as the Virgin, immediately behind the parapet. The reflection in the convex mirror belies such a spatial organization, however (Figures 9.1 and 9.14). Here, the Virgin and the donor are seated at *perpendicular* sides of a table rather than a parapet. Memling also used remarkably different perspective systems for the two wings of his diptych. On the right wing, the wall behind the *Virgin and Child and Maarten*

FIGURE 9.2

Hans Memling, *Virgin and Child and Maarten van Nieuwenhove* (1487). Wood panels, 52.5 × 41.5 cm each (including the original frames). Municipal Museums, Bruges, Hospitaalmuseum Sint-Janshospitaal; image courtesy President and Fellows, Harvard College. The two hinged panels appear to depict a single, unified space, since the Virgin's red robe and the stone parapet with the carpet in the foreground continues over both panels. In the unified space of the full diptych, the donor sits to the Virgin's left and is turned towards her. The convex mirror, on the other hand, reflects a different spatial organization. Here, the Virgin and the donor are seated at two perpendicular sides of the same table. In the mirror's reflection, both figures are silhouetted against bright window-like features, and it has been suggested that these features must represent the beholder's space, as seen from the depicted space back through the picture frames. In 2006 it was discovered that the top left corner of the Virgin and Child was intensively reworked by Memling. X-ray and infra-red imaging reveals that the initial composition showed a similar window as the one on the right, and that the stained glass window with Van Nieuwenhove's coat of arms, the window shutters, and the convex mirror were all added later, painted over a continuing landscape.

van Nieuwenhove is depicted in a relatively steep perspective, and its vanishing point is actually positioned on the left wing. The composition of the Virgin and Child, on the other hand, is in full frontal perspective, and the painter's vantage point is located directly in front of the Virgin, on the panel's vertical center line. Hence, the ideal viewing point of the entire diptych in its fully opened position is not located in front of the center of the diptych, but in front of the center of the left wing with the Virgin and Child. De Vos and others have suggested that the perspective system for Memling's full diptych 'falls into place' when the beholder stands in front of the Virgin and Child while the right wing is placed at a more or less right angle (i.e., around 90°) from the left wing [55].

We test to see whether the final reflected image, as it appears in the painting, does or does not conform to a simple computer graphics model of the tableau. Our technical question, then, is whether the mirror in the left panel of the diptych was painted with the subjects present, or instead added afterwards. This, in turn, comes down to the question of whether the image is highly consistent with the presence of a model, or inconsistent with the presence of a model. In the former case, the figural reference would likely give correspondence; in the latter case, the artist likely worked from memory or imagination. We note in passing that a similar question applies to van Eyck's Arnolfini portrait, specifically whether the room was *fictive* (in the artist's imagination) or actually present. Despite some inconsistencies between the direct and dewarped reflected images [4], the overall consistency between the *full room*, as drawn directly, and the reflection of the room, warped in the convex mirror was high [31]. It would have been a pioneering artistic and visual accomplishment of the highest order if van Eyck could paint a fictive wedding room, and paint the distorted image in the mirror of a fictive room with the spatial consistency we find through the computer reconstructions. As such, it seems far more likely that van Eyck worked from a real, physical room as referent than that he painted an imaginary or fictive room.

Recent technical examination of the diptych by means of infrared reflectography and X-radiography of the Memling diptych found further evidence that the reflections in the mirror do not depict an actual situation since the entire window behind the proper right shoulder of the Virgin was dramatically altered from its initial design. The stained glass window with Van Nieuwenhove's coat of arms, the window shutters, and the convex mirror were all added later by Memling, painted on top of an earlier depicted window with mullions with a view on a landscape, which echoed the present window behind the Virgin's proper left shoulder. The analogous question of praxis is more problematic for the mirror behind the Virgin in the left panel of Memling's *Virgin and Child and Maarten van Nieuwenhove* (1487). Notice that the reflected image does not display the warping that is most prominent around the perimeter of convex mirrors such as the van Eyck *Arnolfini*, the Metsys *Money Lender*, and the Christus *Saint Eligius* mirrors; nor does the dark gray color of the reflected image correspond to the Virgin's red gown as it otherwise should [10]. In short, there are strong indications that the mirror was not part of the original design, but an afterthought by the artist.

9.3   Computer Vision Analysis

We have created a three-dimensional virtual environment in Maya (a three-dimensional commercial modeling software [14]) where the Mary is located (Figure 9.3). We assume that the background of the painting (windows, wooden wall, etc.) is located on the left side wall of the virtual room, the background wall in the painting. This wall also contains the convex mirror. Furthermore, we assume that Mary is located somewhere between the background wall and the camera (observer). The relative positions of camera, Mary, and background wall are constrained since the scene viewed from the camera must match the one in the diptych; in other words, their relative positions must be compatible with what we see in the diptych. Moreover, the amount of curvature in the (reflected) image of straight objects is another constraint.

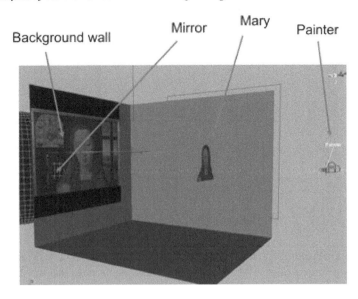

FIGURE 9.3

We modeled the full virtual environment for the left panel of the diptych as a flat back wall, a planar Mary, and a protruding convex mirror whose radius of curvature is estimated so as to conform with other image data. The artist's viewing position is indicated by the camera at the right.

Figure 9.3 shows the virtual setup for our computer graphics analysis. The overall scale is irrelevant to our investigations, but there are several other unknown sizes and positions, such as the position of the artist ("camera"), position and size of Mary, and facial size of the convex mirror. Moreover, we modeled the mirror as a section of a sphere, whose radius of curvature—its bulginess— is *a priori* unknown.

These unknown sizes and positions are constrained and such constraints allow us to infer their relative values. For instance, Figure 9.4 shows the effect of the radius of curvature, R, on the angle of view of the scene, and hence the relative sizes of objects. The theoretical analysis presented in [27] provides the tools for estimating a range of possible values of curvature given the measurements on the observer image plane (that is, the painting), location and orientation of the mirror, and position of the reflected scene (see Section 9.4 for details). Measured quantities can be, for instance, the location of feature points such as corner points in the image plane. There is another parameter we can account for. Not only does the mirror curvature R affect the scale of the reflected scene, but it also induces a deformation on the reflected scene: straight lines are reflected as curves and the shorter the radius of curvature the more severe the deviation. This is, then, another constraint on our configuration. The analytical expressions in [27] allow predicting values for R given a geometrical configuration (observer view direction, distance of the mirror from the observer, distance of the reflected scene from the observer). The best value for R may be then inferred by minimizing the deviation of the predicted position of feature points (e.g., corner points) from their measurements in the painting. Figure 9.5 shows some of the sample configurations we explored. We found that the one that best matched the image data (the distortion in the convex mirror and the relative sizes of objects and positions) is configuration (d), at the lower right. We adjusted the radius of curvature of the mirror to make the dewarped or rectified image as rectilinear as possible (Figure 9.6). This provides the relative size of the (back) of Mary's head which in turn depends upon her distance from the mirror. Notice in the detail of the original (left of Figure 9.13) that the reflected image does not display the warping that is most prominent around the perimeter of convex mirrors such as the van Eyck *Arnolfini*, the

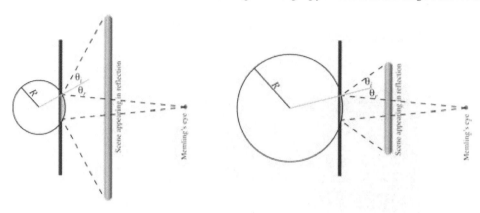

FIGURE 9.4

Left: A small radius of curvature, R, leads to a wide angle of view, that is, a wide angle of view in reflection. *Right*: A large radius of curvature leads to a small angle of view, that is, a narrow angle of view in reflection. Of course, the angle of incidence equals the angle of reflection (i.e., $\theta_i = \theta_r$).

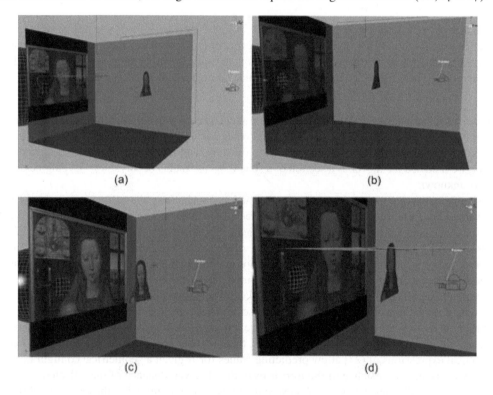

FIGURE 9.5

We adjusted the relative positions and sizes of objects to create different configurations of the scene; four are shown here. The distance values are: 53.7, 17.9, 11.4 and 9 for panel a, b, c and d respectively. The curvature $1/c$ values are: 16, 8.75, 6.85 and 6 for panel a, b, c and d respectively. Numerical values of distance and curvature are represented in *Maya* standard units. Notice that all these quantities can be expressed only up to an unknown scale. We found that the one that best matched the image data (the distortion in the convex mirror and the relative sizes of objects and positions) is configuration (d), at the lower right.

FIGURE 9.6
The dewarped or rectified view, found through adjusting the radius of curvature of the virtual model mirror and additional photo-editing. The figure shows the back of the Mary.

Metsys *Money Lender*, and the Christus *Saint Eligius* mirrors; nor does the dark gray color of the reflected image correspond to the Virgin's red gown as it otherwise should. In short, there are strong indications that the mirror was not part of the original design, but an afterthought by the artist.

9.4 Modeling Reflections Off a Mirror Surface

In this section we summarize our main theoretical tools for studying the relationship between the mirror's location and shape, and corresponding observations. For more details see [27]. Estimating the 3D shape of physical objects is one of the most useful functions of vision. Texture, shading, contour, stereoscopy, motion parallax, and active projection of structured lighting are the most frequently studied cues for recovering 3D shape. These cues, however, are often inadequate for recovering the shape of shiny reflective objects, such as a silver plate, a glass goblet, or a well-washed automobile, for it is not possible to observe their surfaces directly, rather only what they reflect. Yet, the ability of recovering the shape of specular or highly reflective surfaces is valuable in many computer vision applications such as industrial metrology of polished parts, medical imaging of moist or gelatinous tissues, digital archival of artworks and heritage objects, remote sensing of liquid surfaces, and diagnostic of space metallic structures, among others.

Although specular surfaces are difficult to measure with traditional techniques, specular reflections present an additional cue that potentially may be exploited for shape recovery. A curved mirror produces distorted images of the surrounding world (Figure 9.7). For example, the image of a straight line reflected by a curved mirror is, in general, a curve. It is clear that such distortions are systematically related to the shape of the surface. Is it possible to invert this map, and recover the shape of the mirror from its reflected images? The general inverse mirror problem is clearly

FIGURE 9.7
A curved mirror produces distorted images of the surrounding world. We explore the geometry linking the shape of a curved mirror surface to the distortions produced on a scene it reflects.

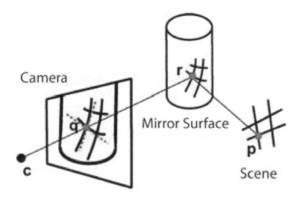

FIGURE 9.8
Setup: A (calibrated) camera is observing the planar scene reflected on the surface of the smooth mirror surface.

under-constrained: by opportunely manipulating the surrounding world, we may produce a great variety of images from any curved mirror surface, as illustrated by the anamorphic images that were popular during the Renaissance (Figure 9.7(b)) or as shown in Figure 9.5. This inverse problem may become tractable under the assumption that some knowledge about the structure of the scene is available.

In [27] we presented the analytical relationship linking the local shape of a curved mirror surface to the distortions produced on a scene it reflects. We call this step direct map. Also, we showed that, by taking advantage of this analytical relationship, local information about the geometry of the surface may be recovered by measuring the local deformation of the reflected images of a planar scene pattern. We call this step inverse map or 3D recovery step. In the inverse map, we assume that the surface is unknown, the camera is calibrated, and the geometry and position of the scene pattern is known in the camera reference system (Figure 9.8). These relationships and constraints are used in the analysis presented in section 9.3. In this section we analyze in further details such relationships, and demonstrate that these can be used to recover location geometry of a reflective surface.

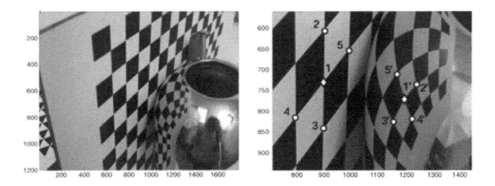

FIGURE 9.9
Mapping between scene points 1, 2, 3, 4, 5 and their corresponding reflections 1', 2', 3', 4', 5'.

9.4.1 Direct Map

We start from the observation that the mapping from a scene line to its reflected curve in the camera image plane (due to mirror reflection) not only changes orientation of the scene line, but also *stretches* its length, modifying the local scale of the scene line. Such a deforming mapping can be easily illustrated by the grid points (1, 2, 3) and their corresponding reflected points (1', 2', 3') on the mirror surface (Figure 9.9). As shown in [27], from this direct map it is possible to derive the analytical expressions for the local geometry in the image (namely, orientation and local scale of reflected image curve at a given point - (e.g., 1')) as a function of the position, orientation and curvature of the mirror surface. This result can be generalized to the case of a scene patch of arbitrary texture, and can be obtained by the direct mapping between the scene patch (around a given point p_o) to the corresponding reflected patch in the image plane (around q_o) as function of the local parameters (location, orientation, and curvature) of the mirror surface (around r_o). Here the point r_o is the reflection of p_o and q_o is its image (Figure 9.10).

9.4.2 Inverse Map

If local shape of the mirror surface is unknown, we can use a number of measurements in the image to estimate shape properties. Typical measurements are local position, orientation, and local scale of a reflected scene patch in the image plane. For instance, such local image measurements may be computed at a point (e.g., 1' in Figure 9.9, or q_o in Figure 9.10) from its four neighboring reflected points (e.g., 2', 3', 4', 5'). As shown in [27], by comparing these measurements with their corresponding analytical expressions, we obtain a set of constraints, which leads to the solution for the surface position r_o. Once r_o is found, normal, curvature and third-order local parameters of the surface around r_o can be found in closed-form solution. Such relationships are critical for predicting values for R (i.e., Memling's mirror curvature radius) given a geometrical configuration (observer's view direction, distance of the mirror from the observer, distance of the reflected scene from the observer). As explained in Section 9.3, the best value for R may be then inferred by minimizing the deviation of the predicted position of feature points (e.g., corner points) from their measurements in the painting.

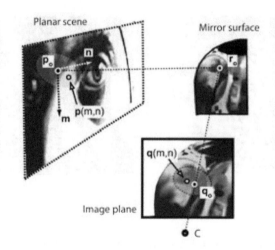

FIGURE 9.10

The direct mapping: The analytical expression describing the mapping from an arbitrary planar patch in the scene (around a generic point p_o) to the corresponding specular reflection in the image. The mapping is a function of the local geometry of the mirror surface around the reflection point r_o.

9.4.3 Experimental Validation

In this section we show that the theoretical results presented in the previous section can be successfully used to estimate shape properties of mirror surfaces from just one single image. We compared reconstruction results with ground truth 3D models. As we shall see in Section 9.5, these results confirm that the analysis introduced in Section 9.3 can be indeed potentially employed to recover shape properties of the Memling's mirror. Theoretical results in [27] are validated by recovering local surface parameters of real mirror objects. A Kodak DC290 digital camera with $1,792 \times 1,200$ resolution was used to capture an image of a mirror surface reflecting a planar grid. The corners of the grid served as mark points and were used to measure position, orientation, and local scale of reflected grid. Notice the grid can be replaced by any pattern as long as local position, orientation, and scale can be extracted from the reflected image. We validated our 3D reconstruction technique with three simple shapes (plane, sphere and cylinder) where ground truth measurements were available (Figures 9.11, and 9.12). In summary, the analysis in [27] gives the ability to:

- Describe the mapping from an arbitrary planar patch in the scene to the corresponding specular reflection in the image. The mapping is a function of the local geometry of the mirror surface.

- Recover local information about the geometry of the surface (i.e., position, curvature, and higher order shape information - up to 3rd order) by measuring the local deformation of the reflected images of a planar scene pattern.

  Moreover, the analysis is appealing in that:

- A static monocular camera can be used (as shown in Figures 9.11 and 9.12). With only one image we can calibrate the camera, the scene, and recover the shape of an unknown mirror surface.

- No initial conditions about the surface position and shape are required.

- Local shape can be recovered in closed form solution.

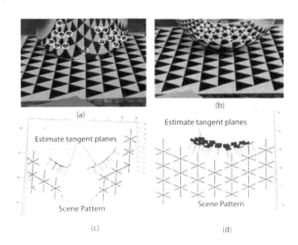

FIGURE 9.11
Qualitative reconstruction results: (a) A cylindrical mirror is placed over the grid pattern. We reconstructed 20 surface points and normals. The reconstructed points are highlighted with white circles. (c) Top view of the reconstruction. Each reconstructed surface point is depicted by its tangent plane and normal. (b) A spherical mirror placed over the grid pattern. (d) Top view of the reconstruction.

Quantity	GT	Mean Error	Standard Dev.	N
Point position	-	−0.48 mm	1.15 mm	12
Normal orient.	-	1.5×10^{-4} rad	6.5×10^{-4} rad	12
Radius	64.9 mm	3.3 mm	7.0 mm	20
Diameter	131.5 mm	0.86 mm	8.5 mm	17

FIGURE 9.12
Quantitative reconstruction results: First column shows the geometrical quantity whose reconstruction accuracy was evaluated. Subsequent columns show the corresponding ground truth value, the mean error and its standard deviation, respectively. The last column shows the number of points reconstructed.

FIGURE 9.13
Left: Mirror in the original painting. *Right*: Original painting with rectified mirror image. Image courtesy President and Fellows, Harvard College.

For these reasons, we view our analysis as a powerful tool for studying the mirror surface in the Memling's diptych. The case of a full uncalibrated scene and camera introduces additional degrees of freedom. Partial lack of calibration can be accommodated by integrating additional cues and some form of prior knowledge on the likely statistics of the scene geometry (as discussed in Section 9.3).

9.5   Results

Once we had the configuration that was most commensurate with the visual information in the painting and the rules of optics, we studied the image of Mary reflected in the convex mirror and the image we would expect had Memling accurately portrayed such a tableau. First, we could rectify the image in the mirror, that is, give the virtual view from the position of the mirror *back* into the space of the tableau. Figure 9.13 shows such a rectified view.

Visual inspection demonstrates that the reflection painted by the artist shows a more dramatic bending of the linear structure on the left side as the right column of Figure 9.14 shows. This bend is not compatible with the best mirror curvature predicted by our analysis as the left column of Figure 9.14 shows. It appears that the artist exaggerated the curvature in order to make the bulging effect of the spherical mirror more compelling. For instance, see the line feature highlighted in red in the lower panels of Figure 9.14. Other minor incompatibilities can be noticed as well. For instance, notice the amount of fore-shortening of the reflected figures–the Mary and the van Nieuewenhove on the right: the best mirror curvature predicted by our analysis would induce a higher amount of fore-shortening in the two figures than those measured in the original painting.

Specifically, the rectified reflected image of the Mary is not compatible with the foreground figure of the Mary. Figure 9.15 shows the rectified reflected image of the Mary. If we overlay it with the foreground image of the Mary, we notice that the two figures do not overlap perfectly; the red region in the figure is the difference. Although we do not assume that the artist would have been photographically accurate throughout a painting, the "error" in the reflected image seems too large to have arisen had Memling had the model as referent.

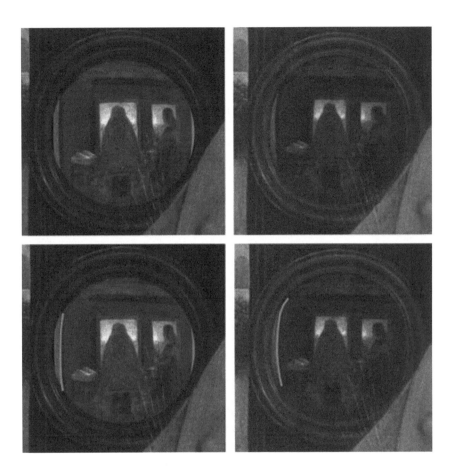

FIGURE 9.14 (SEE COLOR INSERT)

Left column: Image of Mary reflected in the convex mirror we would expect had Memling accurately portrayed such a tableau. Curvature and scene configuration are those of panel (c) in Figure 9.5. *Right column*: Image of Mary reflected in the convex mirror in Memling's original painting. Visual inspection shows that reflection painted by the artist exhibits a more dramatic bending of left side of the wall (e.g., notice the linear structure highlighted in red in the right-bottom panel). This bend is not compatible with the best mirror curvature predicted by our analysis. Notice the lower curvature of the same linear structure highlighted in red in the left-bottom panel.

FIGURE 9.15
Each figure shows the overlap of the rectified (and left-right reversed) reflection of Mary and the scaled direct view. The right panel highlights the difference in shape between these two images - region of non-overlapping that seems too large to have arisen if Memling had worked from an actual sitter as referent.

9.6   Conclusions

In conclusion, we have found two major inconsistencies or incompatibilities between the image depicted in the convex mirror and the image we would expect if the room and Mary were accurately rendered:

- curvature of the mirror and amount of distortion that reflected lines should have in order to be compatible with that curvature.

- the shape of the Mary's figure and its reflected counterpart.

These findings are in line with the idea that the mirror was indeed an afterthought by Memling (or the patron), and that the painted reflection was most probably not painted following an actual model. We hope that our findings will aid further art historical research into the actual usage of this and other diptychs in devotional practice. Our work extends that of Criminisi and his colleagues who opened new vistas through dewarping the reflections in convex mirrors in Renaissance art [4]. While that previous work highlighted differences between northern and Italian approaches to realism (the northern based somewhat more on close observation [2], the Italian based somewhat more on reasoned construction [1]), our work sheds light on religious praxis in the early Renaissance. More broadly, though, our work further demonstrates that new computer methods can shed light on problems in the history of art, and are most likely to profit from collaboration between computer vision experts, deeply familiar with both the power and limitations of image algorithms, and humanistic art scholars, who understand the art historical context and questions addressed [31].

Acknowledgments

We thank the Municipal Museums, Bruges, Hospitaalmuseum Sint-Janshospitaal for permission to reproduce the Memling diptych. Dr. Savarese and Andrey DelPozo were supported, in part, by grant 0413312 from the National Science Foundation. We also profited from the comments from several reviewers.

Bibliography

[1] Laurie Schneider Adams. *Italian Renaissance art*. Westview Press, Boulder, CO, 2001.

[2] Svetlana Alpers. *The art of describing: Dutch art in the seventeenth century*. University of Chicago Press, Chicago, IL, 1984.

[3] Antonio Criminisi. *Acccurate visual metrology from single and multiple uncalibrated images*. ACM Distinguished Dissertation Series. Springer-Verlag, London, 2001.

[4] Antonio Criminisi, Martin Kemp, and Sing-Bing Kang. Reflections of reality in Jan van Eyck and Robert Campin. *Historical Methods*, 3(37):109–121, 2004.

[5] Antonio Criminisi and David G. Stork. Did the great masters use optical projections while painting? Perspective comparison of paintings and photographs of Renaissance chandeliers. In Josef Kittler, Maria Petrou, and Mark S. Nixon, editors, *Proc. 17th International Conference on Pattern Recognition*, volume IV, pages 645–648, 2004.

[6] Richard O. Duda, Peter E. Hart, and David G. Stork. *Pattern classification*. John Wiley and Sons, New York, NY, 2nd edition, 2001.

[7] Sven Dupré, editor. *Early Science and Medicine: A Journal for the Study of Science, Technology and Medicine in the Pre-modern Period: Optics, instruments and painting 1420–1720: Reflections on the Hockney-Falco Thesis*, volume 10, no. 2. Brill Academic Publishers, Leiden, The Netherlands, 2005.

[8] David Falk, Dieter Brill, and David Stork. *Seeing the light: Optics in nature, photography, color, vision and holography*. Wiley, New York, NY, 1986.

[9] Ernst H. Gombrich. *Art and illusion: A study in the psychology of pictorial representation*. Phaidon, London, 1960.

[10] John Oliver Hand, Catherine A. Metzger, and Ron Spronk. *Prayers and portraits: Unfolding the Netherlandish diptych*. Yale University Press, New Haven, CT, 2006.

[11] David Hockney. *Secret knowledge: Rediscovering the lost techniques of the old masters*. Viking Studio, New York, NY, 2001.

[12] David Hockney and Charles M. Falco. Optical insights into Renaissance art. *Optics and Photonics News*, 11(7):52–59, 2000.

[13] David Hockney and Charles M. Falco. Quantitative analysis of qualitative images. In *Electronic Imaging*. SPIE Press, Bellingham, WA, 2005.

[14] http://usa.autodesk.com.

[15] Vincent Ilardi. *Renaissance vision from spectacles to telescopes*. American Philosophical Society, Philadelphia, PA, 2007.

[16] Mohammad Irfan and David G. Stork. Multiple visual features for the computer authentication of Jackson Pollock's drip paintings: Beyond box-counting and fractals. In *SPIE electronic imaging: Image processing: Machine vision applications II*, volume 7251, pages 72510Q1–11. SPIE/IS&T, 2009.

[17] Chris R. Johnson, Ella Hendriks, Igor J. Berezhnoy, Eugene Brevdo, Shannon M. Hughes, Ingrid Daubechies, Jia Li, Erik Postma, and John Z. Wang. Image processing for artist identification. *IEEE Signal Processing magazine*, 25(4):37–48, 2008.

[18] Micah Johnson, David G. Stork, Soma Biswas, and Yasuo Furuichi. Inferring illumination direction estimated from disparate sources in paintings: An investigation into Jan Vermeer's *Girl with a Pearl Earring*. In *Computer image analysis in the study of art*, volume 6810, page 68100I. SPIE/IS&T, 2008.

[19] Thomas Ketelsen, Olaf Simon, Ina Reiche, Silke Merchel, and David G. Stork. Evidence for mechanical (not optical) copying and enlarging in Jan van Eyck's *Portrait of Niccoló Albergati*. In *Optical Society of American Annual Meeting*, paper FWX6, Optical Society of America, 2004.

[20] Roberta Lapucci. *Caravaggio e l'ottica/Caravaggio and optics*. Privately published, Florence, IT, 2005.

[21] Pamela O. Long. *Openness, secrecy, authorship: Technical arts and the culture of knowledge from antiquity to the Renaissance*. Johns Hopkins U. Press, Baltimore, MD, 2001.

[22] Christoph Lüthy. Reactions of historians of science and art to the Hockney thesis: Summary of the European Science Foundation's conference of 12–15 November, 2003. *Optical Society of American Annual Meeting*, Rochester, NY (Abstract), 2004.

[23] Rosamond E. Mack. Lotto: A carpet connoisseur. In David Alan Brown, Peter Humfrey, and Mauro Lucco, editors, *Lorenzo Lotto: Rediscovered master of the Renaissance*, pages 58–67. National Gallery, Washington, DC, 1997.

[24] Peter Nillius and Jan-Olof Eklundh. Automatic estimation of the projected light source direction. In *IEEE Conference on Computer Vision and Pattern Recognition (CVPR01)*, volume 1, pages 1076–1083, 2001.

[25] M. Dirk Robinson and David G. Stork. Aberration analysis of the putative projector for Lorenzo Lotto's *Husband and Wife*: Image analysis through computer ray-tracing. In David G. Stork and Jim Coddington, editors, *Computer image analysis in the study of art*, volume 6810, page 68100H. SPIE/IS&T, Bellingham, WA, 2008.

[26] S. Savarese, R. Spronk, D. G. Stork, and A. DelPozo. Reflections on praxis and facture in a devotional portrait diptych: A computer analysis of the mirror in Hans Memlings *Virgin and Child and Maarten van Nieuwenhove*. In SPIE, editor, *20th Annual Symposium Electronic Imaging*, 2008.

[27] Silvio Savarese, Min Chen, and Pietro Perona. Local shape from mirror reflections. *International Journal of Computer Vision*, 64(1):31–67, 2005.

[28] Sara J. Schechner. Between knowing and doing: Mirrors and their imperfections in the Renaissance. In Sven Dupré, editor, *Early Science and Medicine: A Journal for the Study of Science, Technology and Medicine in the Pre-modern Period: Optics, instruments and painting 1420–1720: Reflections on the Hockney-Falco Thesis*, volume X, no. 2, pages 137–162. Brill Academic Publishers, Leiden, The Netherlands, 2005.

[29] David G. Stork. Color and illumination in the Hockney theory: A critical evaluation. In *Proceedings of the 11th Color Imaging Conference (CIC11)*, volume 11, pages 11–15, Scottsdale, AZ, 2003. Society for Imaging Science and Technology.

[30] David G. Stork. Did Jan van Eyck build the first "photocopier" in 1432? In Reiner Eschbach and Gabriel G. Marcu, editors, *SPIE electronic imaging: Color imaging IX: Processing, hardcopy and applications*, pages 50–56, Bellingham, WA, 2004. SPIE.

[31] David G. Stork. Optics and realism in Renaissance art. *Scientific American*, 291(6):76–84, 2004.

[32] David G. Stork. Optics and the old masters revisited. *Optics and Photonics News*, 15(3):30–37, 2004.

[33] David G. Stork. Were optical projections used in early Renaissance painting? A geometric vision analysis of Jan Van Eyck's Arnolfini portrait and Robert Campin's Mérode Altarpiece. In Longin J. Latecki, David M. Mount, and Angela Y. Wu, editors, *SPIE electronic imaging: Vision geometry XII*, volume 4675, pages 23–30, Bllingham, WA, 2004. SPIE.

[34] David G. Stork. Asymmetry in "Lotto carpets" and implications for Hockney's optical projection theory. In Bernice E. Rogowitz, Thrasyvoulos N. Pappas, and Scott J. Daly, editors, *SPIE electronic imaging: Human vision and electronic imaging X*, volume 5666, pages 337–343, Bellingham, WA, 2005. SPIE.

[35] David G. Stork. Did early Renaissance painters trace optical projections? Evidence pro and con. In Longin J. Latecki, David M. Mount, and Angela Y. Wu, editors, *SPIE electronic imaging: Vision geometry XIII*, volume 5675, pages 25–31, Bellingham, WA, 2005. SPIE.

[36] David G. Stork. Did Georges de la Tour use optical projections while painting *Christ in the Carpenter's Studio*? In Amir Said and John G. Apolstolopoulos, editors, *SPIE electronic imaging: Image and video communications and processing*, volume 5685, pages 214–219, Bellingham, WA, 2005. SPIE.

[37] David G. Stork. Did Hans Memling employ optical projections when painting *Flower Still-Life*? *Leonardo*, 38(2):57–62, 2005.

[38] David G. Stork. Optique et réalisme dans l'art de la renaissance. *Revue Pour la Science*, 327(1):74–86, 2005.

[39] David G. Stork. Spieglein, spieglein and der wand. *Spektrum der Wissenschaft*, 322(2):58–61, 2005.

[40] David G. Stork. Tracing the history of art: Review of Early Science and Medicine: Optics, instruments and painting, 1420-1720: Reflections on the Hockney-Falco theory. *Nature*, 438(7070):916–917, 2005.

[41] David G. Stork. Computer vision, image analysis and master art, Part I. *IEEE Multimedia*, 13(3):16–20, 2006.

[42] David G. Stork. Mathematical foundations for quantifying shape, shading and cast shadows in realist master drawings and paintings. In Gerhard X. Ritter, Mark S. Schmalz, Junior Barrera, and Jaakko T. Astola, editors, *SPIE electronic imaging: Mathematics of data/image pattern recognition, compression and encryption with applications IX*, volume 6314, pages 63150K1–K6, Bellingham, WA, 2006. SPIE.

[43] David G. Stork. Aberration analysis of the putative projector for Lorenzo Lotto's *Husband and Wife*. *Optical Society of American Annual Meeting*, San Jose, CA (Abstract), 2007.

[44] David G. Stork. Imaging technology enhances the study of art. *Vision Systems Design*, 12(10):69–71, 2007.

[45] David G. Stork and Jim Coddington, editors. *Computer image analysis in the study of art*, volume 6810. SPIE/IS&T, Bellingham, WA, 2008.

[46] David G. Stork, Jim Coddington, and Anna Bentkowska-Kafel, editors. *Computer vision and image analysis in the study of art*, volume 7531. SPIE/IS&T, Bellingham, WA, 2010.

[47] David G. Stork and Marco Duarte. Computer vision, image analysis and master art, part III. *IEEE Multimedia*, 14(1):14–18, 2007.

[48] David G. Stork and Marco Duarte. Revisiting computer image analysis and art. *IEEE Multimedia*, 14(3):108–109, 2007.

[49] David G. Stork and Yasuo Furuichi. Image analysis of paintings by computer graphics synthesis: An investigation of the illumination in Georges de la Tour's *Christ in the Carpenter's Studio*. In *Computer image analysis in the study of art*, volume 6810, pages 68100J1–12. SPIE/IS&T, 2008.

[50] David G. Stork and M. Kimo Johnson. Computer vision, image analysis and master art, Part II. *IEEE Multimedia*, 14(3):12–17, 2006.

[51] David G. Stork and M. Kimo Johnson. Estimating the location of illuminants in realist master paintings: Computer image analysis addresses a debate in art history of the Baroque. In *Proceedings of the 18th International Conference on Pattern Recognition*, volume I, pages 255–258, Hong Kong, 2006. IEEE Press.

[52] Richard P. Taylor. Order in Pollock's chaos. *Scientific American*, 287(66):116–121, 2002.

[53] Christopher W. Tyler. "Rosetta stone?" Hockney, Falco and the sources of "opticality" in Renaissance art. *Leonardo*, 37(5):397–401, 2004.

[54] Christopher W. Tyler and David G. Stork. Did Lorenzo Lotto use optical projections when painting *Husband and Wife*? In *Optical Society of American Annual Meeting*, (Optical Society of American 2004).

[55] Dirk De Vos. *Hans Memling: The complete works*. Harry N. Abrams, New York, NY, 1994.

10

Virtual Restoration of Antique Books and Photographs

Filippo Stanco

University of Catania
`Email: fstanco@dmi.unict.it`

Alfredo Restrepo Palacios

Universidad de los Andes
`Email: arestrep@uniandes.edu.co`

Giovanni Ramponi

University of Trieste
`Email: ramponi@units.it`

CONTENTS

10.1 Introduction

Antique photographs and paper documents constitute an immense patrimony that is distributed among thousands of private and public libraries, museums, and archives all over the world. Despite care in their conservation, they are based on fragile materials, and hence they are easily affected by environmental agents. In the cases of photographs and documents, frequently the paper support presents cracks, scratches, holes, and added stamps or text; moreover, chemical reactions between the paper and some microorganisms produce visible stains, and humidity and water cause blotches that change the aspect of the picture or document [55]. Likewise, color photographs and other flat art works, such as tapestries and mosaics, are subject to fading with time.

In fact, digital reproductions are a powerful tool to share this patrimony and to preserve the original materials, and a large amount of intellectual and financial resources have been allocated for this purpose. The digital copy of a document should be as close as possible to the original, with all its textural and physical extra-textual features (including not only glosses and all manuscript signs, but also stains, chromatic alterations, and so on), which are part of the object's history and are of the utmost importance from a philological point of view. On the other hand, by digitization and subsequent virtual restoration, the usability of the object is improved providing an image of it that can be better in several aspects. This procedure also allows for the elaboration of an excellent virtual copy, for instance, of a single extant copy of an edition, especially when it is affected by structural or chromatic pathologies. Two more advantages derive from a restoration operation of this virtual type: it allows the work to be read and otherwise observed by vast numbers of users without affecting the original document traumatically or irreversibly. The use of the computer allows for the greater part of the operations of restoration to be simulated, supplying instruments and materials that will help the "official restorer" in planning future work and appraising the final result.

Virtual restoration has proved to be convenient for various reasons. First, the digital processing is reversible and the original art work is minimally affected so that novel restoration techniques can be attempted without risks. Moreover, the restoration process can be partially automated, enabling the treatment of a larger number of works in a shorter time without further engaging dedicated personnel. Finally, after the development of the restoration tools, the cost of the process is small, so that any professional photographer or small museum can afford it.

In this chapter we report the most important algorithms to digitally detect and restore typical damages that photographs suffer such as foxing, water blotches, fading, and glass cracks; also, defects that books suffer such as yellowing and foxing. We report as well on the state of the art of the quality evaluation methods. In fact, many automatic restoration techniques make use of quality metrics to guide the choice of several parameters. Quality assessments based on measures of the contrast contents of the image are quite common.

10.1.1 Photographic Prints

Since its origin, photography has been considered an important way to document reality. For this reason, its diffusion was fast and wide. In these first one hundred and fifty years the technologies, the materials of support, and the conservation techniques have changed significantly. Despite care in their conservation, the first photographic prints were based on fragile materials, and hence they are easily affected by bad environmental conditions.

There are many common defects in photographic prints. The term "foxing" was used for the first time in the 18th century to indicate the scattered reddish-brown (the color of a fox) spots on the surface of paper in old books [5, 21, 26, 59]. The same technical word was introduced in photography to refer to a similar chemical damage on the prints. Foxing is characterized by a dark-brown center and an area where the color is smoothed (Figure 10.1). In the center all the original information is covered by the stain, and hence it is considered as lacking. The area around the center, on the contrary, can include residual original information that should be enhanced. The causes of foxing are not completely understood; it probably depends on joined fungal activity and metal-induced degradation. The paper used in the oldest prints has microorganisms that can remain latent for decades awaiting conditions appropriate for growth. Moreover, airborne spores can attach to the paper, creating colonies of foxing. Another element that seems to accelerate foxing is the presence of

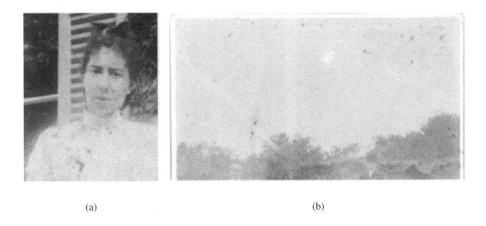

(a)   (b)

FIGURE 10.1 (SEE COLOR INSERT)
(a), (b) Examples of foxing.

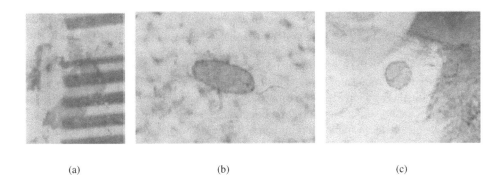

(a)   (b)   (c)

FIGURE 10.2 (SEE COLOR INSERT)
(a), (b) and (c) Examples of water blotches.

iron in the paper. Indeed, if the relative humidity is below 50% and we use modern paper without iron, the foxing is strongly reduced.

Another aggressive and, at the same time, frequent menace to photographic prints comes from water, which can permeate portions of the paper and produce very visible stains on the picture [59]. The result is the so-called *water blotch* (Figure 10.2), which is often characterized by having a vaguely round shape, a color darker than the neighborhood due to the dust which is attracted in the paper texture, and an even darker border where the dust accumulates. An important peculiarity of this type of damage, which differentiates it from many other defects which a print can show, is that the blotch does not completely destroy the content of the picture in the affected area: such a region is darker, but the image details are still visible, at least partially.

Old photographic prints often suffer from the presence of cracks due to inadequate handling. As a result of mechanical stress, the superficial photosensitive emulsion can fracture, exposing the underlying white paper support; the paper in turn can suffer damage from humid and warm air and from polluting agents, changing its color to a yellowish hue [17, 18]. In some cases the support itself can be broken (Figure 10.5).

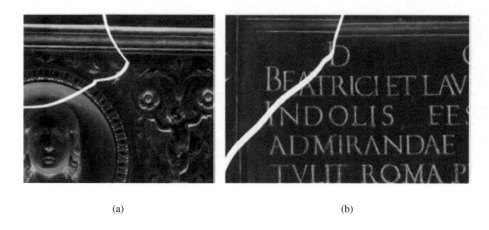

(a)   (b)

FIGURE 10.3
(a), (b) Original fragmented glass negatives.

In the second half of the 19th century photographs were still taken on a glass plate coated by a wet collodion emulsion with silver halide. The first dry gelatin process was proposed by R. L. Maddox in 1871 [29]; this change turned photography into an industry, since photographic materials could now be made available to the user beforehand, rather than be prepared on-site. A few years later G. Eastman introduced a celluloid support to substitute for the glass plate, but the older glass support was, however, used for several more decades, especially in Europe [28]; this fact may be considered lucky with respect to the preservation of original documents for even more than one century now: glass in fact is a chemically more stable support than celluloid. However, many glass plates available in large photographic archives are fractured and cannot be used for reproduction (Figure 10.3).

10.1.2   Antique Books

Digital reproductions of paper documents are a powerful tool to share this patrimony and to preserve the original materials. Large intellectual and financial resources have already been devoted to this purpose. Examples of supported projects in the field abound; without claiming to compile an exhaustive list, we may mention the Universal Library Project [30], the Digital Libraries Initiative [24], or the EU-funded BRICKS Digital Library project [23]. The digital copy of a document should be as close as possible to the original, with all its physical extra-textual features (including glosses and all manuscript signs, but also stains, chromatic alterations, and so on), which are part of the object's history and can be of the utmost importance from a philological point of view [60]. Finally, digital reproduction and restoration represent the only possible salvation for severely damaged books, like the ones in the prestigious *Herzogin Anna Amalie Bibliothek* in Weimar (Germany), recently hit by fire [22].

Also for the antique documents, a typical defect is the "foxing" [21, 25–27, 47] whose origin has been discussed in Section 10.1.1. Physical restoration based on aqueous methods is often insufficient, and the use of a laser with suitable wavelength has been proposed [34].

Moreover, with aging the paper changes its color and becomes yellowish (Figure 10.4). This deterioration depends on the acids introduced in the manufacture of paper as well as on those absorbed from the environment [47]. Despite the fact that this phenomenon can be today avoided by storing the documents with controlled humidity and air pollution [34], there exists a huge amount of already damaged documents.

(a)   (b)

FIGURE 10.4
(a), (b) Original manuscript images with a damaged background.

There are other damages for antique document paper. For example there are documents whose paper has become yellowish or brownish, in which humidity has made the paper undulated, or when the sheet was folded and bending remnants are visible in the scanning process. With respect to the ink, decoloring can be present, or the writing on the back of the sheet may be visible on the front by transparency.

10.2 Detection of the Defects

This section illustrates some techniques used to detect the defects on different types of support; it is organized to parallel the different subjects presented in the Introduction. Thus, we shall subsequently deal with the detection of foxing and of water blotches on photographic prints or antique books, then with cracks on printed photographs, and finally with the case of glass fragmentation in the oldest photographs.

10.2.1 Foxing

Due to the characteristics of this type of defect, the detection of foxing is based on color [59]. The detection process aims to define two different portions in each stain: the central one, opaque, in which the aggression has completely destroyed the original data (phase I), and a semi-transparent periphery where some image details are still recognizable (phase II). To achieve this result, the input image is first converted to the YC_bC_r color space via a conventional transformation [31]; then the histogram of the C_r component is evaluated. It is important to note that both phases in the detection process are completely automated.

Phase I. The first phase of foxing detection determines the locations of the spots. As mentioned above, the foxing damage is related to the presence of iron; hence, the predominant color is red. Usually, vintage photographic prints are gray or sepia, and the foxing stains have a color completely out of the color range of the image. Based on this, in the histogram of the red chrominance matrix C_r the foxing pixels are represented by small bins on the right tail. If the image is grayscale, C_r is null (128), while if it is sepia it presents a histogram in which most bins are located in the right half (128 - 256) of the possible range. It may be observed that in presence of foxing artifacts the typical histogram of C_r has a tail on the right formed by a set of bins with small entries having almost uniform amplitude that is not present on the left. Moreover, the peak of the histogram is located in the left portion of the set of nonzero bins of the histogram. A detection procedure can search for all the connected image parts represented by the bins on the right tail.

If the histogram has the structure presented above, one can search for all the bins representing the damaged pixels. Starting from the right of the histogram, let us mark as foxing this bin B_n. Denote with $h(B_i)$ the height of the bin B_i. For $i = n, n-1, \ldots, 2$ if $|h(B_i) - h(B_{i-1})| \leq Th_{f1}$ then the bin B_{i-1} is marked as foxing; otherwise the procedure is stopped. The center of the last bin marked as foxing gives the value a that is used to perform a thresholding over the matrix C_r. The matrix I_{F1} is a map where the coordinates of foxed pixels are represented as a 0 value.

This detection step may extract isolated points; they do not represent relevant damaged areas, and should be expunged using, for example, a simple 3×3 median filter.

Phase II. The previously detected areas can be expanded by finding all the pixels where the original information is only partially affected by foxing. They are characterized by a lighter coloring than the center of the foxing and their position is near the reddish-brown spot. Therefore they can be searched for starting from the previous detection map I_{F1}. The output of this phase is a new map I_{F2} that initially is equal to I_{F1}. If Ω is a foxing stain detected in phase I, define the following sets of pixels:

$$\partial\Omega = \{(x,y) \mid (x,y) \ is \ 8-connected \ to \ \Omega \ \wedge \ (x,y) \notin \Omega\} \tag{10.1}$$

$$N_{xy} = \{8-connected \ neighborhood \ of \ (x,y)\} \tag{10.2}$$

and

$$N_{xy}^{\Omega} = \{N_{xy} \subset \Omega\} \tag{10.3}$$

For each pixel $(x,y) \in \partial\Omega$ denote as $\overline{C}_r(N_{xy}^{\Omega})$ the average red chrominance value in N_{xy}^{Ω}. If

$$|C_r(x,y) - \overline{C}_r(N_{xy}^{\Omega})| \leq Th_{f2} \tag{10.4}$$

then the pixel is labeled as foxed and stored in I_{F2}. This procedure is repeated until there is at least one new pixel labeled as damaged in the previous iteration.

If the chrominance changes in C_r are uniform, the detection phase can overestimate the number of damaged pixels. To avoid this situation, when a pixel is defined as damaged its $C_r(x,y)$ value can be changed to the average value $\overline{C}_r(N_{xy}^{\Omega})$. In other words, the border values of Ω are propagated over the new pixels that are labeled as damaged.

10.2.2   Water Blotches

Due to the scarcity of scientific literature specifically devoted to the detection of water blotches, it is convenient to start by referring to an analogous but different problem: gaps in artworks. Usually, a gap is characterized by approximately uniform gray levels. In [4] the detection system operates iteratively: it checks whether the gray level of a pixel differs significantly from the uniform range, and hence it decides if this pixel can be classified as belonging to the crack. In [15] the classification is performed optimizing a function based on the luminance uniformity and on the magnitude of the gradient values in the region.

The first specific detection method for water blotches was introduced in [58]. The algorithm first performs an edge detection, and then starting from the selected point the algorithm analysis the surrounding area using a 3×3 window; the samples within the window which do not belong to the border extend the detection map. This method needs a thresholding of the edge detection result. If the threshold is high it could happen that the edges do not delimit a closed area. Conversely, if the threshold is conservative too restricted a portion of the blotch is selected, and multiple user intervention is necessary to complete the detection.

The algorithm for water blotches detection used in [59] combines automatic segmentation algorithms with the user interaction [54]. The user selects one point inside the blotch and all the pixels that belong to that region are added to the detection. This procedure is repeated until the selection is completed. The issue is to reduce the user interaction increasing the performance of the segmentation. A segmentation process partitions an image into its constituent parts or objects. The number of regions obtained can be higher than the segmentation that a user would make based on his/her own perceptive ability. The algorithm reduces these undesired regions using as input for the segmentation a preprocessed image Y_r, that is obtained applying n_r times a *rational filter* (RF) [48] over the luminance component Y of the image in the YC_bC_r color space [31]. The RF enhances the image attenuating small image variations while it preserves edges. Similarly to other nonlinear operators [19, 39], the RF modulates the coefficients of a linear lowpass filter in order to limit its action in presence of luminance changes; for each pixel $Y(i,j)$ the output of the filter is obtained according to:

$$Y_r(i,j) = Y(i,j) \quad + \frac{Y(i-1,j)+Y(i+1,j)-2Y(i,j)}{k(Y(i-1,j)-Y(i+1,j))^2+A}$$
$$+ \frac{Y(i,j-1)+Y(i,j+1)-2Y(i,j)}{k(Y(i,j-1)-Y(i,j+1))^2+A} \tag{10.5}$$

where k and A are parameters that control the filter and take positive values [48]. The resulting image Y_r tends to have still well-marked large edges, while values inside a region are made homogeneous; hence it is the ideal input for a segmentation algorithm. The image Y_r is segmented using *opening by reconstruction* followed by a *closing by reconstruction* [66] that better preserves the shape of the objects. The image Y_s obtained performing opening and closing by reconstruction over the image Y_r is the segmented version of the original image Y. If t is the number of regions R_i inside Y_r such that $\cap_{i=1}^{t} R_i = \emptyset$, it is simple to assert that

$$Y_r = \bigcup_{i=1}^{t} R_i \tag{10.6}$$

At this time a user intervention is requested to select the region that corresponds to the blotch. This region is stored in the map I_B. If the detection is not exhaustive, the user adds other regions and I_B is updated. The RF reduces the insignificant details and permits use of a small radius r in the morphological operators. The lower is the radius r, the higher is the possibility to preserve small significant details.

10.2.3 Cracks

The main cause of cracks is folding, which causes the mechanical breaking of the gelatin and other visible defects in the paper photograph [52]. In the neighborhood of the main crack, the breaking of the gelatin creates a textured pattern of white short segments or microcracks which possess a certain amount of periodicity. In the digitized version of the photograph, geometrically, the main characteristics of cracks are scale and direction while, chromatically, their main characteristic is their achromaticity, of greatest use in the case of sepia prints. A crack detection method can be based on the initial input of the user, who selects a starting and an ending point of the crack, on a set of directional filter banks, on an operator that analyzes the chromatic component of the data, and finally on a tool able to fuse information coming from this and other sources.

Crack detection is a delicate and important step; the literature on crack detection is scarce and to the best of our knowledge completely absent in the case of crack detection in digitized photographic prints. Related work is described in [17, 18], where the problem of virtual restoration of cracks in paintings is addressed. Anywhere, in the case of paintings the cracks are black and have a cause different from cracks in paper photographs that makes their geometric structure different. The detection of pavement cracks is addressed in [68]; it can be performed on the basis of the contrast between small adjacent image blocks or cells. Crack cells are clustered when they fall along a linear string; if further requirements regarding width, length, and orientation of cells are met, the cells are to be joined into one crack. This method benefits from the fact that pavement usually has a simple and identifiable texture; however, photographs are much richer in detail and structured contents. Another approach consists in looking for shortest multiple paths in an image; they can be found also using a constrained expanded trellis, as in [62], where several constraints which help regulate the shape of the paths are incorporated and reduce the computational costs.

Cracks usually go from one side in the frame of the photograph to another and tend to lie along straight lines. A rupture of the paper usually leads to a black region in the scanned image while the exposed paper leads to a white region. In addition, if the digitized version is obtained without a flattening of the photo, in the case of a broken paper, the 3D geometry around the crack may produce different types of reflection of the light used to acquire the photo; this in particular may produce out-of-range intensities at the image sensor. Stereo acquisition could in principle be used to exploit three-dimensional features of cracks for further processing, but this approach would require complex algorithms and is usually not followed in photographic archives. A straight line joining the extremes of the crack will probably have the same direction as the one of the cracks, but it is likely to lie on one side of the crack only, most of the time, so that actual crack pixels cannot be detected with precision. Missed crack pixels evident to the user should be pointed out to the algorithm. The folding of a photograph produces as well a series of parallel 'microcracks' that run on both sides along the main crack. These microcracks usually do not imply a change of hue but a set of image discontinuities, observable at a high-enough resolution. In [52] to detect them, the thick line connecting the user-given crack end point is selected; this subimage is Fourier transformed and frequencies corresponding to high energy contents are found. Next, a Gabor filter bank is used to detect crack pixels. Gabor filtering serves a twofold purpose: using a bank of filters, both the direction and the distance between microcracks is estimated.

In the approach hereby described, it is necessary that the user provides two crack end points; then, a straight line is drawn that connects them. This line is thickened producing a ribbon region that contains the crack as well as a large amount of undamaged neighboring image pixels. An image sample is shown in Figure 10.5 where the localized crack region is in red. The ribbon region is windowed and further processed, first of all taking its Fourier transform: see Figure 10.6(b). The Fourier transform provides information regarding the crack in the frequency domain; namely, the frequencies corresponding to peaks of high energy in the spectrum both indicate the direction and are related to the space between microcracks; the direction of the crack should nearly correspond to that inferred from the user-given crack end points, and the magnitude of the frequency corresponds to the spacing between microcracks. Next, the energy peak frequencies are used to choose a filter bank at a collection of magnitude frequencies from 0 to π and a collection of directions that safely cover the direction of the peak frequencies. The outputs of the filter bank provide information that is used to infer the crack pixels; the so detected crack pixels are corroborated with the pixels detected using color information.

FIGURE 10.5
Digitized photographic print (size 3,112 × 2,297 pixels). In red is the crack region.

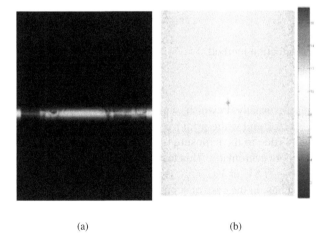

(a)   (b)

FIGURE 10.6
(a) Windowed crack. Gaussian window with $\sigma = 70$ pixels. (b) FFT of the image in (a).

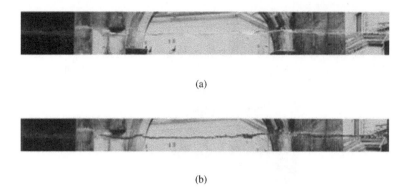

(a)

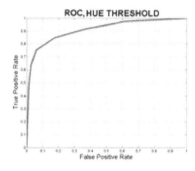

(b)

FIGURE 10.7
(a) Crack region. (b) Result of the crack detection method that uses color.

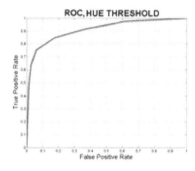

FIGURE 10.8
ROC plot for the hue-based detection method.

Use of Color

Black-and-white pictures are usually brownish, sepia as an extreme. A sepia print typically has a nearly constant hue. Also, a black and white print may have some chromaticity. This can be taken advantage of since both white (due to the exposure of the paper) and black (due to a break of the paper) cracks are usually far more achromatic. This fact is readily revealed by placing the image in a hue-explicit color space such as HSV or YC_rC_b. As it turns out, when it applies, this method has excellent ROC performance. Thus, in the case of sepia prints, many pixels off a certain small range of hues near the red hue corresponding to the sepia tint are safely catalogued as being crack pixels. In particular, the results which can be obtained from the previously described test image are shown in Figure 10.7, where detected crack pixels are shown in blue and where some pixels that do not correspond to cracks but that are off the windowed region are also labelled blue. The image is moved from the RGB to the HSV domain, and a hue threshold is selected in order to label crack pixels. The quantitative performances of this method can be evaluated by a plot of its ROC, obtained by varying the hue threshold. The result is shown in Figure 10.8. It can be noticed that, if one privileges small false detection rates as argued above, a possible optimal point of the curve is the one having

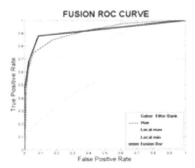

FIGURE 10.9
ROC curve for the combination of the detectors.

probabilities of correct detection ("true positive") and of false alarm ("false positive") respectively equal to 0.63 and 0.02.

Use of a Gabor Filter Bank

At a global level, the overall direction of the crack sets a radial line in the frequency plane along which the energy of the crack should be concentrated. The texture properties of the microcracks are to be exploited as the distance to the origin. A filter bank that covers a set of directions near the main direction and a set of magnitude frequencies is used. It reveals the crack pixels with a good performance level. As pointed out in [12] directional filters, Hough and Radon transforms can be used to detect straight lines and other features of known shape, such as microcracks [32, 52].

Fusion of the Detection Maps

Four binary maps of the area of interest are available at this point, coming respectively from the multi-resolution and multi-orientation Gabor bank, the color analysis, the processed local minima, and the processed local maxima. The first two methods permit one to detect the overall crack structure with a small rate of false positives; the latter two methods permit one to detect different crack subcomponents, with an accuracy which is a function of the selected binarization threshold. The four maps need to be fused to get a final single crack map. Different approaches could be used for this purpose; a simple scheme is based on the likelihood ratio of the data coming from the different operators. In this case, optimal signal detection theory leads to a decision-fusion scheme which uses the local decisions as input to the fusion processor and forms the likelihood ratio of such local decisions. The assumption should be made that the different detectors produce statistically independent maps; this is not true in general, but it has been shown to be an effective hypothesis [69]. The statistical independence-based likelihood ratio can thus be expressed as

$$\lambda(\mathbf{X}) = \frac{P(X_1|H_1)P(X_2|H_1)P(X_3|H_1)P(X_4|H_1)}{P(X_1|H_0)P(X_2|H_0)P(X_3|H_0)P(X_4|H_0)} \tag{10.7}$$

where the X_i are the detector outputs, while H_1 represents the event that the pixel belongs to the crack, and H_0 that the pixel belongs to the photograph. λ is a discrete random variable that takes sixteen different values as expressed in [52].

The ROC for the decision processor is then given as:

$$PD = \sum_{\lambda_i \geq \beta} P(\lambda = \lambda_i|H_1) \,, \quad PF = \sum_{\lambda_i \geq \beta} P(\lambda = \lambda_i|H_0) \tag{10.8}$$

where PD is the true positive rate, PF is the false positive rate, and $i = 1, ..., 16$. An example of ROC is reported in Figure 10.9.

10.2.4   Fragmented Glass Support

The glass plates are acquired with a scanner after having placed their fragments close to each other. Despite the scrupulousness in this operation, it may happen that gaps between fragments in the final digital image are present. Locating the gaps in a fragmented glass plate is an easy task to be performed at scanning time [61]. Gaps originate in areas that are much lighter than the rest of the image, which can be detected by gray-level thresholding. In a second step, each acquired value should be assigned to a given fragment; this result is achieved by connecting pairs of pixels and checking cases in which fragment borders are crossed.

To the best of our knowledge there is only one reference [61] in the open literature specific to the glass plate acquisition and restoration problem, but at least two different scientific contexts exist, the methods of which can be borrowed for the problem at hand. The former is the problem of reconstruction of ancient collapsed wall paintings and mosaics [11,44]. The second context is the solution of a jigsaw puzzle [67]. Both are complex problems because the relative positions of the parts are not known, even if no significant deformations are present.

A method for the virtual restoration of fragmented photographic glass plates is [61]. Since the content of the photographed scene is extremely variable, and it may happen that some fragments represent almost uniform portions of the image, it is more convenient to start from the contour of each fragment and determine its shape. The detection to find where the gaps are located is easy considering how the glass negatives are digitized. Scanners use an illumination source during the acquisition process. The areas without glass do not arrest the light; as a consequence in the acquired image they originate in areas that are much lighter than the rest of the image. In the algorithm described in this section this feature is used to quickly detect the pixels on the gaps. From the input image I, a new image I' is built, where each pixel is labeled "white" if it is in the gap or "black" if it is outside the gap. This new matrix is obtained by performing a thresholding with threshold Th_1 very high.

The algorithm needs some information about the position of the points that belong to the pieces but are close to the border. To extract the positions where these points are located, let us divide I' in overlapping blocks of size 2×2 and then for each block let us apply the rule:

```
if  (#White_pixels ≥ 1 and #White_pixels < 4)
        Black_pixels become Gray_pixels
end
```

The pixels labeled as "gray" belong to the border. In Section 10.4, BF_1 and BF_2 indicate respectively the border of fragments Fr_1 and Fr_2.

10.3   Virtual Restoration of Antique Photographic Prints Affected by Foxing and Water Blotches

Techniques for effectively eliminating foxing and water blotches are not known in the open literature. The restoration process is almost the same for both types of defects due to the fact that, as mentioned in the Introduction, in both cases the affected area (or at least a portion of it) still possesses some residual image information, which is exploited for best results [59]. For clarity, a block scheme of the procedure is shown in Figure 10.10. The input image I is a color image, represented by its R, G and B components.

The restoration process is performed over each RGB color plane. The process is composed of

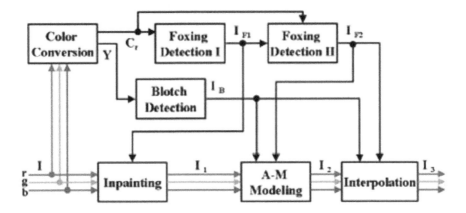

FIGURE 10.10
Flow chart of the algorithm.

three different steps: inpainting, additive-multiplicative (A-M) modeling, and interpolation. The first step is devoted only to the opaque portion of the foxed areas, the remaining ones are used to restore areas affected by both types of aggression. The three restoration steps described above operate without requiring any user intervention.

10.3.1 Inpainting

Inpainting algorithms [3, 8, 13, 38, 41] propagate both the gradient direction and the colors of a band surrounding the hole inside the area to be filled in. Its basic aim is to replace the unrecoverable image data under the opaque foxing layer with values which show good continuity, with respect to the luminance of the area exterior to the stain. Isophote (region with the same level lines) directions are obtained by computing at each pixel along the inpainting contour a gradient vector and by rotating the resulting vector by 90 degrees. This intends to propagate the information while preserving the edges. After few iterations of the inpainting process, the algorithm performs an anisotropic diffusion run to preserve boundaries across the inpainted region.

We use an algorithm that propagates the colors but does not use the gradient direction. If Ω is a single foxing blotch in the map I_{F1}, we denote with $\widetilde{\partial\Omega}$ all the pixels of Ω that are in the border:

$$\widetilde{\partial\Omega} = \{(x, y) \in \Omega \mid \exists (p, q) \in N_{xy} : (p, q) \notin \Omega\}. \tag{10.9}$$

For each pixel $(x, y) \in \widetilde{\partial\Omega}$ we extract the neighborhood N_{xy} as defined in Equation 10.2. Then, we assign to $I_1(x, y)$ the average of the pixels in N_{xy} and not belonging to Ω. The procedure ends when all the pixels in Ω are considered.

The new image I_1 presents more homogeneous foxing spots. They are still visible even if their saturation is reduced. This happens because in this step we have replaced the flawed regions using their outlines which are partially affected by foxing.

10.3.2 Additive/Multiplicative Model

After the inpainting has been performed, the key property of the remaining restoration algorithm must be its ability to exploit the image information still available in the damaged areas. For this purpose, we have selected the A-M model first introduced in [63] for the restoration of scratches in old films.

The model parameters are derived based on the image values in an uncorrupted area surrounding the defect [56]; then, the damaged pixels are substituted with the luminance values yielded by the model. It should be noticed that the same blotch or foxing area may contain different types of image details; in this case a single A-M model is not sufficient and the area is split into different portions where the most effective parameters are selected. More precisely, if

$$\Omega' = \{(x,y) \mid I_{F2}(x,y) = 0 \vee I_B(x,y) = 0\} \tag{10.10}$$

is the corrupted region to restore, as mentioned in [56] a suitable model to describe Ω' in the image I_1 can be the following:

$$I_1(\Omega') = \alpha J(\Omega') + \beta \tag{10.11}$$

where J is an ideal uncorrupted version of the input image I_1, α and β are the parameters to be estimated.

If the variance and the mean (respectively denoted as $var[.]$ and $M[.]$) operators are applied to Equation 10.11, we obtain:

$$\begin{cases} var[I_1(\Omega')] = \alpha^2 var[J(\Omega')] \\ M[I_1(\Omega')] = \alpha M[J(\Omega')] + \beta \end{cases} \tag{10.12}$$

Equation 10.12 can be used to estimate the values α and β. However, the variance and the mean of the uncorrupted image J are unknown. In order to solve Equation (10.11), we approximate $J(\Omega')$ with $I_1(\widetilde{\Omega}')$, where $\widetilde{\Omega}'$ is an uncorrupted area around the blotch. Therefore the approximate values of α and β, $\widetilde{\alpha}$ and $\widetilde{\beta}$ respectively can be obtained by solving Equation (10.12). Subsequently, each pixel in the corrupted regions can be corrected as follows:

$$I_2(\Omega') = (I_1(\Omega') - \widetilde{\beta})/\widetilde{\alpha}. \tag{10.13}$$

where I_2 is the restored image. After this step the area inside the blotch is restored and it appears free of artifacts. The uncorrupted area $\widetilde{\Omega}'$ with width W is automatically extracted. To avoid using pixels that are too close to the blotch Ω', and hence are unreliable, $\widetilde{\Omega}'$ is automatically shifted by D pixels away from the contour. This shifting ensures that an erroneous detection of the defect border does not affect the accuracy of the final restoration.

10.3.3 Interpolation

Indeed, especially in homogeneous image zones, the contour of the processed area can be visible after the two previous steps. This is avoided by a straightforward linear interpolation, operating in the direction orthogonal to the outline of the defect area.

The final output image I_3 is obtained by copying the corresponding values from I_2, for all points which do not belong to the damage border line. For the latter type of points, a linear 1-D interpolation is performed across the border line. It involves a vector of length $2L + 1$. First we perform a simple interpolation and then we assign values at all the possible pixels not considered in the previous steps. More precisely, the luminance gradient is evaluated for each pixel P_i in the contour. Then, an array $\{P_k : i - L \leq k \leq i + L\}$ of $2L + 1$ pixels centered in P_i is considered: the data are chosen along the gradient direction in P_i so that L samples belong to the corrupted region and L belong to the uncorrupted one. If we denote with P_{start} the first pixel of the array and with P_{end} the last one, a conventional linear interpolation can be performed according to the distance between the pixels. If $d(P_j) = \frac{P_j - P_{end}}{P_{start} - P_{end}}$, with $j = 1, \ldots, 2L + 1$, is the normalized distance of each pixel P_j in the array from the P_{start} position, the new intensity values are:

$$I_3(P_j) = d(P_j)I_2(P_{start}) + (1 - d(P_j))I_2(P_{end}) \tag{10.14}$$

(a)   (b)

FIGURE 10.11
(a) Image in Figure 10.1(a) after restoration. (b) Image in Figure 10.1(b) after restoration.

Due to the fact that the gradient orientation can be very different even for neighboring pixels, the interpolation process may skip some samples. To each pixel of this type, we assign a gray level value corresponding to the average of its already interpolated neighbors in a 3×3 mask. Obviously, if the damaged area is split into n different regions, according to the rule in Section 10.3.2, the interpolation step does not take into account pixels that belong to another part of the stain and its related undamaged area. More precisely, if the blotch area Ω_a is associated with $\widetilde{\Omega}_b$, the interpolation will not use the pixels in Ω_i with $i \neq a \wedge i = 1, \ldots, n$ and in $\widetilde{\Omega}_j$ with $j \neq b \wedge j = 1, \ldots, n$.

Sometimes, Ω' can be located in areas with significant details (edges, border of prints, etc.). In these cases we split Ω' into different blocks in order to avoid artifacts. Figure 10.13(b) shows how Ω' and $\widetilde{\Omega}'$ of Figure 10.13(a) are split if $n = 2$: the corresponding areas have the same color.

10.4 Restoration of the Fragmented Glass Plate Photographs

The algorithm proposed in this section can be summarized in three steps [61]: the first one reduces luminance problems; the second one performs a roto-translation to align the different pieces; the last one refines possible residual gaps.

Step 1 The "gray" pixels in I' correspond to pixels in I that have color artifacts not found in their neighbors. This is confirmed by the high values that are found in these positions in the chrominance matrix. Moreover, the areas near the contour are lighter than the rest of the image. We suppose that this depends on the light of the scanner that interacts with the fractured glass, originating a "prism effect" in the border and increasing the luminosity of the areas close to the gap. Our algorithm works in preprocessing to reduce this artifact.

We suppose that the lighter pixels stay inside an area M near the border of width m. To adjust them we use all the available information near the pixel. More precisely: for each pixel in $I(i, j) \in M$ we adjust the luminance according to this rule:

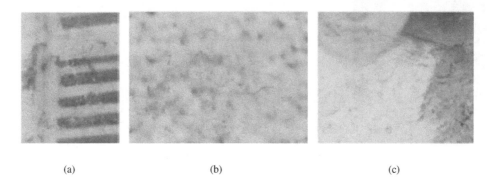

(a)   (b)   (c)

FIGURE 10.12
Examples of restoration of water blotches. (a) Image in Figure 10.2(a) after restoration; (b) Image in Figure 10.2(b) after restoration; (c) Image in Figure 10.2(c) after restoration.

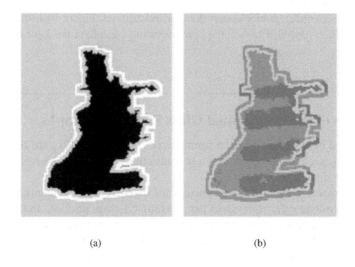

(a)   (b)

FIGURE 10.13
(a) Damaged area Ω' (black) and undamaged area $\widetilde{\Omega}'$ (white); (b) Ω' and $\widetilde{\Omega}'$ are split and the corresponding areas have the same color.

```
N(i,j) = 3 × 3   neighborhood  centred  in   I(i,j)
S={I(p,q) ∈ N(i,j) : I'(p,q)  is  black}
if  I(i,j)  is  gray
    I''(i,j)= the  nearest  pixel  I(p,q) ∈ S
else
    if  |I''(i,j) − I(p,q)| < Th₂
        I''(i,j)= the  nearest  pixel  I(p,q) ∈ S
    end
end
```

Where Th_2 is a suitable threshold, set experimentally. The new image I'' obtained with the correction gives final visual results that are more realistic if it is used as input image.

Step 2 In order to obtain a good match between the pieces of the picture, it is possible to divide the registration-like problem in different parts: rotation and translation. The first copes with the estimation of the rotation angle between BF_1 and BF_2, two binary images where the border lines assume the value 1 with respect to the background 0. The rotation angle α between BF_1 and BF_2 can be estimated as follows:

- Denote as BF_1' the centered version of the curve in BF_1.

- Define $BF_1'^{\delta}$, the image obtained by rotating BF_1' in a suitable subset A of angles, δ, from 0 to 359.

- In order to reduce the computational cost, the decimated images $BF_1'^{\delta,d}$ and $BF_2^{\delta,d}$ have been considered instead of the original ones, $BF_1'^{\delta}$ and BF_2^{δ}.

- For each value $\delta \in A$ the cross-correlation of the two images $BF_1'^{\delta}$ and BF_2^{δ} is evaluated.

- The maximum value M_δ of each cross-correlation matrix - which indicates, for each angle, the best match between the two images - is stored in a vector M_{corr}.

- The position of the maximum value of M_{corr} determines in A the angle β that gives a first estimate for the correct value for α. To increase the precision we repeat the search procedure using the original images, within an interval of angles centered in β. The values of the peak in the cross-correlation matrix are stored in the vector M_{corr}^{β}.

- The maximum value of M_{corr}^{β} is the best match, and it determines the rotation angle α which we are looking for.

- We rotate the fragment Fr_1 and the border image BF_1 by α degrees in order to obtain New_Fr_1 and New_BF_1.

The second part is the estimation of the displacement between the two pieces of the photographic glass plate. We use the well-known phase-correlation technique that is exploited in various motion estimation algorithms [35, 53]. According to the properties of the Fourier transform, it is possible to estimate the displacements between two images as follows:

- Denote $F_1 = \mathcal{F}(New\_BF_1)$ and $F_2 = \mathcal{F}(BF_2)$ as the Fourier transform of the two binary images New_BF_1 and BF_2.

- Estimate the following ratio:

$$\rho = \frac{F_1 \cdot F_2^*}{|F_1| \cdot |F_2|} \tag{10.15}$$

where with $()^*$ we indicate the conjugate operator.

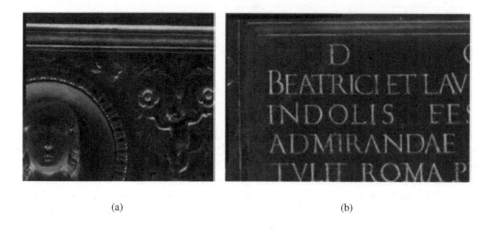

(a)   (b)

FIGURE 10.14
(a) Image 10.3(a) after restoration; (b) Image 10.3(b) after restoration.

- Estimate ϕ as the inverse Fourier transform of the ρ map:

$$\phi = \mathcal{F}^{-1}(\rho) \tag{10.16}$$

Due to numerical precision $|\phi|$ has to be evaluated instead of ϕ. In an ideal case, the ϕ map contains a Dirac delta in the position (dx, dy) where dx and dy denote the horizontal and the vertical displacement between the two input images. In a real case it is sufficient to consider the maximum peak of the map. Now we shift the fragment New_Fr_1 by dx in the horizontal direction and by dy in the vertical one.

Step 3 The image fragments are now very close one to each other. It could happen however that some parts of the fragments are lost before the scanning. For this reason our algorithm eliminates possible pixels that are still in the gap as follows:

```
for each pixel P ∈ the gap G
    N = 3 × 3 neighborhood centred in P
    P = average of P(i,j) with (i,j) ∈ N   and P(i,j) ∉ G
end
```

The image produced by this step is the final restored image. Figure 10.14 reports some results from real negative glass plates taken from Fratelli Alinari archive.

10.5   Restoration of Yellowing and Foxing in Antique Books

A typical processing chain for a page to become a digital document is not easily defined, since its steps are strongly dependent on the purpose of the procedure [60]. After the scanning process, which must be performed accurately avoiding UV illumination and limiting visible light exposure and mechanical stress on the book's binding [72], the document may be segmented into its text / graphic / background components; it may be compressed, separately or jointly, or it may be binarized and input to an Optical Character Recognition (OCR) system [9,45,64]. Owing to the humidity or unintentional water interaction, the pages in old books can be deformed and take local wavelike shape. When these

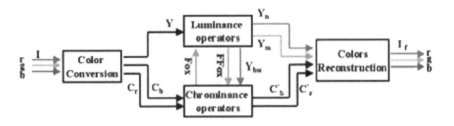

FIGURE 10.15
Block scheme of the algorithms.

books are acquired, the deformations produce projective distortion (skew) in the final image and different colors in its background. If the distortion is significant and produces irregular text lines, it cannot be ignored. In this case, different warping algorithms can be applied [10, 71]. An important element of the overall treatment is the restoration phase. To be generally applied without incurring the criticism of the paleographic expert, the latter has to be performed granting that the processed document gets rid of the defects which are due to aging, but does not lose its original characteristics such as the texture of the paper and the different shades of the ink [37]. There are several algorithms to approach the digital restoration of antique documents. In [6,7,64], a Markov Random Field (MRF) model is used to describe the properties of the boundaries of text characters. The algorithm optimizes this model in order to produce an output where the text is enhanced. This process requires high computational cost and it is usually realized using approximations. To recover ancient text in an Archimede's manuscript, multispectral imaging techniques have been applied [16,65]. The images are acquired with four different wavelengths. The results of these acquisitions depend on the ink properties. Usually, one wavelength reveals the text, while the other ones are helpful to build an image where the text is enhanced. Despite the good results of these techniques, they can be rarely employed due to the multispectral capturing procedure that is very expensive. Different algorithms have been proposed in literature to virtual automated restoration [2, 33, 70]. In [2] a restoration algorithm for the Pahlavi or Middle-age Persian manuscript is provided. It is based over the morphological analysis and connected component concept to segment lines, words, and characters in overlapped Pahlavi documents, in order to preprocess the text before OCR application. Also [33, 70] are based over a morphological analysis of the text but in this case the best parameters are searched for in the neighborhood. In this section a new algorithm for virtual automated restoration of old (since 14th century) paper affected by foxing and yellowing deterioration processes is described. The proposed algorithm can also be used to improve the performance of an OCR. The operators used for eliminating foxing and for paper enhancing have a similar structure [60]. Both algorithms first work over the luminance matrix Y, then adjust the chrominance matrices C_b and C_r (Figure 10.15). The chrominance operations depend on the luminance ones; this happens because the human eye is more sensitive to the details in the luminance field than to those in the chrominance fields.

10.5.1 Foxing

The method we propose automatically detects the foxing stains and then produces a restored version of a page. Detection is based on the chromatic properties of the defect; then, the main aim of the restoration phase is to substitute damaged areas with a color close to the one of the paper background without affecting the image information. To this purpose a detail image is extracted first, typically

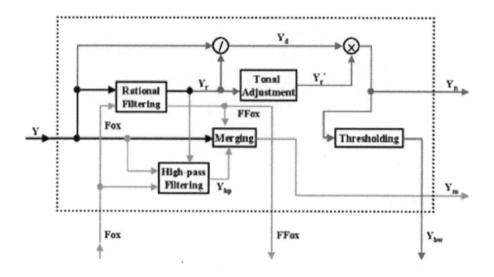

FIGURE 10.16
Block scheme of the luminance operators.

containing printed or handwritten text and miniatures; the rest of the image is processed to eliminate foxing from homogeneous areas; finally the two image components are merged. The chrominance information is modified only in areas affected by foxing and devoid of details.

As expressed in Section 10.2.1, since the foxing damage is formed by a set of reddish-brown spots, we extract their position analyzing the chrominance matrix C_r related to the red. Foxed pixels have the highest values in C_r; we set Max as the maximum value of C_r, and we label as foxed all the pixels whose C_r component has a value greater than $Max - s$, with s a user-selected value. The matrix Fox is a map where the coordinates of foxed pixels are represented as a 1 value.

Image Detail Extraction. First of all, we produce a smoothed image Y_r from the original luminance Y applying n times the RF in Equation 10.5. A detail image Y_{hp} is hence obtained as follows:

$$Y_{hp}(i,j) = Y(i,j) - Y_r(i,j) + K \tag{10.17}$$

where K is a shifting parameter which is set as follows. Let t be a threshold selected using Otsu's method [42]. The values of Y that do not belong to foxing areas ($Fox(i,j)) = 0$) and are not part of the text ($Y(i,j) > t$) are stored in the set M; then,

$$K = mean[M] = mean[\{Y(i,j)|Fox(i,j) = 0 \wedge Y(i,j) > t\}]. \tag{10.18}$$

The detail image Y^{hp} will be used in the *merging* step.

Processing of Foxed Areas. The binary image Fox is the map of foxed areas. As expressed in the preceding text, it has value 1 for the pixels that belong to the foxing, and 0 otherwise. In this step we create a new map that is no longer a binary image but shows smooth transitions between 1 and 0. The slope of the transition is a direct function of the slope of the edges in the original image. To build this new map $FFox$ we use the above RF applied over Fox using the image Y as an edge sensor. More precisely, we use n times Equation 10.5 over Fox where in the denominator Fox is substituted by Y:

$$FFox(i,j) = Fox(i,j) \quad + \frac{Fox(i-1,j)+Fox(i+1,j)-2Fox(i,j)}{k(Y(i-1,j)-Y(i+1,j))^2+A}$$
$$+ \frac{Fox(i,j-1)+Fox(i,j+1)-2Fox(i,j)}{k(Y(i,j-1)-Y(i,j+1))^2+A} \quad (10.19)$$

Merging. In this step we create a new luminance image where the foxing stains are restored. It is formed by merging the shifted highpass image in the foxing areas with the original luminance value in the remaining parts of the image. This is achieved via the combination:

$$Y_m(i,j) = Y_{hp}(i,j)FFox(i,j) + Y(i,j)(1 - FFox(i,j)) \quad (10.20)$$

Color Filtering. As expressed above, the most important characteristic of foxing is its typical color. This means that also the chrominance matrices need to be treated. Our algorithm corrects the foxed areas in the matrices C_r and C_b using the following rule:

$$C'_x(i,j) = \begin{cases} median_x & if\ FFox(i,j) > 0 \\ C_x(i,j) & elsewhere \end{cases} \quad (10.21)$$

where $median_x = median\{C_x(i,j)|FFox(i,j)=0\}$ with $x = \{b,r\}$. The obtained matrices C'_b, and C'_r are the new chrominance matrices for our algorithm. These components, together with Y_m obtained in the previous step, are used to produce the RGB components of the final image I_f.

10.5.2 Page Enhancement

The proposed method to digitally reduce the yellowing of the paper makes the background aspect of the paper more homogeneous, reducing the humidity-induced local alterations; at the same time, parts of the characters which are discolored are restored [60]. Like in the previous section, the algorithm divides the image into text and background. The text is enhanced while the background is adjusted in color. A rational filter is used for detail - background separation. A sigmoidal mapping changes the background luminance; its chromatic components are smoothed via median filtering, in areas determined by a binarized version of the background luminance.

Rational Filter. The luminance component Y of the input image is processed by an RF. The image result, Y_r, contains the most significant background and character structures in the original page. Small but important details in the page are lost in Y_r, but are preserved in our procedure by evaluating the pixel-by-pixel ratio between the original and filtered images:

$$Y_d(i,j) = Y(i,j)/Y_r(i,j) \quad (10.22)$$

The image Y_d is multiplied back at the end of the tonal adjustment process, as indicated in Figure 10.16.

Tonal Adjustment. We want to modify the values of Y_r in order to enhance the contrast between the text and the background. To this purpose we adjust the tonal range of the image using a nonlinear mapping having a sigmoidal shape [60]. To enhance the differences between the text and the background, we use a curve that flattens both dark and bright image areas, as indicatively shown in Figure 10.17. We can control the contrast of the image by increasing or decreasing the slope of its central portion, and moving horizontally its center [57].

The equation experimentally determined as best for the different classes of processed documents is

$$y(x) = -4x^5 + 15x^4 - 20x^3 + 10x^2 \quad (10.23)$$

The output image is more contrasted than the input one. It is multiplied by Y_d, as indicated in Figure 10.16, to obtain Y_n, and the result is used to adjust the chrominance matrices. Y_n is also used in the RGB color space conversion rather than the original Y.

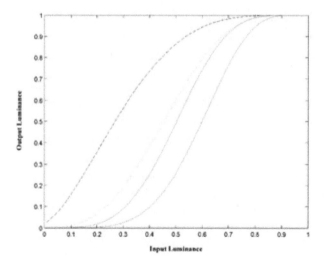

FIGURE 10.17
Examples of tone mapping curves.

Thresholding. In the image Y_n the algorithm automatically searches for a threshold that divides the luminance histogram in two separate classes. We use again Otsu's method of threshold selection. Since the histogram of Y_n shows two separate peaks depending on the tones correction described above, the threshold t selected by Otsu's method is effective [42]. The image Y_{bw} is the binary version of Y_n; its pixels take a value of one only in the positions in which $Y_n(i,j) \geq t$.

Color Filtering. Based on the information derived from the luminance, we then correct the chrominance matrices. We modify only the chrominance values of the pixels that belong to the background and leave all the remaining pixels unchanged. The binary image Y_{bw} is helpful for this operation. Let $M_x = \{C_x(i,j)|Y_{bw}(i,j) = 1\}$, with $x = \{r, b\}$, be the set of chrominance values that belong to the background. The new chrominance matrices are defined as follows:

$$C'_x(i,j) = \begin{cases} median(M_x) & if\ Y_{bw}(i,j) = 1 \\ C_x(i,j) & elsewhere \end{cases} \tag{10.24}$$

The matrices C'_b, and C'_r are the new chrominance matrices for our algorithm. They are used with Y_n to obtain the final image I_f.

In Figure 10.18(b) the restoration of the image in Figure 10.18(a) is reported. In this case, it can be noticed that the foxing stain in the last line of text is removed and the paper deformation has disappeared.

10.5.3   OCR in Antique Documents

A further possible application for the operator is the enhancing of OCR algorithms which detect characters in document images [45]. Commonly diffused OCR systems give better results if the text is black on a white uniform background or vice versa. Obviously, they are based on the assumption that the original image is a high-quality document. Pages from an antique book are not an ideal input for any OCR algorithm: as described above, they are usually characterized by a nonuniform background, and they may be severely damaged. These problems, adding to the faulty recognition of particular characters used in antique printing, produce outputs that are often illegible and far

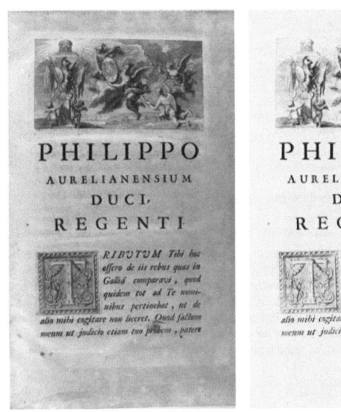

(a)   (b)

FIGURE 10.18 (SEE COLOR INSERT)

(a) An example of background with a variation of colors. Also a foxing stain is visible in the last line of text. (b) Image in Figure 10.18(a) after restoration.

	%
Before our preprocessing	40
After our preprocessing	25

TABLE 10.1
Percentage of mistakes in OCR scan.

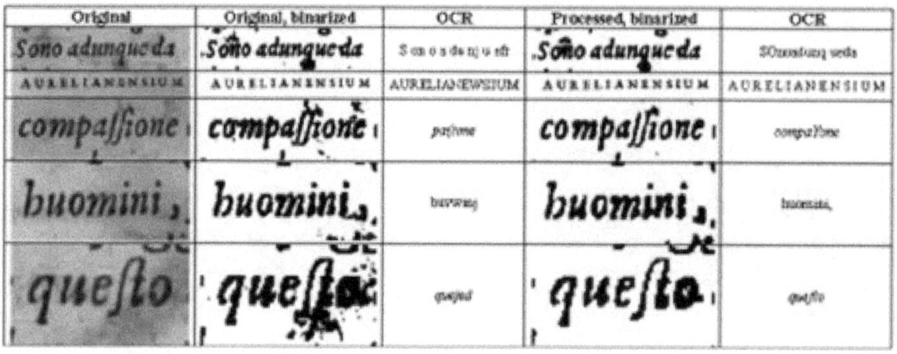

FIGURE 10.19
OCR results with original and processed images.

from reality. An important task is the enhancement of the digital document preserving the text and eliminating background alterations [1, 40, 43]. Since an important part of the OCR is the process of binarization, we propose to use the techniques described in the previous sections to cope with this task. In particular, we suggest creating a binary image Y'_{bw}, analogous to Y'_{bw} as defined in Section 10.5.2 but with different parameters, and using this image as the input of the OCR system. This binary image is obtained eliminating the foxing where necessary, and applying the luminance operator for paper enhancement. For this purpose, we use a high-slope curve during the tonal adjustment. It is a 7th-degree polynomial:

$$y(x) = -20x^7 + 70x^6 - 84x^5 + 35x^4 \tag{10.25}$$

Table 10.1 shows the comparison of the OCR results obtained using a common software tool (ABBYY FineReader 8.0 Professional) with and without the preprocessing method. Without preprocessing the average recognition error rate is 40%; this is a definitely poor result, and is due to the bad quality of the images and to the fact that some characters present in the antique documents are not included in the OCR software set. With the preprocessing the error rate dropped to 25%: still significantly worse than what is expected in conventional office usage, but much better than the figure which the OCR itself can yield, and easily improvable by extending the characters set. Figure 10.19 shows some image details with and without preprocessing. The preprocessing not only reduces the number of mistakes, but it makes the image better readable even by a human observer, and it has lower computational complexity than other techniques (e.g., based on MRF [64]).

10.6 On Image Quality

Estimates of image quality are used for the assessment of the performance of image restoration techniques (as in the *benchmarking of restoration algorithms*) and in the guidance to the choice of the value of a parameter in an adaptive image-restoration technique (as in *optimal-parameter choice*). More generally, image-quality measurement is an important issue in fields such as medical imaging, security, attention, image restoration, image retrieval, etc. Each of these fields uses somewhat different criteria for a definition of image quality.

Consider images of old photographs, aged paintings, old murals, and antique tapestries; in each case, the luminance and color saturation components tend to fade away, with a corresponding loss of contrast. Other than fidelity, the two main aspects of image quality are readability and aesthetics. Readability refers to the easiness with which a human observer interprets an image while aesthetics is an even more subjective, although important aspect. The use of color in antique photographs is rare (even though color photography dates from 1861, when J. C. Maxwell took his *Tartan Ribbon* photograph). Nevertheless, color can be an issue regarding both the readability and the aesthetics of images of antique documents; consider for example the restoration of old books, when the yellowing is to be corrected, or the restoration of black and white photographs that have become brownish over time; however, in the case of antique, virtual document restoration, *color contrast* is usually not an issue, except in the cases of color illustrations such as the case of vignettes.

The preservation of edges and, more generally, of local contrast, has been a constant worry in the history of the field of image processing; likewise, in the field of virtual, antique document restoration, image quality is heavily dependent on the luminance and contrast contents of the image. In fact, time alters most types of pigment, desaturating colors; also, the optics of the devices that capture, transmit, reproduce, and display images tend to decrease the contrast (at out-of-focus regions) and hardly ever increase it. As tools in this nascent field of virtual restoration, it has been found to be a useful approach to consider global as well as local, and joint histograms of variables such as luminance, contrast, and color saturation. The main theme of this section is the distribution of local contrast as a function of local luminance.

10.6.1 On the Measurement of Local Luminance and Local Contrast (Statistics of Location and Dispersion)

Pixelwise luminance is the basic datum used for the computation of the *local luminance* and the local *contrast* of an image. Local luminance is computed as a *location estimate* while contrast is computed as a *dispersion estimate*. A statistic gives, unexpensively and quickly, condensed information regarding some aspect of an image. A statistic can be computed at different levels of locality/globality: pixel, window, image, image data base.

Local Luminance

Local luminance is obtained from pixelwise luminance on the basis of a window. We assume pixelwise luminance to be coded in the interval [0, 1]. A local luminance measurement is obtained by means of the application of an *estimator of location* to a windowed *sample* of (pixelwise) luminances; an estimator of location is a statistic that estimates the *central* value of the luminances in the window. Among the estimators of location, we have the sample mean or average, the sample median and the midrange; these can be seen as linear combinations of order statistics, or *L-statistics*. As customary, the *i-th order statistic* of a sample $\{x_i\}$ is denoted as $x_{(i)}$. For a window of odd size N, we denote the median as $\nu := x_{\left(\frac{N+1}{2}\right)}$ and the midrange as $\mu := \frac{x_{(1)}+x_{(N)}}{2}$.

The eye is sensitive to very bright and very dark regions even of the size of one pixel and thus the midrange is an important measure of local luminance. Different estimators of location and dispersion have different advantages; in particular, if the noise is an issue, it is known that the average, median, and midrange are the maximum likelihood estimators of the mean for populations with Gaussian, Laplacian, and uniform distributions, respectively [49].

Local Contrast

Local (luminance) contrast is obtained by the application of *estimators of dispersion* to windowed samples of pixelwise luminances. Of them, we mainly use the sample variance, its square root (the sample deviation), and the quasiranges (which include the range). Other measures of dispersion are given by the median deviation, the variation from the central pixel and the total variation. Another important measure of local contrast is given by Michelson contrast.

The range ρ is given by $x_{(N)} - x_{(1)}$. Again, because the eye is sensitive to very bright and very dark points in a region of an image, the range is an important estimator of contrast. In general, we denote the quasiranges as $\chi_r := x_{(N-r+1)} - x_{(r)}$, $1 \leq r \leq N/2$. For $r = 1$ we get the range ρ; for r given by the integer part of $N/4$, we have the *quartile quasirange* or *interquartile range*. The quartile quasirange is a measure of the contrast of the central half part of the data. The classical sample variance σ^2 is given by $\frac{1}{N} \sum_1^N (x_i - \bar{x})^2$; for normal populations, an unbiased estimator is obtained dividing the sum above by $N - 1$ instead of N. The sample deviation (also called *rms contrast* [46]) is given by $\sigma := \sqrt{\sigma^2}$. The median deviation is denoted as $\upsilon := median\,\{|x_i - \nu|\}$; as noted in [20], the median deviation is the estimator of dispersion analogous to the median as estimator of location (the "most robust both with regard to gross error sensitivity and to breakdown point.") The *variation from the central pixel* is denoted as $\iota := \frac{1}{N} \sum_1^N |x_i - \bar{x}|$ and the *total variation* of the sample is denoted as

$$\tau := \frac{1}{N^2 - N} \sum_{i=1}^{N} \sum_{j=1}^{N} |x_i - x_j|. \tag{10.26}$$

One other main measure of contrast is that of *Michelson contrast* [46]

$$MC = \frac{max - min}{max + min} \tag{10.27}$$

where max and min are the maximum $x_{(N)}$ and the minimum $x_{(1)}$ of the intensities in the window. The range of the logarithm of intensities is also an interesting measure of local contrast.

Interesting for what follows, Mante et al. [36] have reported a *statistical independence between local luminance and contrast* in natural images using as local luminance λ a weighted sum $\sum w_i \lambda_i$ of pixelwise luminances λ_i; and as contrast, the function

$$GC = \sqrt{\sum w_i \frac{(\lambda_i - \lambda)^2}{\lambda^2}} \tag{10.28}$$

dependent on the weighted contrast of each pixel in the window, with respect to the local luminance λ; the weights w_i decrease with the distance from the center of the context.

Pointwise contrast can be defined in terms of pixelwise and local luminance

$$C(x, y, s) = \frac{I(x, y) - M(x, y, s)}{M(x, y, s)} \tag{10.29}$$

where $I(x, y)$ is the image, $M(x, y, s)$ is the mean of the intensity I in a region $\Omega_{x,y}$ centered in x, y, and s is a scale (or resolution) parameter. This definition pretty much agrees with Weber's law and explicitly includes the scale parameter s. We point out that pointwise contrast is usually a signed magnitude while local contrast is usually unsigned (or taken to be nonnegative).

10.6.2 On the Relationship between Local Contrast and Local Luminance

Scatter plots of local contrast versus local luminance characterize images in important ways and can be used to assess the quality of an image. A statistic that further distills the information in one such scatter plot is the correlation coefficient of local contrast and local luminance. Depending on the estimators used to measure local luminance and contrast, it usually happens that, at very high and very low luminances, the local contrast is restricted to have low values; the (ν, ρ) pair is an exception to this, but we do not consider it in detail here.

Rather, we focus on the distributions of the sample deviation as a function of the sample average, and of the range as a function of the midrange. Local contrast, being nonnegative, is bounded from below in the scatter plots by the line *local contrast* = 0. As we show below [50], the pair $(\bar{x}, \sigma)$ lives on a region bounded by the horizontal axis from below, within the interval [0, 1], and by a piecewise quadratic curve from above. Likewise, the midrange - range pair (μ, ρ) lives in an isosceles triangle [51] with base the horizontal (midrange) axis in the interval [0, 1], and with a third vertex with coordinates $(\mu, \rho) = (0.5, 1)$. The pair median -range has considerably more room of action since the range may take the value 1 for values 0 and 1 of the median; in fact, the whole rectangle [0, 1] $\times$ [0, 1] is possible for the pair (ν, ρ).

The clustering of points against some parts of the boundary of the luminance-contrast space may indicate an overexposed photograph. Scanning and other forms of digitization of images usually scale the points of a luminance-contrast scatter plot down to a proper subset; if there was a clustering against the boundary of the space in the original, in the scanned version there may appear a clustering against a phantom line, parallel to the boundary; in fact, a phantom boundary may appear. Observe for example the image and scatter plot in Figure 10.23.

The (μ, ρ) Pair

Both the range and the midrange are linear combinations of the min $x_{(1)}$ and the max $x_{(N)}$. Besides any properties of robustness that the range and the midrange may have as estimators of dispersion and location, their use allows for an intuitive reading of plots of dispersion versus location; for example, the line $\rho = 2\mu$ corresponds to $min = 0$ while the line $\rho = 2 - 2\mu$ corresponds to the line $max = 1$.

Clearly, the pair (min, max) lives in the triangle with vertices (0, 0), (0, 1) and (1, 1) (i.e., the triangle bounded by the lines $min = 0$, $max = 1$ and the line $min = max$.) The pair (μ, ρ) is obtained transforming the pair (min, max) via the linear transformation $R^2 \rightarrow R^2$ with matrix

$$\begin{bmatrix} 1/2 & 1/2 \\ -1 & 1 \end{bmatrix} \tag{10.30}$$

Therefore, the pair (μ, ρ) lives in the triangle with vertices (0, 0), (0.5, 1) and (1, 0). Part of the usefulness of the (μ, ρ) space comes from the intuitiveness of the (min, max) space; there are also analytical advantages to the (min, max) pair, for example for independent and identically distributed data, the theoretical joint distribution of the max and the min (and therefore that of the range and the midrange) are easier to deduce than that of the sample variance and sample mean.

In the (μ, ρ) plane, the lines through the origin with slope k correspond to the equation $min = \frac{2-k}{2+k}max$ while the lines through the point (1, 0) with slope k' correspond to the equation $max = \frac{2-k'}{2+k'}min + \frac{2k'}{2+k'}$. For example, the line $\rho = \mu$ corresponds to the line $min = 0$ and the line $\mu = 1 - \rho$ to the line $max = \frac{1}{3}min + \frac{2}{3}$. Likewise, the lines of slope k'' through the point $(\mu, \rho) = (0.5, 0)$ correspond to the equation $(k'' - 2)max + (k'' + 2)min = k''$; for example, the lines of slopes 2 and -2 correspond to $min = 0.5$ and $max = 0.5$, respectively.

In Section 10.6.4, using the lines of ρ=0.5, min=0.5, and max=0.5, we subdivide the (μ, ρ) triangle into four triangles, for the purpose of classifying the points in the scatter plots; the triangle with vertices (0, 0), (0.5, 1), and (1, 0) of the (midrange, range) plane is subdivided into the four triangles, A with vertices $\{(0.25, 0.5), (0.5, 1), (0.75, 0.5)\}$, B with vertices $\{(0, 0), (0.25, 0.5), (0.5, 0)\}$, C with vertices $\{(0.25, 0.5), (0.5, 0), (0.75, 0.5)\}$, and D with vertices $\{(0.5, 0), (0.75, 0.5), (1, 0)\}$. The percentages of the points of the scatter plot that fall in each of the regions A, B, C, and D allow for a compact description of the type of image. In Section 10.6.4, based on the relative weights of these regions for a given image, we derive a four-letter descriptor that characterizes the image in the joint dimensions of luminance and contrast.

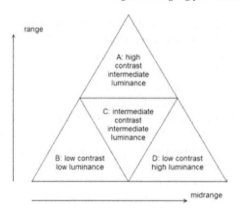

FIGURE 10.20
A partition of the μ ρ-space into the regions A, B, C, and D.

The (μ, ρ) pair lives in a triangle; to find the boundary of the $(\bar{x}, \sigma)$-space requires more work but it is very interesting to note the resemblance. The pair $(\bar{x}, \sigma)$ is restricted to have abscissae in the interval $[0, 1]$; it is bound also, from above and from the sides, by a certain curve that we derive next. Note that if $\bar{x}=0$ or $\bar{x}=1$ then necessarily $\sigma = 0$, since the sample must be constant in those cases; we also have the fact, derived below, that the maximum possible value of σ occurs for $\bar{x}=0.5$.

The $(\bar{x}, \sigma)$ Pair
To get the upper bound curve for the scatter plot (σ, μ), we first derive an alternate expression for the sample variance. Let n be the window (sample) size and let $[x_1, x_2, ...x_n]$ be the corresponding sample data.

$$\sigma^2 = \frac{1}{n^2} \sum_{1 \leq i < j \leq n} (x_i - x_j)^2 \tag{10.31}$$

This expression determines a function of the variables x_i which we want to maximize/minimize, subject to the restrictions $0 \leq x_i \leq 1$ and $\frac{1}{n} \sum_{i=1}^{n} x_i = \bar{x}$, for each $\bar{x}$. Next we find the maximum of $\frac{1}{n^2} \sum_{1 \leq i < j \leq n} (x_i - x_j)^2$ subject to such restrictions, as a function of $\bar{x}$.

First, note that the gradient of the function $f(x) := n^2 \sigma^2 = \sum_{1 \leq i < j \leq n} (x_i - x_j)^2$ is given by $\nabla f(x) = 2[(n-1)x_1 - \sum_{j=2}^{n} x_j, \ldots, (n-1)x_i - \sum_{j \neq i} x_j, \ldots, (n-1)x_n - \sum_{j=1}^{n-1} x_j]$. Next note that $\nabla f(x)$ is normal to the line $\Phi := \{x \in R^n : x_1 = x_2 = ... = x_n\}$ since $\nabla f(x).[1, ..., 1] = 0$.

Lemma. In the restricted domain, the maximum of f occurs at points farthest from the line Φ. This is achieved as follows. Given $\sum_{i=1}^{n} x_i = n\mu = $ k $+ \alpha$; where $k = \lfloor n\mu \rfloor$ and $\alpha = n\mu - k$; with k an integer and α between 0 and 1, the max of f is achieved by letting k of the variables x_i be equal to one, another equal to α and the remaining equal to zero.

Then, the maximum of f is given by $\max(f) = k(1-\alpha)^2 + (n-k-1)\alpha^2 + k(n-k-1)1^2 = (n-1)\alpha^2 - 2k\alpha + nk - k^2$

This function (of the maximum of f in the graph of f as a function of μ) is symmetric around $\mu = 0.5$, with a maximum there and roughly increasing for $\mu \in (0, 0.5)$ although actually piecewise parabolic between each pair of consecutive values k/n and $(k+1)/n$ (these parabolic segments include the vertex of the parabola and are initially decreasing and then increasing.) For large n the roughness of the curve decreases and approaches a monotonic curve. See for example the right-hand side of Figure 10.21, where linear interpolation has replaced the parabolic segments.

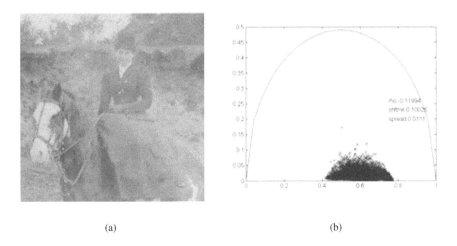

(a)   (b)

FIGURE 10.21
(a) Original image. (b) Corresponding $(\bar{x}, \sigma)$-scatter plot.

10.6.3   Effect of γ-Correction on Scatter Plots of Local Contrast versus Local Luminance

The visualization of the effects of gamma correction on the corresponding local contrast - local luminance plots of the images helps explain not only the change in luminance that the images undergo but also the changes in contrast and the relationships between the changes in contrast and the changes in luminance. As gamma increases (gamma larger than one) a flow of the points in the plane occurs; mainly, points in region D move to regions A and C, points in regions A and C move to region B, and points in region B concentrate (towards the point of zero lightness and zero contrast) and remain in region B; conversely, for values of gamma smaller than one, points in region B move to regions A and C, points in regions A and C move towards region D and points in region D remain in that region, concentrating to the point of largest intensity and zero contrast. Figure 10.23 shows an extreme case of a light (faded) image; after gamma correction, the contrast-luminance plot is more uniformly distributed and the visual quality of the image is improved. The effect of γ-correction on $\mu - \rho$ scatter plots can be seen as a result of the the way the points (min, max) change; thus, consider first the effect of raising to a given power the max and the min. Since the function $f(x) := x^\gamma$ (for positive γ) is a monotonic nondecreasing function, the ordering of the points of the sample is respected; therefore the pair (min, max) becomes the pair (min^γ, max^γ). On the other hand, the difference $x^\gamma - x$ for values of x in the interval [0, 1] is zero at 0 and 1 and maximal at $x = (1/\gamma)^{\frac{1}{\gamma-1}}$; for example, for $\gamma = 0.5$, the max of the difference in absolute value occurs at $x = 0.707$ while for $\gamma = 2$ the max of the abs of the difference occurs at $x = 0.5$. So, for intermediate values of the max and the min, the shift is maximal while for the extreme values the shift is null. The displacement of points in the (min, max) plane results from a vectorial composition so that the shift is in the NE direction for values of gamma smaller than one and in the SW direction for values of gamma larger than one. Translating this to the $\mu\rho$ plane, for γ larger than one there is a movement towards the right for values of gamma smaller than one and towards the left for values of gamma larger than one. On one extreme, for very large values of γ, the points will accumulate near the vertex $(\mu, \rho) = (0, 0)$ and on the other extreme, for values of γ positive but near 0, will be near the vertex $(\mu, \rho) = (1, 0)$. As a function of γ, the range is low for very large and very small (positive) values and has a maximum at an intermediate value of gamma. The midrange on the other hand decreases monotonically with γ.

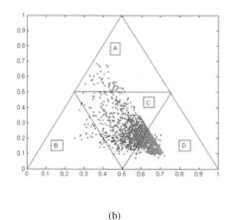

(a)   (b)

FIGURE 10.22
Image "Pisa," gamma corrected with $\gamma=3$, and $\mu\rho$-plot. See the original in Figure 10.23.

region	min	max	median	average
A	0	16	1	3.04
B	0	89.3	16.9	27.03
C	0	64.09	17.7	22.59
D	1	98.4	40.71	43.94

TABLE 10.2
Statistics corresponding to image set considered.

10.6.4 Word Descriptors

For a given image, we compute the number of points in the scatter plot that fall on each of four triangles A, B, C, and D that subdivide the basic $\mu\rho$ triangle; the four regions correspond to *high-contrast/medium-luminance, low-contrast/low-luminance, medium-contrast/medium-luminance* and *low-contrast/high-luminance,* respectively. Variety seems to be a characteristic of good-quality images and we expect such images to have the four regions well represented. Based on the resulting percentages of occupancy, we derive a four-letter word descriptor of the image as follows. Using the percentages, in a decreasing fashion we order the names of the corresponding regions A, B, C, and D; a descriptor of the image is obtained. There are 4!=24 possibilities for the descriptor; nevertheless, some are rather unlikely. As an example, and in order to have an initial reference, based on a small image set of 35 images we computed the min, max, average, and median values of the percentages corresponding to each region A, B, C, and D. The results are summarized in Table 10.2.

We further elaborate on the definition of the word descriptor. By using lowercase or uppercase letters, for each region, respectively, depending on whether or not the percentage is below the corresponding median (observe the third column in Table 10.3), a more sophisticated descriptor result. Nevertheless, it requires the use of a reference data base.

Table 10.3 gives statistics regarding a subset of four images that were subjectively classified as good and in Table 10.4 the four images classified as bad. The corresponding descriptors are CdAb, CBdA, BCdA, and Bdca, for the good images, and Dacb, Da'b'c' (primed letters indicate absence of points), ACdb, and CdbA, for the bad images.

Under gamma correction, the image Pisa underwent a change in word descriptor from Dacb to

region	min	max	average
A	2.4	16	8.20
B	6.4	55.64	33.56
C	21.35	38.7	31.92
D	12.56	38.4	25.98

TABLE 10.3
Statistics corresponding to four images considered of good quality.

region	min	max	average
A	0	55.81	16.76
B	0	12.02	3.31
C	0	60.58	24.50
D	1	9.77	30.68

TABLE 10.4
Statistics corresponding to four images considered of bad quality.

DCab; its quality improved by a decrement of the luminance and an increment of the texture contents. The image description changed from BcdA to Bdca; it became less dark and got more detail as well. Texture1 (ACdb) has higher contrast than Texture2 (CdbA); in each case, local contrast occurs above certain minimal, positive levels, also regions B and D (light and dark regions) are underrepresented; this in opposition to good-quality images which often include regions that are nearly constant (Figure 10.22).

Texture images tend to have a strong C region as well. Globally, texture images tend to have an intermediate luminance and a medium to large contrast. The effects of i.i.d. noise on luminance-contrast plots can be studied theoretically with some distribution analysis. The joint distribution of the max and the min of a sample of n data from an underlying population with probability density function f and cumulative distribution function F is given by [14] $f_{mn,mx}(a,b) = n(n-1)[F(b) - F(a)]^{n-2}f(a)f(b)$. For an underlying uniform distribution U[0,1], with F(t)=t and f(t)=1, the joint density becomes, $f_{mn,mx}(a,b) = n(n-1)[b-a]^{n-2}, a \in [0,1], a \leq b \leq 1$ which has a maximum at (min, max) = (a, b) = (0, 1) and expectation (E[min], E[max]) = $(\frac{1}{n+1}, \frac{n-2}{n-1})$; thus, translating the result to the (midrange, range) plane, the largest likelihood will correspond to the point (0.5, 1) and the cloud of points will have a center of mass at $(\mu, \rho) = (\frac{n^2-3}{2(n+1)}, \frac{n^2-2n-1}{n^2-1})$. A strong A region may be an indication of noise or fine textures in the image. Image regions containing white noise will normally have an intermediate luminance, of a value depending on their mean value, and a contrast that will depend on the strength and distribution of the noise: impulsive noise (with a heavy tailed distribution) will have a larger contrast than uniform noise; higher range values are probable for larger windows. Textured regions having a scale-dependent behavior depend more on the size of the window. For faded images, the distribution of the points in the $\mu\rho$-triangle is highly concentrated near the point (1, 0), an indication of an image with overall high intensity and low contrast. Gamma correction causes a migration of the points, according to a certain flow, that makes the scatter plot more uniformly distributed on the one hand and also makes the overall average luminance near 0.5, on the other; images with good coverage of the $\mu\rho$-triangle and global average luminance near 0.5 look well, in general. Dark images tend to live in the B region while light images tend to live in the D region. We argue that luminance-contrast plots give important information regarding image quality, in the sense that visual quality requires variety in the combination of local contrast and local intensity in the image. In particular, images that live only in the B and D regions only are to be considered of poor quality. Figure 10.23 shows a case of a light (faded) image.

Broadly, we conclude that the letters A and C should be at the end of the descriptor, otherwise

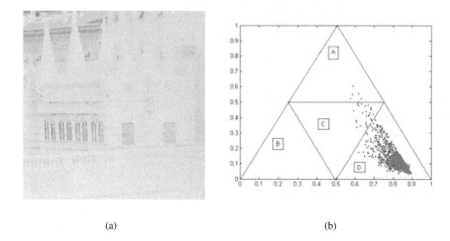

(a)   (b)

FIGURE 10.23
Original image "Pisa" (size 512×512 pixels) and $\mu\rho$-plot.

the image is likely to have excessive detail. The regions B and D should be balanced; in fact, at least 20% of the points should fall on each of the regions B and D, otherwise the image is too light or too dark; also, a 5% minimum on the A and C regions is highly desired. At least two upper-case letters should be present: this will depend on the underlying data set used. In addition, the absence of points in any of the regions is an indicator of lack of variety and of bad image quality.

10.7 Conclusions

In this chapter we first provided a description of problems that typically affect photographs and antique books; we then presented some methods used for the detection of such defects and for the virtual restoration of the affected documents. Some comments on image quality, as far as these types of images are concerned, have been added. It is important to observe that the field we have addressed is far from mature; few authors study it, and many algorithms and techniques are borrowed from similar but not identical application areas. Moreover, in general the automatic detection of damaged areas in imaged material is a difficult task, due to the superposition of the defects with the peculiar structure of the object. Also adequate models for the damage, which would permit one to build dedicated restoration algorithms, are often missing. For all these reasons, we hope that this collection of otherwise dispersed information will trigger further research and improvements.

Acknowledgments

We thank Alinari 24 ORE s.p.a. for its support and useful hints, and for providing the digitized photographs used in our experiments.

    We wish to thank Digital Codex and Biblion Centro Studi sul Libro Antico Onlus for providing the digitized documents used in the experiments. The original documents belong to the Redentoristi Library at S. Maria della Consolazione in Venice, Italy.

Bibliography

[1] Majdi Ben Hadj Ali. Background noise detection and cleaning in document images. In *Proceedings of International Conference on Pattern Recognition*, pages 758–732, 1996.

[2] S. Alirezaee, H. Aghaeinia, M. Ahmadia, and K. Faez. An efficient restoration algorithm for the historic middle-age persian (pahlavi) manuscripts. 3:2114–2120, 2005.

[3] C. Ballester, M. Bertalmio, V. Caselles, G. Sapiro, and J. Verdera. Filling-in by joint interpolation of vector fields and gray levels. *IEEE Transactions on Image Processing*, 10(8):1200–1211, 2001.

[4] M. Barni, F. Bartolini, and V. Cappellini. Image processing for virtual restoration of artworks. *IEEE MultiMedia*, 7(2):34–37, 2000.

[5] T. D. Beckwith, W. H. Swanson, and T. M. Iiams. Deterioration of paper: the cause and effect of foxing. *University of California Publications in Biological Science*, 1(13):299–356, 1940.

[6] L. Bedini, A. Bozzi, and A. Tonazzini. Digital technique for character recognition in old documents. *ERCIM News: http://www.ercim.org/*, (28), January 2004.

[7] L. Bedini, G. M. Del Corso, and A. Tonazzini. A preconditioning technique for edge-preserving image restoration. In *Proceedings of IEEE International Conference on Information Intelligence and Systems*, pages 519–526, 1999.

[8] M. Bertalmio, G. Sapiro, V. Caselles, and C. Ballester. Image inpainting. In *SIGGRAPH 2000*, pages 417–424, 2000.

[9] A. Bozzi and A. Sapuppo. Computer-aided preservation and transcription of ancient manuscript and old printed documents. *ERCIM News: http://www.ercim.org/*, (19), October 1995.

[10] M. S. Brown and W. Brent Seales. Image restoration of arbitrary warped documents. *IEEE Transactions on Pattern Analysis and Machine Intelligence*, 26(10):1295–1306, 2004.

[11] M. Capek and I. Krekule. Alignment of adjacent picture frames captured by a CLSM. *IEEE Transactions on Information Technology in Biomedicine*, 3(2):119–124, June 1999.

[12] M. J. Carlotto. Enhancement of low-contrast curvilinear features in imagery. *IEEE Transactions on Image Processing*, 16(1), 2007.

[13] A. Criminisi, P. Pérez, and K. Toyama. Region filling and object removal by exemplar-based inpainting. *IEEE Transactions on Image Processing*, 13(9):1200–1212, 2004.

[14] H. A. David. *Order Statistics, 3rd ed.* Wiley, 2003.

[15] A. De Rosa, A.M. Bonacchi, V. Cappellini, and M. Barni. Image segmentation and region filling for virtual restoration of art-works. In *Proceedings of ICIP 2001*, number 1, pages 562–565, 2001.

[16] Roger L. Jr Easton, Keith T. Knox, and William A. Christens-Barry. Multispectral imaging of the archimedes palimsest. In *Proceedings of IEEE 32nd Applied Imagery Pattern Recognition Workshop (AIPR03)*, 2004.

[17] I. Giakoumis, N. Nikolaidis, and I. Pitas. Digital image processing techniques for the detection and removal of cracks in digitized paintings. *IEEE Transactions on Image Processing*, 15(1), pp. 178-188, 2006.

[18] I. Giakoumis and I. Pitas. Digital restoration of painting cracks. In *Proceedings on IEEE Internationa Symposium ISCAS '98*, volume 4, 1998.

[19] S. Guillon, P. Baylou, M. Najim, and N. Keskes. Adaptive nonlinear filters for 2D and 3D image enhancement. *Signal Processing*, (67):237–254, 1998.

[20] F. R. Hampe. The influence curve and its role in robust estimation. *JASA*, 69(346):383–393, 1974.

[21] http://palimpsest.stanford.edu/don/dt/dt1434.html.

[22] http://www.anna-amalia bibliothek.de.

[23] http://www.brickscommunity.org.

[24] http://www.dli2.nsf.gov.

[25] http://www.lowyonline.com/spr01_foxing.html.

[26] http://www.octavo.com/collections/essays/foxing.html.

[27] http://www.pbs.org/wgbh/pages/roadshow/speak/foxing.html.

[28] http://www.photogallery.it.

[29] http://www.rleggat.com/photohistory.

[30] http://www.ul.cs.cmu.edu.

[31] A. K. Jain. *Fundamentals of Digital Signal Processing*. Prentice Hall Inc., 1989.

[32] J. K. Kamarainen, V. Kyrki, and H. Kälviäinen. Invariance properties of gabor filter-based features - overview and applications. *IEEE Transactions on Image Processing*, 15(5), 2006.

[33] Tapas Kanungo and Qigong Zheng. Estimating degradation model parameters using neighborhood pattern distributions: An optimization approach. *IEEE Transactions on Pattern Analysis and Machine Intelligence*, 26(4):520–524, 2004.

[34] Z. Kollia, E. Sarantopoulou, A.C. Cefalas, S. Kobe, and Z. Samardzija. Nanometric size control and treatment of historic paper manuscript and prints with laser light at 157 nm. *Appl. Phys., A, Mater. Sci. Process*, 79:379–382, 2004.

[35] L. Lucchese and G. M. Cortelazzo. A noise-robust frequency domain technique for estimating planar roto-translations. *IEEE Transactions on Signal Processing*, 48(6):1769–1786, June 2000.

[36] V. Mante, R. A. Frazor, V. Bonin, W. Geisler, and M. Carandini. Independence of luminance and contrast in natural scenes and in the early visual system. *Nature Neuroscience*, 8(12):1690–1697, 2005.

[37] E. Margin. Proposal for a classification of the different aspects of inks on manuscripts dating from the XVIIth to the beginning of the XIXth century. *www.isyreadet.net*.

[38] S. Masnou and J.-M. Morel. Level lines based disocclusion. In *Proceedings of ICIP 98*, pages 259–263, 1998.

[39] S. K. Mitra, H. Li, I. Lin, and T.Yu. A new class of nonlinear for image enhancement. In *Proceedings of IEEE International Conference on Acoustics, Speech, and Signal Processing ICASSP-91*, number 4, pages 2525–2528, 1991.

[40] H. Negishi, J. Kato, H. Hase, and T. Watanabe. Character extraction from noisy background for an automatic reference system. In *Proceedings of ICDAR*, pages 143–146, 1999.

[41] M. M. Oliveira, B. Bowen, R. McKenna, and Y. S. Chang. Fast digital image inpainting. In *Proceedings of International Conference on Visualization, Imaging and Image Processing VIIP 2001*, pages 261–266, 2001.

[42] N. Otsu. A threshold selection method from gray-scale histogram. *IEEE Transactions on System, Man, Cybernetics*, (SMC-8):62–66, 1978.

[43] H. Ozawa and T. Nakagawa. A character image enhancement method from characters with various background images. In *Proceedings of Second International Conference on Document Analysis and Recognition*, pages 20–22, 1993.

[44] C. Papaodysseus, T. Panagopoulos, M. Exarhos, C. Triantafillou, D. Fragoulis, and C. Doumas. Contour-shape based reconstruction of fragmented, 1600 BC wall paintings. *IEEE Transactions on Signal Processing*, 50(6):1277–1288, June 2002.

[45] T. Pavlidis and S. Mori. Optical character recognition. *In Proceedings of IEEE*, 80(7):1027–1209, July 1992.

[46] E. Peli. Contrast in complex images. *Journal Opt. Soc. Am. A*, 7(10):2032–2040, 1990.

[47] H. J. Porck and R. Teygeler. Preservation science survey: An overview of recent developments in research on the conservation of selected analog library and archival materials. Technical Report ISBN 1-887334-80-7, Council on Library and Information Resources, 2000.

[48] G. Ramponi. The rational filter for image smoothing. *IEEE Signal Processing Letters*, 3(3):63–65, March 1996.

[49] A. Restrepo and A. C. Bovik. On the statistical optimality of locally monotonic regression. *IEEE Transactions on ASSP*, 42(6):1548–1550, 1994.

[50] A. Restrepo and G. Ramponi. Filtering and luminance correction of aged photographs. In *Proceedings of IST/SPIE E.I. Sci. and Techn.*, 2008.

[51] A. Restrepo and G. Ramponi. Word descriptors of image quality based on local dispersion-versus-location distributions. In *Proceedings of EUSIPCO 08*, Lausanne, Switzerland, 2008.

[52] A. Restrepo, E. Fogar, and G. Ramponi. On the detection of cracks in photographic prints. *In Proceedings of Image Processing: Algorithms and Systems VII, IS&T SPIE, Symposium on Electronic Imaging*, 2008.

[53] P. M. B. Van Roosmalen. *Restoration of Archived Film and Video*, PhD Thesis. Universal Press, 1999.

[54] F. Stanco and G. Ramponi. Detection of water blotches in antique documents. In *Proceedings of 8th COST 276 Workshop*, Trondheim (Norway), May 2005.

[55] F. Stanco, G. Ramponi, and A. de Polo. Towards the automated restoration of old photographic prints: a survey. In *Proceedings of IEEE-EUROCON 2003*, September 2003.

[56] F. Stanco, G. Ramponi, and L. Tenze. Removal of semi-transparent blotches in old photographic prints. In *Proceedings of 5th COST 276 Workshop*, October 2003.

[57] F. Stanco, G. Ramponi, and L. Tenze. A method for improving the visual quality of digitized antique books. In *Proceedings of 7th COST 276 Workshop,* Ankara (Tukey), November 2004.

[58] F. Stanco, L. Tenze, and A. De Rosa. An improved method for water blotches detection and restoration. In *Proceedings of IEEE ISSPIT*, pages 457–460, December 2004.

[59] F. Stanco, L. Tenze, and G. Ramponi. Virtual restoration of vintage photographic prints affected by foxing and water blotches. *JEI Journal of Electronic Imaging*, 14(4), December 2005.

[60] F. Stanco, L. Tenze, and G. Ramponi. A technique to correct yellowing and foxing in antique books. *IET Image Processing*, 1(2):123–133, June 2007.

[61] F. Stanco, L. Tenze, G. Ramponi, and A. de Polo. Virtual restoration of fragmented glass plate photographs. In *Proceedings of IEEE-MELECON 2004*, pages 243–246, May 2004.

[62] C. Sun and B. Appleton. Multiple paths extraction in images using a constrained expanded trellis. *IEEE Transactions on Pattern Analysis and Machine Intelligence*, 27(12), 2005.

[63] L. Tenze and G. Ramponi. Line scratch removal in vintage film based on an additive/multiplicative model. In *Proceedings of IEEE-NSIP*, 2003.

[64] A. Tonazzini and L. Bedini. Character segmentation in highly blurred ancient printed documents. In *Proceedings of IEEE International Conference on Image Analysis and Processing*, pages 836–841, 1999.

[65] A. Tonazzini, L. Bedini, and E. Salerno. Image analysis on the archimedes palimpsest. *ERCIM News: http://www.ercim.org/*, (58), July 2004.

[66] L. Vincent. Morphological grayscale reconstruction in image analysis: Applications and efficient algorithms. *IEEE Transactions on Image Processing*, 2(2):176–201, April 1993.

[67] H. Wolfson, E. Schonberg, A. Kalvin, and Y. Lamdan. Solving jigsaw puzzles by computer. *Ann. Oper. Res.*, 12(1-4):51–64, 1988.

[68] Y. Huang, B. Xu. Automatic inspection of pavement cracking distress. *Journal of Electronic Imaging*, 15(1), Jan/Mar 2006.

[69] L. W. Nolte, Y. Liao, and L. M. Collins. Precision fusion of ground-penetrating radar and metal detector algorithms: A robust approach. *IEEE Transactions on Geoscience and Remote Sensing*, 45(2):398–409, Feb. 2007.

[70] Y. Yang, K. Summers, and M. Turner. A text image enhancement system based on segmentation and classification methods. In *Proceedings of the 1st ACM workshop on Hardcopy document processing (HDP '04)*, pages 33–40, New York, NY, USA, 2004. ACM Press.

[71] L. Zhang, Z. Zhang, and T. Xia. Restoring warped document images through 3D shape modeling. *IEEE Transactions on Pattern Analysis and Machine Intelligence*, 28(2):195–208, 2006. Senior Member: Chew Lim Tan.

[72] Z. Zhang and C.L. Tan. Restoration of images scanned from thick bound documents. In *Proceedings of International Conference on Image Processing*, volume 1, pages 1074–1077, 2001.

11

Advances in Automated Restoration of Archived Video

Anil Kokaram, Francois Pitie, David Corrigan
Trinity College
Email: `anil.kokaram@tcd.ie`

Domenico Vitulano, Vittoria Bruni, Andrew Crawford
Istituto per le Applicazioni del Calcolo "M. Picone" - Consiglio Nazionale delle Ricerche
Email: `vitulano@iac.rm.cnr.it, bruni@dmmm.uniroma1.it`

CONTENTS

Since 1995 there has been an explosion in the exploitation and availability of digital visual media. Digital television and Internet broadcasting continue to merge and viewing figures over the Internet have continued to grow. YouTube/Google and social networking sites have completely changed the way we think about visual media broadcasting while HD television and high-quality mobile displays have changed how often we access and demand content. Therefore the demand for good-quality content has never been higher and hence holders of visual archives (film and video) have found themselves in a completely different commercial landscape for their services as well as their media. Material in archives is generally in poor condition with pictures corrupted by dirt, grain, tearing, and so on. Automated restoration systems provide the only mechanism for exploiting that demand. Before 2000, research in restoration was therefore limited to applications in heritage, but since 2000 a wider commercial requirement has moved automated restoration algorithm development from a fringe activity to an enabling technology in an emerging industry [1]. Companies such as MTIFilm,[1] Lowry Digital [1], and Snell and Wilcox[2] provide restoration software and hardware, with all post-production houses (e.g., Framestore London, Cinecite UK, ILM) including *Dirt Busting* software as a matter of course in their arsenal of film treatment tools.

A small subset of defects are illustrated in Figures 11.1-11.6 to show the wide ranging complexity of the problems. Figure 11.1 shows *Dirt and Sparkle* which occurs when material adheres to the film due to electrostatic effects (for instance) and when the film is abraded as it passes through the transport mechanism. It is also referred to as a *Blotch* in the literature. The visual effect is that of

---

[1] www.mtifilm.com
[2] www.snellgroup.com

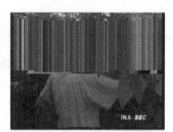

FIGURE 11.1
Dirt and Sparkle

FIGURE 11.2
Film Grain Noise is due to the mechanism for the creation of images on film. A piece of dirt is indicated on the image.

FIGURE 11.3
Betacam Dropout manifests due to errors on Betacam tape.

FIGURE 11.4
Vinegar Syndrome often results in a catastrophic breakdown of the film emulsion.

FIGURE 11.5
Kinescope Moire is an aliasing effect.

FIGURE 11.6
Film Tear

bright and dark flashes at localized instances in the frame. The image indicates where a piece of Dirt is visible. Figure 11.2 shows *Film Grain Noise* which is a common effect and is due to the mechanism for the creation of images on film. It manifests slightly differently depending on the different film stocks. The image shows clearly the textured visible effect of noise in the blue sky at the top left. Blotches and noise typically occur together and are the main form of degradation found on archived film and video. A piece of Dirt is indicated on the image. *Betacam Dropout* (shown in Figure 11.3) manifests due to errors on Betacam tape. It is a missing data effect and several field lines are repeated for a portion of the frame. The repeating field lines are the machine's mechanism for interpolating the missing data. *Vinegar Syndrome* often results in a catastrophic breakdown of the film emulsion and is shown in Figure 11.4. This example shows long strands of missing data over the frame. *Kinescope Moire* (Figure 11.5) is caused by aliasing during Telecine conversion and manifests as rings of degradation that move slightly from frame to frame. Finally, *Film Tear* in Figure 11.6 is simply the physical tearing of a film frame sometimes due to a dirty splice nearby.

This chapter cannot begin to deal in detail with the huge array of algorithms developed for restoration. We instead choose to address the defects that affect the bulk of archived material: *Dirt, Lines, Shake, Flicker,* and *Noise*. The reader is given a brief introduction to the principal issues and major results in each case, followed by pointers to more detailed exposition of the algorithms. The chapter ends with a discussion of the industry in restoration that has evolved from this body of work.

11.1 Dirt and Missing Data

The most common problem in archived film is *dirt* and *sparkle* or *blotches* that manifest as small regions of dark or white pixels. This occurs when film patches have become obscured due to dust or dirt sticking to the film, or completely obliterated due to abrasion of the film material (sparkle). Digital media also suffer from missing data artifacts when bits are corrupted causing loss of blocks or loss of entire frames. Line scratches are also caused by film abrasion and can result in missing data, but often data is still available in the defected area although obscured.

A key difference between missing data as it appears in real footage and *speckle* degradation is that blotches are almost never limited to a single isolated pixel. Therefore simple spatial order statistic filters with small window sizes cannot remove the distortion, while larger window filters will remove too much detail. Assuming that the missing data does not occur in the same location in consecutive frames, most successful schemes repair the damaged region by copying the relevant information from previous or next frames. This relies on the heavy temporal redundancy present in an image sequence. Because this redundancy is prevalent only along motion trajectories, motion estimation has become a key component of missing data treatment systems.

Historically, the approach has been to develop a method to detect the defect [2], then to correct it by some kind of spatio-temporal interpolation activity [3–6]. The traditional concept in detection is to assume that any set of pixels that cannot be located in next and previous frames must represent some kind of impulsive defect and should be removed. This requires some kind of matching criterion and could be dealt with via a model-based approach [7, 8] or several clever heuristics [7–10]. More complicated systems explore the interaction between the motion estimation and missing data detection/correction stages [11–13].

It is worth mentioning that infrared scans of degraded film have been successfully used for detecting blotches and line scratches. With IR scanners such as [14] the defects show up clearly as patches of grayscale and uncorrupted film regions show up as completely dark. In fact infrared techniques have been recently used to build a ground truth to employ in the objective evaluation of the detection results [15, 16]. However, IR can detect image regions which are not perceived as blotches and IR is not suitable if the film medium is not available. This implies that fully digital detection of blotches is still important. In what follows we chart the various solutions up to the present day.

11.1.1 Simple Detection

Consider the simplest image sequence model as follows.

$$I_n(\vec{x}) = I_{n-1}(\vec{x} + \mathbf{d}_{n,n-1}(\vec{x})) + e(\vec{x}) \tag{11.1}$$

where the luminance at pixel site $\vec{x} = [i, j]$ in frame n is $I_n(\vec{x})$, and the two component motion vector mapping site $\vec{x}$ in frame n into frame $n - 1$ is $\mathbf{d}_{n,n-1}(\vec{x})$. The model error follows a Gaussian distribution $e(\cdot) \sim \mathcal{N}(0, \sigma_e^2)$. The model is therefore consistent with luminance conserving, translational motion in the sequence. The key point here is that image sequences obeying the model show luminance constancy along the motion trajectory. Hence if the backward (for instance) Displaced Frame Difference (DFD) at a pixel site, $|\Delta_b(\vec{x})| = |I_n(\vec{x}) - I_{n-1}(\vec{x} + \mathbf{d}_{n,n-1})|$ is high, there is probably some defect in the sequence (e.g., missing data) or occlusion/uncovering.

Hence the earliest work on designing an automatic system to "electronically" detect dirt and sparkle flag a pixel as missing if the forward and backward pixel difference was high. This was undertaken by Richard Storey at the BBC [9, 17] as early as 1983. The natural extension of this idea by incorporating motion was presented by Kokaram around 1993 [2, 7] that allowed for motion-compensated differences. Hence the very simplest detector of missing data flags a pixel as missing when $(|\Delta_f|(\vec{x}) > T) \text{AND} (|\Delta_b(\vec{x})| > T)$, where Δ_f, Δ_b are the forward and backward motion-compensated frame differences. That detector can be improved dramatically if the sign of the DFD is also taken into account, since we can expect the signs to be the same at sites of corruption. That scheme was called the SDIp (Spike Detection Mask).

The SDI detectors clearly are not pixelwise smooth, and missing data hardly ever corrupts a single pixel. To solve these problems spatial and temporal information needs to be incorporated into the detector. In 1996, Nadenau and Mitra [10] presented such a scheme, which used a spatio-temporal window for inference: the Rank Order Detector (ROD). It is generally more robust to motion estimation errors than any of the SDI detectors although it requires the setting of three thresholds. It uses some spatial information in making its decision. The essence of the detector is the premise that blotched pixels are outliers in the local distribution of intensity.

Defining a list of pixels as follows,

$$p_1 = I_{n-1}(\vec{x} + \vec{d}_{n,n-1}(\vec{x}) + [0\ 0])$$
$$p_2 = I_{n-1}(\vec{x} + \vec{d}_{n,n-1}(\vec{x}) + [0\ 1])$$
$$p_3 = I_{n-1}(\vec{x} + \vec{d}_{n,n-1}(\vec{x}) + [0\ -1])$$
$$p_4 = I_{n+1}(\vec{x} + \vec{d}_{n,n+1}(\vec{x}) + [0\ 0])$$
$$p_5 = I_{n+1}(\vec{x} + \vec{d}_{n,n+1}(\vec{x}) + [0\ 1])$$
$$p_6 = I_{n+1}(\vec{x} + \vec{d}_{n,n+1}(\vec{x}) + [0\ -1])$$
$$I_c = I_n(\vec{x}) \tag{11.2}$$

where I_c is the pixel to be tested, the algorithm may be enumerated as follows.

1. Sort p_1 to p_6 into the list $[r_1, r_2, r_3, \ldots, r_6]$ where r_1 is minimum. The median of these pixels is then calculated as $M = (r_3 + r_4)/2$.

2. Three motion-compensated difference values are calculated as follows:

If $I_c > M$			If $I_c \leq M$		
e_1	$=$	$I_c - r_6$	e_1	$=$	$r_1 - I_c$
e_2	$=$	$I_c - r_5$	e_2	$=$	$r_2 - I_c$
e_3	$=$	$I_c - r_4$	e_3	$=$	$r_3 - I_c$

3. Three thresholds are selected t_1, t_2, t_3. If any of the differences exceeds these thresholds, then a blotch is flagged as follows

$$b_{\text{ROD}}(\vec{x}) = \begin{cases} 1 & \text{if } (e_1 > t_1) \text{ OR } (e_2 > t_2) \text{ OR } (e_3 > t_3) \\ 0 & \text{otherwise} \end{cases}$$

where $t_3 \geq t_2 \geq t_1$. The choice of t_1 is the most important. The detector works by measuring the "outlierness" of the current pixel when compared to a set of others chosen from other frames. The choice of the shape of the region from which the other pixels were chosen is arbitrary.

11.1.2   Better Modeling

These heuristics have one theme in common. They use some prediction of the underlying true image, based on information around the site to be tested. The difference between that prediction and the observed value is then proportional in some way to the probability that a site is corrupted. In the SDIp, the prediction is strictly temporal. In the ROD the prediction is based on an order statistic; other approaches use morphological filters to achieve the same goal. Poor predictions imply high probability of corruptions. Joyeux, Buisson, Decenciere, Harvey, Boukir et al. have been implementing variants of these techniques for film restoration since the mid-1990s [8, 18–23]. Joyeux [8] points out that these techniques are particularly attractive because of their low computational cost. This idea is more appropriately quantified with a Bayesian approach to the problem. Since 1993 [2,5,7,12,13,24–27] this approach was adopted not only to detect missing data but also to reconstruct the underlying image and motion field. Reconstruction of the motion field is very important indeed since temporal prediction is highly dependent on correct motion information.

Assume that we can identify missing pixels by examining the presence of temporal discontinuities. To do this, associate a variable $s(\vec{x})$ with each pixel in an image. That variable can have six values or states $b00, b01, b10$. The variable b is binary, indicating whether a pixel site is corrupted, $b = 1$, or not, $b = 0$. The remaining digits encode occlusion hence 00 indicates no occlusion, while 01, 10 imply pixels that are occluded in the next and previous frames respectively. Our process for missing pixel detection requires one to estimate $s(\cdot)$, and those pixels with state $b = 1$ are to be reconstructed. The field $b(\vec{x})$ is the missing data detection field. The degradation model is then a mixture process with replacement and additive noise as follows:

$$G_n(\vec{x}) = (1 - b(\vec{x}))I_n(\vec{x}) + b(\vec{x})c(\vec{x}) + \mu(\vec{x}) \tag{11.3}$$

where $\mu(\cdot) \sim \mathcal{N}(0, \sigma_\mu^2)$ is the additive noise, and $c(\vec{x})$ is a field of random variables that explain the actual value of the observed corruption at sites where $b(\vec{x}) = 1$.

Three distinct (but interdependent) tasks can now be identified in the restoration problem. The ultimate goal is image estimation (i.e., revealing $I_n(\cdot)$ given the observed missing and noisy data). The missing data detection problem is that of estimating $b(\vec{x})$ at each pixel site. The noise reduction problem is that of reducing $\mu(\vec{x})$ *without* affecting image details. From the degradation model of (11.3) the principal unknown quantities in frame n are $I_n(\vec{x})$, $s(\vec{x})$, $c(\vec{x})$, the motion $\vec{d}_{n,n-1}$ and the model error $\sigma_e^2(\vec{x})$. These variables are lumped together into a single vector $\boldsymbol{\theta}(\vec{x})$ at each pixel site $\vec{x}$. The Bayesian approach infers these unknowns conditional upon the corrupted data intensities from the current and surrounding frames $G_{n-1}(\vec{x})$, $G_n(\vec{x})$, and $G_{n+1}(\vec{x})$. For the purposes of missing data treatment, it is assumed that corruption does not occur at the same location in consecutive frames, thus in effect $G_{n-1} = I_{n-1}$, $G_{n+1} = I_{n+1}$.

Proceeding in a Bayesian fashion, the conditional may be written in terms of a product of a likelihood and a prior as follows:

$$p(\boldsymbol{\theta}|I_{n-1}, G_n, I_{n+1}) \propto p(G_n|\boldsymbol{\theta}, I_{n-1}, I_{n+1})p(\boldsymbol{\theta}|I_{n-1}, I_{n+1}) \tag{11.4}$$

This posterior may be expanded at the single pixel scale, exploiting conditional independence in the model, to yield

$$
\begin{aligned}
&p(\theta(\vec{x})|G_n(\vec{x}), I_{n-1}, I_{n+1}, \theta(-\vec{x})) \\
&\propto p(G_n(\vec{x})|\theta(\vec{x}), I_{n-1}, I_{n+1})p(\theta(\vec{x})|I_{n-1}, I_{n+1}, \theta(-\vec{x})) \\
&= p(G_n(\vec{x})|I_n(\vec{x}), c(\vec{x}), b(\vec{x})) \\
&\quad \times p(I_n(\vec{x})|\sigma_e(\vec{x})^2, \vec{d}(\vec{x}), s(\vec{x}), I_{n-1}, I_{n+1}) \\
&\quad \times p(s(\vec{x})|S)p(c(\vec{x})|C)p(\vec{d}(\vec{x})|D)p(\sigma_e(\vec{x})^2)
\end{aligned}
\tag{11.5}
$$

where $\theta(-\vec{x})$ denotes the collection of θ values in frame n with $\theta(\vec{x})$ omitted, and C, D, S, denote local dependence neighborhoods around $\vec{x}$ (in frame n) for variables b c, d, and s respectively.

Solution

Although that Bayesian expression seems quite complicated, it is possible to exploit marginalization cleverly to simplify the problem at each site and then yield surprisingly intuitive iterative steps. These steps involve local pixel differencing interspersed with simple motion refinement. The key step is to realize that the variables of this problem can be grouped into two sets. The first set contains a pixel state, $s(\vec{x})$ (defined previously), its associated "clean" image value $I_n(\vec{x})$, and finally the corruption value $c(\vec{x})$ at each pixel site. The second set contains a motion and error variable $[\vec{d}_{n,n-1}, \vec{d}_{n,n+1}, \sigma_e^2]$ for each block in the image. In other words, the variables can be grouped into a *pixelwise* group $[s, c(\vec{x}), I_n]$, that varies on a pixel grid, and a *blockwise* group $[\vec{d}_{n,n-1}, \vec{d}_{n,n+1}, \sigma_e^2]$, that varies on a coarser grid. Exploiting marginalization we can factor the distributions as follows (using the pixelwise triplet for illustration)

$$p(s, i_n, c | I_{n-1}, I_{n+1}, S, C) = p(i_n | c, s) p(c | s, C) p(s | S, C, I_{n-1}, I_{n+1})$$

$$\text{Where} \quad p(c | s, C) \propto p(c, s | C) = \int_{i_n} p(s, i_n, c | I_{n-1}, I_{n+1}, S, C) di_n$$

$$\text{and} \quad p(s | S, C, I_{n-1}, I_{n+1}) = \int_c p(c, s | C) dc \tag{11.6}$$

To solve for these unknowns, the ICM (Iterative Conditional Modes) algorithm [28] is used. At each site, the variable set that maximizes the *local* conditional distribution *given* the state of the variables around, is chosen as a suboptimal estimate. Each site is visited with variables being replaced with ICM estimates. The maximization is performed through the factorization above, using the marginalized estimate of each variable in turn. Copious details can be found in Kokaram [13] and the reader is directed there for more information.

11.1.3 Further Refinement

The reader would be forgiven for thinking that the Bayesian approach outlined above encapsulates the essentials of the missing data problem in video sequences and is completely successful at removing degradation. Sadly, its great weakness lies in the underlying image sequence prediction model that is used for the data likelihood. In fact, it is quite often in a real image sequence that motion blur, self-occluding, and non-rigid motion (e.g., cloth, transparent/translucent objects, and specular reflections) all contrive to cause the breakdown of that model. Therefore, while the method above works well in many cases, when it fails, it fails in unusual ways. Removal of single frame specular highlights are possible, and increased blur at the boundaries of motion blurred regions also occur. Furthermore, in a real application, a user may wish to remove dirt but leave the noise intact. In that case the decision-making process that rejects dirt can leave "blobs" of discontinuous material which are not in the right place. Note, however, that when using a spatiotemporal linear prediction (3DAR) model as the image sequence modeling equation, these effects are reduced in the sense that erroneously interpolated areas tend to be smoother.

This problem was called *Pathological Motion* in the Aurora[3] project of 1995-1999. P. van Roosmaalen [4], Raphael Bornard [29], and Andre Rares [30–32] were the first to address the issue. In each case the idea that worked was to turn off any dirt detection when PM was detected. The most reliable PM detector used the density of discontinuity detection to flag regions as showing unusually high self-occluding activity and hence most likely to be false alarms. The PM detection itself was a post-process after detecting temporal discontinuities of any type. To put it another way, we do not expect there to be blotched pixels in the same location in successive frames. If that is observed, then it is likely that something is wrong with the detection process and it is better to turn off the restoration than to risk damage to good pictures. A simpler and practical formulation was developed by Kent

---

[3]EU Funded 5th Framework research project

FIGURE 11.7 (SEE COLOR INSERT)
Top row: showing four frames from four different sequences exhibiting pathological motion. Second row from top: PM/blotch detection in red/green respectively. Third row: restoration employing standard detection without PM detection. Incorrectly reconstructed areas are highlighted in cyan. Last row: restoration with PM detection embedded in the blotch removal process. Note the better preservation of the PM regions while still rejecting blotches.

et al. [33]. That is based on a simple global motion estimator which is used to detect foreground regions. The detector takes the conservative approach of reducing the probability of detecting a blotch in foreground regions. The Kent algorithm compromises performance for computational load. The idea of using PM detection to turn off dirt removal has been very successful in general and has been generalized in work by Corrigan et al. [34]. In that work, Corrigan uses two features for PM detection: (1) repeated discontinuities detected in the same position in consecutive frames, (2) unusually divergent motion fields. These constraints were initially built for the discontinuity estimation step and as a result the process does not treat PM detection as a post-process (as in the case of Bornard, Rares, and Roosmalen) but integrates it with the blotch detection optimization in a unified framework. Figure 11.7 shows examples of four sequences in which PM occurs. These examples represent typical problems: clothing, periodically rotating objects, fast moving and warping objects (birds), and motion blur (in the chopper blades). The third row in that figure shows what happens when a standard blotch removal process [2] is used to detect and remove the blotches with a 3D median filter. The highlighted regions exhibit very poor reconstruction since the motion information is not correct at all in PM areas. The second row shows the result of Corrigan's process, highlighting in red areas detected as PM and in green areas detected as blotches. We see in the last row how much improved the picture reconstruction is because of the reduction of false alarms in PM areas.

Vlachos et al. [15] have attempted to create more robust systems by performing a spatial discontinuity detection first followed by a temporal validation step. This reduces the computation involved and works well for small high-contrast missing areas. In [16], two uncorrelated blotch detectors are fused to improve performance, based on the assumption that the conflict between the two detectors gives false alarms. The detectors employ different features: one is temporal while the other spatial. In contrast with the work in [15], the temporal detection is applied first, leaving the validation of the detection results to the spatial detector. The SROD detector [4] is employed in the first step, looking for outliers of the intensity distribution of each analyzed pixel. The distribution is the one extracted from adjacent frames in the image sequence. A modified morphological spatial filter is used in the validation step, in order to make it suitable also for semi-transparent objects. Fusion is achieved by machine learning algorithms.

Despite these improvements, blotches which are poorly contrasted still evade detection. This has led recent work to consider variants of the corruption model and this is discussed next.

11.2   Semi-Transparent Defects

Many missing data defects do not affect the image in a catastrophic way. In fact many blotches and certainly line scratches still retain part of the original information in the defected area [13, 35–37]. These are *semi-transparent* defects. Examples are shown in Figure 11.8.

The mixture model of Equation 11.3 is therefore not appropriate for these defects. Semi-transparent blotches can appear as irregularly shaped regions (see Figure 11.8), with variable color and intensity. They are commonly caused by contact with moisture so that underlying image details remain in the affected area, whereas the color and/or average intensity is changed. Note that even with opaque blotches the edge of the blotch region shows a measure of transparency. Despite the huge amount of work done for the restoration of opaque blotches (missing data defects), very little has been done specifically for semi-transparent blotches. As pointed out in [16], semi-transparency, large size and low visibility are the main reasons for which standard methods fail in the processing of that kind of degradation. Pure in-painting methods [38] that synthesize information may seem not suitable for this task [39, 40]. Even for semi-transparent blotches, temporal information alone cannot completely characterize the defect. For this reason, some authors have paid particular attention to the

(a)   (b)

FIGURE 11.8 (SEE COLOR INSERT)

Examples of semi-transparent defects. The slightly visible semi-transparent blotch is circled in the right image.

spatial characterization of the defect. The spatial result can be then validated using the inter-frame information for avoiding further ambiguities—see [15, 16] for a thorough discussion.

In 2008 Bruni et al. [41] were the first to approach the problem of detection of semi-transparent defects by exploiting observations about the HVS response to temporal discontinuities. They consider that the sudden reaction of human visual system (HVS) to the presence of a semi-transparent blotch corresponds to the projection of the degraded image G into a new space where semi-transparent blotches become the most visible objects in the scene. The projection operator P depends on the physical event that generated the blotch. For instance, for blotches caused by contact with moisture, a suitable projection space is the saturation component in the HSV color space, since moisture causes a sort of miscellanea of colors. Moreover, human eyes perceive semi-transparent blotches as uniform areas even though it is not so. It means that blotches emphasize their visibility with respect to the remaining part of the scene at particular resolutions. A low pass filter simulates this effect: it removes redundant frequencies that are not perceived by HVS, providing the visual homogeneity of the degraded region (see Figure 16.12). The optimum level of resolution r is a trade-off between the enhancement of the degraded region and the preservation of its geometrical shape and size. Finally, the recognition and selection of the most visible regions in the previously defined projection space $P_r[G]$ is performed through the definition of suitable distortion measures, that account for both global and local visibility of objects in the whole scene, as human eyes do. Conventional contrast definitions generally account for pixel-wise measures by considering the analyzed region as opaque objects over a uniform background. Unfortunately, this is not the case of semi-transparent blotches since they preserve and inherit background features. The proposed region-based distortion measure is the following

$$D(\Omega_T) = \frac{1}{|\Omega_T|} \sum_{(x,y) \in \Omega_T} D_1(x,y) D_2(x,y), \qquad \Omega_T = \{(x,y) \in \Omega : \ P_r[G](x,y) > T\}$$

(11.7)

where T is a threshold value, Ω is the image domain, Ω_T is the visible region whose intensity value over-exceeds T, $|\Omega_T|$ is its size, while (x,y) is a point in Ω_T. D_1 measures the change of the perception of a given object with respect to the fixed background, before and after its transformation through a threshold based operator, i.e.,

$$D_1 = \frac{P_r[G] - \tilde{G}}{M} \tag{11.8}$$

where M is the mean of the degraded image $P_r[G]$, $\tilde{G}$ is the result of the clipping transformation on $P_r[G]$ with $\tilde{G} = T$ if $P_r[G] > T$, otherwise $\tilde{G} = P_r[G]$. In other words, it evaluates how an object of intensity $P_r[G]$ changes if it is substituted for the threshold value T. D_2 measures the change of the contrast of the same object over different backgrounds, i.e.,

$$D_2 = \frac{P_r[G](M_T - M)}{M M_T} \tag{11.9}$$

where M_T is the mean of the non-clipped region of $P_r[G]$.

The optimal threshold $\overline{T}$ is the one such that $D(\Omega_T)$ is maximum (i.e., the point where occurs a good separation between the foreground and the background of the image). It is the maximum contrast for the image which is able to separate different objects of the image without introducing artifacts. In fact, from that point on, the clipping operator selects pixels belonging to both degradation and background.

The advantages of that kind of approach are its capability of: (1) detecting all blotches in the scene, even the slightly visible one or the ones masked by the underlying image content; (2) fine tuning all the involved thresholds and/or parameters since they are adaptively computed according to both local and global perception-based measurements; (3) being independent of the shape and size of the defect. These features make it a good candidate for spatial detector.

11.2.1   Reconstruction

Removal of semi-transparent defects in still images has been addressed since 2005 [37, 40, 41], and that is the topic of Chapter 10 of this book and will not be discussed further here. Handling semi-transparency in reconstruction for video sequences requires explicit modeling of the corruption. In 2009 Mohammed et al. [42] approached this problem from the standpoint of matting. Their work is related to work in 2007 by Crawford et al. [39] and the essential equation is in fact a non-binary version of Equation 11.3. Hence we write instead

$$G_n(\vec{x}) = (1 - \alpha(\vec{x}))I_n(\vec{x}) + \alpha(\vec{x})\Lambda + \mu(\vec{x}) \tag{11.10}$$

In this equation, α takes the place of b and is non-binary, but also varying between 0 and 1, while the corruption variable Λ, replaces $c(\vec{x})$ and is constant. Setting Λ to 255 (for an 8 bit) image, then allows a blotch to be modeled as a continuous mixture between the underlying original image and a bright white corruption. In the matting problem, observed objects are modeled as a mixture of hidden foreground and background layers and the challenge is to extract the best *alpha* matte as well as foreground and background object layers. The proposed model is identical to the matting problem except that one of the layers is known, i.e., Λ. Hence the semi-transparency restoration problem is posed as the solution of the above equation for α, I_n at each pixel site that is corrupted, i.e., the extraction of an α mattte and a background layer I_n. In their work, Mohammed et al. employ the previous and next frames to model the color distribution of the background layer I_n. Smoothness of the background layer is an important prior and is imposed using the gradients of the surrounding frames in that region as well as an MRF (Markov Random Field). A Graph Cuts algorithm allows the solution for the variables. This is a very interesting model and in theory can be incorporated into the joint Bayesian framework introduced previously in this chapter. It appears to be the most suitable model for these artifacts and will no doubt feature heavily in future work.

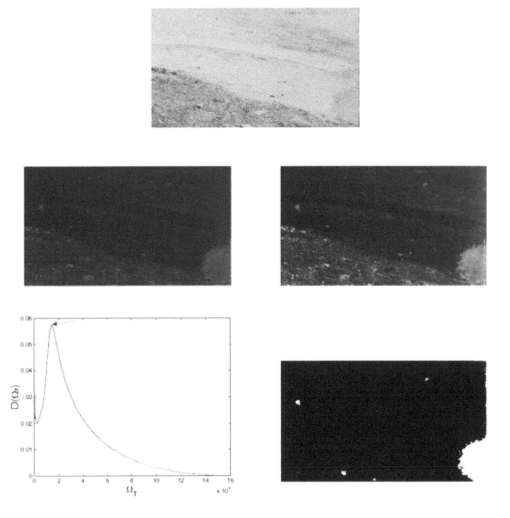

FIGURE 11.9

From top to bottom, left to right: Original semi-transparent blotch; its saturation component; the smoothed saturation component at the optimal resolution; global distortion as in Equation (11.7); detection mask.

FIGURE 11.10
Line scratches manifest in much archived footage. They also occur due to accidents in film developing. The color of the scratch depends on which side of the film layer it occurs. It is often the case that not all the image information inside the scratch is lost. They are a challenge to remove because they persist in the same location from frame to frame.

11.3 Line Scratches

Another very common form of degradation in film is line scratches. These are caused during developing or due to material being stuck in the film gate and abrading or smearing material over many consecutive frames. Figure 11.10 shows some examples. Early work by Kokaram [7,43] on detection and removal of this defect concentrated on spatial detection and reconstruction. Detection was limited to straight scratches and relied on the horizontally impulsive nature of the artifact (it generally is very narrow and well contrasted) combined with the longitudinal correlation. The degradation model was additive and is as follows.

$$G_n(x) = I_n(x) + \sum_{p=1}^{P} L_p(x - x_p) \tag{11.11}$$

where G_n is observed degraded image, I_n is the original clean image, and there are P lines distributed around location centres x_p with an intensity cross section $L_p(\cdot)$ that varies horizontally only. The observed intensity profile of a line scratch is fairly regular and is brighter near the center of the scratch, getting darker near its extremities. Kokaram proposed a damped sinusoid to model this profile. However, considering that the line scratch can probably be explained as a diffraction effect through the narrow vertical slit created by the abrasion of the film emulsion, a more appropriate model is a *sinc* function. The width of the observed scratch depends on the width of the slit, while the brightness of the observed scratch changes according to the depth of the scratch on the film material. Since the scratch does not penetrate the film material, original image information persists and so the result is a semi-transparent artifact.

In 1996 Kokaram built the line detector by using the Hough Transform to isolate line candidates that are narrow and bright vertical lines. These candidates were then validated using the damped sinusoid model to reject false alarms. Four years later, Bretschneider et al. [44] employed the vertical detail component of the wavelet transform of the image to detect the line. A *sinc*-like shape was assumed for the horizontal line projection L_p, i.e., $L_p(x - x_p) = b_p sinc(\frac{3\pi(x-x_p)}{2w_p})$ in Equation 11.11, and only the vertical details of the degraded and clean images G, I are used for the degradation model. Given that line visibility is an important aspect for detection, Bruni et al. [36,45] in 2004

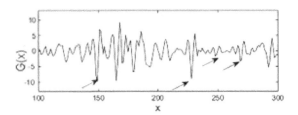

FIGURE 11.11
Horizontal cross-section of an image containing line scratches (indicated by arrows).

introduced aspects of human perception into the degradation model as follows

$$G(x) = (1 - (1 - \gamma_p) \exp^{-\frac{2}{w_p}(x - x_p)}) I(x) + (1 - \gamma_p) L_p(x - x_p), \quad \forall p$$

$$\text{where } L_p(x - x_p) = b_p sinc^2 (\frac{\pi(x - x_p)}{w_p}) \tag{11.12}$$

The normalized parameter γ_p balances the semi-transparency of the defect by taking into account its visibility with respect to the surrounding image content, according to Weber's law [46]. In particular, γ_p approaches zero in case of important scratches, while it converges to 1 whenever the scratch is masked by its context in the scene. In that way, even slightly visible defects are represented and they are detected only if their contrast value is up to the just noticeable threshold [36, 46], see for example rightmost scratches in Figure 11.11. The resulting algorithm is computationally very fast, making its use in film analysis practical. Moreover, visibility laws allow both to avoid false alarms (or at least limit their number) and to tune the employed perception-based threshold. The same arguments remain valid whenever color films are considered, as shown in [47], where the color of the observed scratch is related to the depth of the slit on the film material. Note that blue scratches were treated in 2005 [48]. In that work intense blue scratches are detected as maxima points of the horizontal projection of a suitable detection mask created by thresholding hue, saturation, and value amplitudes ranges.

Since 1999 Besserer, Joyeux et al. [49–51] took a different approach to detection, exploiting the temporal coherence of the line scratch over several frames. Besserer [49] presented an excellent tracking algorithm (using a Kalman tracker) that connected horizontally impulsive line elements which persist across several frames at a time. Figure 11.12 shows the representation used. Their work yields very high quality detection masks and it seems sensible that a combination of the ideas of Bruni et al. and Besserer et al. are the way forward.

However, removing line scratches is notoriously difficult. While convincing spatial interpolation can be achieved in a single frame, over several frames any error in reconstruction is clearly seen since it is correlated with the same position in many frames. Example based texture synthesis, famously introduced by Efros et al. [52] can achieve very good spatial reconstructions, but temporally the result is poor if simply repeated on multiple frames. Most of the proposed approaches assume the absence of the original information in the degraded region, see for instance [7, 44, 51, 53–57]. Therefore, they propagate neighboring clean information into the degraded area. The neighboring information can be

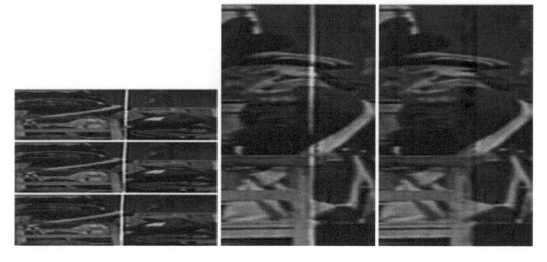

FIGURE 11.12
Left: The image sequence representation useful for detecting line scratches. Center: Frame with scratch. Right: Reconstruction of the line scratch in the region of local motion. Errors in reconstruction are much more visible in a sequence. Work on temporally coherent interpolation is important.

found in the same frame [7, 44, 53, 54] or also in the preceding and successive frame exploiting the temporal coherency, as done in [51, 55, 56]. The propagation of information can be performed using in-painting methods, as in [53, 54], or interpolation schemes. In [7], an autoregressive filter is used for predicting the original image value within the degraded area. A cubic interpolation is used in [58], by also taking into account the texture near the degraded area (see also [57] for a similar approach), while in [44] a different interpolation scheme is used for low and high frequency components. Finally, in [55] each restored pixel is obtained by a linear regression using the block in the image that better matches the neighborhood of the degraded pixel. Figure 11.12 shows the problem of poor temporal consistency in the region of local motion. The autoregressive interpolator of Kokaram et al. [43] was used here.

A smaller class of restoration approaches assumes the presence of the original information in the degraded area. An additive multiplicative model is employed in [59]. In that work, the image content in the degraded area is modified using a linear model in order to match with the mean and variance of the original surrounding information. In [48] blue scratches are removed by comparing their contribution in the blue and green color channels with the one in the red channel, since a blue scratch is assumed to be negligible in the red channel. Visibility laws are used in [45, 47] to also guide the restoration process. The idea is very simple: the contribution of a line scratch must be suitably attenuated till it is no more visible in the scene. It means that the contrast value between the degraded area and the surrounding region must be small enough to perceive the whole area as a uniform one. A Wiener filter based shrinkage is then applied to the degraded region, where the defect is modelled as noise, while the original image is derived by the inversion of the equation model (Equation11.12). This approach works only because the horizontal shape of the scratch is well defined, i.e., it is the $sinc^2$ shape in Equation 11.12.

Any successful approach to line removal, however, must enforce temporal coherence of the interpolated region. An extension to the ideas of Efros to spatiotemporal synthesis for line scratch removal was introduced by Bornard et al. [29] and was quite successful. It relied on using local motion around the line scratch to fetch useful information outside that region in the next and previous frames. They used global motion only and hence there were difficulties with objects showing local motion. The problem is how to reconstruct the underlying missing data including reconstructing the motion convincingly over several frames. Recent work by Irani et al. [60], Kokaram et al. [61], and Sapiro et al. [62] has considered the problem of object removal in image sequences. While Kokaram takes a purely motion-based approach, Irani and Sapiro both use a combination of more sophisticated methods relying on substituting 3D cubes of data from other parts of the sequence or 3D inpainting. These ideas could be applied to the removal of line scratches. However, the nature of the defect is more severe than the cases considered for object removal thus far, in that motion around the region may not be periodic or may be highly varying over short distances. Line scratch removal technology in industry therefore still requires manual review of manipulated pictures.

11.4 Global Defects

The previous defects have all been local in nature. Blotches and lines affect only certain areas of each frame, leaving other areas untouched. Random brightness fluctuation (called *flicker*) is common in archived sequences however, as is shake or warping of each frame. These artifacts affect the entire image. Shake occurs in archived film because of (1) warping of the film medium and (2) worn guide holes in the film transport area causing each frame to be in different locations relative to the scanning or display equipment. Flicker is caused by the degradation of the medium (ageing of the film stock), varying exposure times, or curious effects of poor standards conversion. Varying exposure time is common to hand-cranked footage, but also happens with mechanized cameras in early films or most recently with personal 8mm. The flicker artifact is still a problem today, even with the use of digital cameras. One frequent source of modern flicker comes from radiometric calibration issues. Consider for instance time-lapse sequences, where each film frame is taken at long interval rates, or sequences made from multiple cameras, like in the "inbetweening" special effect used in *The Matrix* (1999). These kinds of sequences usually flicker when played back because each frame has been captured under different lighting conditions and calibration settings.

Of these two artifacts, shake has received a huge amount of attention in the literature, but not for archival purposes. In fact it is because of the rise of handheld digital video cameras that handheld video shake has become a common problem and hence attracted widespread attention. Shake in archive material can be much more difficult to remove than handheld camera shake however, since nonlinear warping can indeed occur that will have little to do with the camera motion. Furthermore, in archival material, shake (or warp) and flicker often occur together, making the removal process that much more difficult. Kokaram et al. [63] in 2003 were the earliest to address these two defects in tandem if only to note that warp removal is best done after flicker removal. Work in removing camera shake can indeed be applied to archival material, and the underlying idea in both cases is to estimate the global motion of the scene in its entirety, and then filter out the random global motion components. Simple FIR (Finite Impulse Response) filters were used in [63] together with an affine model for the global motion component. Recent work by Liu et al. in 2009 [64] presents what amounts to a breakthrough in handheld camera shake removal. In that work, the 3D trajectory of the camera is estimated using structure from motion ideas, and the camera path is then smoothed in the scaled 3D space. The main contribution is the development of image warps which do not overdistort the scene when balancing the stabilization. Handheld camera shake remains a very active and important industrial topic for mobile phone video capture, but there remains a gap in the application of those ideas to archival material. Missing areas and brightness flicker will confuse a structure from motion algorithm (for instance) and the fact that the observed image instability is not solely due to camera position alone (i.e., physical distortion of the image medium is also an issue), means that there is still work to be done here. Very interestingly in 2009, Robinson et al.[4] have commercialized a

---

[4]The Foundry (www.thefoundry.co.uk)

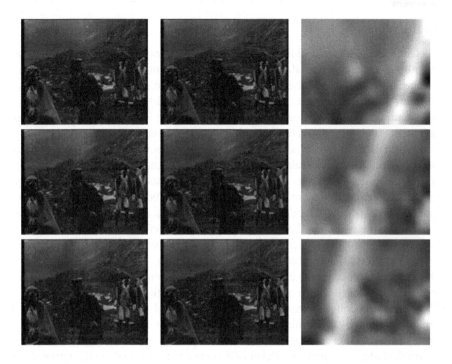

FIGURE 11.13
Example of strongly localized archive flicker. On the left, the original sequence (showing a strong black diagonal swathe moving across the image from right to left), on the middle, the deflickered sequence, and on the right the corresponding flicker map.

solution to the problem of CMOS rolling shutter. In modern CMOS (Complementary Metal-Oxide Semiconductor) devices the frame scan time is much slower than with CCD arrays. As a result a visible warp of the image can occur when the capture device is moved during capture. This is related to the archival problem in the sense that the observed "shake" is then a combination of camera motion *and* image warp. Their solution is based on detailed optic flow analysis of the scene but no further details are available currently. We do not address the warp/shake problem further in this section except to note that this area remains open as far as archive restoration is concerned.

Flicker is considered as a global artifact as it affects the entire frame but the brightness fluctuations also frequently present some spatial variations across each frame. The variations are usually very smooth but may appear as localized structures, like the black diagonal on the right of the frame in the second row (leftmost column) in Figure 11.13. These structures are a typical manifestation of severe flicker on archive footage. An extreme case of localization appears in modern footage in the case of *in-scene* flicker. This occurs when fluorescent lights are out of synchronization with the acquisition rate and the camera is not in the same place as the lights. If the fluorescent lights are inside a room for instance and the camera is outside the room, then only the door is flickering whereas the rest of the scene is non-flickering. As visible in Figure 11.14, due to the complexity of the scene, the flicker requires a pixel-wise granularity.

Deflicker techniques generally consist of two stages: (1) the flicker model estimation and (2) the flicker compensation that aligns the brightness level.

Flicker Models

In the general sense, the flicker artifact between two images u and v can be modeled as a mapping t on the grayscale component that depends on the pixel position $\mathbf{x} = (x, y)$:

$$v(\mathbf{x}) = t(u(\mathbf{x}), \mathbf{x}) + \epsilon(\mathbf{x}) \tag{11.13}$$

The outlier term $\epsilon(\mathbf{x})$ accounts for the image disparities due for instance to motion or missing data. The mapping at a particular pixel is usually modeled as being linear $t(u(\mathbf{x}), \mathbf{x}) = a(\mathbf{x})u(\mathbf{x}) + b(\mathbf{x})$ in modern footage [4, 63, 65] and nonlinear in old footage [66–68]. Ideally the mapping t should be estimated at every pixel site but this would require too much computation. Since the flicker artifact is spatially smooth, the solution adopted in the literature is to interpolate the mapping using 2D polynomial [4, 65], cosine [63], or spline [66] functions. Using splines for instance yields the following interpolation [66]:

$$v(\mathbf{x}) = \sum_i w(\mathbf{x} - \mathbf{x}^{(i)}) \, t^{(i)}(u(\mathbf{x})) \tag{11.14}$$

where $\mathbf{x}^{(i)}$ is the 2D position of the ith control point on the image, $t^{(i)}$ the estimated mapping at this control point, and $w(\mathbf{x})$ the interpolating 2D spline.

The problem to estimate the flicker mapping is that not all of each image pair can be matched. The parts that cannot be matched are due to occlusions/uncovering because of motion or simply due to missing data (blotches/dirt/dropout) in the case of degraded film and video material. To cope with this problem, Roosmalen [4] and Yang [69] suggest detecting occluding areas based on spotting large intensity differences that cannot be explained by flicker alone. Parameter estimation is then performed only on the blocks in which there are no outliers detected. Estimates for the "missing blocks" are then generated by some suitable interpolation algorithm. Unfortunately, this method for detecting outliers fails in the presence of heavy flicker degradation.

Flicker Compensation

The second step is to find the mapping that needs to be applied on each frame to compensate for the fluctuations and thus align the brightness levels. This can be done by estimating the flicker between the current frame u_n and the last restored frame u_{n-1}^R and apply the mapping onto u_n. The brightness levels are then *locked* to match the levels of the first frame. To avoid error accumulations,

Roosmalen [4] relaxes the brightness stabilization by constructing the restored frame as a mixture of the locked frame $u_n^L(x, y)$ and the observed image $u_n(x, y)$:

$$u_n^R(x, y) = k\, u_n^L(x, y) + (1 - k)\, u_n(x, y) \tag{11.15}$$

where k is a forgetting factor usually set between 0.85 and 0.9. There is thus a trade-off between the amount of deflicker that can be handled and the propagation of errors in the restored sequence. The key solution to flicker compensation is actually to consider the problem as a filtering problem [63,66]. Instead of locking the brightness to the previous frame, it should be sought to average the brightness levels between the current frame and its past and future neighboring frames. The filtering idea [66] can be simplified as follows:

$$u_n^R(x, y) = \frac{1}{2T + 1} \sum_{i=-T}^{T} t_{n,n+i}(u_n(\mathbf{x}), \mathbf{x}) \tag{11.16}$$

where $t_{n,n+i}$ is the estimated flicker mapping that aligns the brightness levels of the current frame u_n to match the neighboring frame u_{n+i}. The number of neighboring frames T can be up to seven frames forward and backward. This compensation method results in more stable brightness alignment and is also less dependent on a perfect flicker estimation since the compensation depends on several estimations.

11.4.1 Modeling

Since the earliest work on deflickering by Roosmalen et al. in 1999, many deflicker algorithms specifically targeted smooth instances of flicker. For non-parametric models, the limitation arises from the manipulation of local histograms [66, 70] which are not designed to be used at pixel resolution. For parametric methods, the limitation comes from the amount of reliable data required for the mapping estimations. Because the parametric models involve more than one parameter per mapping, the problem becomes under-determined on flat areas and intractable at pixel resolution. The generic solution proposed in [4] is to use the smoothness assumption and interpolate the mapping on these flat areas from more reliable neighboring mappings. In 2006 however, Pitie et al. [71] introduced a model that was able to deal with in-scene flicker, or flicker that varies much more quickly across the image. A good example of this is shown in Figure 11.14, and it is that work that has been the most successful at removing flicker of many different kinds. That algorithm is in use today by The Foundry.

The basis of the idea is that, instead of looking for a complicated smoothness prior, a flicker model is established with only one parameter per pixel and is still able to handle nonlinear flicker distortions. To arrive at the model, consider that a pixel is only affected by a percentage of the original flicker source:

$$v(\mathbf{x}) = \alpha(\mathbf{x})\, (t_0(u(\mathbf{x})) - u(\mathbf{x})) + u(\mathbf{x}) \tag{11.17}$$

This model can be understood by assuming that in *in-scene* flicker, the light which is the source of flicker has a global impact of $t_0(u(\mathbf{x}))$, but that due to the scene geometry, a particular point can only receive a percentage $\alpha(\mathbf{x})$ of this light. The complexity of the problem is then dramatically reduced because the mapping estimation is done globally, whilst the local variations are modeled with only one parameter α. Thus, provided that the flicker can be derived from only one source, the problem comes down to the estimation of one parameter per pixel, which yields a fully determined problem. In practice this model needs to be re-parameterized before it can be useful. For details the reader is directed to Pitie et al. [71]. Figures 11.13 and 11.14 show results of deflickering with this idea as well as the resulting flicker map $\alpha(\cdot)$.

FIGURE 11.14

Example of *in-scene* flicker localization due to out of sync fluorescent lighting. On the left, the original sequence (note the fluctuations in the color of the window from frame to frame), in the middle, the deflickered sequence, and on the right the corresponding flicker map. The scene geometry is too complex to consider the flicker as having smooth variations.

11.5   An Evolving Industry

In a sense, the industry for digital visual manipulation is divided into two categories. The television broadcasters are mainly interested in real-time processing, while post-production houses (both film and video) who edit and apply effects to movies, are more interested that the picture is as good as possible. That means users in post-production tend to prefer general purpose hardware running various software tools for editing and compositing and hence are usually not concerned with real-time processing. Interestingly enough, it is television broadcast hardware producers that had started in the late 1980s to produce restoration systems. The BBC used their hardware (Debra) for dirt and noise reduction in-house in the late 1980s and it was real time built out of discrete logic devices (it was quite large indeed). Digital Vision www.digitalvision.se launched a PC-based system for color correction and dirt/noise reduction in 1988 that exploited dedicated hardware, again allowing real-time operation. Around 1997, Snell and Wilcox launched *Archangel*, dedicated hardware specifically targeting motion-compensated, real-time noise and dirt removal. Significantly, *Archangel* was the first hardware-based restoration system to arise from an EU Research Project: AURORA (1994-1998). Terranex Systems teranexlive.dimentians.com is a more recent arrival (~2002) using a massively parallel array of processors on a chip to create single hardware units that allow real-time noise/dirt/scratch concealment. The algorithms implemented in hardware tend to be deterministic in nature; clever use of motion-compensated filtering combined with simple decision making over multiframes. S&W were the first to introduce a real-time hardware line scratch removal system which was very successful.

It was in 1997 that the first software-based restoration systems emerged for film post-production. Four appeared almost simultaneously. Lowry Digital was the first to use a massive network of Apple Macs to achieve fast throughput and high-quality film restoration. They used their own software-designed systems and essentially were a high-quality post-production unit for restoration. DUST, established in France, took a similar approach. MTIFilm (www.mtifilm.com) was the first dedicated software system to appear for restoration, marketed as a restoration system to post-production houses and film studios. They therefore hold the accolade of being the first to design a professional restoration interface in software that showed a timeline, before/after and so on. DaVinci/Digital Revival, emerging from a collaboration with Cambridge University in 1996, followed shortly after offering software for restoration on networks of Linux machines. HS-ART released their Diamant[5] software restoration system about the same time. Diamant was also a result of an EU collaboration and driven by the needs of archivists; it is perhaps the only self-contained software system for editing and automated restoration available today. The Diamant product introduced an interesting innovation to show a representation of the entire movie, using horizontal projections of each image stacked horizontally. This proved surprisingly useful for restoration since flicker and shake in particular can easily be spotted with this representation. It is worthwhile to note that film scanner manufacturers like Philips, Sony, Imagica, and Thompson all incorporate some level of software-based dirt and noise reduction in their systems today.

Since 2003, however, the software restoration space has become more interesting. The increasing speed of PCs and the large repositories of video found in communities like *GoogleVideo* and *YouTube*, implies that the need for video manipulation has become more mainstream. Software plug-ins that enable restoration for consumer and professional software platforms like *After Effects*, *Flame*, *Shake*, and *FinalCutPro*, are now available from Adobe Systems, Autodesk, The Foundry, RedGiantSoftware, and GreenParrotPictures. The PixelFarm has recently also joined the bandwagon of post-production software manufacturers that have seen the growing niche of restoration systems attractive. The Foundry and RedGiantSoftware together with GreenParrotPictures were the first to use algorithms derived from the use of Markov Random Field priors in dustbusting and motion interpolation. The continuing advance of high-definition television in 2009 seems also to be driving the demand not only for restoration but also for resolution improvement. HDTV sets are widely available, as is HD broadcasting, and this is being coupled with the rapid proliferation of the Blu-ray HD format. Hence viewers can see defects much more readily than before and this drives the need for better quality pictures. The growth in the industry therefore seems clear and there is no reason to suspect that the demand for restoration will lessen in the near future.

Bibliography

[1] T. Perry, "Ian Caven: He had to be in pictures," *IEEE Spectrum*, vol. 42, pp. 24–25, February 2005.

[2] A. Kokaram, R. Morris, W. Fitzgerald and P. Rayner, "Detection of missing data in image sequences," *IEEE Transactions on Image Processing*, vol. 4, pp. 1496–1508, November 1995.

[3] A. Kokaram, R. Morris, W. Fitzgerald and P. Rayner, "Interpolation of missing data in image sequences," *IEEE Transactions on Image Processing*, vol. 4, pp. 1509–1519, November 1995.

[4] P. M. van Roosmalen, *Restoration of Archived Film and Video* (ISBN 90-901-2792-5). PhD thesis, Technische Universiteit Delft, The Netherlands, September 1999.

[5] P. V. M. Roosmalen, A. Kokaram and J. Biemond, "Fast high quality interpolation of missing data in image sequences using a controlled pasting scheme," in *IEEE Conference on Acoustics Speech and Signal Processing (ICASSP '99)*, vol. IMDSP 1.2, pp. 3105–3108, March 1999.

[6] G. Arce, "Multistage order statistic filters for image sequence processing," *IEEE Transactions on Signal Processing*, vol. 39, pp. 1146–1161, May 1991.

---

[5]http://www.hs-art.com/diamant/index.html

[7] A. C. Kokaram, *Motion Picture Restoration: Digital Algorithms for Artefact Suppression in Degraded Motion Picture Film and Video.* Springer Verlag, ISBN 3-540-76040-7, 1998.

[8] L. Joyeux, S. Boukir, B. Besserer and O. Buisson, "Reconstruction of degraded image sequences: application to film restoration," *Image and Vision Computing*, no. 19, pp. 503–516, 2001.

[9] R. Storey, "Electronic detection and concealment of film dirt," *SMPTE Journal*, vol. 94, pp. 642–647, June 1985.

[10] M. J. Nadenau and S. K. Mitra, "Blotch and scratch detection in image sequences based on rank ordered differences," in *Proceedings of the 5th International Workshop on Time-Varying Image Processing and Moving Object Recognition*, September 1996.

[11] A. C. Kokaram, "Advances in the detection and reconstruction of blotches in archived film and video," in *IEE Seminar on Digital Restoration of Film and Video Archives (Ref. No. 2001/049)*, pp. 7/1–7/6, January 2001.

[12] A. Kokaram and S. J. Godsill, "MCMC for joint noise reduction and missing data treatment in degraded video," *IEEE Transactions on Signal Processing, Special Issue on MCMC*, vol. 50, pp. 189–205, February 2002.

[13] A. C. Kokaram, "On missing data treatment for degraded video and film archives: a survey and a new Bayesian approach," *IEEE Transactions on Image Processing*, vol. 13, pp. 397–415, March 2004.

[14] DIGITALICE, "http://motion.kodak.com/it/it/motion/products/lab_and_post_production/dice.htm."

[15] J. Ren and T. Vlachos, "Segmentation-assisted detection of dirt impairments in archived film sequences," *IEEE Transactions on Systems, Man and Cybernetics, Part B*, vol. 37, pp. 463–470, April 2007.

[16] S. Tilie, I. Bloch and L. Laborelli, "Fusion of complementary detectors for improving blotch detection in digitized films," *Pattern Recognition Letters*, vol. 28, pp. 1735–1746, May 2007.

[17] R. Storey, "Electronic detection and concealment of film dirt," *UK Patent Specification No. 2139039*, 1984.

[18] E. D. Ferrandière, *Motion picture restoration using morphological tools*, pp. 361–368. Kluwer Academic Publishers, May 199.

[19] E. D. Ferrandière and J. Serra, "Detection of local defects in old motion pictures," in *VII National Symposium on Pattern Recognition and Image Analysis*, pp. 145–150, April 1997.

[20] E. D. Ferrandière, *Mathematical morphology and motion picture restoration*. John Wiley and Sons, New York, 2001.

[21] O. Buisson, *Analyse de séquences d'images haute résolution, application à la restauration numérique de films cinématographiques.* PhD thesis, Université de La Rochelle, France, December 1997.

[22] E. D. Ferrandière, *Restauration automatique de films anciens.* PhD thesis, Ecole des Mines de Paris, France, December 1997.

[23] O. Buisson, B. Besserer, S. Boukir and F. Helt, "Deterioration detection for digital film restoration," in *Proceedings of IEEE International Conference Computer Vision and Pattern Recognition*, vol. 1, pp. 78–84, June 1997.

[24] R. D. Morris and W. J. Fitzgerald, "Detection and correction of speckle degradation in image sequences using a 3D Markov random field," in *Proceedings International Conference on Image Processing: Theory and Applications (IPTA '93)*, Elsevier, June 1993.

[25] A. Kokaram and S. Godsill, "Joint detection, interpolation, motion and parameter estimation for image sequences with missing data," in *IEEE International Conference on Image Processing*, pp. 191–194, October 1997.

[26] A. Kokaram, "Parametric texture synthesis for filling holes in pictures," in *IEEE International Conference on Image Processing 2002*, September 2002.

[27] A. Kokaram, "Practical MCMC for missing data treatment in degraded video," in *European Conference on Computer Vision, Workshop on Statistical Methods in Video Processing* (ISBN 0-9581044-0-9), pp. 85–90, Monash University, Australia, June 2002.

[28] J. Besag, "On the statistical analysis of dirty pictures," *Journal of the Royal Statistical Society B*, vol. 48, pp. 259–302, 1986.

[29] R. Bornard, E. Lecan, L. Laborelli and J.-H. Chenot, "Missing data correction in still images and image sequences," in *ACM Multimedia*, December 2002.

[30] J. B. A. Rares and M. J.T. Reinders, "Statistical analysis of pathological motion areas," in *The 2001 IEE Seminar on Digital Restoration of Film and Video Archives*, (London, UK), January 2001.

[31] J. B. A. Rares and M. J.T. Reinders, "Complex event classification in degraded image sequences," in *Proceedings of ICIP 2001 (IEEE)*, ISBN 0-7803-6727-8, (Thessaloniki, Greece), October 2001.

[32] J. B. A. Rares and M. J.T. Reinders, "Image sequence restoration in the presence of pathological motion and severe artifacts," in *Proceedings of ICASSP 2002 (IEEE)*, (Orlando, Florida, USA), May 2002.

[33] B. Kent, A. Kokaram, B. Collis and S. Robinson, "Two layer segmentation for handling pathological motion in degraded post production media," in *IEEE International Conference on Image Processing*, pp. 299–302, October 2004.

[34] A. K. D. Corrigan and N. Harte, "Pathological motion detection for robust missing data treatment," *EURASIP Journal on Advances in Signal Processing*, vol. 2008, April 2008.

[35] F. Stanco, G. Ramponi and A. D. Polo, "Towards the automated restoration of old photographic prints: A survey," in *IEEE EUROCON, Ljubljana, Slovenia, September 2003*, 2003.

[36] V. Bruni and D. Vitulano, "A generalized model for scratch detection," *IEEE Transactions on Image Processing*, vol. 13, pp. 44–50, January 2004.

[37] F. Stanco, L. Tenze and G. Ramponi, "Virtual restoration of vintage photographic prints affected by foxing and water blotches," *Journal of Electronic Imaging*, vol. 14, December 2005.

[38] M. Bertalmio, L. Vese, G. Sapiro, V. Caselles and S. Osher, "Simultaneous structure and texture image inpainting," *IEEE Transactions on Image Processing*, vol. 12, pp. 882 –889, August 2003.

[39] A. J. Crawford, V. Bruni, A. C. Kokaram and D. Vitulano, "Multiscale semitransparent blotch removal on archived photographs using Bayesian matting techniques and visibility laws," in *Proceedings of IEEE ICIP '07*, (S. Antonio, Florida), 2007.

[40] A. Greenblatt, K. Panetta and S. Agaian, "Restoration of semitransparent blotches in damaged texts, manuscripts and images through localized, logarithmic image enhancement," in *Proceedings ISCCSP '08*, (Malta), 2008.

[41] V. Bruni, A. Crawford, A. Kokaram and D. Vitulano, "Perception measures for digital detection and restoration of semi-transparent blotches," in *Mobile Multimedia/Image Processing, Security, and Applications 2008*. Edited by Agaian, Sos S.; Jassim, Sabah A. *Proceedings of the SPIE*, Volume 6982, pp. 69820J-69820J-11 (2008) SPIE Defense Security, (Orlando Florida), March 2008.

[42] M. Ahmed, F. Pitić and A. Kokaram, "Extraction of non-binary blotch mattes," in *IEEE Conference on Image Processing (ICIP'09)*, (Cairo, Egypt), October 2009.

[43] A. Kokaram, "Detection and removal of line scratches in degraded motion picture sequences," in *Signal Processing VIII*, vol. I, pp. 5–8, September 1996.

[44] T. Bretschneider, O. Kao and P. Bones, "Removal of vertical scratches in digitised historical film sequences using wavelet decomposition," in *Proceedings of Image and Vision Computing* (New Zealand), 2000.

[45] V. Bruni, A. Kokaram and D. Vitulano, "Fast removal of line scratches in old movies," in *Proceedings of ICPR'04*, (Cambridge, UK), 2004.

[46] S. Winkler, *Digital video quality - vision models and metrics*. John Wiley and Sons, 2005.

[47] V. Bruni, P. Ferrara and D. Vitulano, "Removal of color scratches from old motion picture films exploiting human perception," *EURASIP Journal on Advances in Signal Processing*, vol. 2008, July 2008.

[48] L. Maddalena and A. Petrosino, "Restoration of blue scratches in digital image sequences," Tech. Rep. ID 21, Italian National Council of Researchers, Napoli, Italy, December 2005.

[49] B. Besserer and C. Thire, "Detection and tracking scheme for line scratch removal in an image sequence," *Lectures Notes in Computer Science*, vol. 3023/2004, pp. 264–175, 2004.

[50] L. Joyeux, O. Buisson, B. Besserer and S. Boukir, "Detection and removal of line scratches in motion picture films," in *Proceedings of CVPR'99*, (Fort Collins, Colorado, USA), 1999.

[51] L. Joyeux, S. Boukir and B. Besserer, "Film line removal using Kalman filtering and Bayesian restoration," in *Proceedings of IEEE WACV'2000, Palm Springs, California*, 2000.

[52] A. A. Efros and T. K. Leung, "Texture synthesis by non-parametric sampling," in *Proceedings of the IEEE International Conference on Computer Vision (ICCV)*, vol. 2, pp. 1033–1038, September 1999.

[53] M. Bertalmio, G. Sapiro, V. Caselles and C. Ballester, "Image inpainting," in *Computer Graphics, SIGGRAPH 2000*, 2000.

[54] S. Esedoglu and J. Sheno, "Digital inpainting based on the Mumford-Shah-Euler image model," *European Journal of Applied Mathematics*, vol. 13, pp. 353–370, 2002.

[55] M. Gulu, O. Urhan and S. Erturk, "Scratch detection via temporal coherency analysis and removal using edge priority based interpolation," in *Proceedings of IEEE International Symposium on Circuits and Systems, 2006*, 2006.

[56] M. Haindl and F. Filip, "Fast restoration of colour movie scratches," in *Proceedings of ICPR 2002*, (Quebec, Canada), August 2002.

[57] L. Rosenthaler and R. Gschwind, "Restoration of movie films by digital image processing," in *Proceedings of IEE Seminar on Digital Restoration of Film and Video Archives 2001*, 2001.

[58] G. Laccetti, L. Maddalena and A. Petrosino, "Parallel/distributed film line scratch restoration by fusion techniques," *Lectures Notes in Computer Science, Springer Berlin*, vol. 3044/2004, pp. 525–535, September 2004.

[59] L. Tenze and G. Ramponi, "Line scratch removal in vintage film based on an additive/multiplicative model," in *Proceedings of IEEE-EURASIP NSIP-03*, (Grado, Italy), 2003.

[60] Y. Wexler, E. Shechtman and M. Irani, "Space-time video completion," in *2004 IEEE Computer Society Conference on Computer Vision and Pattern Recognition (CVPR'04)*, vol. 1, pp. 120–127, 2004.

[61] A. Kokaram, B. Collis and S. Robinson, "Automated rig removal with Bayesian motion interpolation," *Proceedings of the IEE Journal on Vision, Image and Signal Processing*, vol. 152, pp. 407–414, August 2005.

[62] K. Patwardhan, G. Sapiro and M. Bertalmio, "Video inpainting of occluding and occluded objects," in *IEEE International Conference on Image Processing (ICIP)*, vol. 2, pp. 69–72, 2005.

[63] A. C. Kokaram, R. Dahyot, F. Pitié, and H. Denman, "Simultaneous luminance and position stabilization for film and video," in *Visual Communications and Image Processing (VCIP)*, (San Jose, California USA), January 2003.

[64] F. Liu, M. Gleicher, H. Jin and A. Agarwala, "Content-preserving warps for 3D video stabilization," in *ACM Transactions on Graphics SIGGRAPH 2009*, August 2009.

[65] T. Ohuchi, T. Seto, T. Komatsu and T. Saito, "A robust method of image flicker correction for heavily-corrupted old film sequences," in *Proceedings of the 2000 International Conference on Image Processing (ICIP'00)*, September 2000.

[66] F. Pitié, R. Dahyot, F. Kelly and A. C. Kokaram, "A new robust technique for stabilizing brightness fluctuations in image sequences," in *2nd Workshop on Statistical Methods in Video Processing In Conjunction with ECCV 2004 Prague*, (Prague, Czech Republic), May 2004.

[67] T. Vlachos, "Flicker correction for archived film sequences," *IEEE Transactions on Circuits and Systems for Video Technology*, vol. 14, pp. 508–516, April 2004.

[68] G. Forbin, T. Vlachos and S. Tredwell, "Flicker compensation for archived film using a spatially-adaptive non-Linear model," in *Proceedings of the IEEE International Conference on Acoustics, Speech, and Signal Processing (ICASSP'05)*, May 2006.

[69] X. Yang and N. Chong, "Enhanced approach to film flicker removal," *Proceedings of SPIE Applications of Digital Image Processing XXIII*, vol. 4115, pp. 39–47, 2000.

[70] P. Schallauer, A. Pinz and W. Haas, "Automatic restoration algorithms for 35mm film," *Videre: Journal of Computer Vision Research*, vol. 1, pp. 60–85, Summer 1999.

[71] F. Pitié, B. Kent, B. Collis and A. Kokaram, "Localised deflicker of moving images," in *IEE European Conference on Visual Media Production (CVMP'06)*, (London), December 2006.

12

Computational Analysis of Archaeological Ceramic Vessels and Their Fragments

Andrew R. Willis

University of North Carolina at Charlotte

`Email: arwillis@uncc.edu`

CONTENTS

12.1 Introduction

Since the dawn of mankind our ancestors have created objects to serve in everyday life. Inevitably, these objects will erode, fragment, and break apart due to numerous causes which include so-called "acts of God" (floods, earthquakes, etc.), war, neglect, carelessness, and many more. During field excavations, archaeologists and anthropologists rarely find intact artifacts. More often, they find fragments of incomplete relics and much work is spent studying individual fragments and trying to find correlations between fragments. The geometry of the uncovered fragments alone provide important archeological information and reconstructed objects, i.e., being able to know approximately or exactly the whole object given one or more of its fragments, provides even more information.

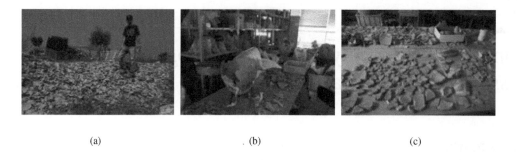

(a)   . (b)   (c)

FIGURE 12.1
(a-c) provide an example of the large number of ceramic fragments found at a typical archaeological site. Near the city Herzliya-Pituach, Israel, the Apollonia-Arsuf site contains a multitude of Greek, Byzantine, Muslim, and Crusader ceramics. Some finds such as those in (a) are too numerous to store in museums and must remain on site while others such as those in (b) and (c) have more archaeological significance and are painstakingly assembled by hand off-site.

Reconstructions for individual fragments are often accomplished by matching the fragment into a library of shape templates for objects that are commonly found on the site, for instances where more than one fragment is available, researchers match the fragments to one another to reassemble the artifact from its fragments. For flat surfaces such as stone tablets, this problem is similar to that of solving a large jigsaw puzzle where each piece is irregularly shaped. For free-form surfaces such as sculpture and pottery, the problem becomes a 3D jigsaw puzzle. More generally, use can be made not just of fragment geometry but also of fragment surface appearance since fragments often have natural or human-imposed patterns on their surfaces. In this chapter we discuss the development of automatic reconstruction systems capable of coping with the realities of real-world geometric puzzles that anthropologists and archaeologists face on a daily basis. Such systems must do more than find matching fragments and subsequently align them; these systems must be capable of simultaneously solving an unknown number of puzzles where all of the puzzle pieces are mixed together in an unorganized pile and each puzzle may be missing an unknown number of pieces.

12.1.1   Significance of Generic Artifact Reconstruction

Systems capable of automatically reconstructing objects from their fragments can greatly aid in the study of many civilizations and have potential applications in other domains such as forensic science where they may be used to reconstruct broken bones or shattered glass and then estimate the point of impact and perhaps the energy incident to the glass structure. The key contributions of such programs are that they are huge time-savers for archaeologists who may otherwise spend many hours assembling broken artifacts by hand. They also endow archaeologists with tools for quickly making precise measurements on fragments and also on reconstructed objects, thus permitting interpretations not heretofore possible. In fact, in most cases, broken vessels and artifacts are not reassembled unless the fragments are visually similar and discovered in a context that is local in both time and space. Automated reconstruction systems working from large databases of digitized fragments could uncover numerous partial or complete reconstructions of artifacts that may have been excavated during different years of the same excavation, or possibly from different sites altogether. In this way, reconstruction systems not only save researchers time but, given a sufficient database of fragments, also have the capacity to reconstruct artifacts which would have otherwise remained as an incoherent pile of fragments.

12.1.2 Significance of Ceramic Vessel Reconstruction

Ceramic pot fragments (referred to as *sherds*) are often the most important source of information from which archaeologists make inferences about the society that existed at a given site (see Figure 12.1). The primary use archaeologists make of the hundreds of thousands of sherds found at a site is to establish typological and chronological sequences of vessel types that developed during the course of a site's occupation. Indicative sherds, such as neck or base fragments or decorated pieces, allow them to infer the shape and type of the vessel in question. Indicative and non-indicative sherds are also counted to estimate the total number of vessel types used during a certain period in a specific location. Pottery sherds and reconstructed vessels are the best indicators of a site's chronology, the population's socio-economic standards, and the extent of local or international trade. Estimating the vessel type, exact shape, size, and volume from a portion of the neck relies heavily on the shape reconstruction offered by classifying the vessel type (see Section 12.2.3). In such cases, subtle variations in estimated curvatures, especially for large vessels, may precipitate volume estimation errors of up to 30%. Such errors may be significant enough, if repetitive for one site or every site, to promote false typological and chronological conclusions. Completely or partially reconstructed vessels provide more accurate information on the exact vessels used on the site and can help prevent such difficulties. For example, Apollonia-Arsuf, a Mediterranean maritime city in Israel that flourished between the 6th century BCE and the 13th century CE, was heavily involved in trade. One important aspect of this commerce involved the transportation of liquids in amphorae (see Figure 12.4 for an example of a Roman amphora and its vessel type). The standard volume of these changed by about 10% from the 5th to 6th century AD. Such subtle variations in volume are impossible to discriminate from only one or a few necks. Hence, complete or partial reconstructions provide a far more accurate assessment of the period that a specific type of amphora was in use, and the numbers of these used. This in turn allows archaeologists to determine changing economic trends such as inflation or economic crises. Another example from [1], discusses how a pottery analysis system allowed archaeologists to discern subtle variations present in collections of "torpedo" jars. The long cylindrical vessels were used to carry trade items during the Iron Age in Israel and Phoenecia and controversy surrounded the manufacturing origin of the jars which was independently claimed to occur in Tyre and Hazor. Detailed computational analysis of the jar shape made light of consistent morphological differences in the jars found at each cite and supported a third hypothesis that claimed that jars were separately manufactured at both sites, an insight offered due to accurately quantifying the shape of these vessels.

While geometric reconstruction can better quantify the number, shape, typology, and volume of ceramics at a site, matching of fragment pieces may provide more complete restoration of decorative patterns and inscriptions on fragment surfaces that have been separated when the artifact broke apart. Examples of decorative patterns include inscriptions, figurative images, and surface patterns (e.g., geometric patterns, floral patterns, etc.). Completion of these patterns can provide more accurate dating of the artifact and, by context, can provide important information about the site as a whole.

12.1.3 Traditional Puzzle Solving and Artifact Reconstruction

Initial work on puzzle-solving may be traced back to the problem of computationally-based automatic reconstruction of jigsaw puzzles which has been discussed as far back as the mid-1960s [2] with some additional work in the '80s [3, 4]. However, numerous works on the problem have emerged in recent years and current systems are capable of automatically assembling jigsaw puzzles with hundreds of pieces [5, 6].

The problem of reassembling *real-world* archaeological puzzle pieces, i.e., reassembling free-form broken objects from measurements of their fragments, has received little attention until recently. As shown in Figure 12.2, this problem is more difficult than assembling commercially produced jigsaw puzzles. A major complication is that fragments are irregular and can match along any subset

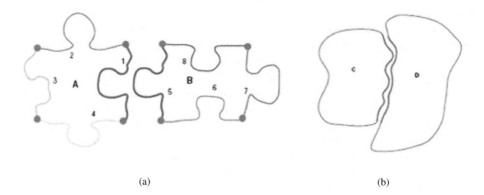

(a)   (b)

FIGURE 12.2 (SEE COLOR INSERT)

Two figures from [7] demonstrate the differences between reassembling commercial jigsaw puzzles and reconstructing broken artifacts. As shown in (a), jigsaw pieces have readily identifiable corners (red dots) allowing programs to easily separate portions of the boundary (shown as different color curves) that will match with some other unique puzzle piece. Additionally, each boundary segment is a smooth planar curve having an *isthmus* or neck that is highly indicative for finding that unique matching puzzle piece. Figure (b) shows two hypothetical fragment boundaries. Note here that the problem is made much more difficult as corners are not easily identifiable and may not indicate the beginning or end of a curve that will uniquely match some other fragment. Worse, any portion of a boundary curve may match to any other fragment and the curve itself may match equally well with numerous similar boundaries from other fragments.

of their complete boundary. Additionally, typical real world puzzles introduce a vast variety of possible sources of confusion. Some of these include: (1) physical degradation of the fragments due to chipping and, if exposed to the elements, erosion; (2) the number of puzzles being solved may not be known (i.e., we are given a collection of fragments coming from an unknown number of objects); (3) fragments may be missing as they may have yet to be discovered or may have been destroyed by some phenomenon.

12.2   Artifact Reconstruction Systems: Basic Components and Concepts

While artifact reconstruction systems are currently present only in laboratory environments, continuing research promises to develop systems that may be used by archaeologists in the field in the immediate future. These systems will be particularly useful to excavations where the influx of ceramic fragments from daily excavation is so large that it is prohibitively time-consuming and expensive to devote careful consideration of each and every fragment (see Figure 12.1(a) for an appreciation of the volume of sherds that may be found at a site).

Such cases are more common than one might expect and conventional methods for coping with the overwhelming amount of pot sherds is to catalog merely the *indicative* sherds, i.e., those sherds whose geometry and appearance provide sufficient information to uniquely determine the typological class (i.e., vessel type from which the sherd originates). Indicative sherds tend to come from regions of the vessel that show significant shape variation such as rim sherds, base sherds, or other geometrically interesting portions to the vessel such as the spout, handle, etc. Other sherds are categorized as *non-indicative* as there is not enough observable evidence to determine the type of vessel from which the sherd originates. These sherds tend to come from slowly varying portions of the vessel surface typically associated with the body of the vessel and, as such, are commonly referred to as *body sherds*. Given the relatively small amount of information that can be directly extracted from these sherds, typical approaches seek to record statistical information such as the size, frequency, and appearance associated with groups of body sherds found at a given location. This data can be used to approximate useful information such as the number of intact vessels or the total volume of the vessels at a given site location via statistical methods.

In some rare cases a group of fragments are found in close proximity, which indicates that the group of fragments are likely to come from one or a small group of vessels. In these rare cases, the fragments are saved and, if there is a sufficient number of sherds, time, and interest, the vessel may be reconstructed by piecing the fragments back together by hand.

The labor intensive process of excavating, cleaning, and cataloging sherds can easily dominate the day-to-day effort of an excavation team, especially if there is an effort to reconstruct sherd finds into vessels. Tel-Aviv University has a full-time ceramics expert dedicated year-round to such work (Figure 12.1(b,c) shows the ceramics laboratory). A system capable of efficiently capturing accurate sherd information and performing semi-automatic or fully-automatic operations on the captured data can assign quantities to what was previously approximate or empirical observations for both indicative and non-indicative fragments. As detailed in Section 12.1.2, this can improve the accuracy for many aspects of archaeological ceramic analysis such as typological classification [1, 16], volume estimation [17], or complete vessel reconstruction [18]. A system designed for this purpose requires three key components:

1. A sherd digitization station; this is a data-capture device that takes real-world sherds as input and generates a virtual 3D model of the sherd surface geometry and appearance as output (see Section 12.2.1).

2. A sherd database; digitized models will be merged into a fragment database. The digitized

sherd will be cataloged along with excavation-specific data fields which specify other useful information such as the context, pose, period, locus, etc.

3. A computational server for sherd analysis; this is a computer that will access the fragment database and provide quantitative analysis of the digitized sherds (see Section 12.3 and Section 12.4).

As this chapter is dedicated to computational reconstruction, we concentrate on the digitization process (1) and algorithms for computational analysis and reconstruction that would be executed on the computational server (3).

12.2.1 Digitizing Artifacts

Computational methods for object reconstruction must have digital representations of the fragments to be assembled. There are a vast variety of potential techniques for obtaining 2D and 3D fragment measurements. Measurement technologies for ancient artifacts are almost all non-contact (i.e., optical measurement systems). For ceramic vessel reconstruction, digitization is usually accomplished using commercially available 3D triangulation-based laser-scanners that now produce highly accurate measurements with (x, y, z) errors < 0.25 mm (see Figures 12.5, 12.6, for examples). Yet, there are numerous image-based techniques from computer vision literature that are capable of capturing the 3D geometry from images. Popular methods here include stereoscopic reconstruction [19] and photometric stereo (shape-from-shading) [20]. The ultimate output of these measurement devices is a digital representation, i.e., digitized version, of the fragment as a 3D surface mesh.

While measurements for flat fragments can usually be captured with one or two images from a calibrated camera, generating a complete model of a curved sherd fragment is often laborious and time consuming. Laser scanning technologies typically recover surface depth measurements through triangulation using a laser line emitter and high-resolution image sensor with known relative positions and orientations (see Figure 12.3(a,b)) [21]. The result is a *range image,* i.e., a collection of measurements which have the form $z = f(x, y)$, where z is the object depth for an (x, y) location in the sensor image plane. By changing the relative orientation of the sensor and fragment, different range images are captured, each of which provides a partial description of the fragment surface (Figure 12.3c,d). The complete fragment model is generated by aligning the measured range images to a common coordinate system using techniques such as [22–24] and subsequently merging the range images using techniques such as [25] to produce a virtual 3D model of the real-world fragment (Figure 12.3e).

12.2.2 Approaches for Computational Sherd Analysis

Current techniques for computational analysis of ceramic sherds fall into two distinct categories: (1) *typological recognition,* i.e., reconstruction of the vessel from typological classification [26], and (2) *vessel reconstruction,* i.e., piecing together a broken vessel by matching together its broken sherds. Typological recognition requires two inputs: (1) a digitized sherd model and (2) a typological database, i.e., a set of potential pot vessel shapes where it is presumed that the fragment originates from some portion of one of the vessels stored in the database. This is a *vessel recognition problem* (i.e., given a fragment model we seek to *recognize* the vessel from which the fragment came). Vessel reconstruction takes as input a collection of digitized sherd models and seeks to estimate the shape of the unknown vessel(s) from which these fragments came by automatically reassembling the sherds in a manner analogous to 3D puzzle-solving. This is a *vessel reconstruction problem* (i.e., fragments are matched together to uncover the a priori unknown global shape of the unknown vessel).

Each approach has inherent benefits and drawbacks. A key benefit to vessel recognition is the potential for recovering the global vessel geometry from a single sherd by recognizing the typological

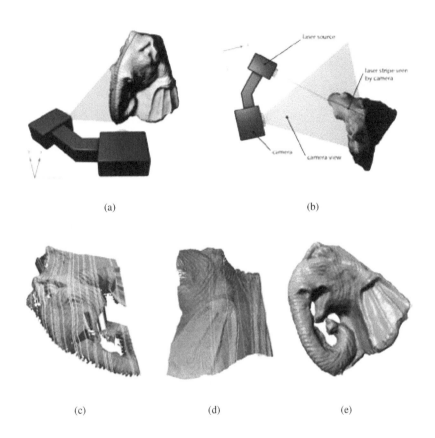

(a) (b)

(c) (d) (e)

FIGURE 12.3

(a,b) show views of a typical triangulation-based laser scanner used for artifact digitization. A "stripe" of laser light is swept across the surface of the object and the laser stripe projected onto the object surface is observed by an imaging sensor. Using an a priori known relationship between the emitted laser stripe and the observing sensor, a range estimate is calculated for each surface point via triangulation yielding a range image: $z = f(x, y)$. By changing the relative position of the sensor and object, multiple range images may be measured (c,d) which are subsequently merged to generate a complete model of the 3D object (e).

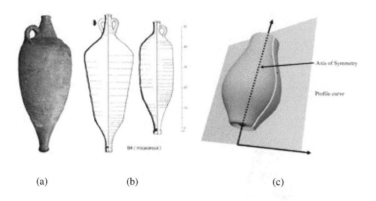

(a)   (b)   (c)

FIGURE 12.4

(a) An amphora (from p. 102-104 of *Roman Pottery in Britain* (1996).) with rounded shoulder, narrow neck, and beaded rim produced between the 1st and 6th century A.D. (b) A typological representation of this artifact. (c) A mathematical representation for the geometric shape of the vessel via a central axis and an associated profile curve.

class of the sherd. However, such approaches can be unreliable for sherds that are small or non-indicative, which results in multiple potential typological classifications. In addition, archaeologists need to hand-specify all plausible typological varieties that can occur at each site, which precludes accurate classification of "surprise finds" (i.e., sherds coming from unexpected vessels that may not be in the typological database). Further, the typological database provides normative models for the shape of each vessel and small, potentially significant (see Section 12.1.2 for an example) variations from these normative models may be ignored. Key benefits to vessel learning are the ability to build vessels having arbitrary typology without prior knowledge of their shape and the ability to integrate information from multiple sherds to develop a more accurate model of the unknown shape unique to each individual vessel. Vessel reconstructions also provide users specific instructions detailing how to reconstruct *actual artifacts* that may have otherwise remained as an unorganized pile of sherds. Reconstrucion also requires less input from the archaeologists since they do not need to specify a typological database. Yet, in many cases the typology of vessels at a site may be well-known to archaeological researchers. Ignoring this information can make vessel learning unnecessarily complex and computationally costly. In addition, the "puzzle-solving" process can require more user intervention since there may be multiple potential reconstructions for a single vessel that must be manually discarded or validated by an archaeologist.

12.2.3 Computerized Typology for Automated Sherd Classification

Typological classification seeks to cluster artifacts into discrete "types." Individual types are determined by the archaeologists and often the shape of the vessel is a fundamental discriminative feature associated with each type. The shape of ceramic vessels is specified in terms of a *profile curve*, i.e., the curve generated by tracing the cross-section of the vessel when intersected by a plane that contains the vessel's axis of symmetry (see Figure 12.4). Due to the rotational symmetry of vessels made on a potter's wheel, the profile curve is typically plotted as a 2D planar curve with respect to the vertical axis of symmetry. The 3D vessel surface may then be recovered by rotating the profile curve around the vertical axis. Pot surface patterns and asymmetric features are also provided in standard vessel type drawings. This additional information is typically provided using stylistically

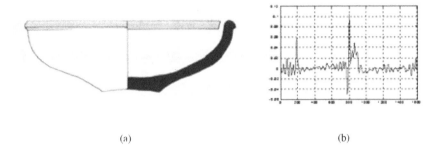

(a)   (b)

FIGURE 12.5
In these figures from [1], the authors represent the typological shape of the profile curve (a) using the curvature signature, $\kappa(s)$, generated by computing curvature as a function of the traveled arc-length. (b) shows the resulting curvature signature that results by traversing the profile curve starting at the bottom of the base and proceeding around the outside of the profile curve in a counter-clockwise direction. Finite difference methods are commonly applied to compute the curvature signature but often require smoothing, i.e., low-pass filtering, to attenuate noise in the computed curvature values, which improves the robustness and accuracy of typological profile curve matching.

distinct annotations that are incorporated into the typological drawing, e.g., outlined shape and not a solid shape such as that in Figure 12.4(b), and include things such as handles, spouts, holes, etc. [27].

Typological classification represent the profile curve with a specific *shape model*, i.e., a computational description of the profile curve shape. This shape model is estimated for each element in the typological database. Classification of new sherds is accomplished by first estimating the axis of symmetry either manually or automatically and subsequently computing the profile curve with respect to this axis. A shape model is then extracted from the profile curve and compared against the profile curves of the database. An error functional is defined that compares the extracted sherd profile curve against the profile curves of each typological class in the database and provides an overall matching score as output. Large values of the similarity function are indicative of typological classes that share a shape similarity with the unclassified sherd. The complete typological profile curve may then be used to reconstruct the entire vessel given the typological classification of the sherd. In [1], the authors compare the curvature signature as shown in Figure 12.5 using the error functional (12.1). This functional compares two curved denoted α and β by parameterizing these curves in terms of their arc-length, s, and computing a weighted sum of squared differences between the curvatures, $\kappa_\alpha(s)$ and $\kappa_\beta(s)$, over the extent of these curves. Low values of this functional indicate candidate typologies for a given unclassified sherd.

$$d_{\alpha\beta} = \sqrt{\int_{s_{min}}^{s_{max}} (\kappa_\alpha(s) - \kappa_\beta(s))^2 w(s) ds} \tag{12.1}$$

As the profile curve matching is geometric, shape models often originate from classical literature on differential geometry such as [28] and there is a considerable body of work within the computer-aided design and computer graphics community that discusses applications of these classical theories to discrete curves and polygonal surfaces (see Figure 12.5). In particular, many researchers represent 2D shapes using a *curvature signature,* $\kappa(s)$, which represents a planar curve by curvature as a function of the curve arc length. For discretely sampled curves this representation is obtained by computing local curvatures of the fragment boundary when sampled at uniform intervals of arc-length. The resulting discrete 1-dimensional function is referred to as a *curvature signature,* $\kappa(s)$,

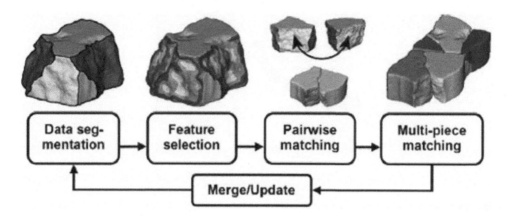

FIGURE 12.6
This figure from [10] shows the generic process for object reconstruction. It consists of four steps (from left to right): (1) classify the boundaries of interest for matching, (2) compute a shape model for these boundaries in terms of features, (3) search for correct pairwise matches of the computed features, and (4) implement a merge and update procedure that merges pairwise matches into larger multi-piece configurations.

which has been extensively studied for curve matching with seminal work on the topic available in [29]. Examples of its use in 2D reconstruction for matching curves are numerous [1, 9, 29] and its generic extension to 3D space-curve matching is also a key contributor for 3D puzzle-solving approaches [10, 22, 30, 31]. Provided that samples are taken sufficiently small intervals, this representation benefits from two theorems of differential geometry: (1) *the fundamental existence and uniqueness theorem*, which ensures that, for planar curves, $\kappa(s)$ is equivalent to the original boundary curve up to an unknown position and orientation [32], and (2) two planar curves are congruent if and only if they have the same curvature signature $\kappa(s)$ [28]. Since the curvature signature is an intrinsic property of the boundary, it is coordinate-free (i.e., it is invariant to Euclidean transformation). This allows researchers to match two profile contours by directly comparing their curvature signatures (i.e., no alignment between the two curves is required) [1].

Unfortunately, the position- and orientation-free representation afforded by the curvature signature is countered by the inherently noisy values obtained when computing curvatures using standard methods such as finite-different methods. These effects are widely acknowledged in the literature and nearly all techniques using the curvature signature (or other differential quantities such as torsion) perform low-pass filtering, i.e., curve smoothing, to control the estimated curvature variability [1, 9, 22, 29, 33, 34]. Another important source of variability is any bias present in the estimation technique. Details on estimation approaches and their relative merits are widely available in both 2D [35, 36] and 3D [36–41].

12.2.4   Computerized Vessel Reconstruction by Fragment Matching

Using digitized fragments as raw data, sherd assembly approaches follow a four-step process as summarized in Figure 12.6 to reconstruct vessels from measurements of their sherd fragments. Raw data for a typical sherd surface may have 0.5M-2M (x, y, z) points. Considering a single broken artifact may consist of 10-30 fragments, direct manipulation of the raw data has a huge computational cost. Reconstruction approaches typically start by subdividing the fragment into matchable subsections, i.e., segmenting the boundary into regions to be matched (Figure 12.6, step 1).

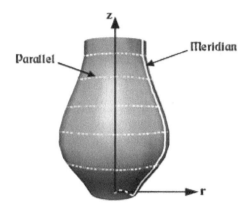

FIGURE 12.7
Geometry of a surface of revolution

For each identified boundary subsection, a compact representation of the fragment shape is computed in terms of features taken from the fragment surface (Figure 12.6, step 2). Estimated features from portions of each fragment are then compared against features from other fragments to identify candidate pairwise matches between fragments (Figure 12.6, step 3). Finally, an assembly rationale is implemented through a computational search procedure which updates and merges pairwise matches into large configurations of aligned fragments until a stopping criterion is met (Figure 12.6, step 4). While this chapter is dedicated to methods for reconstructing ceramic pottery, the reader can refer to [42], which provides a survey of methods for generic 3D object reconstruction (i.e., reconstruction of objects that may not have the special symmetry properties typically found in ceramic vessels).

12.3 Computational Models for Vessels and Their Fragments

All of the proposed sherd analysis methods require a mathematical, i.e., computational, representation for the shape of the vessel and its fragments. The representations used are based on the assumption that the vessel is axially symmetric, i.e., the surface of the vessel may be well approximated by a surface of revolution, with special exceptions for pot components such as handles, spouts, and surface decorations which violate this assumption. Reliable and accurate extraction of a central axis and profile curve model from sherd data is then an important problem for both automatic typological classification and automatic vessel reconstruction. Several solutions to this problem have been proposed in the literature, each approach takes as input a digitized 3D sherd model and seeks to generate an estimate of the central axis, or, an axis and profile curve as output.

12.3.1 Modeling Ceramic Vessels as Surfaces of Revolution

A surface of revolution $S \in \mathbb{R}^3$ is obtained by revolving a planar profile curve $\alpha \in \mathbb{R}^2$ about a line $1 \in \mathbb{R}^3$. α is called the *profile curve* and 1 the *axis* of S. When the z-axis is taken as the axis of revolution with profile curve $\alpha(z)$, the surface S may be represented parametrically as (12.2).[1]

---
[1]The adopted notation and terminology for surfaces is that used in classic texts such as [43, 44].

$$S(\theta, z) = (\alpha(z)\cos\theta, \ \alpha(z)\sin\theta, \ z) \tag{12.2}$$

With this parameterization, the curves $z = constant$ are *parallels* of S and the curves $\theta = constant$ are *meridians* of S. The profile curve characterizes how the radius and height of the surface change for a fixed meridian (see Figure (12.7)). In Equation (12.2), the radius function, $r = \alpha(z)$, is a single-valued function of z. For archaeological sherds, this presents a problem since profile curves that are multi-valued with respect to the z-axis commonly occur. Examples include sherds which come from pot bases and rims. Figures 12.8(d) and 12.8(f) represent typical examples of sherds which have multi-valued profile curves. For this reason, the profile curve model from (12.2) needs to be exchanged for a model that can accommodate these situations, e.g., a spline [45] or implicit polynomial [46, 47]. The problem of interest is: given an unorganized set of 3D measured points of a small patch of a larger axially symmetric surface, estimate a surface geometry model for the patch. This geometry is completely specified by an axis in 3D and a 2D profile curve with respect to that axis. Consequently, estimation of an axially symmetric surface model reduces to the problem of estimating the parameters of the axis, $\mathbf{I}$, and profile curve, α.

12.3.2   Estimating the Vessel Axis from Digitized Sherds

One approach for surface estimation is to solve for the unknown vessel axis and then compute the profile curve using the estimated pot axis. [48, 49] provide two distinct approaches for finding the axis of symmetry for a surface of revolution from 3D surface measurements. [49] uses the geometric property that the line defined by a point, $\mathbf{p}_i$, on a surface of revolution and its associated surface normal, $\mathbf{n}_i$, must intersect the axis of symmetry. One may then pose the axis estimation problem as one of determining the 3D line having minimum distance to the set of lines given by each measured surface point and associated surface normal. Solving this problem in a Euclidean coordinate space requires optimization of a non-linear error functional. Yet, when the problem is formulated by using Plücker coordinates, i.e., a 6-dimensional space where Euclidean lines may be represented as points, a linear solution may be found.

To do so, each measured surface point and associated normal is used to define a 3D line as represented by a Plücker coordinate. The union of all these lines form a linear complex and the approach forms a optimization function that seeks to find the line of minimal distance (in Plücker coordinate space) to this linear complex. The function turns out to be a positive semi-definite quadratic form that may be directly minimized as a generalized eigenvalue problem whose solution is taken to be the unknown axis of revolution. The Plücker coordinate associated with the line defined for each surface point/normal pair is given by (12.3).

$$\mathbf{l}_i = (\overline{t_i}, t_i) = (\mathbf{p} \times \mathbf{n}_i, \mathbf{n}_i) \tag{12.3}$$

This collection of lines defines a linear complex and we seek to find sherd axis which is taken as the line having minimum mean squared distance to the lines of the complex. A non-Euclidean distance metric between this the unknown line $\mathbf{x} = (\overline{x}, x)$ and the ith line in the linear complex, $\mathbf{l}_i$, is defined using the permuted inner product for Plücker vectors (12.4).

$$d(\mathbf{l}_i, \mathbf{x}) = \overline{t_i} \cdot x + t_i \cdot \overline{x} \tag{12.4}$$

When the lines $\mathbf{l}_i$ and $\mathbf{x}$ intersect $d(\mathbf{l}_i, \mathbf{x}) = 0$. Additionally, we must constrain our solution to be a Euclidean line which requires us to enforce the Plücker relation: $\overline{x} \cdot x = \mathbf{p}_i \times \mathbf{n}_i \cdot \mathbf{n}_i = 0$. Since the permuted inner product (12.4) can be negative, we would like to minimize the sum of the squared values of $d(\ )$ over all data points. The value of $\mathbf{x} = (\overline{x}, x)$ which minimizes this functional is taken as the axis estimate (12.5) for a linear complex involving N lines.

$$\widehat{\mathbf{x}} = \min_{x,\overline{x}} \sum_{i=1}^{N} (\overline{t_i} \cdot x + t_i \cdot \overline{x})^2 \tag{12.5}$$

Equation (12.5) can be re-written into a quadratic matrix-vector form (12.6).

$$F(\mathbf{x}) = \mathbf{x}^t \mathbf{L} \mathbf{x} \tag{12.6}$$

where $\mathbf{L}$ is the scatter matrix of the linear complex Plücker coordinates; $\mathbf{L} = \sum_i \mathbf{l}_i \mathbf{l}_i^t$. A linear solution to (12.6) may be obtained by requiring the solution to have a unit-length direction vector which is also a constraint needed to avoid the trivial zero vector solution. Since this constraint is expressible in terms of a quadratic equation in axis line: $\mathbf{x}^t \mathbf{D} \mathbf{x} = \|x\| = 1$, where $\mathbf{D} = diag(0, 0, 0, 1, 1, 1)$, the solution to (12.6) reduces to a generalized eigenvalue problem (12.7).

$$(\mathbf{L} - \lambda \mathbf{D})\mathbf{x} = 0 \tag{12.7}$$

The solution, $\widehat{\mathbf{x}}$, is taken to be the line described by the Plücker vector associated with the smallest eigenvalue given by singular value decomposition of the matrix $\mathbf{L}$. Since this solution may be computed linearly, it is computationally efficient. In fact, several methods typically use this solution as a starting point for non-linear minimization. In practice, this method was found to work well particularly when the angle between the surface normal and the axis of symmetry are close to orthogonal. Acute angles between the surface normal and unknown axis typically come from points at the base of the vessel and can create difficulties in finding stable solutions.

In [48] authors choose to solve for the set of spheres that make second order contact with the measured surface at each surface point. Specifically, for each measured surface point, the authors seek the parameters of the sphere having the same tangent plane as the surface and also having the same radius of curvature in the direction of a surface parallel. Such spheres are said to have second order contact with the surface and are commonly referred to as osculating spheres, i.e., spheres that "kiss" the surface at each point. Using the constraint that each of these spheres must have a center that lies on the unknown axis of symmetry, a non-linear error functional is derived (12.8) that expresses the unknown parameters of the fragment axis as a function of the measured surface point, $\mathbf{p}_i$, and normal, $\mathbf{n}_i$, data.

$$f(\mathbf{p}_0, \mathbf{v}) = \sum_{i=1}^{N} \left\| (\mathbf{p}_i - \mathbf{p}_0) \times \mathbf{v} - \frac{\|(\mathbf{p}_i - \mathbf{p}_0) \times \mathbf{v}\|}{\|\mathbf{n}_i \times \mathbf{v}\|} (\mathbf{n}_i \times \mathbf{v}) \right\|^2 \tag{12.8}$$

where $\mathbf{p}_0$ is a point on the unknown vessel axis and $\mathbf{v}$ is a vector in the direction of the unknown vessel axis. The method is non-linear due to added constraint that requires the surface of revolution to have a constant radius for a given height along the axis, due to the added constraint that not enforced in [49].

12.3.3 Estimating Profile Curves from Digitized Sherds

The methods of Section 12.3.2 emphasize the extraction of the axis alone as a necessary predecessor for estimation of the profile curve. Since the sherd profile curve is estimated with respect to vessel axis, extraction of the profile curve using such techniques involves (1) estimating the axis line and (2) estimating a profile curve given the axis estimate from the previous step. There are several limitations to this approach: (1) axis estimates are not always highly accurate; this is particularly true for small sherds and sherds that do not have indicative shapes, e.g., body sherds. Surface data from such sherds may locally appear planar or spherical, which can create unstable results (planes and spheres are geometries that have no well-defined axis of symmetry). This can be partially attributed to the fact that analysis for each of these methods is based on *local* features of the sherd surface. The linear

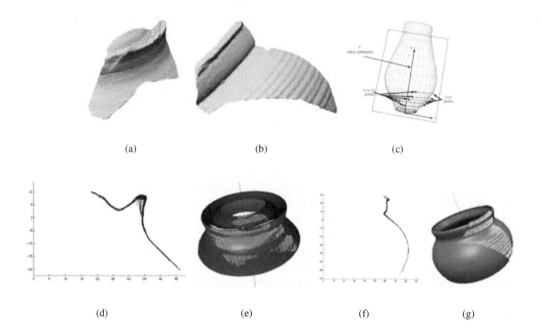

(a)   (b)   (c)

(d)   (e)   (f)   (g)

FIGURE 12.8

(a,b) shows two surface scans of archaeological sherds excavated at the Great Temple site of Petra, Jordan [14]. Only the outer surface data has been captured; we refer to such scans as *eggshell* models of the sherd surface. (c) Several sherd surface points (in blue) are projected into a plane containing the estimated vessel axis (in red). (d,f) Sherd surface data from (b,c) has been projected into the (r, z) plane as shown in (c) using an estimated vessel axis. (e,g) The estimated 3D axially symmetric algebraic surface is shown along with the 3D sherd surface data (superimposed in white).

complex of [49] requires only the position and tangential information, i.e., *information provided by the data and its first-order differential (tangent-plane)*. The osculating spheres of [48] extends this local analysis to include information from the second-order differential, *i.e., the principal curvatures*.

Use of these differential surface properties requires that the underlying sherd surface be continuous [49] and that the first derivative of the surface is also continuous [48]. A logical extension of these methods is to require that *all* the sherd data lie along a single smooth, i.e., differentiable, surface. In this case one seeks to find a single *global* surface model that well approximates *all* of the measured sherd data. In [50, 51] authors suggest such a method where the goal is to determine the surface shape and pose (position and orientation) of the axially symmetric vessel that best approximates the measured sherd surface data.

Algebraic models are typically used as a representation for vessel and sherd surface shapes as they provide a great degree of flexibility for shape representation and can be computed quickly. Specific algebraic models used by researchers include a spline curve [45], where knots for the spline are judiciously chosen to fall at locations of high curvature along the profile, or an implicit polynomial curve as discussed in [46, 47].

Unfortunately, the standard methods for surface fitting, i.e., fitting as a linear least-squares estimation problem [52], is not appropriate since we require that the fit surface be axially symmetric. In fact, fitting axially symmetric surfaces in general position to the sherd surface data requires highly non-linear constraints on the polynomial coefficients, even for polynomials having modest shape-representation capabilities (degree 3 or higher). For this reason, profile curve estimation is decomposed to three steps:

1. Estimate the pose of the vessel, i.e., the position and direction of the vessel axis (see Section 12.3.2).

2. Specify a plane that contains the estimated axis and rotate the measured sherd data around the estimated axis into the plane.

3. Fit a 2D-curve, the profile curve, using the in-plane coordinates of the sherd surface points as data.

Note that this procedure for estimation of the axially symmetric surface is *completely equivalent* to fitting a 3D axially-symmetric surface to the measure sherd data but does not require enforcement of the highly non-linear constraint equations that are encountered when one attempts to fit an axially symmetric surface of degree 3 or higher directly to a collection of (x, y, z) surface data.

Step 2 of the process above may be quickly and easily accomplished by defining a cylindrical coordinate system based on the vessel axis. The coordinate system origin, $\mathbf{o}$, may be arbitrary. In practice, this is taken as the point on the axis closest to the mean of the measured sherd surface data. The coordinate system z-axis, $\mathbf{z}$, is taken as the unit vector in the direction of the estimated axis and the coordinate system r-axis, $\mathbf{r}$, is taken as an arbitrary unit vector perpendicular to the z-axis. Rotation of each surface data point, $\mathbf{p}_i$, and surface normal, $\mathbf{n}_i$, into the plane is then accomplished by applying the standard equations of a Euclidean to cylindrical coordinate system change:

$$r_i = \sqrt{(p_i - o)^t (I - zz^t)(p_i - o)} \qquad z_i = (p_i - o)^t \cdot z \qquad (12.9)$$

$$n_i^p = (n_i \cdot r, n_i \cdot z) \qquad (12.10)$$

This projection of the data into a plane is shown graphically in Figure 12.8(a). The θ-component of the measured data is discarded since the surface fitting error is invariant to changes in this parameter, i.e., error between the data and the estimated profile curve will not change as the sherd rotates around the vessel axis at a fixed height. This projection (12.3.3) takes each measured data point, p_i, from

xyz-space to rz-space, which is referred to as radius (r) - height (z) space. A polynomial may then be fit to the data point scatter observed in radius-height space (see Figure 12.8(b)) using standard fitting techniques.

[51] demonstrates profile curve estimation results using the Gradient-1 fitting algorithm described in [47]. This is accomplished by minimizing the curve-fitting objective function (12.11) with respect to the unknown profile-curve coefficients of the curve $\alpha(r, z)$.

$$e_{grad1} = \sum_{i=1}^{I} \left(\alpha^2(r_i, z_i) + \lambda \left\| \mathbf{n}_i^p - \nabla \alpha(r_i, z_i) \right\|^2 \right), \qquad (12.11)$$

where $\alpha(r, z)$ is a 2D implicit polynomial curve of degree d that has $[(d+1)(d+2)/2]$ unknown coefficients and is the set of points satisfying (12.12).

$$\alpha_d(r, z) = \sum_{0 \le j+k \le d; \ j,k \ge 0} a_{jk} r^j z^k = 0. \qquad (12.12)$$

Fortunately, (12.11) may be solved via standard linear least-squares minimization. The fitting approach provides an additional penalty term to the usual least-squares objective function that requires not only that the surface pass through the measured surface data, but also that the measured surface normals agree with the normals of the fit polynomial at each surface point which can significantly improve the accuracy of the polynomial fit. Appropriate values for the free-parameter λ must be determined by the user but will depend upon the relative strength of the noise present in the measured points and surface normals. [51] prescribes a value $\lambda = 0.01$ to generate their experimental results.

The scalar residual, e_{grad1}, is a measure of asymmetry resulting from the axis/profile-curve fit. The important thing to notice is that the objective function (12.11) seeks to minimize the approximate *orthogonal distance* of the measured data to the unknown *surface*. This approximation will generate improved models where the surface is smooth, which is often the case for archaeological sherds (see Figure 12.8 for examples of such sherds). It is also interesting to note that the approximation can be done without computing any local derivatives information from the measured data which makes the estimation procedure robust to noisy sherd data and enables estimates to be computed when there are no measured surface normals (here the term $\left\| \mathbf{n}_i^p - \nabla \alpha(r_i, z_i) \right\|^2$ from (12.11) is zero).

Note that the methods proposed here use *eggshell surfaces,* i.e., only the geometry of the outer surface of the sherd is used for estimation of the sherd axis and profile (see Figure 12.8). The authors justify this simplification for two reasons: (1) due to the manufacturing process, the interior and exterior contours of the vessel are often very similar, and (2) use of only the outer surface can greatly expedite the labor-intensive process of digitizing the excavated sherds as discussed in Section 12.2.1.

12.3.4   Simultaneously Solving for the Axis and Profile Curve

Research in [50, 51] outlines approaches that simultaneously solve for both the axis of symmetry (i.e., the pose of the vessel, and the shape of the sherd surface). The approach is very similar to that specified in Section 12.3.3 and incorporates an outer loop that allows errors observed in the profile curve fit to refine the unknown parameters of the axis. This is accomplished by using the gradient of the fitting error, i.e., the fitting residual e_{grad1} to update the vessel axis (or equivalent pose). This additional step allows the axis position and orientation to be refined based on observed fitting errors. In practice, this method for extracting the sherd surface model is particularly useful for body sherds which present difficult axis estimation problems due to their subtle surface variations. Here, the initial estimates of the vessel axis may be far from the best solution and the iterative optimization provided by the outer loop allows for a much improved estimated of the global surface geometry (axis and profile) albeit convergence to the solution may be slow (this is due to the fact that the residual error e_{grad1} changes very slowly, i.e., the error surface is nearly flat).

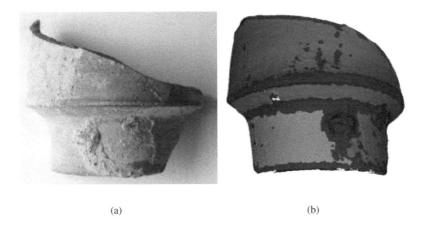

(a) (b)

FIGURE 12.9 (SEE COLOR INSERT)
(a) An archaeological pot fragment excavated in Petra, Jordan. (b) A segmentation of the fragment surface in (a) from [53] obtained by dividing the surface into regions that are well approximated by a single quadratic surface patch. (b) Four quadratic surface patches indicated in orange, red, green, and brown. Blue surface points are those points that do not lie close to any of the estimated quadratic patches and typically include blemishes on the sherd surface created by calcification, chipping, and weathering. This asymmetric surface data can make automatic estimation of the vessel axis and profile curve problematic, and exclusion of these regions can greatly improve the accuracy and robustness of sherd surface estimation approaches.

12.3.5 Dealing with Asymmetries in Archaeological Ceramic Sherds

While many archaeological sherds are excavated with little or no surface blemishes, it is not uncommon to observe local asymmetric variations in the sherd surface due to calcification, dirt, chipping, or other deposits commonly found on archaeological sherds. These asymmetries create outliers that can adversely effect the accuracy of the final solution for the vessel estimation approaches of Section 12.3.2 and Section 12.3.3. Standard techniques for coping with outliers can be included in the approach by finding points having unusually high fitting error ([48] proposes applying this outlier-removal scheme at each iteration of their non-linear minimization). Yet outlier detection of this kind may be ineffective; especially when there are large populations of similarly valued outliers. Such situations can (and do) happen for real-world fragments such as that shown in Figure 12.9(a).

[53] proposes a method to automatically locate these outliers by approximating smooth symmetric portions of the sherd surface with one or more quadratic surface patches. For each surface patch, a quadratic implicit polynomial is fit to the sherd surface data and the unknown axis location and orientation is directly computed from the polynomial coefficients. In contrast to past approaches which are either local (Section 12.3.2) or global (Section 12.3.3), this approach identifies a collection of surface patches, i.e., clusters of surface points that appear to lie on a single quadratic patch. As one can see from Figure 12.9(b), these patches typically form a patchwork over the sherd surface. This method is a compromise between local and global approaches and, to some extent, can benefit from the positive aspects of each approach. The large number of surface points in each patch make sherd surface estimates from these patches robust to local variations such as measurement noise, which is a weakness of local approaches (Section 12.3.2). Each surface patch is restricted to be quadratic, i.e, to individually have simple shape; the patchwork offered by the collection of patches can then represent very complex shapes which is a weakness of global approaches (Section 12.3.3) that tend to smooth out fine details and can be insensitive to important local shape variations. A

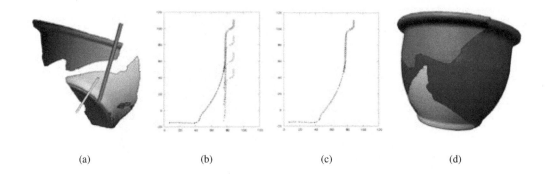

(a)   (b)   (c)   (d)

FIGURE 12.10

(a) Two fragments from a contemporary pot in arbitrary 3D orientation with their estimated axes rendered shown as cylinders. (b) The profile curve of one fragment is shifted vertically to compute t_z, the best height for which the two fragment profile curves match. (c) The aligned axis/profile-curves for the fragment pair is shown as a set of connected 2D (r, z) points. Point sample locations are indicated with crosses "+" and boxes "□" for the two fragment models. The two aligned profile curves for the fragment pair are merged to generate a global model of the pot. (d) The reconstructed 3D axially symmetric surface generated by merging the axis and profile curve estimates for the two sherds. The data of the two sherds has been superimposed. Of note here is that we may reconstruct the complete pot shape from fragments which may not share a common boundary as is the case in (d).

by-product of this approach is also the ability to identify clusters of points that are not suitable for surface estimation, i.e., outliers that occur when asymmetries exist on the sherd surface (see Figure 12.9(a,b)).

12.3.6   Vessel Reconstruction by Matching Axis/Profile Curve Models

Since individual sherds typically provide only a fraction of the complete vessel profile, their axis and profile curve estimates provide only *a portion* of the complete vessel profile. In contrast to typological classification, an alternative approach for vessel reconstruction is to match together sherd profile curves. Figure 12.10 shows results from [13] which proposes a method to do this. The method extracts a model of the sherd axis and profile curve for all sherds that may be fit together. Sherds that share similarly shaped regions of their profile curve are then merged into a larger profile curve (see Figure 12.10). As shown in Figure 12.11, this method is capable of reconstructing archaeological vessels when none of the vessel fragments touch. Solutions such as this may be particularly difficult for human puzzle solvers to piece together since manual reconstruction is often dependent upon matching sherds together by their common boundary.

12.4   Vessel Reconstruction by Sherd Matching: 3D Puzzle Solving

For the purpose of clarity and conciseness, we frame the problem of sherd matching for ceramic vessel reconstruction (commonly known as *3D puzzle-solving*) strictly as a geometric estimation

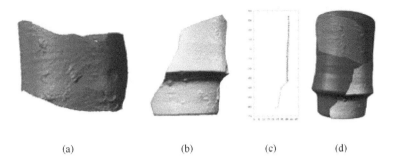

(a)   (b)   (c)   (d)

FIGURE 12.11
In this figure taken from [13], (a) and (b) are surface measurement meshes from archaeological
fragments of a Nabatean cup which are to be aligned. (c) shows the aligned mean vectors of the
estimated axis/profile-curve distributions for the fragment pair. (d) shows the 3D axially symmetric
surface generated by merging the two distributions. It is shown as a brown surface mesh with the
aligned fragment data superimposed.

problem, i.e., we do not consider other important features such as color, patterns, context, etc. (see
Section 12.5.1 for a brief discussion on these topics).

We begin by denoting the global geometry of the fractured object as a large collection of unknown
variables, Ω, which may be decomposed into two subsets: (1) Ω_{outer}, those variables that characterize
the outer boundary of the object before being fractured, i.e., the *intact boundary*, and (2) Ω_{inner}, those
variables that characterize the boundaries generated when the object was fractured, i.e., the *fracture
boundaries*. A third set of variables describe locations where three or more fragment boundaries have
broken apart and are referred to as *fracture junctions* or simply *junctions* and denoted $\Omega_{inner\perp}$ (see
Figure 12.12). A fourth set of variables describe special junctions between the intact boundary and the
fracture boundary and are referred to as *outer-surface junctions* and denoted $\Omega_{outer\perp}$. Together these
variables account for all the usable geometric information for geometric reconstruction techniques.
Measurements from fragments allow us to estimate some subset of these unknown variables. Our
task in geometric reconstruction is to define a computational method to merge these estimates to
uncover not only the shape of the global intact surface, but also how that surface was fractured when
it broke apart in antiquity by computing an estimate for Ω.

Table 12.1 specifies terminology for generic puzzle-solving by the dimension of the geometric
primitives being estimated: (1) 2D-reconstruction, i.e., the reconstruction of flat objects such as stone
tablets or frescoes and (2) 3D-reconstruction, i.e., the reconstruction of 3D objects (objects that are
not flat). When reviewing this table, note that typically Ω_{outer} and Ω_{inner} are a single contiguous
planar curve or surface patch. However, in some cases, a single fragment may have two distinct outer
surface patches such as that in the leftmost image of Figure 12.6. Here, the cyan-colored surface
regions show two disjoint surface patches that together define Ω_{outer}.

12.4.1 The Bayesian Formulation

As mentioned in Section 12.2.4 reconstruction of artifacts by fragment matching consists of four
steps. The first step is segmentation of the surface; for the purposes of our discussion, we again
deal with eggshell surfaces, i.e., only the sherd outer surface (see Section 12.3.3 for details on
eggshell surfaces). Fortunately, this step is not a challenge as the boundary of sherds for eggshell
surfaces correspond to junctions between fracture surfaces and the outer surface, i.e., measurements
of $\Omega_{outer\perp}$ from Table 12.1. Surface points away from the boundary are then measurements from the

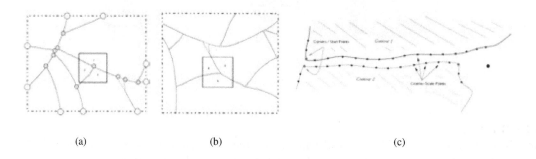

(a)   (b)   (c)

FIGURE 12.12 (SEE COLOR INSERT)

In these modified figures from [7], we define the geometric features used by various approaches for assembly. In (a), the outer contour of the square ceramic tile, i.e., the *intact boundary,* Ω_{outer}, is shown in red; *fracture boundaries,* Ω_{inner}, are shown in blue; *vertices,* $\Omega_{inner\perp}$, are shown as green points; and *outer vertices,* $\Omega_{outer\perp}$, are shown as yellow points. Note that most automatic assembly algorithms including [8–11,22], only consider the fracture boundaries, Ω_{inner}, and perhaps the vertices, $\Omega_{inner\perp}$. The black boxes in Figures (a,b) show "T" and "Y" junctions respectively that ceramics tend to generate when fractured. These junctions are typically high curvature locations along the fragment fracture boundary and are important for reconstruction as they often denote origin points for fracture boundaries that match to only one other fragment (c).

Geometric Variable	Generic Term	2D Reconstruction		3D Reconstruction	
		Geometric Primitive	Referred to as:	Geometric Primitive	Referred to as:
Ω_{outer}	outer boundary	planar curve	outer curve	surface patch	outer surface
Ω_{inner}	fracture boundary	planar curve	fracture curve	surface patch	fracture surface
$\Omega_{inner\perp}$	fracture junction	2D point	vertex	3D space-curve	break curve
$\Omega_{outer\perp}$	outer fracture junction	2D point	outer vertex	3D space-curve	outer break curve

TABLE 12.1

Variable definitions and a generic terminology that applies to puzzle-solving methods for both 2D and 3D artifact reconstruction (see Figure 12.12).

sherd outer surface, Ω_{outer}. The second step of the reconstruction is to extract features for surface reconstruction. For reconstruction of axially symmetric vessels these features consist of: (1) the fragment outer surface, Ω_{outer}, as fit by a surface of revolution in terms of a central axis and a profile curve with respect to that axis (see Section 12.3 for details), and (2) the outer break curves, $\Omega_{outer\perp}$, which denote locations along the sherd outer surface where two sherds broke apart (see Figures 12.12 and 12.13). $\Omega_{outer\perp}$ is modeled as sequences of 3D points with respect to the pot axis, and their (x, y, z) coordinates constitute the variables of $\Omega_{outer\perp}$. Similar to the approach in [7], the outer break curve segments come together at junctions. As these junctions are 3D points constrained to lie on the 2D axially symmetric surface, we call these locations *outer vertices* which typically correspond to high curvature locations along the outer fragment surface boundary (see Figures 12.12 and 12.13). Note, in Figure 12.13(b), the profile curve is modeled as an algebraic curve of degree 6, i.e., as the zero set of a 6th degree implicit polynomial. The axis/profile-curve model that is common to both sherds i and j when the sherds are aligned is denoted $\Omega_{outer(i,j)}$, and the portion of the outer break-curves, $\Omega_{outer\perp}$, along which fragments i and j broke apart, is denoted $\Omega_{outer\perp(i,j)}$.

Because pot reconstruction in [18, 54] is treated as Bayesian Maximum A-Posteriori (MAP) estimation, the matching performance functional is determined. For a pair of fragments having their datasets in some "aligned" position, the matching performance functional depends on the outer break curve data for each fragment and the outer surface data for each fragment. Note that only the portion of outer-break curve along which the two fragments match is used. Denote this data by $C_{(i,j)}$. Similarly, for the two outer-surface data sets, one for each fragment, only that data that is pertinent to the patch of axis/profile-curve shared by the two fragments is used. Denote this data set by $\mathcal{D}_{(i,j)}$. In words, the matching performance functional for the aligned pair of fragments, i and j, is *P(break-curve data, outer-surface data — outer break-curve variables, axis/profile-curve variables)*, which, in equation form is

$$P(C_{(i,j)}, \mathcal{D}_{(i,j)} | \Omega_{outer\perp(i,j)}, \Omega_{outer(i,j)}) \tag{12.13}$$

Equation (12.13) assumes the fragment data alignment is fixed, then the values used for $\Omega_{outer\perp(i,j)}$ and $\Omega_{outer(i,j)}$ are the values for which (12.13) is maximum, i.e., the MLE for these. In (12.13) it is assumed the data for the two fragments are aligned. Alignment means a Euclidean transformation $\mathbf{T}_{(i,j)}$ (3D translation and rotation) of one of these datasets with respect to the other. Hence, $C_{(i,j)}$ and $\mathcal{D}_{(i,j)}$ depend on $\mathbf{T}_{(i,j)}$, and the conditioning should, therefore, also be on $\mathbf{T}_{(i,j)}$, and the resulting (12.13) should then be maximized with respect to $\mathbf{T}_{(i,j)}$. Then the matching performance functional for a pair of fragments is

$$P(C_{(i,j)} | \widehat{\Omega}_{outer\perp(i,j)}, \widehat{\Omega}_{outer(i,j)}, \widehat{T}_{(i,j)}) P(\mathcal{D}_{(i,j)} | \widehat{\Omega}_{outer(i,j)}, \widehat{\mathbf{T}}_{(i,j)}) \tag{12.14}$$

where (12.13) factored into a product because $C_{(i,j)}$ and $\mathcal{D}_{(i,j)}$ are conditionally independent. What are the explicit expressions for the two factors in (12.14)? It is assumed that a curve data point is a 2D circularly symmetric Gaussian perturbation, within a plane, of a point on the curve. The point on the curve is that which is closest to the data point, and the plane is the plane perpendicular to the curve at the point on the curve. It is assumed that a surface data point is an independent identically distributed Gaussian perturbation, along a line perpendicular to the surface at the point on the surface closest to the data point. For example, assume a plane specified by point $\mathbf{p}_0$ and unit vector $\mathbf{w}$, then the probability density for the surface measurement point $\mathbf{p}$ is given as:

$$P(\mathbf{p} | \mathbf{w}, \mathbf{p}_0) = \frac{1}{\sqrt{2\pi\sigma_s^2}} \exp\left(-\frac{1}{2\sigma_s^2} \left[\mathbf{w}^t(\mathbf{p} - \mathbf{p}_0)\right]^2\right) \tag{12.15}$$

There is a reasonable rationale for these models [54]. Then each factor in (12.14) is a product of Gaussian probability density functions, and (12.14) is a constant multiplying an exponential which has an exponent which is a sum of two quadratic forms (energy functions). One energy function is a measure of how well the two outer break curve-data sets, one from each curve, represent a

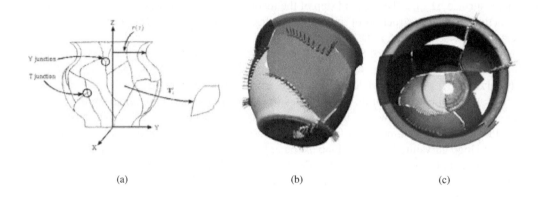

(a)   (b)   (c)

FIGURE 12.13
Figures (a) shows the geometric features used to model broken axially symmetric pots and (b,c)
show the result of a pot reconstruction result. (a) shows the outer surface features for the pot, Ω_{outer},
which specify a surface of revolution as represented by a 3D-line, the pot central axis, and a planar
profile-curve with respect to that axis. (b,c) show two views of a 13 fragment pot reconstruction
where 10 of the pieces were provided to the system for reconstruction. Small spheres on the pot
surface in (b,c) denote locations of matched boundaries.

common curve, and the other energy function is a measure of how well the two outer surface-data
sets represent a common surface.

12.4.2   Searching for the Solution

Initial approaches to this problem attempted to compute the geometry of the pot as a maximum
likelihood estimation problem where we seek to find the collection of matched outer surface break
curves and global axis/profile-curve geometry that maximizes the probability of all the measured
fragment data. To align a pair of fragments, the authors align a pair of outer vertices (points of high
curvature on the outer break-curve) one on each fragment, and then align the fragments keeping
these vertices in close proximity. Since there are roughly 4 vertices per fragment, if there were
100 fragments, the number of alignments required to compare a pair of fragments is 32, and for
all 5,000 pairs of fragments, the number of alignments is 160,000. The search algorithm starts
with all of these possible configurations of two fragments, arranges them in a stack by order of
decreasing probably or equivalently, increasing energy, and then grows configurations by adding
one fragment, at each recursion, to the top configuration in the stack and then keeping the resulting
configuration at the top of the stack or moving it to a new position if appropriate. In this way, high
probability configurations are grown and configurations will automatically grow for different pots.
Note, configuration probabilities or energies must be appropriately normalized here because different
configurations involve different numbers of data points.

Results for this approach are shown in Figure 12.13. While this method works, due to the
computational complexity of the search problem, i.e., finding correct fragment matches, and the
computational cost of each fragment comparison, this method can handle only small numbers of
fragments. The computational killer here is that this approach processes the raw range data each
time two or more surface fragments are aligned and the number of data points for each fragment is
typically in the many thousands. A recent variation on this approach, [13], uses compact probability
distributions for the fragment geometry rather than repeatedly using the same raw data, thus speeding
up the matching process and allowing for quicker or more comprehensive searches for compatible

matches. Also, rather than starting alignments by aligning break-curve vertices, it is orders of magnitude faster to first align axis/profile-curves for groups of fragments (see Figure 12.11). A final comment is that use of the global structure of axial symmetry provides robustness to erosion and chipping of the outer break-curves. This is because the axis/profile-curve estimates are insensitive to these deformations, and the axis/profile-curve estimates provide such powerful information for fragment reassembly that perfect fitting of outer break-curves where pot fragments come together is no longer critical.

12.5 Current Trends in Computational Artifact Reconstruction

Despite the advances in computational techniques for ceramic analysis there remains much work both in terms of theoretical research and in terms of pragmatic technologies that archaeologists can deploy. Current methods also fall short of using all of the available information from the digitized fragments which can be very useful for reconstruction. Breakthrough developments capable of significantly boosting the performance and accuracy of computational ceramic analysis will need to incorporate more than simply geometric measurements.

12.5.1 Going beyond Geometry: Textures and Patterns on Sherd Surfaces

In [7,15,55], the authors include information obtained from patterns on the outer fragment surface to augment their matching performance functional for reconstructing ceramic tiles. Figure 12.14(a-c) shows how patterns apparent on an inscribed Roman marble tablet discovered in Petra, Jordan, are used for improved fracture curve matching in [7]. In (a), the author computes the intensity profile of the fracture curve along a curve inset a small distance from the boundary (shown in blue). The matching performance functional used in this case is a combination of the differences in matched pixel luminance (Figure 12.14(a)), the difference in matched gradient magnitudes (Figure 12.14(b)), and a curve continuation metric called the *Euler Spiral Completion Curve* that seeks to complete curves across the fracture in the directions of the gradient vectors shown in (Figure 12.14(c)) while minimizing the curvature variation of the estimated completion curve.

In [15,55], the authors use in-painting and texture synthesis methods to extend the patterns on the outer surface of fragments to predict those patterns that appear on matching pieces (see Figure 12.14(d,e)). The mean and variance of the pixel values within small windows around the image boundary are then computed as the fragment features. Note that these are intrinsic features of the fragment, like curvature, and are also Euclidean invariant. These features are then matched with those of other fragments by transforming one piece with respect to the other and subsequently taking a weighted Euclidean distance between features at corresponding locations where the weighting coefficients decrease for the matched features close to the boundary of the extended region, i.e., matches closest to the original un-extended fragment boundary have the highest weights. This matching procedure is then a weighted cross-correlation of the image features along the extended boundary which must be optimized with respect to the variables of the unknown transformation matching the fragments. The fast Fourier transform FFT is then used to efficiently solve for the translation variables of the transformation followed by an optimization on the unknown one-parameter rotation to determine the best relative alignment of the fragments. The search procedure is a best-first approach that searches through the fragment matches exhaustively to compute a global solution which may change depending upon the chosen order of fragments to match as shown in Figure 12.14(f,g).

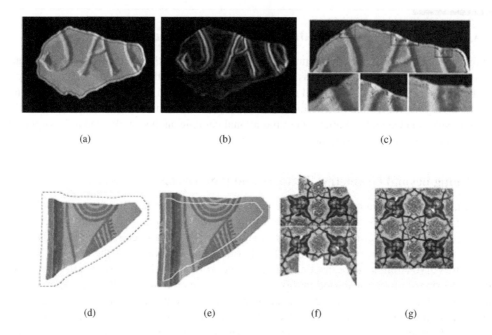

(a)   (b)   (c)

(d)   (e)   (f)   (g)

FIGURE 12.14

Figures (a-c), taken from [7], demonstrate one technique for matching fragments using patterns apparent on the fragment outer surface. This technique matches three features from the fragment surface along a slightly inset boundary (shown as the blue curve in (a)). The features are: (a) pixel luminance values along the boundary, (b) the gradient magnitude along the boundary, and (c) the energy in a continuation curve that spans the fracture in the direction of the image gradient vectors. Figures (d-g), taken from [15], show a second approach for appearance-based fragment matching that uses texture in-painting to extend the surface pattern on the piece (d) to predict the patterns that appear on matching pieces (e) (see text for details).

12.5.2 Discussion and Future Work

Of the 2D reconstruction systems discussed, only [7, 56] provide results for archaeological fragments and these include only small groups of matched fragments, i.e., pairs. Of the 3D reconstruction systems discussed [10, 11, 13], all present reconstructions of real-world artifacts although [11] is not automatic and performs reconstruction with a large amount of outside help from expert users. Reconstructions including four archaeological pot fragments from the Great Temple site are presented in [13] and three fragments from the Forma Urbis Romae project are presented in [10]. While this may initially seem discouraging, one must remember that one major contributing factor for such small numbers of matched archaeological data is, in part, difficulties in quickly and easily collecting and digitizing the fragments. At present, practical reassembly requires a collection of fragments which have been excavated near one another so that the likelihood of many pieces having come from the same object is high. Development of quick and accurate scanning technologies already under development promises to streamline the reassembly process, making these computational aids applicable in real-world archaeological settings.

A wide variety of approaches for ceramic vessel analysis and reconstruction have been investigated using just about every conceivable piece of geometric information. Yet, we are surprised to note that much information available from the outer surface remains unused in current approaches. [7, 50] are examples which use such information by incorporating continuity constraints across matched outer surface boundaries to construct matches that are both continuous and smooth across the joined outer surface boundary. Yet, no current methods incorporate higher order geometric features in the vicinity of the outer surface boundary. Since artifact outer surfaces are typically smooth, we feel that these features may be computed accurately and be informative enough for fragment matching to improve the current state-of-the-art. Use of this information is applicable to both free-form 3D reconstruction and 2D tablet reconstruction problems. As noted in the third approach explored by the Forma Urbis Romae project, being able to identify the outer boundary of fragments constrains the potential matches for these fragments considerably, which is a big performance boost when searching for potential candidate matching fragments. This is an area that will likely receive more consideration as scientists develop the next generation of artifact reconstruction systems.

Another aspect to this problem requiring attention is the development of an information-theoretic basis for searching compatible matches. Current methods either have an extremely difficult search problem with many false positives [7, 50] or sufficiently discriminative features to quickly find the solution with a simple matching procedure [10]. Making expeditious use of computed features requires knowledge of how much discriminative information a given feature contains, a central concept in machine learning. Seminal work in this regard is offered in [33] where the authors analyze the information content of 2D ceramic tile fracture boundaries in terms of their curvature signatures. Using the discrete sine transform, the authors propose a method for measuring the amount of information present in a 2D boundary curve. Search procedures that concentrate on high-information features identified using techniques such as this also show promise for improving the speed, accuracy, and reliability of reconstruction systems. A related aspect of the problem is the development of a generative model for cracks and fractures as proposed in [57] or for erosion. Such work may lead to material-specific fracture boundary models capable of succinctly representing and robustly estimating fracture boundaries and fracture boundary junctions, i.e., the vertices and break curve junctions that provide crucial matching features and performance speed-ups to some computational reconstruction systems.

While approaches to date are sophisticated and computationally efficient, none of the published works claim to be ready for deployment in the field as a tool for archaeologists. In fact, it is surprising to note that the Forma Urbis Romae project, using brute force computational methods and expert users, has been very successful in discovering twenty new matches for a reconstruction problem that humans have been studying for hundreds of years [11]. This encourages the use of current reconstruction systems in a cybernetic (i.e., a collaborative human-and-computer) context. Such systems would use computerized artifact reconstruction to complement manual reconstruction, exploiting the powers of each to generate a system that is more efficient and accurate than either could be independently. Here, the computer can quickly identify a small set of candidate matches that the human user can manually prune to aid the computer in obtaining the correct global solution.

12.5.3 Conclusion

The state-of-the art in ceramic analysis and reconstruction techniques and concepts have been discussed. The detailed quantities that can be captured using accurate digitization sensors and computer-aided analysis is already beginning to provide new insights to the archaeological community. As these technologies and research progress, the impact of such systems will increase and allow archaeologists to do more detailed analysis on what is typically an overwhelming wealth of sherd information. Our treatment of computational sherd analysis started with an overview of approaches in this area followed by a description that details how sherds may be digitized for computational ceramic analysis. Vessel profile curve analysis was discussed as it applied to the problem of computational classification of sherds to ceramic typologies. Solutions to this problem can be of great assistance to archaeologists, especially for discriminating subtle variations in vessel morphology that may have archaeological significance. The underlying mathematical models required to represent sherd surfaces and algorithms capable of estimating these models were discussed. Finally, a description was provided that details methods for 3D puzzle-solving (i.e., uncovering the shape of ceramic vessels by matching digitized sherds). Several extensions to the geometric matching problem that use patterns apparent on the fragment outer surface that show promise to extend analysis beyond that of matching strictly geometry were also discussed. The models needed for solving these problems are new and challenging, and most involve 3D which is largely unexplored by both the archaeological and pattern recognition community. We would like to thank the authors of [1,7,10,15,58] for granting permission to use their figures for this publication and the National Science Foundation for funding this work through NSF ITR Grant #0205477.

Acknowledgments

Work presented in this chapter was supported in part by a fellowship offered by Analog Devices and grants BCS-9980091, IIS-0205477, and IIS-0808718 from the National Science Foundation. The support of my family, friends, co-workers, and especially my wife Julianna has made the completion of this work a reality. The pottery assembly system reflects discussions with a diverse group of people from several disciplines including Engineering (David Cooper and Benjamin Kimia), Applied Math (David Mumford), and Archaeology (Martha Sharp Joukowsky). Without these helpful discussions this work would not have been possible.

Bibliography

[1] A. Gilboa, A. Karasik, I. Sharon, and U. Smilansky, "Towards computerized typology and classification of ceramics," *Journal of Archaeological Science*, vol. 31, no. 681–694, 2004.

[2] H. Freeman and L. Garder, "A pictorial jigsaw puzzles: The computer solution of a problem in pattern recognition," *IEEE Trans. Electron. Comput.*, vol. 13, pp. 118–127, 1964.

[3] H. Wolfson, E. Schonberg, A. Kalvin, and Y. Lambdan, "Solving jigsaw puzzles using computer vision," *Annals of Operations Research*, vol. 12, pp. 51–64, 1988.

[4] G. Radack and N. Badler, "Jigsaw puzzle matching using a boundary-centered polar encoding," *Computer Graphics and Image Processing*, vol. 19, pp. 1–17, 1982.

[5] D. Goldberg, C. Malon, and M. Bern, "A global approach to automatic solution of jigsaw puzzles," in *Proc. of Conf. on Computational Geometry*, pp. 82–87, 2002.

[6] S. Li, *Markov Random Field Modeling in Image Analysis*, ch. 7. Springer, 2001.

[7] J. McBride, "Archaeological fragment re-assembly using curve matching," Master's thesis, Brown University, September 2002.

[8] G. Papaioannou, E.-A. Karabassi, and T. Theoharis, "Reconstruction of three-dimensional objects through matching of their parts," *PAMI*, 2002.

[9] J. Stolfi and H. Leitão, "A multiscale method for the reassembly of two-dimensional fragmented objects," *PAMI*, 2002.

[10] Q.-X. Huang, S. Flöry, N. Gelfand, M. Hofer, and H. Pottmann, "Reassembling fractured objects by geometric matching," *ACM Transactions on Graphics*, vol. 25, no. 3, pp. 569–578, 2006.

[11] D. Koller, J. Trimble, T. Najbjerg, N. Gelfand, and M. Levoy, "Fragments of the city: Stanford's digital forma urbis romae project," *Journal of Roman Archaeology Suppl.*, vol. 61, pp. 237–252, 2006.

[12] W. Kong and B. B. Kimia, "On solving 2D and 3D puzzles using curve matching," in *Proc. of CVPR*, (Hawaii, USA), IEEE, Computer Society, December 2001.

[13] A. Willis and D. Cooper, "Estimating a-priori unknown 3D axially symmetric surfaces from noisy measurements of their fragments," in *Proc. of the 3rd Intnl. Symposium on 3D Data Processing, Visualization, and Transmission*, pp. 334–341, 2006.

[14] M. S. Joukowsky, *Petra Great Temple - Volume 1: Brown University Excavations 1993-1997*. Providence, RI, USA: Petra Exploration Fund, 1999.

[15] M. C. Sağiroğlu and A. Erçil, "A texture based matching approach for automated assembly of puzzles," in *The 8th International Conference on Pattern Recognition*, vol. 3, pp. 1036–1041, 2006.

[16] M. Kampel and R. Sablatnig, "Automated segmentation of archaeological profiles for classification," in *ICPR*, vol. I, pp. pp. 57–60, 2002.

[17] S. Tosovic, M. Kampel, and R. Sablatnig, "Combining shape from silhouette and shape from structured light for volume estimation of archaeological vessels," in *ICPR*, vol. I, pp. pp. 364–367, 2002.

[18] A. Willis and D. Cooper, "Bayesian assembly of 3D axially symmetric shapes from fragments," in *Proc. of the Conf. on Computer Vision and Pattern Recognition*, vol. I, pp. 82–89, 2004.

[19] Y. Ma, S. Soatto, J. Kosecka, and S. S. Sastry, *An Invitation to 3-D Vision*. Springer, 2003.

[20] T. Malzbender, B. Wilburn, D. Gelb, and B. Ambrisco, "Surface enhancement using real-time photometric stereo and reflectance transformation," in *Eurographics Symposium on Rendering*, pp. 245–248, June 2006.

[21] S. Rusinkiewicz, O. A. Hall-Holt, and M. Levoy, "Real-time 3D model acquisition.," in *SIG-GRAPH*, pp. 438–446, 2002.

[22] P. Besl and N. McKay, "A method for registration of 3-D shapes," *PAMI*, vol. 14, no. 2, pp. 239–256, 1992.

[23] A. Fitzgibbon, "Robust registration of 2D and 3D point sets," in *Proc. of the British Machine Vision Conference*, 2001.

[24] S. Rusinkiewicz and M. Levoy, "Efficient variants of the ICP algorithm," in *Intl. Conf. on 3D Digital Imaging and Modeling*, pp. 145–152, 2001.

[25] G. Turk and M. Levoy, "Zippered polygon meshes from range images," in *Proc. of SIGGRAPH*, pp. 311–318, 1994.

[26] M. Kampel and R. Sablatnig, "Virtual reconstruction of broken and unbroken pottery," in *Proc. of the Conf. on 3D Digital Imaging and Modeling*, pp. 318–325, 2003.

[27] R. H. Smith, "An approach to the drawing of pottery and small finds for excavation records," *World Archaeology*, vol. 2, pp. 212–228, October 1970.

[28] H. W. Guggenheimer, *Differential Geometry*. Dover Publications, Inc., 1977.

[29] H. J. Wolfson, "On curve matching," *IEEE Transactions on Pattern Analysis and Machine Intelligence*, vol. 12, no. 5, pp. 483–489, 1990.

[30] G. Ucoluk and I. H. Toroslu, "Automatic reconstruction of broken 3-D surface objects," *Computers & Graphics*, vol. 23, pp. 573–582, August 1999.

[31] B. Kimia and J. McBride, "Archaeological fragment re-assembly using curve matching," in *CVPR : Workshop on Computer Vision Applications in Archaeology*, 2003.

[32] M. M. Lipschutz, *Schaum's Outline of Theory and Problems of Differential Geometry*. McGraw-Hill, 1969.

[33] J. Stolfi and H. Leitão, "Measuring the information content of fracture lines," *International Journal of Computer Vision*, vol. 65, no. 3, pp. 163–174, 2005.

[34] W. Kong, "On solving 2D and 3D puzzles using curve matching," Master's thesis, Brown University, May 2002.

[35] S. Hermann and R. Klette, "A comparative study on 2d curvature estimators," in *Intl. Conf. on Computing: Theory and Applications (ICCTA)*, pp. 584–589, 2007.

[36] S. Manay, B.-W. Hong, D. Cremers, A. J. Yezzi, and S. Soatto, "Integral invariants for shape matching," *IEEE Trans. Pattern Anal. Mach. Intell.*, vol. 28, no. 10, pp. 1602–1618, 2006.

[37] X. Tang, "A sampling framework for accurate curvature estimation in discrete surfaces," *IEEE Transactions on Visualization and Computer Graphics*, vol. 11, no. 5, pp. 573–583, 2005.

[38] S. Rusinkiewicz, "Estimating curvatures and their derivatives on triangle meshes," in *3DPVT '04: Proceedings of the 3D Data Processing, Visualization, and Transmission, 2nd International Symposium*, pp. 486–493, 2004.

[39] S. Petitjean, "A survey of methods for recovering quadrics in triangle meshes," *ACM Comput. Surv.*, vol. 34, no. 2, pp. 211–262, 2002.

[40] M. Desbrun, M. Meyer, P. Schröder, and A. H. Barr, "Implicit fairing of irregular meshes using diffusion and curvature flow," in *SIGGRAPH '99: Proceedings of the 26th Annual Conference on Computer Graphics and Interactive Techniques*, pp. 317–324, 1999.

[41] G. Taubin, "A signal processing approach to fair surface design," in *SIGGRAPH '95: Proceedings of the 22nd Annual Conference on Computer Graphics and Interactive Techniques*, pp. 351–358, 1995.

[42] A. Willis and D. B. Cooper, "From ruins to relics: Computational reconstruction of ancient artifacts," *IEEE Signal Processing Magazine*, vol. 25, pp. 65–83, July 2008.

[43] D. J. Struik, *Lectures on Classical Differential Geometry*. NY: Dover, 1950.

[44] E. Kreyszig, *Differential Geometry*, pp. 128–129. Dover, 1963.

[45] M. Kampel, R. Sablatnig, et al., "Fitting of a closed planar representing a profile of an archaeological fragment," pp. pp. 263–269, 2001.

[46] G. Taubin, "Estimation of planar curves, surfaces and nonplanar space curves defined by implicit equations with applications to edge and range image segmentation," *IEEE Trans. on Pattern Anal. Machine Intell.*, vol. 13, pp. 1115–1138, November 1991.

[47] T. Tasdizen, J. P. Tarel, and D. B. Cooper, "Improving the stability of algebraic curves for applications," *IEEE Trans. on Image Proc.*, vol. 9, pp. 405–416, March 2000.

[48] D. Mumford and Y. Cao, "Geometric structure estimation of axially symmetric pots from small fragments," in *Proc. of Int. Conf. on Signal Processing, Pattern Recognition, and Applications*, pp. 92–97, 2002.

[49] H. Pottmann, M. Peternell, and B. Ravani, "An introduction to line geometry with applications," *Computer-Aided Design*, vol. 31, pp. 3–16, 1999.

[50] A. Willis, D. Cooper, et al., "Bayesian pot-assembly from fragments as problems in perceptual-grouping and geometric-learning," in *ICPR*, vol. III, pp. 297–302, 2002.

[51] A. Willis, D. Cooper, et al., "Accurately estimating sherd 3D surface geometry with application to pot reconstruction," in *CVPR Workshop: ACVA*, June 2003.

[52] G. Taubin, "Estimation of planar curves, surfaces and nonplanar space curves defined by implicit equations with applications to edge and range image segmentation," *PAMI*, vol. 13, no. 11, pp. 1115–1138, 1991.

[53] Y. Sui and A. Willis, "Using Markov random fields and algebraic geometry to extract 3D symmetry properties," in *Fourth International Symposium on 3D Data Processing, Visualization and Transmission (3DPVT)*, (Atlanta, GA), June 2008.

[54] A. Willis, D. Cooper, et al., "Assembling virtual pots from 3D measurements of their fragments," in *VAST International Symposium on Virtual Reality Archaeology and Cultural Heritage*, pp. 241–253, 2001.

[55] M. C. Sağiroğlu and A. Erçil, "A texture based approach to reconstruction of archaeological finds," in *The 6th International Symposium on Virtual Reality, Archaeology, and Cultural Heritage*, pp. 137–142, 2005.

[56] C. Aras, "Hindsite: A robust system for archaeological fragment re-assembly," Master's thesis, Brown University, May 2007.

[57] B. Desbenoit, E. Galin, and S. Akkouche, "Modeling cracks and fractures," *The Visual Computer*, vol. 21, no. 8, pp. 717–726, 2005.

[58] P. Tyers, *Roman Pottery in Britain*, p. 102–104, Routledge, 1996.

13

Digital Reconstruction and Mosaicing of Cultural Artifacts

Efthymia Tsamoura, Nikos Nikolaidis, Ioannis Pitas

Aristotle University of Thessaloniki
Email: etsamour@csd.auth.gr, nikolaid@aiia.csd.auth.gr, pitas@aiia.csd.auth.gr

CONTENTS

13.1 Introduction

Reconstruction of broken or torn artifacts of archaeological, historical, or cultural importance is an indispensable tool in the hands of researchers in these fields. Indeed, ceramic pot fragments are an important source of information from which archaeologists can draw inferences about the civilization that existed at a given archaeological site (e.g., its chronology, the population's socioeconomic standards, etc.). For murals and mosaics (see Figure 13.1), accurate reconstructions provide information to archaeologists about the iconography, stylistic and drawing tool developments, etc. In the case of stone tablets and other artifacts that bear text inscriptions (e.g., papyrus), reconstructions that allow epigraphists to read part of the entire text provide important information about the society's organization and the scientific and cultural evolution (e.g., poetry, drama) of ancient civilizations such as Greek, Persian, Egyptian, etc.

During the last years, researchers in the fields of image processing and analysis and computer vision have dealt with computer-based approaches for the reconstruction of such artifacts, providing a valuable helping hand to archaeologists, historians, and restoration experts. There are many advantages in using computers for the reconstruction of fragmented artifacts. Some of them are listed below [53]:

- Computer-aided reconstruction (either automatic, or human supervised) may lead to huge time savings for archaeologists who spend many hours assembling broken artifacts by hand.

- In most cases, the reconstruction of broken ancient artifacts is extremely difficult for many reasons, e.g., the fragments are highly spoiled (e.g., from environmental and other effects), or are fragmented into many tiny pieces. Furthermore, fragments from many different pots, murals, or mosaics may be found mixed during the excavations. Unless the fragments are visually similar and discovered in a context that is local in both time and space, archaeologist very rarely proceed to reconstruction. Thus, computer-aided approaches may help in the reconstruction of objects that would have otherwise remained in pieces.

- The correct computer-based reconstruction of artifacts provides the ability to conduct precise measurements on reconstructed objects. These measurements can sometimes provide a better insight for historical events and historical evolution. When such measurements are performed by archaeologists using empirical techniques they usually have a large margin of error (see [53] for an example).

- Finally, the development of computer-based 3D fragmented objects reconstruction methods may have potential applications in other domains such as anthropology or forensic science (e.g., in the reconstruction of a human skeleton from broken bones).

A number of automatic or semiautomatic reconstruction methods have been proposed in the literature. Some of them were actually used in projects for the reconstruction of fragmented objects. A human-guided method for the reassembly of murals at the Akrotiri archaeological site in the Greek island of Thera (Santorini), dated back to 1600 BC, was presented in [31]. The murals of Thera were preserved because they were covered by pumice from the eruption of a volcano. The walls that were decorated with the murals collapsed before the volcanic eruption, due to strong earthquakes. Thus, a single painting is usually scattered into many fragments, sometimes mixed with the fragments from other murals, rendering the restoration of the murals from their constituent fragments a very painstaking and time-consuming process. The method has been applied to two fragmented murals consisting of 262 fragments. Another important project, that dealt with the problem of supervised reconstruction of 3D objects, was Stanford's Digital Forma Urbis Romae project [21]. The project aimed to reconstruct the Seventh Marble of Rome or Forma Urbis Romae (211 AD). This enormous map of Rome, measuring 18×13 meters, was carved onto 150 marble slabs installed on a wall of an hall of the Templum Pacis. The map was destroyed in the 5th century, resulting in 1,186 pieces. Since its rediscovery in 1562, scholars have focused on joining the fragments and reconstructing this great monument, but this was a very difficult puzzle to work with since the map is drastically incomplete (only 10 percent of the map survives). In addition, many fragments are huge and heavy, while others are so small that their carved surfaces hardly provide any identifiable information. Finally, a project dealing with the reassembly of documents of archaeological importance is "The Sinaitic Glagolitic Sacramentary (Euchologium) Fragments" project [19]. The project deals with the recording, investigation, and edition of two very important medieval Slavonic manuscripts, discovered in 1975 in St. Catherine's monastery on Mount Sinai, namely the "Euchologii Sinaitici pars nova" and the "Missale (Sacramentarium) Sinaiticum" manuscripts. The folios of these manuscripts are degraded due to water.

Another example of fragmented objects where computer assistance would be valuable is ripped-up documents. This problem arises in archival study and investigation science. Documents may be torn by hand or shredded by a machine called a shredder. Similar to archaeological fragmented objects, manually ripped-up documents may also suffer damages at several levels, such as torn edges, moisture, obliteration, charring, and shredding. Thus, their manual reconstruction may require days or even months, depending on the number of fragments and their conservation state. Another problem that occurs when reassembling documents by hand lies in their manipulation. The physical reconstruction of a document modifies the original document, since products like glue and adhesive tape are added into it. Thus, the computer-aided virtual reconstruction of ripped-up documents may reduce or eliminate the time-consuming manual effort, and also lead to reassembled documents that exhibit small degradation with respect to the original.

A special case of fragmented 2D objects, which are not related to cultural heritage and its preservation, are the jigsaw puzzles. A jigsaw puzzle is a tiling puzzle that requires the assembly of numerous small, oddly shaped, interlocking pieces, that, when properly assembled, form a picture without gaps and overlaps. Jigsaw puzzles were originally created by painting a picture on a flat, rectangular piece of wood, and then cutting that picture into small pieces with a jigsaw, hence the name. John Spilsbury, a London mapmaker and engraver, is credited with creating the first jigsaw puzzles around 1760. The characteristics of jigsaw puzzles, (e.g., that each puzzle piece is perfectly cut and well preserved) are different from those of archaeological fragmented artifacts. However, puzzle reassembly and 2D object reconstruction methods have many common characteristics. In general, reassembly of puzzles is considered less difficult than reconstruction of ancient artifacts or paper documents, due to the following reasons [31]:

- There is no a priori knowledge about the picture content drawn on the surface of the unbroken object.

- The size and the shape of the fragments vary dramatically in contrast to what happens in jigsaw puzzles.

- Gaps may exist between adjacent fragments of a fragmented object, due to degradation. Furthermore, one or more fragments may be missing.

- The fragments are often highly degraded from various factors (e.g., environmental effects), in contrast to jigsaw puzzle pieces.

- There is no unique solution concerning the matching of two fragments. As will be discussed later, there is no criterion to confirm that the matching of two fragments is correct.

In summary, the types of fragmented objects whose reconstruction has been studied in the corresponding literature can be classified into the following two broad categories:

- 2D objects, such as 2D puzzles, paper documents, paper images, mosaics, murals.

- 3D objects, symmetric or asymmetric, such as tiles, bowls, amphorae or other types of ceramic pots.

Another problem that is closely related to that of 2D fragmented object reconstruction is that of mosaicing or stitching. Mosaicing is the process of reconstructing or re-stitching a single, continuous image from a set of overlapping sub-images. Image mosaicing is essential for the creation of high-resolution digital images of architectural monuments and works of art (especially of those with considerable dimensions like frescoes and large-size paintings) for archival, digital analysis, and restoration purposes [25,38]. In such applications, the required very high resolution image acquisition stresses the limits of acquisition devices. To overcome this obstacle, digitization procedures that utilize the acquisition of different, overlapping views, sometimes with the aid of sensor arrays and positioning mechanisms, are used. The acquired sub-images (Figure 13.2) are subsequently processed by a mosaicing algorithm.

Several image mosaicing techniques have been proposed in the literature [10, 12, 23, 25, 33, 35, 38, 40, 44, 47, 58] since mosaicing is also important in other areas that include the creation of high-resolution large-scale panoramas for virtual environments, image-based rendering, consumer photography, medical imaging [12, 47], aerial [58], satellite, and underwater [35] imaging, etc. The mosaicing process is comprised of two steps. The first step involves the estimation of the optimal transformation (translation, rotation, scaling. etc.) that aligns each sub-image with respect to each neighboring one. If one assumes that only translational camera motion takes place during the acquisition then only the relative displacement has to be evaluated. This step is the most computationally intensive part of the entire process. In the general case of a set of $M_1 \times M_2$ sub-images, optimal mosaicing (i.e., mosaicing by concurrently searching for the optimal position of all sub-images), would require a search in a m-dimensional space, where $m = 2(2M_1M_2 - M_1 - M_2)$, the term in parenthesis representing the number of all pairs of neighboring sub-images. Obviously the computational cost becomes prohibitive, as the number of sub-images increases. The second step of the mosaicing process utilizes displacement information found in the previous step in order to combine each pair of neighboring sub-images with invisible seams and thus reconstruct the whole image (Figure 13.15).

In this chapter we will try to provide a concise review of algorithms proposed so far for the 2D or 3D reassembly of fragmented or torn objects of cultural importance. Section 13.2 deals with the three steps of an object reconstruction procedure, whereas reconstruction methods will be reviewed in Section 13.3. Despite the fact that several methods were developed for the reassembly of shredded documents [37, 41, 51, 52], in this chapter we will review only methods proposed for manually ripped-up (hand-torn) documents, since the former have no application in cultural heritage preservation. Furthermore, although computer-aided reassembly of puzzles (which attracted significant interest since the mid-'80s [13, 17, 22, 56]) is not directly related to the topic of this chapter, we will briefly review a small number of such techniques, since there is a close relation between puzzle reassembly and 2D object reconstruction methods and the two areas share many principles. Subsequently, in Section 13.4 we present an automatic four-step method for the reassembly of torn 2D paper images or fragmented paintings [50]. Finally, in Section 13.5 we will describe two methods that aim to reduce the number of computations required for the evaluation of the sub-image displacements in mosaicing, without affecting significantly the matching error [28]. Readers interested in more details about the computational reconstruction of 2D/3D ancient artifacts and mosaicing techniques can delve into the thorough review and tutorial papers of Willis and Cooper [53] and Szeliski [48] respectively.

Throughout this chapter we will use the term *reassembly* for 2D objects and *reconstruction* for 3D objects. Furthermore, the term *fracture surface* will be used hereafter to denote the boundary surface of a 3D fragment that was created through the fracturing, whereas the term *fracture curve* will be used to denote the boundary of a 2D fragment or the curve delineating the fracture on the outer or inner surface of a 3D fragment (Figure 13.3).

13.2   The Three-Step Object Reconstruction Procedure

The purpose of this section is to present the steps required for the overall reconstruction of an object from its consistent fragments, as well as a comparative study of the methods proposed for each step (e.g., the features and the techniques utilized to identify the matching of two fragments).

Before the execution of a reconstruction method, digitization of fragments must take place. In case of paper documents or color images, the fragments can be either scanned or photographed.

(a)   (b)

FIGURE 13.1
Fragments from Byzantine murals.

FIGURE 13.2
A $M_1 = 3$ by $M_2 = 2$ sub-image acquisition of a painting.

FIGURE 13.3
The outer surface (the one with depictions on it), the fracture surface (the gray irregular surface on the perimeter of the fragment), and the fracture curve (in red) of a 3D fragment.

The same procedure is followed for 3D objects when only 2D data are utilized for reassembly, as in [8,31]. In this case, many conditions must be met when the fragments are photographed, e.g., same illumination conditions, no artificial shadows, fixed distance of the fragment plane from the camera focus, object surface parallel to the photographic glass and minimal photo distortion, etc. In the case of actual 3D object reconstruction, the fragments can be either scanned with a 3D laser scanner, or photographed from different viewpoints and reconstructed in 3D using photogrammetry techniques. It must be noted that in all methods that will be reviewed, all input fragments are assumed to belong to the same broken or torn object, unless it is explicitly stated.

The overall object reconstruction procedure can be either automatic or semi-automatic. In the former category, the object reconstruction process is not guided by users, whose role is limited to the digitization of the input fragments and sometimes the selection of appropriate parameter values. In the latter category [24,31], the user selects the correct matches for pairs of fragments among the ones proposed by the system and/or the appropriate affine transformation in order to correctly align them along their matching segments.

All the proposed methods try to identify fragments that were adjacent in the unbroken object. The adjacent fragments should have high similarity with respect to, e.g., the shape and/or color of fracture curves in case of 2D objects or the fracture surfaces in case of 3D objects. Similarities may also exist in the depictions on the surface of the fragments. Thus, the existing methods perform pair-wise fragments comparisons based on several criteria such as shape, color, or texture in order to identify candidate adjacent fragments.

In general, the reconstruction of an object is a three-step procedure:

- **Fragment preprocessing**. The purpose of this optional step is to reduce the computational burden of the steps that follow. Two different approaches for fragment preprocessing exist. The first reduces the number of pair-wise fragment comparisons that will be performed during the second step, by retaining only the set of (possibly) correct adjacent fragment pairs. The second approach is to reduce the number of matching possibilities between a certain pair of fragments.

- **Matching and alignment of candidate adjacent fragments**. The second operation that takes place is the matching of pairs of fragments, namely the identification of point correspondences between fracture curves or surfaces of the two fragments. If pairs of candidate adjacent fragments have been generated during the first step, then only the matching of these pairs is estimated. Similarly, the matching process is guided by information produced during the preprocessing step, in cases where the possible matching options of two fragments have been already decided, see for example [14]. The identification of the matching of two fragments can be done through several approaches. One approach is to utilize information only from a fragment's fracture curve or

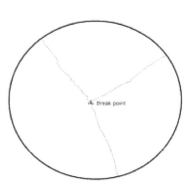

FIGURE 13.4
A break point on a fragmented object. The break points of the three fragments are marked with red dots.

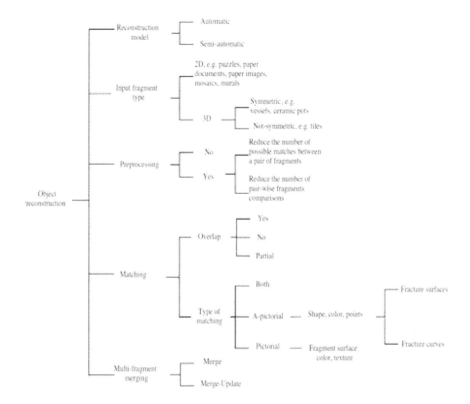

FIGURE 13.5
A schematic representation of the various steps and characteristics of object reconstruction methods and their variations.

surface, while another is to utilize information depicted on the surfaces of a fragment. Regarding the former approach, several alternatives for contour/surface comparisons have been proposed. The majority of them perform comparisons utilizing shape and or color criteria, while other utilize pixels/points from the fragments' contours/surfaces. Usually, some candidate adjacent fragments found in the previous step are discarded if the matching quality does not satisfy certain criteria [50]. The output is a (new) set of candidate adjacent pairs that are appropriately matched.

Many matching methods simultaneously estimate the alignment transformation (i.e., the appropriate affine transformation of one fragment relative to its adjacent one), in order to correctly align and merge them into one. This combined matching and alignment procedure is usually referred to as registration. If the alignment transformation is not found during matching, a separate secondary alignment step is executed. In order to identify this transformation certain criteria are maximized, while information from the matching step is also exploited (see for example [50]). Usually, the affine transformation must satisfy certain constraints (e.g., adjacent fragments must not overlap).

- **Multi-fragment merging**. In this step, the input fragments are appropriately merged in order to reconstruct the overall object. The matching of potentially adjacent fragments and their alignment are utilized to this end. The adjacent fragments that will be merged are selected, among the pairs produced during the previous steps, using heuristic criteria.

13.2.1   Fragment Preprocessing

Most algorithms involve pair-wise matchings in order to discover fragments that were adjacent in the original object. Therefore, for an object consisting of hundreds or thousands of fragments, which is usually the case in archaeological excavations, the cost of these comparisons becomes prohibitively high. Thus, one possible way to speed up the process is to reduce the number of pair-wise comparisons, either manually, by having the users proposing pairs of potentially adjacent fragments, or through more advanced methods. In [50], for example, a matching approach based on color similarity is employed in order to identify possibly adjacent image fragments. Preprocessing approaches of this type are obviously optional and may lead to reconstruction errors (e.g., two adjacent fragments in the unbroken object may be erroneously classified as not adjacent and not retained for further examination). A second approach is to reduce the matching possibilities between two fragments. This is very useful when the input fragments are three-dimensional. For example, Willis et al. utilize the break points of the 3D fragmented objects in order to speed up the matching identification process between any pair of sherds [54]. The break points of an object are the surface points where the object has broken. An example of a break point is shown in Figure 13.4. In reality, a break point on the original object is "split" among the various fragments that meet on it. Utilizing the break points of the fragments, the number of the possible contour matchings between a pair of contour curves is significantly reduced. For more details see Section 13.3. An alternative preprocessing method is introduced in [14]. There, the boundary surfaces of the input fragments are classified into original surfaces, which come from the boundary surface of the unbroken object, and fracture surfaces, which were created when the object broke. After that, only matchings over points on the fracture surfaces are investigated. Preprocessing techniques may be also employed in 2D puzzles reassembly due to the shape characteristics of puzzle pieces. For example, no matching contour segments need to be searched along a linear contour segment of a puzzle piece, since such a segment obviously belongs to the border of the puzzle [57]. Approaches of this category are beneficial for the correct reconstruction of an object, since an original surface cannot be matched with a break-surface in 3D reconstruction. However, the high degradation of some fragments, especially in archaeology, may lead to erroneous classification of original surfaces to break-surfaces and vice versa.

13.2.2 Matching of Candidate Adjacent Fragments

After the optional preprocessing step, the identification of the matching of two input fragments takes place. Matching of contour segments can be performed in two ways [20]. In certain cases, only information (e.g., color and/or shape) from the fracture contours or surfaces is considered. We will call this matching non-pictorial (or a-pictorial). On the other hand, when information (e.g., texture, color) depicted on the surfaces of a fragment is accounted to identify or disambiguate the matching of two fragments, the matching is called pictorial. To the best of our knowledge all the methods that were proposed so far for the reconstruction of 2D and 3D objects, except for some methods for 2D puzzles reassembly such as [27, 39, 57], perform non-pictorial matching. For example, in [57], the candidate adjacent puzzle piece pairs are disambiguated using pictorial criteria related to the Red, Green, Blue (RGB) colors on the surfaces of the pieces. In [39], only pictorial criteria are utilized to identify the matching of puzzle pieces. Each puzzle piece is expanded in a band around the border of the piece by predicting the pictorial information from the surface outwards, using inpainting techniques [7].

The non-pictorial matching methods employ shape (e.g., curvature, torsion) or color features of a fracture curve or surface, positions of 2D/3D points on the fracture curve or surface, or combine shape and/or color features, as well as points. In particular, methods for non-pictorial matching of 3D fragments can be further classified into those that perform matching by employing characteristics or points from the fracture surfaces and those that employ characteristics or points from the fracture curves. Examples of methods in the latter category include [2, 18, 54].

As already mentioned, some methods perform registration (concurrent mapping and alignment) of the contour or surface points, while others perform the alignment of the matched contour or surface points in a subsequent step. For example, in [54] an affine transformation is found that registers the 3D points from the fracture boundaries of two fragments, in [1], the alignment of the matched contour pixels is estimated using a brute force method, while in [50], alignment is estimated using an Iterative Closest Point (ICP) variant [5].

In order to correctly reconstruct the fragmented object, the adjacent fragments must not overlap in the reconstructed object. The methods proposed in the literature, both pictorial and non-pictorial, are divided into three categories with respect to this feature: those that do not allow the overlap of adjacent fragments, those that allow the overlap of a small percentage of the contours or outer surface segments, and those that allow the overlap of adjacent fragments. Obviously this categorization does not apply in semi-automatic object reconstruction methods (e.g., [24, 31]), where users select the correct matching among those proposed by the system, since in this case users are supposed to arrange fragments so that no overlap occurs.

13.2.3 Multi-Fragment Merging

The last step for the overall object reconstruction is multi-fragment merging. Here, the correct adjacent fragments among the candidates produced during the previous step are merged, in order to form the (partially) overall object. The proposed approaches follow either the merge or the merge-update paradigm. In the merge approach, the matched adjacent candidate fragments are iteratively merged into bigger fragments, until one object is produced. In the second category, two or more fragments are merged, forming one or more new and bigger fragments, and subsequently the reconstruction procedure is applied on the new fragments. In particular, pair-wise fragment comparisons are performed in the new set of fragments, namely those that either consist of two or more initial fragments or fragments that are not yet merged with others. The new adjacent candidate fragments are properly aligned and merged and the procedure is repeated. The procedure ends when all input fragments are merged into a single object. The advantage of this method is that some pair-wise matches have a better chance to be discovered after previously found adjacent fragments have been merged into a new bigger fragment. On the other hand, the cost of the overall reconstruction process increases significantly, especially in cases where the input fragments are relatively big and the fragments are digitized in high resolution. Figure 13.5 summarizes the basic characteristics of the overall or partial object reconstruction techniques.

13.3   Approaches for Object Reconstruction

In this section, some automatic and semi-automatic methods for object reconstruction are briefly reviewed. Table 13.1 reports the following for every method: the object reconstruction model that is employed (i.e., automatic or semi-automatic), the type of fragments that it operates on (e.g., paper documents, puzzle pieces, etc.), whether fragment preprocessing is performed or not, and how fragments are merged during the overall object reconstruction step (i.e., through a merge or a merge-update technique). The last three columns of Table 13.1 provide details regarding the properties (see Figure 13.5) of the involved matching techniques (e.g., the fragment features that are utilized, whether fragments overlap in the reconstructed object or not, etc.).

13.3.1   Torn by Hand Document Reassembly

An automatic method for the reassembly of torn by hand paper documents is proposed in [16]. This method does not virtually reassemble the entire paper document, but proposes pairs of possibly adjacent fragments and their matching. Thus, the paper fragments are not aligned, nor checked for overlapping surface segments. The discovery of the matching of two paper fragments is performed using a non-pictorial approach. The fragment contours are initially approximated through polygonal lines. Then, shape features, namely the angle and the Euclidian distance of a contour point with respect to its two neighboring contour points, are estimated from the fragments' contours and a simple point-wise matching approach is introduced; for every pair of contour points, belonging to two different fragments, the corresponding angle and distance features are compared. Each comparison is assigned a value and the sum of these values characterizes the matching. Finally, the merge-based method proposed in [8] is utilized in order to reassemble the entire paper document; in every iteration, the method merges the document fragments with the highest matching similarity, as described above.

A method that targets the same problem is proposed in [59]. A non-pictorial method identifies the matching contour segments of the paper fragments; shape features, namely turning functions [55], are estimated from every fragment contour and a turning function matching algorithm is introduced to discover matching contour segments [59]. The alignment transformation of the fragments is found during the turning function matching procedure using the method proposed in [55]. In a subsequent step, the estimated matchings are disambiguated (i.e., it is verified whether an estimated matching is correct or not) through the gradient projection method. To this end, a *global confidence score* is assigned to each matching. This score reflects the probability that a found matching is correct. The fragments whose matching has confidence score equal to one are merged and the whole procedure starts again, namely the matching contour segments are identified for all pairs of fragments and so on.

13.3.2   3D Objects Reconstruction

Willis et al. [54], in STITCH project (see Chapter 12 for more details), proposed a method for the automatic reconstruction of 3D rotationally symmetric archaeological fragmented objects (e.g., ceramic pots). The matching between every pair of input 3D fragments is identified through a non-pictorial, two-phase approach, which employs both 3D shape features (e.g., the axis of rotation and the profile of sherds) and points from the fracture curves on the outer surface of the fragments.

In the first phase, an affine transformation that aligns the points from the fracture curves of the two fragments is found. The manually identified break points (Figure 13.5) of the 3D fragments are utilized to speed up the matching process. In particular, given a pair of fragments, two of their corresponding break points, one from each fragment, are matched and an affine transformation is found that aligns the two fracture curve segments located in a small neighborhood around each break point, using a Maximum Likelihood Estimation (MLE) approach [9]. If after the alignment the fragments overlap significantly, the matching is rejected. In the second phase, the alignments are refined. This step takes into consideration the points from the fracture curves, the axis of rotation, and the profile curve (i.e., the silhouette of the pot when seen from the side) of the fragments. Each refined matching is then assigned a score. The overall object reconstruction follows the merge-update method, where pairs of fragments are heuristically selected and merged into bigger fragments.

In [18, 30], two automatic methods are proposed for the matching and the alignment of pairs of 3D, free-form, and rotationally symmetric objects, respectively. In [30], the matching is performed through a non-pictorial method whose basic concept is that, given two 3D fragments, the best fit is likely to occur at their relative position and orientation, which minimizes the point-by-point distance between their mutually visible fracture surfaces. For this reason the authors introduce an error measure for the complementary matching between two fragments at a given relative pose, based on this point-by-point distance. Simultaneous matching and alignment is performed by minimizing this measure through simulated annealing [45]. It must be noted that in the resulting matching-alignment the fragments may overlap. The same problem is considered in [18] for fragments coming from a rotationally symmetric object. Fragments that are candidates for matching are pre-aligned by aligning their axis of rotation, thus reducing the matching and alignment problem to that of finding a rotation (around the axis of rotation) and a translation value for one fragment relative to its candidate adjacent one. The best matching-alignment is found by estimating the relative pose that minimizes the point-by-point distance of the points lying on the profile curves of the fragments.

Andrews et al. [2], propose an automatic method for the reconstruction of pairs or triplets of 3D symmetric archaeological fragments. There, no fragment preprocessing is performed. The matching is found through a non-pictorial two-phase method. During the first phase, several matchings-alignments are estimated for every pair of fragments. The 3D points of the fracture curves in the outer and inner surface of the fragments as well as the axis of rotation of the fragments are utilized to this end. In the second phase, these matchings are refined using the quasi-Newton method, and evaluated according to several criteria, namely the angle formed by the fragments rotation axes, the perpendicular distance between the rotation axes, and the distance of the matched fracture curves points. After this phase, one matching is retained for every pair of fragments. The aligned fragments are not checked for overlapping. Finally, in the overall object reconstruction step, a greedy merge strategy selects pairs of fragments to form triplets.

Huang et al. [14] proposed an automatic method for the reconstruction of free-form archaeological fragmented objects. In the preprocessing step, the boundary surface of the input fragments is segmented into a set of surfaces bounded by sharp curves. These are classified into original surfaces, which come from the boundary surface of the unbroken object (and are in general smoother), and fracture surfaces, which were created when the object broke. The segmentation and classification are performed through a multi-scale edge extraction [32] and a normalized graph cut [43] algorithm.

After the preprocessing step, information extracted from the 3D points of the fracture surfaces is utilized for the matching and alignment of the fragments. In particular, several shape descriptors, namely the volume descriptor, the volume distance descriptor, and the deviation descriptor are estimated from the points lying on the fracture surfaces [36]. These descriptors are used to cluster the points lying on the fracture surfaces into feature clusters. Each feature cluster is represented by a set of (1) single-valued parameters and (2) vectors (the principal axes evaluated through principal component analysis on the x, y, z data of the cluster points) and points (e.g., the barycenter of the points in the cluster). The fracture surface matching and alignment is performed utilizing the produced feature clusters. The overall object reconstruction is performed through a merge-update

TABLE 13.1

Characteristics of the reviewed object reconstruction methods. Methods marked with "*" can reconstruct objects that consist of only two fragments. The method marked with "+" identifies only the matching contour segments of two 2D fragments. The method marked with "++" performs fragments registration in a subsequent step. Abbreviations: reg.=registration, reconstruc.= reconstruction, preproc.=preprocessing.

Paper	Reconstruc.	Application	Preproc.	Fragment merging	Type of matching	Overlap	Reg.
Justino [16]	Automatic	Paper documents	No	Merge	Non-pictorial: shape features (angle and distance of a contour pixel, with respect to its two neighboring contour pixels)	Yes	No
Zhu [59]	Automatic	Paper documents	No	Merge-Update	Non-pictorial: shape features (turning functions estimated from the contour pixels)	No	Yes
Willis [54]	Automatic	3D symmetric objects	Yes	Merge-Update	Non-pictorial: shape features (axis/profile curve of a fragment) and 3D points of the contour curve	Partially	No
Papaioannou [30]	*	3D free-form objects	No	*	Non-pictorial: 3D points of the outer surfaces	Yes	Yes
Andrews [2]	Automatic	3D symmetric objects	No	Merge	Non-pictorial: shape features (axis of rotation of a fragment) and 3D points of the inner and outer surface fracture curves	Yes	Yes
Kampel [18]	*	3D symmetric objects	No	*	Non-pictorial: 3D points of the fragments' profile curves	Yes	Yes
Huang [14]	Automatic	3D free-form objects	Yes	Merge-Update	Non-pictorial: shape features (volume descriptor, volume distance descriptor and deviation descriptor) estimated from the fracture surfaces	No	Yes
Lu [24]	Semi-automatic	3D free-form objects	No	Merge-Update	Non-pictorial: shape features (curvature and torsion) estimated from the fracture curves	No	User-def.
Papaodysseus [31]	Semi-automatic	2D objects	No	Merge-Update	Non-pictorial: pixels of the contours	No	Yes
Aminogi [1]	*	2D objects	No	*	Non-pictorial: color and shape features (curvature) estimated from the contour pixels	No	++
Leitao [8]	+	2D objects	No	+	Non-pictorial: shape features (curvature) estimated from the contour pixels	Yes	No
Yao [57]	Automatic	2D puzzle pieces	Yes	Merge	Pictorial (color on narrow strips) and non-pictorial (gap area)	Yes	Yes
Sagiroglu [39]	Automatic	2D puzzle pieces	No	Merge-Update	Pictorial (color and texture)	No	Yes
Nielsen [27]	Automatic	2D puzzle pieces	No	Merge	Pictorial (color)	Yes	No

approach. In particular, the matches and the relative alignment of the fragments are modeled by a graph, [15]. Using a greedy approach, this graph is partitioned and the fragments of each resulting sub-graph are merged into one fragment. After that, the entire reconstruction procedure is repeated using the new set of fragments.

In [24], a human-supervised collaborative reconstruction system is described. The aim of the system is to propose a potential matching between any pair of input fragments. The matching is found by integrating shape features (curvature and torsion) estimated from all 3D points in the fracture curves of the fragments under a cyclic distance algorithm [34]. Then the users select to merge or not the proposed fragments. The fragment alignment is performed interactively (in a VRML environment) by the users, while the object reconstruction procedure follows the merge-update paradigm.

13.3.3 2D Puzzles Reassembly

In [57], a method for automatic 2D jigsaw puzzles reassembly is proposed. This three-step method pre-processes input puzzle pieces in order to limit the matching possibilities between pairs of pieces. In particular, each piece is classified into three categories: corner, boundary, or interior. After that, puzzle piece matching and alignment is performed, in order to identify the correspondences of the pixels on the contours of the puzzle pieces. The matching criterion is to minimize the gap area formed after the pieces' alignment. The best adjacent candidates (those that are best matched and aligned) are further disambiguated using pictorial criteria. In order to judge the quality of the matching of two pieces along one of their sides, the following statistical measure is being evaluated on two narrow strips Φ_1, Φ_2 (one on each piece) that extend from each side towards the interior of the piece:

$$1 - \frac{\sigma_b^2}{\sigma_T^2} \tag{13.1}$$

where

$$\sigma_b^2 = n_1(\bar{P}_1 - \bar{P}_m)^2 + n_2(\bar{P}_2 - \bar{P}_m)^2 \tag{13.2}$$

n_1, n_2 are the number of pixels in regions Φ_1, Φ_2 respectively, $\bar{P}_1$, $\bar{P}_2$, $\bar{P}_m$ are the average R, G, or B color values in Φ_1, Φ_2 and $\Phi_1 \bigcup \Phi_2$, and σ_T^2 is the R, G, or B color channel variance in $\Phi_1 \bigcup \Phi_2$. This measure (taken from [29], which utilizes $\frac{\sigma_b^2}{\sigma_T^2}$) takes values close to one if Φ_1, Φ_2 are very similar and can be merged and close to zero when they are very dissimilar. Only adjacent candidates that result in large (close to one) values of this statistic are retained. In the final step, the puzzle is reassembled using a greedy-like approach. One piece at a time is added to the puzzle. This piece is the best adjacent candidate of one or more pieces added in the previous iteration.

In [39], an automatic approach that exploits only pictorial information for the matching of puzzle pieces is proposed. More precisely, each puzzle piece is expanded in a band around the border of the piece by predicting the pictorial information in this band. Inpainting techniques are used for this task. The method does not perform pair-wise comparisons in order to identify matches [7]. Instead, a global puzzle reassembly procedure takes place. Local mean and variance values are estimated for every pixel of the original and the predicted part of the piece, and a feature image is created by assigning to each pixel a 2-D vector containing these values. Feature images are registered using a correlation maximization approach based on Fast Fourier Transform (FFT). Each registration is assigned a score equal to the correlation between the predicted parts of one piece and the parts (either predicted or original) of the other piece. In this registration the puzzle pieces must not overlap. Each time the best registered- with respect to pieces already added in the puzzle- piece is added.

In [27], a method that makes use of color features is proposed. The method relies on the observation that if two puzzle pieces are adjacent in the puzzle, their pictorial information in an area very close to the matching contours of the two puzzle pieces is approximately the same. The overall puzzle reassembly is performed using the merge-like approaches proposed in [13, 56].

13.3.4   2D Objects Reassembly

In [31], a system for the semi-automatic reconstruction of the murals of the Greek island of Thera is described. The mural fragments are photographed and so the problem of mural reconstruction is mapped to the 2D space. Given a pair of fragments having a certain relative orientation, the system estimates a possible matching (i.e., contour pixels correspondences) using only the pixels from the fragments' contours. This matching is evaluated according to several criteria such as the number of overlapping pixels or the number of pixels lying in the gap after aligning the fragments. The latter is related to the translation of the fragments, since their relative orientation is considered fixed. After that, their relative orientation changes by a small amount and the procedure described above is repeated. The best matching is the one which optimizes the criteria previously described. The user selects the matches-alignments that will be retained and merges the corresponding fragments. The new fragments are added in the set and the procedure is repeated from the beginning.

In [8], an automatic, time-efficient non-pictorial matching approach is presented. The fragments' contours are initially sampled in a coarse scale. The curvature of the coarse contours is estimated and a dynamic programming algorithm is employed in order to identify contour pixel correspondences. Each pair of matching contour segments is assigned a two-term quadratic mismatch cost. The first term estimates the total difference between the curvatures, while the second one is used to penalize matches that are too irregular. A pair of contour segments are considered to match if this mismatch cost is less than a user-defined threshold. The pairs of candidate matching segments are mapped back to the original contours, which are sampled using a finer scale this time, and the dynamic programming technique is employed again.

In [1], the matching of the image fragments is based on both the shape and the color characteristics of the fragments contours. One contour pixel sequence is overlaid on another one and, for each such placement, the curvature and color differences of the corresponding contour pixels are estimated. If their total sum is less than a user-defined threshold, the contour segments are considered to match. After that, their alignment is found using a brute force approach that searches all possible alignment angles between the two fragments.

As it is obvious, the literature on 2D object reassembly is rather limited. A novel contribution towards this direction will be presented in the next section.

13.4   Automatic Color-Based Reassembly of Fragmented Images and Paintings

In this section a novel automatic four-step method for the reassembly of torn 2D paper images is presented. The identification of possibly adjacent image fragments is the first step. It is considered that two fragments were adjacent in the original image if the color similarity of the depictions on their outer surfaces is high. As a result of this step, for each image fragment, the L fragments having the highest color similarity with it are selected. Subsequently, the matching contour segments for every pair of potentially adjacent input fragments are identified. The matching algorithm is non-pictorial and utilizes color information within a dynamic-programming framework [46]. At the end of the second step, for each fragment, its matching contour segments with K other fragments are retained. The third step aims to align each fragment with respect to its adjacent ones along their matching contour segments, using a variant of the ICP algorithm. Once the matching contour segments of pairs of image fragments are identified and aligned, the reassembly of the overall image takes place. To this end, a novel feature, namely the alignment angles found during the previous step, is introduced and a merge-like paradigm is employed. The main steps of the proposed method are shown in Figure 13.6.

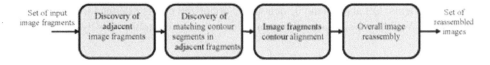

FIGURE 13.6
The steps of the automatic color-based paper image reassembly approach proposed in [50].

13.4.1 Discovery of Adjacent Image Fragments

Let $\mathcal{F} = \{f_1, f_2, \ldots, f_N\}$ be the set of N image fragments. The aim of this step is to speed up the overall image reassembly process by reducing the number of pair-wise fragment comparisons that take place in the second step. It is assumed that the color similarity of the depictions on the outer surfaces of the fragments that were adjacent in the original image is high. Thus, a matching approach based on color similarity is used to identify spatially adjacent image fragments. In particular, color quantization is performed on every image fragment f_k using the Gretag Macbeth Color Checker palette [11]. This palette consists of twenty-four colors chosen so as to represent a variety of naturally occurring colors. After performing color quantization, the normalized color histograms of the image fragments in $\mathcal{F}$ are estimated. These normalized color histograms are compared with the histogram intersection similarity measure [11]:

$$d_{HI}(h_k, h_l) = \sum_{i=1}^{24} \min(h_k(i), h_l(i))(1 - |h_k(i) - h_l(i)|) \tag{13.3}$$

where h_k and h_l denote the normalized color histograms extracted from image fragments f_k and f_l, respectively. $h_k(i)$ denotes the value of the i-th bin of h_k and is defined as the number of pixels having color i divided by the total number of pixels. Once this step is finished, a list of the L most chromatically similar fragments is retained for every image fragment. It must be emphasized that this step is optional and can be skipped, especially if the number of image fragments is rather small.

13.4.2 Discovery of Matching Contour Segments of Adjacent Image Fragments

In this step, a novel algorithm is utilized for identifying the matching contour segments of pairs of input fragments. Note that this step is performed only for the fragment pairs identified in the previous step or for all pairs of fragments, if the first step has been skipped. In both cases, we denote the set of image fragment pairs by $\mathcal{E}$. The utilized fragment matching algorithm is non-pictorial, since it utilizes information exclusively regarding the color of fragments' contours.

A color quantization preprocessing step is initially performed on the contour pixels of the image fragments, in order to avoid comparing directly pixel colors that may contain noise. Kohonen Neural Networks (KNNs) are employed for this purpose [4]. KNNs consist of an input and an output node layer. In the input layer, the number of nodes equals the dimension of input vectors, while the number of nodes in the output layer is equal to the number of produced clusters.

In order to perform color quantization of the contour pixels of the image fragments by means of the KNNs, the following steps take place. Initially, a number of N_p pixels is randomly sampled from the image fragments in $\mathcal{E}$ and mapped to the La^*b^* color space. N_p is empirically set to 25% of the total image fragment pixels [50]. After that, a $[3 \times C]$ KNN is defined, where 3 corresponds to the dimension of the input La^*b^* space and C to the predefined number of color clusters, and the KNN iterative learning procedure is applied on the subset of N_p pixels. After training the network, the color of every contour pixel is represented by its color cluster label c_j, $j = \{1, \ldots, C\}$.

Let $U = [u_i]_{i=1}^n$ and $V = [v_j]_{j=1}^m$ be two sequences that correspond to the contour pixels of two different image fragments. Let also a_i and b_j denote the quantized colors assigned to the pixels u_i and v_j, respectively. We define a similarity function F between two contour pixels by

$$F_{u_i,v_j}[a_i, b_j] = \begin{cases} e > 0, & a_i = b_j \\ d < 0, & a_i \neq b_j \end{cases} \tag{13.4}$$

Our goal is to find a contour pixel mapping function Φ, given U and V, that satisfies the following conditions:

- $\Phi[u_i] = v_k$ and $\Phi[u_{i+1}] = v_l, \forall i, k \leq l \leq m$.

- $\Phi[u_i] \neq \emptyset$.

The first condition implies that more than one contour pixels in U can be mapped to the same contour pixel in V. However, pixels in contour U that are mapped in the same pixel in V must be strictly consecutive. This condition guarantees that no "folding" of the U contour pixel sequence will occur during its matching with V. The second condition ensures that every contour pixel in U is mapped to a contour pixel in V. The algorithm that is used to identify the mapping function Φ is a variant of the Smith Waterman dynamic programming algorithm [46] . Each mapping between segments of input sequences is assigned a score $S > 0$. Large values of S imply high color similarity. A $n \times m$ similarity matrix $\mathbf{H}$ is set up, where the ith row of matrix $\mathbf{H}$ corresponds to u_i, while the jth column corresponds to v_j. $H_{i,j}$ denotes the best mapping score among the pairs of all sub-sequences ending at pixels u_i and v_j. The algorithm gradually fills matrix $\mathbf{H}$ and forms the mapping function Φ. During filling, each matrix cell is assigned the highest possible value, as our purpose is to maximize the mapping score S.

Being a dynamic programming algorithm, the solution to an instance of the problem is given in terms of solutions to its smaller sub-instances. Thus, matrix $\mathbf{H}$ is recursively filled-up using the formula

$$H_{i,j} = \max \left\{ H_{i-1,j-1} + F_{u_i,v_j}(a_i, b_j), H_{i-1,j} + g, H_{i,j-1} + g, 0 \right\}, \tag{13.5}$$

where $g < 0$.

Let Φ be the estimated mapping of the $[u_k]_{k=1}^{i-1}$ and $[v_l]_{l=1}^{j-1}$ subsequences. If $H_{i,j} = H_{i-1,j-1} + F_{u_i,v_j}[a_i, b_j]$, then the relation $\Phi[u_i] = v_j$ is appended to Φ. On the other hand, if $H_{i,j} = H_{i,j-1} + g$ or $H_{i,j} = H_{i-1,j} + g$ is selected, then $\Phi[u_i] = v_{j-1}$ or $\Phi[u_{i-1}] = v_j$ is appended to Φ, respectively. A *gap* is formed when either of the last two cases occur. Generally, a gap is formed when one or more contour pixels of the first fragment are mapped to the same contour pixel of another fragment. The percentage of the mapping gaps is a measure of the dissimilarity between the input contour segments, i.e., a high gap percentage reveals contours with many dissimilarities. The parameter g is negative, in order to penalize mappings with many gaps.

When this stage is completed, we identify an area in $\mathbf{H}$ with high similarity values H_{ij}. Diagonal patterns are preferred, since they correspond to mappings without many gaps. Let H_{e_1,e_2} and H_{s_1,s_2} be the lowest right and highest left borders of this area, respectively. In the implemented modification, H_{e_1,e_2} is the maximum value of $\mathbf{H}$, while $H_{s_1,s_2} = 0$. Then we select the mapping starting from (s_1, s_2) and ending at (e_1, e_2), which is formed during filling $\mathbf{H}$. In summary, the steps in order to identify the matching contour segments of two image fragments are the following:

- Let $U = [u_i]_{i=1}^n$ and $V = [v_j]_{j=1}^m$, be the corresponding contour pixel sequences of two fragments. Quantize the contour pixel colors of the two image fragments using a KNN (note that this step is performed only once for all image fragments in $\mathcal{E}$). Let $[a_i]_{i=1}^n$ and $[b_j]_{j=1}^m$ be the quantized colors sequences assigned to U and V, respectively.

- Initialize $H_{i,j}$ according to $F_{u_i,v_j}[a_i, b_j]$.

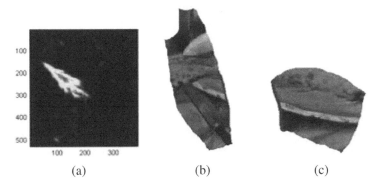

(a)   (b)   (c)

FIGURE 13.7 (SEE COLOR INSERT)
Similarity matrix (a) for the contour sequences of two fragments (b), (c). The correctly matched contour segments are shown in blue in (b), (c).

- Fill **H** according to Equation (13.5).

- Select an area in matrix **H**, as previously discussed. Let H_{e_1,e_2} and H_{s_1,s_2} be the lowest right and highest left borders of this area. Select the mapping starting from (s_1, s_2) and ending at (e_1, e_2).

Details on how to select e, d, and g are presented in [50]. An example of a similarity matrix is presented in Figure 13.7(a). The horizontal axis of the matrix corresponds to the quantized colors of the contour pixels sequence of the image shown in Figure 13.7(b), while the columns correspond to the quantized colors of the contour pixels sequence of the image shown in Figure 13.7(c). High values correspond to contour segments with a high degree of similarity (deep red colored areas). The correctly matched contour sequences are shown with blue color in Figures 13.7(b) and 13.7(c). At the end of the second step, for each image fragment f we retain one matching contour segment with K $(0 < K \leq L)$ other image fragments, producing the set of *true* pairs of adjacent fragments $\{(f, f_i), i = 1, \ldots, K\}$.

13.4.3   Contour Alignment of Fragments

Since the previous step only identifies the matching of the contour segments of image fragments, their alignment transformation must be found. A popular algorithm that is used for the registration of two point sets is the ICP algorithm [5]. The ICP starts with two point sets (in our case pixels from the two contour segments) and an initial guess of their relative rigid body geometrical transformation. It then refines the transformation parameters by iteratively generating pairs of point correspondences and minimizing the sum of the L_2 distances between points from the two point sets. The original form of the ICP algorithm is not robust to outliers coming from errors during the contour segment matching. These outliers may create serious problems to alignment, if not properly handled.

In order to develop a robust contour alignment procedure, three popular, robust to outliers, variants of ICP, namely [6, 42, 49], have being evaluated. The evaluation of these variants was done by applying them for the alignment of correctly matched contour segments (i.e., contour segments that really match in the original image). These contour matchings may contain misidentified pixels or erroneously non-identified contour pixels. The experiments detailed in [50] have shown that the ICP variant proposed in [42], called ICPIF (ICP using Invariant Features), outperforms the other two variants. ICPIF introduces new features, such as the second order moments and the spherical harmonics, in order to improve the correspondence selection. After the computation of the above features for every point of the two contour segments, the L_2 distance, computed during the second step of ICP, is replaced by the weighted sum of the L_2 distance of the two point sets and distances based on the introduced features. The goal is to estimate point correspondences that are not only based on the Euclidean distance but also incorporate shape invariant features.

13.4.4 Overall Image Reassembly

Once the true adjacent image fragments pairs and their matching contour segments are identified and properly aligned, the final step is the reassembly of the overall image. Approaches that utilize only the mapping scores, such as [22], are inadequate for producing the correct overall image, since the mapping scores of the matching contour segments are not always well correlated with the true ones. Thus, a novel solution has been developed, which is based on the alignment angle that best aligns the matching contour segments of two image fragments. This approach follows the merge paradigm.

Consider three image fragments f_i, f_j, and f_k, where f_i has a matching contour segment with f_j and f_k. We denote by θ_j the rotation angle by which the individual fragment f_j must be rotated, in order to be correctly placed inside the overall reassembled image. The alignment angle, by which we must rotate fragment f_i to align it with the matching contour segment of fragment f_j (before fragment f_j is rotated by θ_j), is denoted by θ_{ij}. In order to align fragments f_i and f_j with respect to each other and place them correctly in the reassembled image, the following steps must be performed:

1. Rotate fragment f_j by θ_j to correctly orient it in the assembled image.

2. Rotate fragment f_i by $\theta_{ij} + \theta_j$ to correctly align its matching contour segment with the corresponding matching contour segment of fragment f_j.

This procedure will simultaneously align fragment f_i with fragment f_j and provide its correct orientation inside the entire image. We can then state that the matching contour segments of pairs (f_i, f_j) and (f_i, f_k) are *compatible*, if and only if:

$$\theta_{ij} + \theta_j = \theta_{ik} + \theta_k \tag{13.6}$$

Following this rationale, it is considered that an image is fully reconstructed if all its matching contour segments are compatible according to Equation (13.6); this image is called *valid*.

Based on Equation (13.6), the so-called relative alignment angle is defined, which will direct the multiple fragments merging procedure. The relative alignment angle ϕ_i^j of a fragment f_i with respect to a fragment f_j is evaluated by the formula

$$\phi_i^j = \theta_{ij} + \theta_j, \tag{13.7}$$

where θ_{ij} and θ_j are defined as above. It is clear that for a set of fragments placed in the reassembled image and a fragment f_i having matching contour segments with a subset of the above image fragments (i.e., $I = \{f_1, f_2, \ldots, f_n\}$) the above matching contour segments are compatible if

$$\phi_i^1 = \phi_i^2 = \phi_i^3 = \ldots = \phi_i^n. \tag{13.8}$$

Consequently, if a new fragment i is to be merged with a partially reassembled image that consists of some image fragments, then the number of the new valid partially reassembled images that will be generated is equal to the cardinality r of the relative alignment angle set, $\{\phi_i^l, l = 1, \ldots, r\}$. We have developed a reassembly algorithm that utilizes the aforementioned assumptions to produce a number of possible reassembled images, all being potential solutions to the reassembly problem. The algorithm is initialized by taking M pairs of input fragments that have the highest mapping scores for the corresponding matching contour segments, where $M \leq N * K$ and N is the number of total input image fragments. These pairs are reassembled (merged into one object) to produce a number of M partially reassembled (two-fragment) images that are further extended by inserting

	Reassembly performance	Computation time
Without first stage	63.49%	7.58 min
Including first stage	42.15%	3.42 min
Human	-	17.37 min

TABLE 13.2
Reassembly performance and computational complexity of the proposed method.

one fragment each time. The selection of the initial M fragment pairs is crucial for the algorithm performance, since the involvement of an erroneous input pair would inevitably lead to a wrong image reconstruction. The reliability of the proposed initialization is both intuitively expected and experimentally proven.

The set of M partially reassembled images is iteratively updated in order to include further image fragments. At every step and for every partial image (initially consisting of two image fragments), the fragment having the maximum mapping score with an image fragment that belongs in the partially reconstructed image is added. After computing the relative alignment angles of this fragment with respect to the partially reconstructed images, a number of images, equal to the cardinality of these relative alignment angles, are reassembled for each partial image. Those images compose the new set of partially reassembled images (replacing the previous ones) that will be included in the next iteration. Consequently, in the next step a new fragment will be involved in the update procedure of these images. This iterative algorithm terminates when no more image fragments are left to be inserted into the reassembled images.

The proposed method differs from others found in the literature, as it is based on the alignment angles. This fact renders it sensitive to errors in the estimation of these angles. However, this drawback is ameliorated, by setting a threshold when comparing the relative alignment angles. Thus, two relative alignment angles ϕ_1 and ϕ_2 are regarded to be equal if $|\phi_1 - \phi_2| < \epsilon$. Another "backup" approach is to add manual intervention to the system, and ask the user to select the correct alignment transformation.

13.4.5 Image Reassembly Experiments

The performance of the proposed method was evaluated using 70 paper image prints. Each print had size 25 cm × 20 cm and was torn by hand into $N = 20$ pieces that were scanned at 300 dpi resolution, in order to create a challenging image fragments dataset. An example is shown in Figure 13.8. Experiments were conducted in a quad-core PC with 4GB RAM and the algorithm has been implemented in C++. For the KNN color quantization in the second step, the number N_p of sampled image pixels was set equal to 25% of the total pixels. The utilized Kohonen nets resulted into $C = 50$ color clusters, while the learning procedure took 750 epochs. Nodes in the output layer of the neural network were organized under a random lattice. The parameters of the Smith Waterman algorithm were set to $e = 1$, $d = -0.5$, and $g = -0.5$. At the end of the second step, for each image fragment, matching contour segments with $K = 10$ fragments were retained.

The performance of the method was estimated as follows. Let I be a manually reassembled image and $\mathcal{I} = \{I_1, I_2, \ldots, I_n\}$ be the set of automatically reassembled images generated by the algorithm for this image. The reassembly is defined to be correct when there is a $\tilde{k} \in \{1, \ldots, n\}$ for which $I_{\tilde{k}}$ is declared similar to I by a human observer. In Table 13.2, the reassembly performance, namely the percentage of the test images that were correctly reassembled according to the above definition, and the computational time of the presented approach, for the 70 test images are shown. The last row shows the mean manual reassembly time (the accuracy of manual reassembly was not measured). Eight people of ages between 23 and 31 participated in this experiment. The average time needed for each person to correctly reassemble each fragmented image was evaluated. As expected, the overall

FIGURE 13.8
A torn by hand image used in the experiments.

performance is higher when the first stage is omitted. However, in this case the computation cost is higher, since both the second and the third step are executed for every pair of input image fragments. Table 13.2 shows that the choice whether to employ the first step on the reassembly process depends on the computational cost that is associated within the input image fragments. As the amount of image fragments increases, the higher overall performance of the variation that omits the first step gives way to the higher computational cost that is associated with it. Figure 13.9 shows correctly aligned couples of image fragments produced during the second and the third steps of the proposed algorithm. The image resulting from the reassembly of fragments shown in Figure 13.8 is shown in Figure 13.10. It can be seen that its reconstruction is nearly perfect. The white region in the middle of the reassembled image is due to missing pieces of paper that were not scanned.

In the future, we plan to further improve the performance of the proposed method. Improvements can be introduced in each step of the method. For example, the procedure of the discovery of the spatial adjacent image fragments may be improved by employing not only color but also textual or semantic features. Another possible extension is to utilize both the color and the shape of the fragments contours in order to perform matching. The evaluation of the proposed method in a real-world fragmented image database, such as a database with archaeological data, is worth exploring.

13.5 Reduced Complexity Image Mosaicing Utilizing Spanning Trees

As already mentioned, the mosaicing process is comprised of two steps. The first step involves the estimation of the optimal transformation (translation, rotation, scaling, etc.) that aligns each

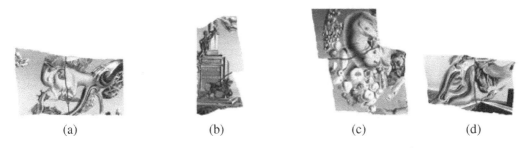

(a) (b) (c) (d)

FIGURE 13.9
New fragments created by aligning and assembling original fragments along their matching contour segments.

FIGURE 13.10
The result of reassembling the image depicted in Figure 13.8.

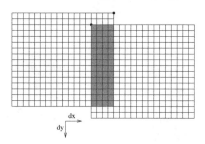

FIGURE 13.11
Two neighboring sub-images and the associated overlap region $W(\mathbf{d})$ (which is identified by the gray area) for a displacement of $\mathbf{d} = [-4\ 2]^T$.

sub-image with respect to each neighboring one, whereas the second step utilizes displacement information found in the previous step in order to combine each pair of neighboring sub-images with invisible seams and reconstruct the entire image.

In this section we will describe two methods that aim to reduce the number of computations required for the evaluation of the sub-image displacements, without significantly affecting the matching error. It is important to note that, despite the fact that the methods are illustrated for the particular case where sub-images are only translated with respect to each other, the proposed matching methodology can be easily applied to more complex situations (e.g., cases that involve camera rotation or zooming during the acquisition of the sub-images). Before dealing with the general case of mosaicing an arbitrary number of sub-images, the case of two images will be studied in the next subsection, since it provides significant insight to the problem.

13.5.1   Two-Image Mosaicing

Let us assume that the displacement vector $\mathbf{d}$ is constrained to take values in the following set:

$$\mathbf{d} \in \{[d_1\ d_2]^T :\ d_i \in \{d_{i_{\min}}, \ldots, d_{i_{\max}}\},\ \ i = 1, 2\}. \tag{13.9}$$

If $I_j(\mathbf{n})$, $j = 1,\ 2$, denotes the intensity of the j-th image at pixel coordinates $\mathbf{n} = [n_1\ n_2]^T \in W(\mathbf{d})$, where $W(\mathbf{d})$ denotes the area where the two images overlap (Figure 13.11), then the matching error $E(\mathbf{d})$ associated with a specific displacement $\mathbf{d}$, can be expressed as follows:

$$E(\mathbf{d}) = \frac{\displaystyle\sum_{\mathbf{n} \in W(\mathbf{d})} |I_1(\mathbf{n}) - I_2(\mathbf{n})|^p}{\|W(\mathbf{d})\|} \tag{13.10}$$

where $\|W(\mathbf{d})\|$ denotes the number of pixels in the overlap area $W(\mathbf{d})$. For $p = 1, 2$ Equation (13.10) expresses the Matching Mean Absolute Error (MMAE) and the Matching Mean Square Error (MMSE), respectively. Subsequently, the optimal displacement value $\mathbf{d}_{opt}$ can be estimated through the following minimization:

$$\mathbf{d}_{opt} = \underset{\mathbf{d}}{\arg\min} E(\mathbf{d}). \tag{13.11}$$

From (13.9) and (13.11) it is obvious that this minimization process requires the evaluation of (13.10) over all possible values of $\mathbf{d}$. Block matching techniques such as 2D logarithmic search, three-point search, and conjugate gradient search [26] can be employed in order to reduce the computational cost associated with the exhaustive minimization procedure of Equation (13.11). These procedures may provide estimates $\hat{\mathbf{d}}_{opt}$ of the optimal displacement value $\mathbf{d}_{opt}$. In our simulations, the 2D logarithmic search was employed. With respect to the error metric, experiments showed that similar results were obtained by using either the MMAE or the MMSE criterion. Thus, MMAE was used since it is faster to compute.

13.5.2 Spanning Tree Mosaicing

If $M_1 \times M_2$ sub-images are to be mosaiced, a displacement matrix $\mathbf{D}$ is involved. This matrix plays a role similar to that of the displacement vector $\mathbf{d}$ mentioned in the previous subsection. The $2M_1M_2 - M_1 - M_2$ columns of $\mathbf{D}$ are two-dimensional vectors, each corresponding to a displacement value between two neighboring sub-images. An expression for the matching quality, similar to the two-image case, can be derived in the multiple image case, by substituting $\mathbf{d}$ with $\mathbf{D}$ in Equation (13.10) and extending the summation over all neighboring images. The optimal value $\mathbf{D}_{opt}$ of the displacement matrix can then be derived as follows:

$$\mathbf{D}_{opt} = \underset{\mathbf{D}}{\operatorname{argmin}} E(\mathbf{D}). \tag{13.12}$$

The minimization in the equation above involves prohibitive computational complexity, since in this case a much larger search space is involved. Indeed, if each of the column vectors in $\mathbf{D}$ is of the form (13.9), $\mathbf{D}$ may assume $((d_{1_{\max}} - d_{1_{\min}} + 1)(d_{2_{\max}} - d_{2_{\min}} + 1))^{2M_1M_2 - M_1 - M_2}$ different values. Thus, computational complexity increases exponentially with respect to the number of sub-images. Moreover, calculation of $E(\mathbf{D})$ poses additional computational problems, since the overlap area W is now a multi-dimensional set.

In order to avoid exhaustive matching, certain constraints can be imposed on the way images are matched. Indeed, a faster method may be devised by performing simple matches only (i.e., matches between an image and one of its neighbors). The proposed method can be illustrated with the aid of a mosaicing example. In Figure 13.12(a) a mosaic of $M_1 = 2$ by $M_2 = 2$ sub-images is depicted. If one associates each image with a graph node and each local matching of two sub-images with an edge, the mosaicing of the four images can be represented by the graph of Figure 13.12(b). Computation of $\mathbf{D}_{opt}$ requires an exhaustive search in 8-dimensional space. To reduce this complexity, the entire mosaicing process is decomposed into simpler steps each involving the mosaicing of two images at a time. Spanning trees can provide a representation of the possible mosaicing procedures, under this constraint. Figures 13.12(c)-(f) illustrate the four different spanning trees that correspond to the graph of Figure 13.12(b). For example, in the case depicted in Figure 13.12(d) three pairwise image matches should be performed: image A to C, D to C, and D to B, while Figures 13.12(c), 13.12(e), and 13.12(f) illustrate the other three possible mosaicing procedures. The final image is the one obtained by the procedure which results in the smallest matching error. Obviously, this procedure is sub-optimal but offers a significant decrease in computational complexity. The number of spanning trees that correspond to a certain graph can be calculated by the matrix-tree Theorem [3]:

Let G be a non-trivial graph with adjacency array $\mathbf{A}$ and degree array $\mathbf{C}$. The number of the discrete spanning trees of G is equal with each cofactor of array $\mathbf{C} - \mathbf{A}$.

Both $\mathbf{A}$ and $\mathbf{C}$ are matrices of size $(M_1M_2) \times (M_1M_2)$. If node v_i is adjacent to node v_j, $\mathbf{A}(i,j) = 1$, otherwise $\mathbf{A}(i,j) = 0$. Additionally, the degree matrix is of the form:

$$\mathbf{C} = \operatorname{diag}(d(v_1), \dots, d(v_{M_1M_2})) \tag{13.13}$$

where $d(v_i)$ denotes the number of nodes adjacent to v_i. The number of trees increases rapidly with the grid size. For example the numbers of spanning trees for grids of size 3×3, 4×4, and 5×5 are 192, 100, 352, and 5.6×10^8 respectively. The proposed spanning tree mosaicing (STM) procedure is outlined below:

1. For each pair of neighboring images calculate the optimal displacement and the associated matching error.

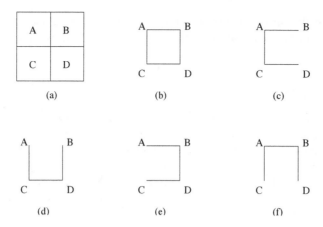

FIGURE 13.12
(a) A mosaic of $M_1 = 2$ by $M_2 = 2$ sub-images, (b) the corresponding graph, (c)-(f) the four possible spanning trees.

2. For each spanning tree that is associated with the specific graph, calculate the corresponding total matching error, by summing the local matching errors which are associated with the two-image matches represented by the given tree.

3. Select the tree associated with the smallest total matching error.

4. Perform mosaicing of two images at a time, following a path on the selected spanning tree.

It should be clarified once again that sub-optimal results are obtained by the STM procedure (i.e., only an approximation $\hat{\mathbf{D}}_{opt}$ of the optimal matrix is computed). However, this is compensated for by the significant speed gains provided by the algorithm.

13.5.3   Sub-Graph Spanning Tree Mosaicing

As already mentioned, the number of trees grows very fast with respect to grid size and thus for large values of M_1 and M_2, STM becomes computationally demanding. The second approach that is proposed, namely the *Sub-graph STM (SGSTM)* may, partially, address this issue. In SGSTM, a graph is partitioned into sub-graphs by splitting the original graph vertically and/or horizontally. A sample partitioning of this type is depicted in Figure 13.13. By splitting the graph vertically and then horizontally, four sub-graphs result. STM can be applied separately to each one of the four sub-graphs of Figure 13.13(d). It can be shown that in this case a total of 194 spanning trees should be examined. Since four images will be produced by the STM process (one for each sub-graph), a further STM step will be required, in order to mosaic these four images into the final image. Thus, 4 more trees should be added to the trees examined in the previous step, for a total of 198 trees. In contrast, an STM of the original image set would require matching error calculations for 100,352 trees.

If the original graph is of size $M \times M$ ($M = 2^\nu$), the image can be gradually mosaiced by decomposing the original graph into an appropriate number of 2×2 sub-graphs, performing STM on each one, decomposing once more the resulting $\frac{M}{2} \times \frac{M}{2}$ graphs and so on, until one image emerges. After mosaicing a partition's sub-graphs, new displacement matrices that correspond to the resulting sub-images should be calculated. The number of 2×2 graphs in this procedure is equal to $\frac{M}{2}\frac{M}{2} + \frac{M}{4}\frac{M}{4} + \ldots + 1 = \frac{M^2 - 1}{3}$. Since four spanning trees exist for a 2×2 graph, the matching error

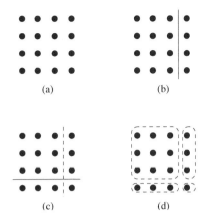

FIGURE 13.13
(a) A graph corresponding to an $M_1 = 4$ by $M_2 = 4$ image mosaic. (b) Vertical split of (a). (c) Horizontal split of (b). (d) Resulting partition of graph.

Method	Error	Time
STM	27.41	2.19
SGSTM	29.60	1.95

TABLE 13.3
STM and SGSTM performance results for image set 1.

of $4\frac{M^2-1}{3}$ trees should be evaluated. Theoretical speedup values of SGSTM over STM are depicted in Figure 13.14. It is obvious that SGSTM introduces large computational savings over the STM approach. Obviously, the quality of SGSTM mosaicing may be inferior to the one provided by STM, since SGSTM involves the examination of a significantly smaller number of possible sub-image matches. However, SGSTM can be utilized for fast visualization of mosaicing results (e.g., mosaic previews).

13.5.4 Experimental Evaluation

Simulations were carried out in order to assess the performance of the proposed methods, on several image sets. In the following, comments and results are presented for two of these sets. The first set consisted of the 6 sub-images of Figure 13.2, which were arranged in $M_1 = 3$ rows of $M_2 = 2$ sub-images, each having resolution of 238×318 pixels. For each one of the $2M_1M_2 - M_1 - M_2 = 7$ pairs of neighboring sub-images, matching errors were calculated using the MMAE criterion. Subsequently, for each one of the 15 spanning trees that correspond to this 2×3 node graph, the total matching error was calculated. The reconstructed images for both STM and SGSTM are depicted in Figure 13.15. The total time required to find all spanning trees that correspond to the given graph, calculate optimal neighboring image displacements, and output the overall matching error of each tree (for both STM and SGSTM) is tabulated in Table 13.3, in seconds. The corresponding matching errors are also presented in this table. In the case of SGSTM, the original 3×2 graph was partitioned into four sub-graphs. Two of them had sizes 2×1, while the last two consisted of one node each. In the SGSTM case, matching error calculation is required for 6 trees, instead of the 15 trees of the STM case. However, the approximately 2.5 theoretical speedup factor of SGSTM over STM is not attained, because quoted time figures take into account the amount of time required to re-compute

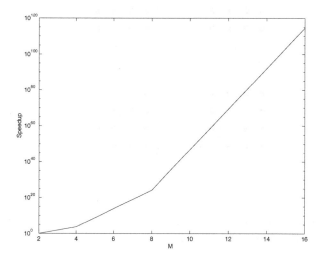

FIGURE 13.14
Theoretical speedup of SGSTM over STM, for graphs of size $M \times M$ ($M = 2^\nu$).

(a)   (b)

FIGURE 13.15
(a) Spanning tree mosaicing (STM) reconstructed image. (b) Reconstructed image after sub-graph spanning tree mosaicing (SGSTM).

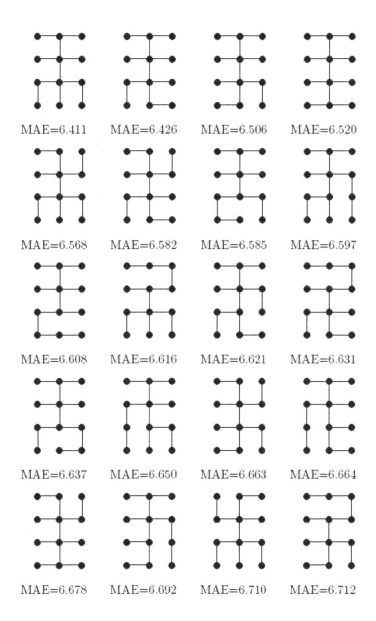

FIGURE 13.16
The twenty spanning trees that produced the lowest MMAE figures.

Method	Error	Time
STM	6.4	782.7
SGSTM	9.7	68.9

TABLE 13.4
STM and SGSTM performance results for image set 2.

displacement matrices. Due to this overhead a speedup factor of 1.12 was recorded. The second set consisted of 12 sub-images, which were arranged on a grid of $M_1 = 4$ rows of $M_2 = 3$ sub-images each. Each sub-image had a resolution of 951×951 pixels. For this graph 2,415 spanning trees exist. Similar to the previous set, for each one of the $2M_1M_2 - M_1 - M_2 = 17$ pairs of neighboring sub-images, matching errors were calculated and the 2D logarithmic search was utilized in order to obtain the optimal displacement. Subsequently, for each one of the 2,415 spanning trees, the total matching error was calculated. The total time that was required to find all spanning trees, calculate optimal neighboring image displacements, and output the overall matching error of each tree in this case is presented in Table 13.4. In the case of SGSTM, the graph was decomposed into four sub-graphs: two of size 2×2 and two of size 2×1, which required the calculation of the total error for 14 trees, compared to the 2,415 of STM. It is evident that, in this image set, SGSTM is more than an order of magnitude faster than STM. More specifically, the speedup factor provided by the SGSTM method over the STM method was 11.4. Of course, this speedup was accompanied by a significant increase in the matching error. The twenty spanning trees that exhibited the lowest MMAE scores for this image set are depicted in Figure 13.16. By studying these spanning trees as well as the MMAE scores we can conclude that the most characteristic feature of the trees with the lowest error figures is that optimal matching begins from the center and proceeded outward. In other words, in these trees, the central nodes of the graph were connected. One possible explanation is that the matching quality of the central nodes is more crucial to the overall mosaicing quality than the matching quality of the other nodes.

13.6   Conclusions

Reconstruction of broken or torn artifacts bearing archaeological, historical, or cultural importance, such as ceramics, murals, mosaics, papyrus or paper manuscripts, stone tablets, etc. is a labor intensive procedure when performed manually by experts. Fortunately, computer-based or computer-assisted approaches for the reconstruction of such artifacts have started to emerge during the last few years. This area of research, which lies within the broader field of image processing and analysis and computer vision, is an extremely challenging one, since it has to deal with issues such as high degradation of fragments, missing fragments, lack of a priori information regarding the correct appearance of the reconstructed object, and so on. As a result, it is expected to attract considerable attention in the years to come, hopefully leading to methods that are highly automated, robust, and of general applicability.

Mosaicing or stitching, namely creation of a a single image from a set of overlapping sub-images, is also of high importance in archival, digital analysis and restoration of cultural heritage artifacts such as murals or large paintings. Being applicable in other fields, such as consumer photography or aerial, satellite, and underwater imaging, mosaicing has attracted the interest of a larger part if the imaging community than the reconstruction of archaeological or cultural artifacts. This has led more solid results, but sufficient room for additional, fruitful research still exists.

In this chapter, we have attempted to provide a concise review of algorithms for the 2D or 3D reassembly of fragmented or torn objects and have also presented algorithms that we have developed for the automatic reassembly of torn 2D paper images or fragmented paintings and the computationally efficient mosaicing of images. We hope that the content of this chapter, along with the other chapters of this book, will stimulate new research in this highly rewarding field, eventually leading to more efficient approaches towards preservation and study of cultural heritage.

Bibliography

[1] F. Amigoni, S. Gazzani, and S. Podico. A method for reassembling fragments in image reconstruction. In *Proceedings of the International Conference on Image Processing (ICIP)*, pages 581–584, September 2003.

[2] Stuart Andrews and David H. Laidlaw. Toward a framework for assembling broken pottery vessels. In *Proceedings of the Eighteenth American Conference on Artificial Intelligence*, pages 945–946, 2002.

[3] N. L. Biggs, E. K. Lloyd, and R. J. Wilson. *Graph Theory 1736-1936*. Clarendon Press, 1986.

[4] Gail A. Carpenter and Stephen Grossberg. *Pattern Recognition by Self-Organizing Neural Networks*. The MIT Press, Cambridge, MA, USA, 1991.

[5] Yang Chen and Gérard Medioni. Object modelling by registration of multiple range images. *Image and Vision Computing*, 10(3):145–155, 1992.

[6] D. Chetverikov, D. Svirko, D. Stepanov, and P. Krsek. The trimmed iterative closest point algorithm. *Proceedings of the 16th International Conference on Pattern Recognition*, 3:545–548, 2002.

[7] A. Criminisi, P. Perez, and K. Toyama. Object removal by exemplar-based inpainting. In *Proceedings of the IEEE International Conference on Computer Vision and Pattern Recognition*, pages 721–728, 2003.

[8] Helena Cristina da Gama Leitao and Jorge Stolfi. A multiscale method for the reassembly of two-dimensional fragmented objects. *IEEE Transactions Pattern Analysis Machine Intelligence*, 24(9):1239–1251, 2002.

[9] Scott R. Eliason. *Maximum Likelihood Estimation: Logic and Practice*. Sage Publications, Inc., 1993.

[10] K. Shoji, F. Toyama, and J. Miyamichi. Image mosaicing from a set of images without configuration information. In *Proceedings of the International Conference on Pattern Recognition (ICPR '04)*, volume 2, pages 899 – 902, 2004.

[11] M.A. Gavrielides, E. Sikudova, and I. Pitas. Color based descriptors for image fingerprinting. *IEEE Transactions on Multimedia*, 8(4):740–748, August 2006.

[12] Y. Gehua and C.V. Stewart. Covariance-driven mosaic formation from sparsely-overlapping image sets with application to retinal image mosaicing. In *Proceedings of the IEEE Conference on Computer Vision and Pattern Recognition (CVPR '04)*, volume 1, pages 804–810, 2004.

[13] David Goldberg, Christopher Malon, and Marshall Bern. A global approach to automatic solution of jigsaw puzzles. *Computational Geometry: Theory and Applications*, 28(2-3):165–174, 2004.

[14] Qi-Xing Huang, Simon Flöry, Natasha Gelfand, Michael Hofer, and Helmut Pottmann. Reassembling fractured objects by geometric matching. *ACM Transacions on Graphics*, 25(3):569–578, 2006.

[15] D. Huber. Automatic three-dimensional modeling from reality. PhD thesis, Carnegie Mellon University, 2002.

[16] Edson Justino, Luiz S. Oliveira, and Cinthia Freitas. Reconstructing shredded documents through feature matching. *Forensic Science International*, 160(2):140–147, 2006.

[17] Alan Kalvin, Edith Schonberg, Jacob T. Schwartz, and Micha Sharir. Two-dimensional, model-based, boundary matching using footprints. *International Journal of Robotics Research*, 5(4):38–55, 1987.

[18] Martin Kampel and Robert Sablatnig. On 3D mosaicing of rotationally symmetric ceramic fragments. In *Proceedings of the International Conference on Pattern Recognition*, volume 2, pages 265–268, 2004.

[19] F. Kleber, M. Lettner, M. Diem, M. Vill, R. Sablatnig, H. Miklas, and M. Gau. Multispectral acquisition and analysis of ancient documents. In *Proceedings of the 14th International Conference on Virtual Systems and MultiMedia (VSMM '08)*, pages 184–191, 2008.

[20] Florian Kleber and Robert Sablatnig. A survey of techniques for document and archaeology artefact reconstruction. In *Proceedings of the International Conference on Document Analysis and Recognition*, pages 1061–1065, 2009.

[21] D. Koller, J. Trimble, T. Najbjerg, N. Gelfand, and M. Levoy. Fragments of the city: Stanford's digital Forma Urbis Romae project. *Journal of Roman Archaeology*, 61:237–252, March 2006.

[22] Weixin Kong and Benjamin. B. Kimia. On solving 2D and 3D puzzles using curve matching. In *Proceedings of the IEEE International Conference on Computer Vision and Pattern Recognition*, volume 2, pages 583–590, 2001.

[23] Anat Levin, Assaf Zomet, Shmuel Peleg, and Yair Weiss. Seamless image stitching in the gradient domain. In *Proceedings of the Eighth European Conference on Computer Vision (ECCV '04)*, pages 377–389, 2004.

[24] Yifan Lu, Henry Gardner, Huidong Jin, Nianjun Liu, Rhys Hawkins, and Ian Farrington. Interactive reconstruction of archaeological fragments in a collaborative environment. In *Proceedings of the Digital Image Computing Techniques and Applications (DICTA '07)*, pages 23–29, 2007.

[25] F. Bartolini, M. Corsini, and V. Cappellini. Mosaicing for high resolution acquisition of paintings. In *Proceedings of the 7th International Conference on Virtual Systems and Multimedia*, pages 39–48, 2001.

[26] A. N. Netravali and B. G. Haskell. *Digital Pictures: Representation and Compression*. Plenum Press, 1988.

[27] Ture R. Nielsen, Peter Drewsen, and Klaus Hansen. Solving jigsaw puzzles using image features. *Pattern Recognition Letters*, 29(14):1924–1933, 2008.

[28] N. Nikolaidis and I. Pitas. Using spanning trees for reduced complexity image mosaicing. In *Proceedings of the European Signal Processing Conference (EUSIPCO '06)*, 2006.

[29] N. Otsu. An automatic threshold selection method based on discriminant and least squares criteria. *IEICE Transactions*, J63-D(4):349–356, 1980.

[30] G. Papaioannou, E. A. Karabassi, and T. Theoharis. Reconstruction of three-dimensional objects through matching of their parts. *IEEE Transactions on Pattern Analysis and Machine Intelligence*, 24(1):114–124, January 2002.

[31] C. Papaodysseus, T. Panagopoulos, M. Exarhos, C. Triantafillou, D. Fragoulis, and C. Doumas. Contour-shape based reconstruction of fragmented, 1600 BC wall paintings. *IEEE Transactions on Signal Processing*, 50(6):1277–1288, 2002.

[32] M. Pauly, R. Keiser, M.H. Gross. Multi-scale feature extraction on point-sampled surfaces. *Computer Graphics Forum*, 22(3):281–290, 2003.

[33] S. Peleg and J. Herman. Panoramic mosaics with videobrush. In *Proceedings of the DARPA Image Understanding Workshop 97*, pages 261–264, 1997.

[34] G. Peris and A. Marzal. Fast cyclic edit distance computation with weighted edit costs in classification. In *Proceedings of the 16th International Conference on Pattern Recognition (ICPR'02)*, pages 184–187, 2002.

[35] O. Pizarro and H. Singh. Toward large-area mosaicing for underwater scientific applications. *IEEE Journal of Oceanic Engineering*, 28(4):651–672, October 2003.

[36] H. Pottmann, J. Wallner, Q.-X. Huang, and Y.-L. Yang. Integral invariants for robust geometry processing. *Computer Aided Geometric Design*, 26(1):37–60, 2009.

[37] Matthias Prandtstetter and Günther R. Raidl. Combining forces to reconstruct strip shredded text documents. In *Proceedings of the 5th International Workshop on Hybrid Metaheuristics (HM '08)*, pages 175–189, 2008.

[38] W. Puech, A. G. Bors, I. Pitas, and J-M. Chassery. Projection distortion analysis for flattened image mosaicing from straight uniform generalized cylinders. *Pattern Recognition*, 34(8):1657–1670, August 2001.

[39] Mahmut Samil Sagiroglu and Aytul Ercil. A texture based matching approach for automated assembly of puzzles. In *Proceedings of the 18th International Conference on Pattern Recognition (ICPR '06)*, pages 1036–1041, 2006.

[40] H. S. Sawhney and R. Kumar. True multi-image alignment and its application to mosaicing and lens distortion. In *Proceedings of the IEEE Conference on Computer Vision and Pattern Recognition (CVPR '97)*, pages 450–456, 1997.

[41] Kulesh Shanmugasundaram and Nasir Memon. Automatic reassembly of document fragments via context based statistical models. In *Proceedings of the 19th Annual Computer Security Applications Conference (ACSAC '03)*, pages 152–159, 2003.

[42] G. C. Sharp, S. W. Lee, and D. K. Wehe. ICP registration using invariant features. *IEEE Transactions on Pattern Analysis and Machine Inteligence*, 24(1):90–102, January 2002.

[43] J. Shi and J. Malik. Normalized cuts and image segmentation. *IEEE Transactions on Pattern Analysis and Machine Intelligence (PAMI)*, 22(8):888-905.

[44] H.-Y. Shum and R. Szelisky. Construction of panoramic image mosaics with global and local alignment. *International Journal of Computer Vision*, 36(2):101–130, February 2000.

[45] P. Siarry, G. Berthiau, F. Durbin, and J. Haussy. Enhanced simulated annealing for globally minimizing functions of many-continuous variables. *ACM Transactions on Mathematical Software*, 23(2):209–228, 1997.

[46] T. F. Smith and M.S. Waterman. Identification of common molecular subsequences. *Journal of Molecular Biology*, 147(1):195–197, 1981.

[47] V. Swarnakar, M. Jeong, R. Wasserman, E. Andres, and D Wobschall. Integrated distortion correction and reconstruction technique for digital mosaic mammography. In *Proceedings of the SPIE Medical Imaging: Image Display*, volume 3031, pages 673–681, 1997.

[48] R. Szeliski. Image alignment and stitching: A tutorial. *Foundations and Trends in Computer Graphics and Vision*, 2(1):1–104, 2006.

[49] E. Trucco, A. Fusiello, and V. Roberto. Robust motion and correspondences of noisy 3D point sets with missing data. *Pattern Recognition Letters*, 20(9):889–898, 1999.

[50] E. Tsamoura and I. Pitas. Automatic color based reassembly of fragmented images and paintings. *IEEE Transactions on Image Processing*, 19(3):680–690, March 2010.

[51] A. Ukovich and G. Ramponi. Feature extraction and clustering for the computer-aided reconstruction of strip-cut shredded documents. *Journal of Electronic Imaging*, 17(1):1–13, 2008.

[52] A. Ukovich, G. Ramponi, H. Doulaverakis, Y. Kompatsiaris, and M.G. Strintzis. Shredded document reconstruction using mpeg-7 standard descriptors. In *Proceedings of the 4th IEEE International Symposium on Signal Processing and Information Technology*, pages 334–337, 2004.

[53] A. R. Willis and D.B. Cooper. Computational reconstruction of ancient artifacts. *IEEE Signal Processing Magazine*, 25(4):165–83, July 2008.

[54] Andrew Willis, Stuart Andrews, Jill Baker, Yan Cao, Dongjin Han, Kongbin Kang, Weixin Kong, Frederic F. Leymarie, Xavier Orriols, Senem Velipasalar, Eileen L. Vote, David B. Cooper, Martha S. Joukowsky, Benjamin B. Kimia, David H. Laidlaw, and David Mumford. Assembling virtual pots from 3D measurements of their fragments. In *Proceedings of the Virtual Reality Archeology and Cultural Heritage Conference (VAST '01)*, pages 241–254, 2001.

[55] H. Wolfson. On curve matching. *IEEE Transactions on Pattern Analysis and Machine Intelligence*, 12(2):483–489, 1990.

[56] H. Wolfson, E. Schonberg, A. Kalvin, and Y. Lamdan. Solving jigsaw puzzles by computer. *Annals of Operations Research*, 12(1-4):51–64, 1988.

[57] Feng-Hui Yao and Gui-Feng Shao. A shape and image merging technique to solve jigsaw puzzles. *Pattern Recognition Letters*, 24(12):1819–1835, August 2003.

[58] Zhaoda Zhu Yong Li, Daiyin Zhu, and Ling Wang. Automatic mosaicing for airborne SAR imaging based on subaperture processing. In *Proceedings of the IEEE International Geoscience and Remote Sensing Symposium (IGARSS '05)*, pages 4644–4647, 2005.

[59] Liangjia Zhu, Zongtan Zhou, and Dewen Hu. Globally consistent reconstruction of ripped-up documents. *IEEE Transactions on Pattern Analysis and Machine Intelligence*, 30(1):1–13, 2008.

14

Analysis of Ancient Mosaic Images for Dedicated Applications

Lamia Benyoussef, Stéphane Derrode

École Centrale Marseille
Email: lbenyoussef@ec-marseille.fr, sderrode@ec-marseille.fr

CONTENTS

14.1 Introduction

Mosaic is a form of art to produce decorative images or patterns out of small components. It requires a great deal of time and labor and it is often very expensive. Mosaic is, in return, far more durable in time than painting, so it can be used in places where painting cannot be of practical use, such as floors. Mosaic moreover allows the realization of light effects that are impossible with other media. The success of the mosaic art form through the ages is at the origin of large collections in several museums; among them we can cite the Great Palace Mosaic Museum (Istanbul, Turkey) and the Musée du Bardo (Tunis, Tunisia).

Recently, questions have arisen about the virtual conservation and management of collections and their accessibility to experts such as archeologists or art historians, and even to the larger public. Requirements concern facilities

- for cataloging collections and distant consultation, and for intelligent retrieval tools (e.g., the extraction of objects with a semantic meaning, such as animal or human, in a complex mosaic scene);

- for virtual conservation (e.g., restoration of tesserae color), reconstruction (e.g., missing patches), and anastylosis of ancient mosaics.

The aim of this chapter is to draw an overview of image processing–based methods dedicated to the specificities of mosaic images for applications to restoration and cataloging of ancient mosaics.

14.1.1   Some Historical Facts about Mosaics

The following description only intends to give some basis information about the history of mosaics. And so, it is quite superficial and partial. Interested readers should consult the numerous books covering the subject; among them we can cite the following books: [11, 16, 26]. The worldwide web also gives access to very complete lists of bibliographical references on ancient mosaics (e.g., [44]).

The earliest mosaics were installed as floors or pavements, occasionally as walls. Later, ceilings were also decorated. Originally, mosaics were used strictly in an architectural context. Mosaicking smaller, portable items such as panels or portraits is a later development, though there are a few rare instances of items such as icons from the late Byzantine era. The first mosaics that we know of are from what is now the Middle East. Temple columns in ancient Babylon (present-day Iraq) had thousands of small clay cones pressed into wet plaster in decorative and geometric patterns. These date from five thousand years ago. From these humble beginnings, mosaic developed into a major art form. It was the Greeks, in the four centuries BC, who raised the pebble technique to an art form, with precise geometric patterns and detailed scenes of people and animals. By 200 BC, specially manufactured pieces, "tessera," were being used to give extra detail and range of color to the work. Using small tesserae, sometimes only a few millimeters in size, meant that mosaics could imitate paintings. Many of the mosaics preserved at, for example, Pompeii were the work of Greek artists. The expansion of the Roman Empire took mosaics further afield, although the level of skill and artistry was diluted. Typically Roman subjects were scenes celebrating their gods, domestic themes, and geometric designs.

With the rise of the Byzantine Empire from the 5th century onwards, centered on Byzantium (present-day Istanbul, Turkey), the art form took on new characteristics. These included Eastern influences in style and the use of special glass tesserae called "smalti," manufactured in northern Italy, made from thick sheets of colored glass. Smalti have a rough surface and contain tiny air bubbles. They are sometimes backed with reflective silver or gold leaf. Whereas Roman mosaics were mostly used as floors, the Byzantines specialized in covering walls and ceilings. The smalti were ungrouted, allowing light to reflect and refract within the glass. Also, they were set at slight angles to the wall, so that they caught the light in different ways. The gold tesserae sparkle as the viewer moves around within the building. Roman images were absorbed into the typical Christian themes of the Byzantine mosaics, although some work is decorative and some incorporates portraits of Emperors and Empresses.

In the west of Europe, the Moors brought Islamic mosaic and tile art into the Iberian peninsula in the 8th century, while elsewhere in the Muslim world, stone, glass, and ceramic were all used in mosaics. In contrast to the figurative representations in Byzantine art, Islamic motifs are mainly geometric. Examples can be seen in Spain at the Great Mosque at Cordoba, the Alhambra Palace in Granada, and Meknes in Morocco (Figure 14.4(b)). In Arabic countries a distinctive decorative style called "zillij" uses purpose-made ceramic shapes that are further worked by hand to allow them to tessellate (fit together perfectly to cover a surface).

In the rest of Europe, mosaic went into decline throughout the Middle Ages, although some tiling patterns in abbeys, for example, used mosaic effects. In the 19th century there was a revival of interest, particularly in the Byzantine style, with buildings such as Westminster Cathedral in London and *Sacre-Cœur* in Paris. The "Art Nouveau" movement also embraced mosaic art. In Barcelona, Antoni Gaudí worked with Josep Maria Jujol to produce the stunning ceramic mosaics of the Guell Park in the first two decades of the 20th century. These used a technique known as *trencadis* in which tiles (purpose-made and waste tiles) covered surfaces of buildings. They also incorporated broken crockery and other found objects, a revolutionary idea in formal art and architecture.

Mosaic still continues to interest artists but also crafts people because it is a very accessible, non-elitist form of creativity. The field is rich with new ideas and approaches, and organizations such as the British Association for Modern Mosaic (BAMM) [2] and the Society of American Mosaic Artists (SAMA) [1] exist to promote mosaic. As a recent example, we can cite the development of algorithms for computer-aided generation of mosaic images from a raster image, simulating ancient or modern styles [6].

14.1.2 Mosaics, Images, and Digital Applications

Mosaics are made of colored tiles, called tessera or tessella, usually formed in the shape of a cube of materials separated by a joint of mortar. The earliest known mosaic materials were small cones of clay pressed into wet plaster. Semi-precious stones such as lapis lazuli and onyx, as well as shells and terra cotta were also used. As the art developed, glass, ceramic, and stone tesserae were the most common materials, along with pebbles. Modernly, any small singular component can be used: traditional materials, glass or ceramic cast or cut into tiles, plus plastic, polymer clay (such as Sculpey or Fimo), beads, buttons, bottle caps, pearls, etc. The physical study of materials used in mosaics continues to be an active research field [10, 17, 38].

Structure Characteristics

Mosaics can be built on different natural ground made of soil or rock, or on top of a previous pavement. The mosaic itself is composed of a variety of foundation or preparatory layers and a layer of tesserae. Also, the mosaic surface exhibits irregular hollows (tesserae) and bumps (mortar) through the scene. Hence, a mosaic can not be considered as a plane surface, as for paintings, but has numerous forms of irregularities, typical of the artwork style:

- *Shape of tesserae.* The shapes of tesserae are irregular, from square shapes to polygonal ones.

- *Organization of tesserae.* Tesserae are not positioned according to a regular lattice. On the contrary, the smart and judicious choice in the orientation, size, and positioning of tesserae characterize the artwork style and exhibit the "general flow" (guidelines) of the mosaic chosen by the mosaicist.

- *Mortar joint.* This positioning makes the joints appear as an irregular network with numerous interconnections throughout the mosaic. Network color intensity, mainly middle gray, is however not uniform through the scene.

- *Color of tesserae.* Because of materials used and their oldness, tesserae of ancient mosaics are generally characterized by pastel colors, with low contrast.

Oldness Characteristics

Other artifacts can be associated to the wear, erosion, and oldness of ancient mosaics (some of them are illustrated by images in Figures 14.1 and 14.2):

- *Disaggregated tesserae.* Some tesserae can display a loss of cohesion of their surfaces, disintegrated into powder or small grains. Also erosion and patinae deteriorations are commonly encountered in studying ancient mosaics.

- *Missing tessera patches.* Most mosaics are not completely preserved and, as a consequence of disaggregation, lacunas generally corrupt mosaic scenes.

- *Color alteration.* Alteration of the mosaic surface characterized by a localized change in color (due to fire damage, graffiti, etc.).

Data Acquisition

For later processing and virtual anastylosis, mosaic scenes must first be digitized using a camera. All peculiarities cited above have a strong and complex impact on acquisition and the way mosaic scenes appear in an image with a limited resolution:

FIGURE 14.1
Excerpt of an ancient Greek mosaic scene (from Paphos, Cyprus), showing missing patches and geometric deformation due to slant acquisition [25].

(a) Disaggregated   (b) Eroded   (c) Altered in color

FIGURE 14.2
Three examples of deteriorated tesserae, from [34].

- *Relief shadows*. During snapshot acquisition, the relief of tesserae generates shadows on mortar which does not appear nearly uniform in intensity all over the image, as it should.

- *Indoor/outdoor acquisition*. Previous item is further accentuated by indoor acquisition with flash. In outdoor acquisition, light is not controlled and one should expect severe contrast variations in acquisitions.

- *Geometric deformation*. Another degradation comes from the snapshot acquisition angle, especially for very large mosaic scenes lying on the floor. This point produces perspective (rotation, scaling , and shearing) deformations of tesserae shape in image and the thinning down of the mortar width (until disappearing).

- *Tessera resolution*. The resolution of tesserae in an image (i.e., the number of pixels to describe a tessera) should not be too low for later processing. On the other hand, a too high resolution will only be able to capture a partial scene.

As a consequence, very large mosaic scenes have to be recorded using multiple overlapping images that are subsequently co-registered in geometry and color to produce the entire digital scene. An interesting complementary approach to camera acquisition is given by laser scanning. This technique, used recently for ancient mosaics analysis [23, 37], can produce a 3D mesh of the mosaic surface and material for a detailed shape analysis.

Hence, dedicated processing methods are required to take into account the specificities of mosaic and the specificities of mosaic images, and to adopt image-processing strategies suited to the tiling organization of tesserae. This way, one can expect better performances than general purpose algorithms. The nature of the artifacts in mosaic images makes segmentation methods based on pixel intensities inefficient (e.g., pixels associated to the mortar interfere with the classification process of tesserae). A strategy better suited to mosaic images is to consider tesserae as indivisible entities with an almost uniform gray-level value.

Section 14.2 gives a brief description of several projects involving image-processing methods for restoration, preservation, and indexation of ancient mosaics. Then we present in detail some fundamental applications of the tessera-oriented point-of-view on mosaic images proposed by the authors, according to the diagram in Figure 14.3. Section 14.3 presents an effective way to extract tesserae, while Section 14.4 exposes two direct applications of this strategy for mosaic image segmentation and coding. Section 14.5 presents an effective way to retrieve the main guidelines of a mosaic which can be helpful for delimitating semantic objects present in a complex scene. Finally, the conclusion in Section 14.6 examines some open issues and suggests future research directions to succeed in providing appropriated tools to museums and experts.

14.2 Recent Image-Processing Projects Concerned with Mosaics

A few projects involving computer science and image processing for mosaic conservation, restoration, or cataloging are reported in the literature. Here is a short description of them.

14.2.1 St. Vitus Cathedral Mosaic Restoration

The more accomplished project concerns the restoration of the St. Vitus cathedral mosaics, in Prague (Czech Republic), reported in [49]. The Golden Gate (the southern entrance to the cathedral) is decorated with a unique work of art: a colored, richly gilded mosaic representing the Last Judgement, see Figure 14.4(a). In 1992 the Office of the President of the Czech Republic and the Getty Conservation Institute [3] began to cooperate to restore and conserve the mosaic.

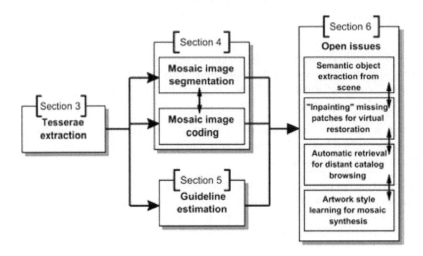

FIGURE 14.3
Organization of chapter with the processing pipeline.

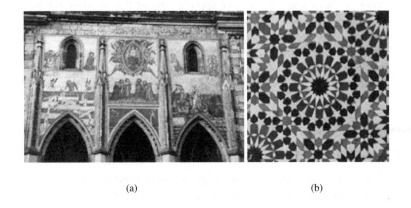

(a)   (b)

FIGURE 14.4
(a) Photo of the "Last Judgement" mosaic in the St. Vitus cathedral (Prague, Czech Republic), from [4]. Reprinted with permission B. Zitova et al. (b) Islamic mosaic (Meknes, Morocco).

Having the historical photo of the mosaic from the end of the 19th century and the photos of the current state, authors study the evolution of the mosaic, which was several times reconstructed and conserved. First they tried to restore the historical photograph - remove noise, deblur the image, increase the contrast. Then, they removed the geometrical difference between images by means of the multi-modal registration using mutual information. Finally, they identified mutual differences between the photos, which indicate the changes on the mosaic during the centuries.

14.2.2 Arabo-Moresque and Islamic Mosaic Pattern Classification

Lot of attention has also been paid to Islamic mosaics, the artwork style of which is illustrated in Figure 14.4(b). The particularities of such mosaics come from the periodicity and symmetry of tile patterns.

A. Zarghili et al., [47, 48] propose a method to index an Arabo-Moresque decor database which is not based on symmetry. They use a supervised mosaicking technique to capture the whole principal geometric information (connected set of polygonal shapes, called "spine") of a pattern. The spine is then described by using Fourier shape descriptors [35] to allow retrieving of images even under translation, rotation, and scale. But, according to [14], the method cannot be automatized and does not allow the classification of these patterns according to any criterion.

In [21], the authors propose image-processing techniques to restore mosaic patterns. Image analysis tools are developed to obtain information about design patterns which are used to recover missing motifs or tesserae. One difficulty was to propose methods robust to the discrepancies between equal object shapes (due to manual artwork, oldness, etc.). The symmetry, once recovered, allows virtual reconstruction by inpainting methods and physical restoration of damaged parts of mosaics.

To study Islamic geometrical patterns, and all periodic patterns such as those encountered in textile patterns [41] or wallpapers, several works are based on the symmetry group theory. In [14], the authors first classify patterns into one of the three following categories: (1) pattern generated by translation along one dimension, (2) patterns which contain translational symmetries in two independent directions (refers to the seventeen wallpaper groups), and (3) "rosettes," which describes patterns that begin at a central point and grow radially outward. For every pattern, authors extract the symmetry group and the fundamental region (i.e., a representative region in the image from which the whole image can be regenerated). Finally, they describe the fundamental region by a simple color histogram and build the feature vector, which is a combination of the symmetry feature and histogram information. The authors show promising experiments for either classification or indexing.

In [15], the authors exploit symmetry and auto-similarity of motifs to design a system to index Arabo-Moresque mosaic images based on the fractal dimension. Mosaics are first segmented automatically using color information. The classification decomposes the initial motif into a set of basic shapes. Contours of those shapes are then characterized by their fractal dimension which gives, according to the authors, a relevant measure of the geometric structure of the tile pattern. Some retrieval performances are also reported.

14.2.3 Roman Mosaics Indexation

Recently, two Content-Based Image Retrieval (CBIR) systems have been proposed to catalog and index Roman mosaic images.

The first one, proposed in [32], details a complete CBIR system which includes (1) object extraction from a complex mosaic scene by using unsupervised statistical segmentation and (2) invariant description of semantic objects using the analytical Fourier-Mellin transform [13, 20]. Similarity between querying mosaic and the database is based on an index constructed from the invariant descriptors and an appropriate metric (Euclidean and Hausdorff).

The second CBIR [27, 28] is a general system to index and retrieve by the content historic document images. While the object annotation in database images is done manually and off-line, the

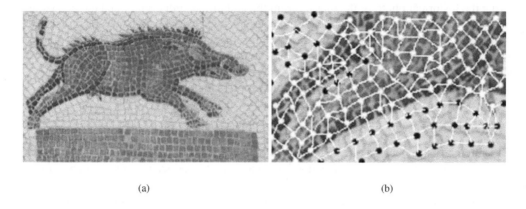

(a)   (b)

FIGURE 14.5
Excerpt of an ancient mosaic showing a wild boar (a) and a zoom on its hind legs (b).

indexation is done automatically using an extended curvature scale space descriptor [33] suitable for concave and convex shapes. The query/retrieval of pertinent shapes from the database starts with a user drawing query (with a computer mouse or a pen) that is compared to entries in the database using a fuzzy similarity measure. The system integrates an XML Database conforming to the MPEG7 standard, and experiments on large databases provided by the National Library of Tunisia and some Tunisian museums are reported.

One key point for such systems to be effective is to improve semantic objects extraction. Pixel-based methods, as the one used in [32], require heavy post-processing to detect shapes of interest from the pixels class. Recently, a new viewpoint on mosaic images, based directly on tesserae, has been proposed [8]. Once tesserae extraction is achieved, this strategy facilitates basic steps toward CBIR applications, as described in the following.

14.3 Tesserae Extraction

A "natural way" to describe mosaics is to consider tesserae in their own instead of groups of isolated pixels, allowing the development of tessera-oriented processing. Indeed, a mosaic image can be seen as an irregular lattice in which a node is a tessera (Figure 14.5). The grid structure has been super-imposed to the zoomed image in Figure 14.5(b) to illustrate the complexity of the neighborhood system, both in the number of neighbors and in their orientation. To adopt this point of view on mosaic images, a robust method is required to extract tesserae from the network of mortar surrounding them.

A way to consider this problem is to deal with the dual problem (i.e., the extraction of the network of mortar) which is close to recurrent problems encountered in several image-processing applications (i.e., in medical imaging: vascular network segmentation from angiographies [12, 46], or in satellite imaging: road extraction in urban scenes [29, 30]).

Several approaches have been proposed. Methods based on contour extraction are widely used and mainly rely on the assumption that the network pixels and neighboring ones have different gray levels in order to compute gradients. But methods based on high-pass filter, such as Harris's corner detector, highlight pixels belonging to the network, not connected components. Higher-level processings detect lines with varying widths [18, 39]. Strategies that track the entire network from

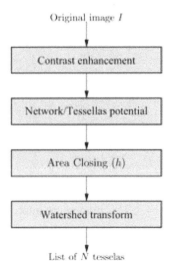

Original image I

Contrast enhancement

Network/Tessellas potential

Area Closing (h)

Watershed transform

List of N tessellas

FIGURE 14.6

Processing chain involved in tessera extraction, using gray-scale morphology tools.

a starting point [7, 45] are difficult to justify in the mosaic image case due to the high number of interconnections in a mosaic network graph.

Numerous methods based on Markov modeling [19, 40] or active contours [36, 42] have also been proposed. These methods are quite efficient but time consuming. In the case of mosaics, these methods are not suited because of the high density of the network to be extracted in images. In [19], a Markov model is applied on a graph of adjacency crests, detected by a Watershed Transformation applied on a criterion image. This criterion image, computed from the original one, exhibits the potential of each pixel to belong to the network.

A study of tesserae extraction has been conducted in [8]. The solution adopted, and preferred over other experimented strategies, is based on gray-level morphology which is suited to the tiling organization of mosaics. The Watershed Transformation (WT) approach [31] appeared interesting for mosaic images since this method is a good compromise between low-level methods (contour detection) and approaches by energy minimization (Markov model or active contours). But to work well, the WT needs to be computed on a criterion image (processed from the original one) that shows tesserae as catchment basins and the mortar network as watershed crests.

The entire algorithm is sketched in Figure 14.6. The goal is to present to the WT a potential image that exhibits tesserae as catchment basins and the network as a crest crossing the entire image. The pre-processings are: (1) contrast enhancement, based on top-hat and bottom-hat transforms; (2) potential criterion computation to exhibit the network as a crest; and (3) area closing [43] to reduce over-detection by WT.

The extraction quality greatly depends on the way mosaic images have been acquired. It should not be taken too far from the mosaic since individual tiles will not be visible: according to experiments, tesserae should appear with a resolution of not less than 10×10 pixels. Parameter h is a threshold used by the area closing operator [43]. Its value has an impact on the number of extracted tiles: a small value gives an over-detection whereas a big one produces an under-detection one. Hence, it should be (roughly) adjusted according to the mean size α of tesserae in the image (in number of pixels), assuming that α is almost constant in a mosaic. Parameter h should be less than α^2 to avoid small tesserae to be deleted by the morphological operator. In experiments, value $h = \alpha^2/2$ gives good results.

Figure 14.7 shows the criterion image obtained by applying the method to the wild boar image.

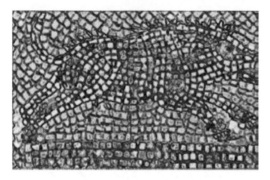

FIGURE 14.7
Criterion image obtained from Figure 14.5. This processing improves the contrat between tessera and the network of mortar.

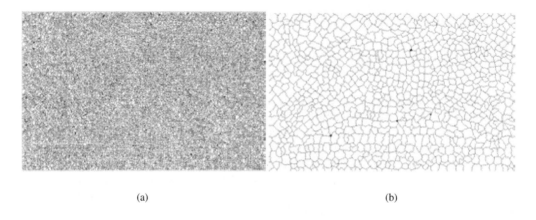

(a)   (b)

FIGURE 14.8
Extraction of tesserae from the criterion image in Figure 14.7, without (a) and with (b) area closing operator.

As can be seen in this example, the network appears in dark. However, tesserae are not uniform in texture and show local gray-level crests that should be deleted before WT in order to avoid over-segmentation. Following [19], an area closing [43] is first computed on the criterion image. This processing gives fewer minima while retaining crest locations as illustrated in Figure 14.8. The crest contours now represent correctly the network, which is confirmed by the zoom in Figure 14.9(a). To determine the width of the network (and not only a one-pixel skeleton as done by WT), which varies through the image, a simple threshold is applied on neighboring pixels of crests: a pixel is aggregated to the crest if its gray value is different by no more than 10% of the skeleton mean gray value. The result of applying such a threshold can be observed in Figure 14.9(b).

A second example of tesserae extraction based on gray-level morphological processing is given in Figure 14.10.

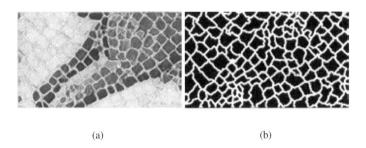

(a)   (b)

FIGURE 14.9
Result of tesserae extraction on the zoom in Figure 14.5(a) in (a), and network/tesserae classification in (b).

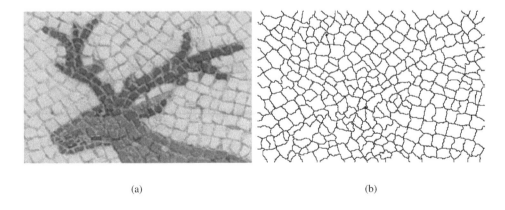

(a)   (b)

FIGURE 14.10
Illustration of the tesserae extraction algorithm on the mosaic image of a deer.

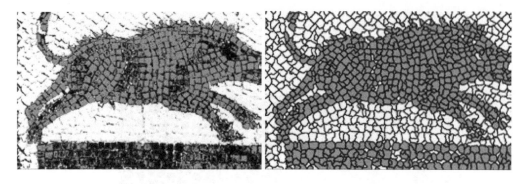

(a) Pixel-based K-means segmentation.   (b) Tile-based K-means segmentation.

FIGURE 14.11
Segmentation comparison of the mosaic image in Figure 14.5(a) with a pixel-based strategy (a) and the tessera-based strategy presented in Section 14.3 (b) Both pixel- and tile-based segmentations make use of an unsupervised K-means classification algorithm, the first one with three classes, and the second one with only two classes since the mortar is already extracted.

14.4 Tessera-Based Segmentation and Coding

This section describes two simple and direct applications of the tessera-oriented representation of ancient mosaic images: segmentation and coding. Especially, these applications preserve the irregularity of the tiles' shape and so the mosaic effect perception. From now on, we assume that all tesserae in a mosaic image have been extracted successfully (i.e., one can access the coordinates of every pixel belonging to the contour of each tile in the mosaic). Each pixel that does not belong to a tessera is assumed to be a background pixel (i.e., a mortar network pixel).

14.4.1 Segmentation of Ancient Mosaic Images

Instead of using all pixels from the tesserae, and since tesserae are almost homogeneous in color, classification can be performed directly on tesserae, not on pixels, by using some dedicated features. Tesserae can be characterized by several features, among them the mean and variance of gray-level values. Also one can take into account the number of pixels in the tessera, using a multi-features classification. All pixels belonging to the mortar network are directly associated to a unique class (they do not participate to the classification). The class color corresponds to the mean color of the mortar pixels.

In Figure 14.11, a simple K-means algorithm based on the mean gray-level value of tesserae was sufficient to get an interesting segmentation. From a semantic point of view, this segmentation outperforms the result obtained by a classical pixel-based K-means strategy. A second example that confirms such interesting behavior is provided in Figure 14.12. Once again, the segmentation quality depends on the robustness of tessera extraction.

14.4.2 Tessera-Based Coding and Lossy Compression

Another direct application of the WT tessera labeling is to develop a compact and efficient coding representation of ancient mosaic images [9]. This kind of module can be useful for rapid thumbnails

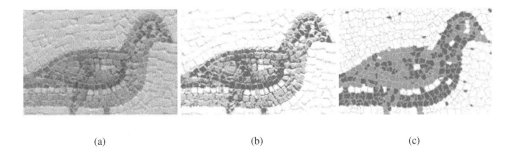

(a) (b) (c)

FIGURE 14.12

K-means segmentation of a mosaic representing a bird (a) with the pixel-based (b) and the tile-based (c) strategies.

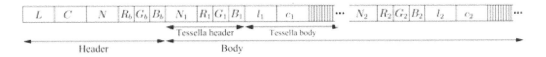

FIGURE 14.13

Sketch of the *cms* file format to compress images of ancient mosaics (without Huffman coding).

transmission in case of distant catalog browsing application. The compression proposed is lossy but preserves the specificities of mosaic images, especially the irregularity of tesserae shape. In a way, the strategy described below is better suited to ancient mosaic images than a general purpose lossy compression algorithm, such as *jpeg*, which is based on square blocks.

We still consider that all tesserae in a mosaic have been extracted successfully (i.e., we can access the coordinates of every pixel belonging to the contour of each tile in the mosaic). The network can be seen as the mosaic image background (i.e., each pixel that does not belong to a tessera is assumed to be a background pixel). Let I be a mosaic image with dimensions $L \times C$, made of N colored tiles. The tessera-based lossy-compression specificity has been designed in order to take into account the two following observations:

- A tile n, $n \in [1, N]$, is usually formed in the shape of a cube of material, mostly with uniform color. Hence a vector of three intensity colors (R_n, G_n, B_n) is enough to describe the color of tessera n. This color can be easily estimated by computing the mean intensity of pixels belonging to the tessera.

- The mortar intensity, mainly middle gray, is not uniform throughout the image because of shadows due to non-flat mosaic surfaces and snapshot acquisition angle. However, the network intensity should appear the same everywhere in the image. Hence, all pixels belonging to the mortar network are coded by the same color vector (R_b, G_b, B_b); b stands for *background*. This value can be estimated by computing the mean intensity of network.

Coding File Structure

The *cms* ("Compressed MoSaic") format is structured as follows (see Figure 14.13) :

- **Header:** The header is made of

  1. The original image size in pixels (2×2 bytes).
  2. The number N of tesserae in the image (2 bytes).

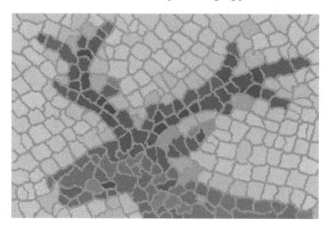

FIGURE 14.14
Result of coding/decoding of the deer image in Figure 14.10(a).

3. The mean color values (R_b, G_b, B_b) of the network (3 bytes).

Hence, 9 bytes are necessary to code the header. Other information such as the name of the original image or a comment can be added.

- **Body:** Each tessera n has the following code structure

  1. *Tessera header* contains the number of pixels N_n of the contour (1 byte), and the mean color values (R_n, G_n, B_n) of the tessera (3 bytes).

  2. *Tessera body* codes the closed-contour using the coordinates (l_1, c_1) of the first pixel (4 bytes) and Freeman chain code (or whatever code) for the subsequent $N_n - 1$ pixels. If no Huffman coding technic is applied, only 3 bits per contour pixels are necessary.

Counting the number of bytes necessary to code a tessera n gives: $8 + \frac{(N_n-1)\,3}{8}$ rounded to the ceil value. The last byte is filled in to start the next tessera coding on a new byte. Hence, 21 bytes are needed to code a tessera with $N_n = 34$.

Mosaic image reconstruction from the compressed file is done by reconstructing the image background using the network color intensity, on which each tessera with its own color is superimposed. The order in which tesserae are saved or read in the body of the *cms* file has no influence on the reconstruction quality, but can be changed for a desired application. For example, one can organize tesserae by grouping tiles in the file according to the semantic object they belong to (e.g., human, animal, etc.). Hence reconstruction will progressively show the different objects present in the mosaic scene. Tesserae dedicated to the background, mainly of uniform color as illustrated in Figures 14.5(a), 14.10(a) and 14.12(a), can be saved at the end of the file since their semantic meaning is, in general, less interesting than the one from objects tesserae. Specific organization of tesserae within the file can also be of interest for mosaic database indexing and text- or content-based retrieval (see Section 14.6). Hence, we get a compact and easy-to-handle lossy-compressed representation of mosaic image. The compression of a mosaic image and its reconstruction from a *cms* file are summed up in Algorithms 1 and 2.

Experimental Results

To evaluate the compression algorithm, we process the deer mosaic image in Figure 14.10(a). The decoding result is shown in Figure 14.14 and must to be compared with the original image. Pixels attributed to the network represent 27.5% of the total number of pixels in the image (255 ×

Algorithm 1 *cms* format coding.

Require: Mosaic image I.
Ensure: Tessera-oriented coding of I in file $\mathtt{f}$.
  Tessera extraction (*cf.* Section 14.3).
  for $n = 1, \ldots, N$ **do**
    Write the number of contour pixels $N_n \leftarrow \mathtt{f}$.
    Write the mean color value $V_n \leftarrow \mathtt{f}$.
    Write the first contour pixel coordinates $l_n, c_n \leftarrow \mathtt{f}$.
    for $i = 2, \ldots, N_n$ **do**
      Compute chain code direction for pixel i.
      Write direction $\leftarrow \mathtt{f}$.
    end for
  end for

Algorithm 2 *cms* format decoding.

Require: Compressed file $\mathtt{f}$.
Ensure: Tessera-oriented decoding of file $\mathtt{f}$ into image I.
  Read file header $\rightarrow L, C, N, R_b, G_b, B_b$.
  Create an image I with dimensions $L \times C$ and background color (R_b, G_b, B_b).
  for $n = 1, \ldots, N$ **do**
    Read tessera header $\rightarrow N_n, R_n, G_n, B_n, l_n, c_n$.
    Set pixel (l_n, c_n) with color (R_n, G_n, B_n).
    for $i = 2, \ldots, N_n$ **do**
      Compute coordinates (l_i, c_i).
      Set pixel (l_i, c_i) with color (R_n, G_n, B_n).
    end for
    Fill in the closed contour with color (R_n, G_n, B_n).
  end for

---

394). The number of tesserae detected is 314. The image produces a *cms* file size of 8,766 bytes, which gives a compression factor of about 11.5 with respect to a raw coding of pixels (e.g., *ppm* file format). The compression ratio depends on the mean tessera size in the mosaic image. The bigger the tiles, the higher the compression ratio is expected to be.

From a qualitative point of view, some remarks can be pointed out. As expected, the shape of the network and tesserae is kept very similar to the original ones. The mean colors used to describe network and tesserae intensities make the decoded image appear pastel with respect to the original one, since artifacts such as shadow or dirt are cleaned. Nevertheless the global aspect of the mosaic is respected and of sufficient quality to understand the scene content. Lacunas and missing patches are not taken into account by the actual algorithm. It considers holes as tesserae but should better associate them to the mortar network since we observe the preparatory layer mortar appears behind. One solution is to fuse the network with tesserae that show mean color close to the mean network color.

14.5   Guidelines Estimation for Mosaic Structure Retrieval

Ancient mosaicists avoided aligning their tiles according to rectangular grids. Indeed, such grids emphasize only horizontal and vertical lines and may distract the observer from seeing the overall picture. Hence, they placed tiles in order to emphasis the strong edges of the subject to be represented, influencing the overall perception of the mosaic [5]. Hence, organization and positioning of tesserae interesting give information for experts since they emphasize the main directional guidelines chosen by the artist (characteristic of the style). This information is of crucial interest for mosaic-dedicated applications such as content-based retrieval of semantic elements or region-based mosaic image coding.

To get directional guidelines, one can first think of using the principal axes of an ellipse-equivalent shape of each tessera, using well-known formulae based on geometrical moments (minor and major axes). However, ancient mosaic tesserae are neither box- nor regular-shaped and principal axes quickly appear not robust enough. One major drawback of such a method is that it does not take into account information of neighboring tesserae, which is of great importance for regularization and for recovering the main guidelines that emphasize the "general flow" of a mosaic.

Hence, an energy-based contextual algorithm for retrieving main directional guidelines in a mosaic has been proposed [8]. The energy to be minimized is constructed by using two key features: the mean gray value and the border direction of each tessera. The optimization is done either by gradient descent or by simulated annealing.

Methodology
Each tessera n is represented by

- Its barycenter (x_n, y_n) computed on the support Ω_n of n.

- The list of its neighboring tesserae: $\mathcal{V}_n = \{v_{n,1}, \ldots, v_{n,T_n}\}$. A neighbor is a tessera that shares at least one pixel of mortar network with n.

It should be noted that the number of neighbors T_n is different from one tile to the other since tesserae are not organized according to a regular grid. Each tessera n is characterized by an energy of configuration which links itself to each of its neighbors $v_{n,t} \in \mathcal{V}_n$. This energy, denoted by $E_{n,t}$ with $t \in [1, \ldots, T_n]$, is the sum of two complementary terms:

- The first term Q is based on the mean gray value of tesserae. It is proportional to the sum of the difference of gray-level means (1) between n and $v_{n,t}$, and (2) between n and the symmetrical

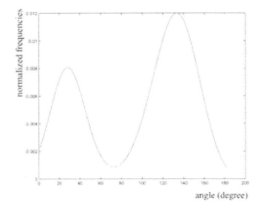

FIGURE 14.15

Plot of the regularized and normalized histogram of the contour orientation of one tile from the boar mosaic image in Figure 14.5(a).

tessera of $v_{n,t}$ with respect to n. This feature favors alignment of tesserae with low contrast, which is a characteristic of directional guidelines.

- The second term R is based on the orientation of tessera contours. We compute the histogram of the orientation of segments constituting the contour of tessera n. This histogram is regularized using a Gaussian kernel, the result of which is illustrated in Figure 14.15. It should be noted that the two modes of the histogram, at approximatively 90^0 each other, correspond to the two ambiguous orthogonal main directions of a square-shaped tile. It is then possible to estimate the p.d.f. at angle $\alpha_{n,t}$ given by the barycenter of n and the one of $v_{n,t}$.

Terms Q and R are normalized to belong to range $[0, 1]$. We can then initialize the "main direction" of a tile (i.e., the direction of the neighboring tessera which gives the highest $Q + R$ value):

$$t_{n,max} = \arg \max_{t \in [1,...,T_n]} E_{n,t} \tag{14.1}$$

The energy of a tessera is then defined as $C_n = (2 - E_{n,t_{n,max}}) + \lambda V_n$, with λ a weighting factor set manually. Term V_n is defined as

$$V_n = \frac{1}{T_n} \sum_{t=1}^{T_n} |\alpha_{n,t_{n,max}} - \alpha_{t,t_{t,max}}| \tag{14.2}$$

which is the normalized sum of absolute difference between the main direction of tile n and the main direction of its neighbor t.

The next step to to minimize the mosaic energy, defined as the sum of C_n for all tesserae in the mosaic. This is done by selecting the tessera n which gives the highest value for V_n. To reduce the contribution of this tessera, we try another main direction and recompute the mosaic energy. At that point, two strategies have been tested:

- Deterministic framework (Gradient Descent): if the mosaic energy reduces then the new main direction is validated, otherwise another main direction is tested. When all directions for this tile have been tested, we repeat the process for the next tile with high V_n value.

- Stochastic framework (Simulated Annealing): a configuration which gives a higher mosaic energy can be validated according to the simulated annealing principle [24]. This strategy allows one to search for the global minimum, which can not be reached with the previous strategy since the mosaic energy function is not convex.

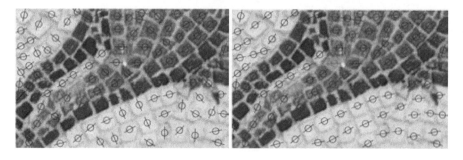

(a) Configuration at initialization.   (b) Configuration after SA.

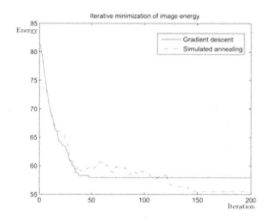

(c) Energy evolution for GD and SA.

FIGURE 14.16
Cartography of tesserae orientation. Each tessera is characterized by its center of mass (circle) and its orientation (segment crossing the circle).

The process is iterated until the mosaic energy is almost constant. The cartography of tessera orientation is made of the main direction of each tile at the last iteration.

Figure 14.16 illustrates the application of the tessera orientation methodology on the detail of the boar image in Figure 14.5(a). From the initial configuration of tesserae (a) we get the final configuration (b) using Simulated Annealing (SA) for optimization. Figure 14.16(c) shows the evolution of the computed energy during iterations of both a gradient descent (GD) algorithm and a simulated annealing one. As expected, SA reaches a lower minimum than GD but to the detriment of numerous additional iterations (approximately 150 for SA versus 50 for GD). Indeed, GD searches for a local minimum and is highly dependent on the initial configuration, whereas SA is expected to reach the global minimum due to its stochastic nature.

The tessera cartography obtained with SA optimization is really satisfying when visually compared to the main directional guidelines of the mosaic. This is especially true for regions at the borders between classes. A second example of cartography is proposed in Figure 14.17, after both simulated annealing and gradient descent optimizations. Results are very similar except for some guidelines where SA gives a better result. Once again, the tessera orientation estimation methodology, which makes use of contextual information, gives regularized results that emphasize the mosaic guidelines. Nevertheless, confusions can be found on areas with homogeneous colors and where

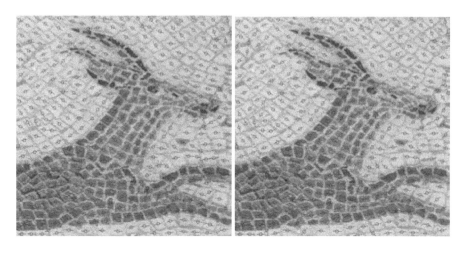

(a) Configuration after SA.   (b) Configuration after GD.

FIGURE 14.17
Another example of tesserae orientation cartography with (a) simulated annealing optimization and (b) gradient descent optimization.

tesserae are square-like shaped. Indeed, for those kinds of tesserae two orthogonal directions are equally probable, which generally gives ambiguous results. However, these areas of uniform color are of limited interest for object-based scene applications, such as mosaic pattern recognition.

Finally, Figure 14.18 shows a failure case in tessera orientation estimation. Indeed, the result will not allow one to find the main directional guidelines in the mosaic; that can be more easily observed in Figure 14.12(a). The main reason comes from an over-detection of tesserae from the extraction step. This behavior is observed on mosaics built with tesserae of different sizes (e.g., large tesserae for the background and small ones for objects or details). Hence the shape of extracted tiles does not correspond to the shape of tesserae and the orientation is corrupted, showing no particular guideline in the mosaic.

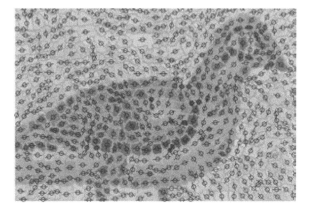

FIGURE 14.18
Result of guidelines detection for the "bird" mosaic in Figure 14.12(a).

14.6   Open Issues and Research Directions in Mosaic Image Analysis

In this chapter, an overview of image processing methods for images of ancient mosaics has been proposed. The analysis of such images is a difficult task due to many reasons, among them the irregularity of shapes, the particular organization of tesserae, and the degradation of tiles due to their oldness. Also, the way snapshot acquisition is performed has a crucial impact on the subsequent processings.

A few works only are reported in the literature regarding the numerous problems that experts (museum curators, archeologists, art historians, etc.) face about possible virtual restoration/lifting, conservation, and cataloging of ancient mosaics. Each artwork style and period requires the development of dedicated processing methods to take into account a priori information of materials, shape, color, etc. An interesting point of view on the problem is to consider, within the processings, that a mosaic is a collection of tiles and not only a series of image pixels. If this tessera-oriented strategy seems suited to numerous cases, improvement regarding the robustness of tesserae extraction is still required, especially when mosaic tiles show a large difference in size and a high level of degradation.

The algorithms described within Sections 14.3 to 14.5 are only a few basic, but crucial, steps toward complete applications expected by experts such as (1) ancient mosaic restoration with virtual "inpainting" tools taking into account main guidelines and tesserae shapes, and (2) multi-site and distant catalog browsing with "intelligent" retrieval capabilities of semantic objects. A simple *html* interface has been designed to illustrate the possible use of basic modules for browsing, *in situ*, a distant database of Roman-style mosaics. A simple query using textual keywords like "human," "face," or "animal" allows one to visualize the corresponding mosaics in a distant database. Figure 14.19 shows an example of this application with the word "bird" as query. Of course, one goal of the project is to replace keywords by self-content indexing using semantic objects extraction and characterization, based on segmentation, main guidelines estimation, and connected components labeling algorithms [22]. The mosaic coding strategy presented above can be used to construct a tile-oriented CBIR application. If the CBIR can work directly on the coded stream

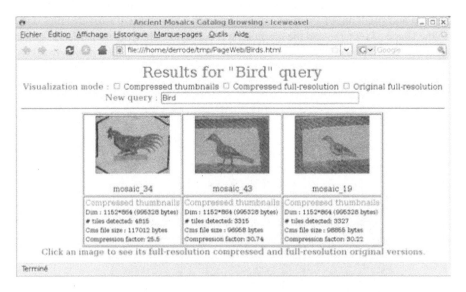

FIGURE 14.19
Example of querying a distant mosaic catalog by keyword.

(without reconstruction), one can expect to reduce drastically the RAM and/or hard disk memory requirements.

The problem of quality of processings for segmentation and lossy coding has not been addressed in this chapter. Particular attention should be paid to this issue in order to quantify the effectiveness of the approach. To that goal, specific quality metrics for tesserae shape and color have to be defined with experts and evaluated against image resolution. Cumulative error impact should then be analyzed along the complete processing chain to evaluate the level of accuracy for the targeted application. Note that fully automatic tools are not always required and applications piloted by users can be of interest.

An interesting open issue is the construction of an invariant representation of semantic mosaic objects which is robust to the tiling organization of shapes. Such a module will help CBIR systems to retrieve mosaic objects independently of their orientation, pose, and size in the scenes. Dedicated registration algorithms to reconstruct a complete mosaic scene from a set of sub-images can also be of interest in case of *in situ* acquisition of a large mosaic for example.

Finally, one very promising research direction is to develop new image-processing methods to automatically learn ancient mosaicists' way of working and style (e.g., tessera color and size, organization, guidelines, etc.). These tools could first be used to virtually fill in missing patches or restore deteriorated patches in a mosaic scene, using tessera-dedicated inpainting methods. They could also be exploited for emulating a mosaic artwork style from a raster input image using computer-aided algorithms, *cf.* [6] for an overview of digital mosaic creation.

Bibliography

[1] http://www.americanmosaics.org/ The Society of American Mosaic Artists.

[2] http://www.bamm.org.uk/ The British Association for Modern Mosaic.

[3] http://www.getty.edu/conservation/ The Getty Conservation Institute.

[4] http://zoi.utia.cas.cz/mosaic.html, Institute of Information Theory and Automation, Academy of Sciences of the Czech Republic.

[5] S. Battiato, G. Di Blasi, G. M. Farinella, and G. Gallo. A novel technique for Opus Vermiculatum mosaic rendering. In *14th International Conference on Computer Graphics, Visualization and Computer Vision (WSCG'06)*, Plzen, Czech Republic, 2006.

[6] S. Battiato, G. Di Blasi, G. M. Farinella, and G. Gallo. Digital mosaic frameworks: an overview. *Computer Graphic Forum*, 26(4):794–812, December 2007.

[7] A. Baumgartner, S. Hinz, and C. Wiedemann. Efficient methods and interfaces for road tracking. In *Photogrammetric Computer Vision (PCV'02)*, page B:28, Graz, Austria, 2002.

[8] L. Benyoussef and S. Derrode. Tessella-oriented segmentation and guideline estimation of ancient mosaic images. *Journal of Electronic Imaging*, 17(4), 2008.

[9] L. Benyoussef and S. Derrode. Représentation orientée tesselle d'images de mosaiques anciennes. In *GRETSI'09*, Dijon, France, 8-11 September 2009.

[10] C. Boschettia, C. Leonellib, M. Macchiarolac, P. Veronesiband, A. Corradib, and C. Sada. Early evidences of vitreous materials in Roman mosaics from Italy: An archaeological and archaeometric integrated study. *Journal of Cultural Heritage*, 9(sup. 1):e21–e26, 2008.

[11] P. Bruneau. *La mosaïque antique*. Presses de l'université de Paris-Sorbonne, 1987.

[12] A. C. S. Chung. Image segmentation methods for detecting blood vessels in angiography. In *IEEE International Conference on Automation, Robotics and Computer Vision (ICARCV'06)*, pages 1–6, Republic of Singapore, 2006.

[13] S. Derrode and F. Ghorbel. Robust and efficient Fourier-Mellin transform approximations for invariant grey-level image description and reconstruction. *Computer Vision and Image Understanding*, 83(1):57–78, 2001.

[14] M. O. Djibril and R. O. H. Thami. Islamic geometrical patterns indexing and classification using discrete symmetry groups. *Journal on Computing and Cultural Heritage*, 1(2):1–14, 2008.

[15] M. O. Djibril, R. O. H. Thami, R. Benslimane, and M. Daoudi. Une nouvelle technique pour l'indexation des arabesques basée sur la dimension fractale. In *Compression et Représentation des Signaux Audiovisuels (CORESA'05)*, Rennes, France, 2005.

[16] K. M. D. Dunbain. *Mosaics of the Greek and Roman World*. Cambridge University Press, 1999.

[17] C. Fiori, M. Vandini, S. Prati, and G. Chiavari. Vaterite in the mortars of a mosaic in the Saint Peter basilica, Vatican (Rome). *Journal of Cultural Heritage*, 10(2):248–257, 2009.

[18] M. A. Fischler, J. M. Tenenbaum, and H. C. Wolf. Detection of road and linear structures in low resolution aerial images using multi-source knowledge integration techniques. *Computer Graphics and Image Processing*, 15(3):201–223, 1981.

[19] T. Géraud and J. B. Mouret. Fast road network extraction in satellite images using mathematical morphology and Markov random fields. *EURASIP Journal on Applied Signal Processing*, 2004(16):2503–2514, 2004.

[20] F. Ghorbel. A complete invariant description for gray-level images by the harmonic analysis approach. *Pattern Recognition Letters*, 15:1043–1051, 1994.

[21] F. A. Gil, J. M. Gomis, and M. Pérez. Reconstruction techniques for image analysis of ancient Islamic mosaics. *Int. Journal of Virtual Reality*, 8(3):5–12, 2009.

[22] R. C. Gonzalez and R. E. Woods. *Digital Image Processing*. Addison-Wesley Publishing Company, 1992.

[23] R. Kadobayashi, N. Kochi, H. Otani, and R. Furukawa. Comparison and evaluation of laser scanning and photogrammetry and their combined use for digital recording of Cultural Heritage. In *ISPRS Congress*, Istanbul, Turkey, 2004.

[24] S. Kirkpatrick, C. D. Gelatt, and M. P. Vecchi. Optimization by simulated annealing. *Science, New Series*, 220(4598):671–680, May 1983.

[25] P. Kratochvil. Greek mosaic from paphos, cyprus, http://www.publicdomainpictures.net/view-image.php?image=4512&picture=mosaiques-grecques&large=1.

[26] R. J. Ling. *Ancient Mosaics*. British Museum, 1998.

[27] W. Maghrebi, L. Baccour, M. A. Khabou, and A. M. Alimi. An indexing and retrieval system of historic art images based on fuzzy shape similarity. In *6th Mexican International Conference on Artificial Intelligence (MICAI 2007)*, pages 623–633, Aguascalientes, Mexico, 2007.

[28] W. Maghrebi, A. Borchani, M. A. Khabou, and A. M. Alimi. A system for historic document image indexing and retrieval based on XML database conforming to MPEG7 standard. In *7th International Workshop on Graphics Recognition. Recent Advances and New Opportunities (GREC'07)*, pages 114–125, Curitiba, Brazil, 2007.

[29] H. Mayer, S. Hinz, U. Bacher, and E. Baltsavias. A test of automatic road extraction approaches. In *Photogrammetric Computer Vision (PCV'06)*, Bonn, Germany, 2006.

[30] J. B. Mena and J. A. Malpica. An automatic method for road extraction in rural and semi-urban areas starting from high resolution satellite imagery. *Pattern Recognition Letters*, 26(9):1201–1220, July 2005.

[31] F. Meyer and S. Beucher. The morphological approach of segmentation: the watershed transformation. In *Mathematical Morphology in Image Processing*, Dougherty E. Editor. Marcel Dekker, New York, 1992.

[32] M. M'hedhbi, R. Mezhoud, S. M'hiri, and F. Ghorbel. A new content-based image indexing and retrieval system of mosaic images. In *3rd International Conference on Information and Communication Technologies: from Theory to Applications (ICT-TA'06)*, pages 1715–1719, Damascus, Syria, 24-28 April 2006.

[33] F. Mokhtarian and A. Mackworth. A theory of multiscale, curvature-based shape representation for planar curves. *IEEE Transactions on Pattern Analysis and Machine Intelligence*, 14(8):789–805, 1992.

[34] Mosaics *In Situ* project - illustrated glossary. Technical report, The Getty Conservation Institute and the Israel Antiquities Authority, 2003.

[35] E. Persoon and K.-S. Fu. Shape discrimination using Fourier descriptors. *IEEE Transactions On Systems, Man and Cybernetics*, 7(3):170–179, 1977.

[36] M. Rochery, I. H. Jermyn, and J. Zerubia. Higher order active contours. *International Journal of Computer Vision*, 69(1):27–42, August 2006.

[37] G. Salemi, V. Achilli, M. Ferrarese, and G. Boatto. High resolution morphometric reconstruction of multimaterial tiles of an ancient mosaic. In *ISPRS Congress*, page B5:303ff, Beijing, China, 2008.

[38] C. S. Salerno, C. Moretti, T. Medici, T. Morna, and M. Verità. Glass weathering in 18th century mosaics: The São João Chapel in the São Roque Church in Lisbon. *Journal of Cultural Heritage*, 9(sup. 1):e37–e40, 2008.

[39] C. Steger. An unbiased detector of curvilinear structures. *IEEE Transactions on Image Processing*, 20(2):113–125, February 1998.

[40] F. Tupin, H. Maitre, J. F. Mangin, J. M. Nicolas, and E. Pechersky. Detection of linear features in SAR images: Application to road network extraction. *IEEE Transactions on Geoscience and Remote Sensing*, 36(2):434–453, 1998.

[41] J. M. Valiente, F. Albert, and J. M. Gomis. A computational model for pattern and tile designs classification using plane symmetry groups. In *The Ibero American Congress on Pattern Recognition (CIARP)*, Havana, Cuba, 2005.

[42] C. M. Van Bemmel, L. J. Spreeuwers, M. A. Viergever, and W. J. Niessen. Level-set-based artery-vein separation in blood pool agent CE-MR angiograms. *IEEE Transactions on Medical Imaging*, 22:1224–1234, 2003.

[43] L. Vincent. Grayscale area openings and closings: Their applications and efficient implementation. In *EURASIP Workshop on Mathematical Morphology and its Applications to Signal Processing*, pages 22–27, Barcelona, Spain, May 1993.

[44] R. Westgate. Ancient mosaics - a bibliography. http://www.cardiff.ac.uk/hisar/people/rw/mosaicbibliog.htm

[45] O. Wink, W. Niessen, and M. A. Viergever. Multiscale vessel tracking. *IEEE Transactions on Medical Imaging*, 23(1):130–133, 2004.

[46] P. J. Yim, G. Boudewijn, C. Vasbinder, V. B. Ho, and P. L. Choyke. Isosurfaces as deformable models for magnetic resonance angiography. *IEEE Transactions on Medical Imaging*, 22(7):875–881, July 2003.

[47] A. Zarghili, N. Gadi, R. Bensliman, and K. Bouatouch. Arabo-Moresque decor image retrieval system based on mosaic representations. *Journal of Cultural Heritage*, 2(2):149–154, 2001.

[48] A. Zarghili, J. Kharroubia, and R. Bensliman. Arabo-Moresque decor images retrieval system based on spatial relationships indexing. *Journal of Cultural Heritage*, 9(3):317–325, 2008.

[49] B. Zitová, J. Flusser, and F. Šroubek. An application of image processing in the medieval mosaic conservation. *Pattern Analalysis and Applications*, 7(1):18–25, 2004.

15

Digital Reproduction of Ancient Mosaics

Sebastiano Battiato, Giovanni Gallo, Giovanni Puglisi

University of Catania
Email: battiato@dmi.unict.it, gallo@dmi.unict.it, puglisi@dmi.unict.it

Gianpiero Di Blasi

University of Palermo
Email: diblasi@dinfo.unipa.it

CONTENTS

15.1 Art and Computer Graphics

Art often provides valuable hints for technological innovations especially in the field of image processing and computer graphics. In this chapter we survey in a unified framework several methods to transform raster input images into good-quality mosaics. For each of the major different approaches in literature the chapter reports a short description and a discussion of the most relevant issues. To complete the survey comparisons among the different techniques both in terms of visual quality and computational complexity are provided.

Non-Photorealistic Rendering (NPR) is a successful area of computer graphics and it is nowadays applied to many relevant contexts: scientific visualization, information visualization, and artistic style emulation [12, 38]. NPR's goals may be considered complementary to the traditional main goal of computer graphics, which is to model and render 3D scenes in a natural (i.e., photorealistic) way. Within NPR the recent approach to digitally reproduce artistic media (such as watercolors, crayons, charcoal, etc.) and artistic styles (such as cubism, impressionism, pointillism, etc.) has been gaining momentum and is very promising [2, 13, 15]. Several denominations have been proposed for this narrower area within NPR: Artistic Rendering (AR) [12] and Computational Aesthetics (CA) [35] are among the most popular ones. None of these names have reached general acceptance within the research community. The authors of this chapter believe that, beyond names, a proper definition of the area that explains in a suitable way its purpose and its aim may be the following: *to reproduce the aesthetic essence of arts by means of computational tools.*

The focus of this chapter is to review the state of the art for the problem of ancient digital mosaic creation [3]. The survey restricts its scope only to the techniques that explicitly bear the "ancient

FIGURE 15.1
Examples of ancient mosaics from [42].

mosaic" name (or equivalent) and that make use of primitives larger than pixels, points, or lines. Stippling [17, 31, 39, 40] and hatching [39, 41] are hence not covered here, although their visual similarity to mosaic naturally leads to approaches for these techniques that closely resemble the mosaic techniques. Mosaics, in essence, are images obtained by cementing together small colored fragments (see Figure 15.1). Likely, they are the most ancient examples of discrete primitive based images. In the digital realm, mosaics are illustrations composed by a collection of small images called tiles. The tiles tessellate a source image with the purpose of reproducing the original visual information rendered into a new mosaic-like style. The same source image may be translated into many strikingly different mosaics. Factors like tile dataset, constraints on positioning, deformations, and rotations of the tiles are indeed very influential upon the final results. As an example, the creation of a digital mosaic resembling the visual style of an ancient-looking man-made mosaic is a challenging problem because it has to take into account the polygonal shape of the tiles, the small size of the tiles, the need to pack the tiles as densely as possible and, not least, the strong visual influence that tile orientation has on the overall perception of the mosaic. In particular, orientation cannot be arbitrary but it is constrained to follow the gestalt choices made by the author of the source picture. Tiles, hence, must follow and emphasize the main orientations chosen by the artist.

The rest of this chapter is organized as follows: in Section 15.2 a brief history of ancient mosaics is summarized; Section 15.3 states the problem of digital mosaic creation into a mathematical framework; in Section 15.4 we present the crystallization mosaic techniques, while Section 15.5 explains the ancient mosaic methods. In Section 15.6 we extend the problem in a 3D environment. Finally Section 15.7 is devoted to final discussions and suggestions for future works.

15.2   History of Ancient Mosaics

The first examples of mosaic go back some 4,000 years or more, with the use of terracotta cones pushed into a background to give decoration [25] as demonstrated by unusual cone-shaped tesserae of various lengths found in ancient Sumerian manufacts. Other relevant examples, realized in almost the same period, can be found in some Egyptian and Phoenician mosaics. After that were different pebble pavements, using different colored stones to create patterns, although these tended to be unstructured decoration (about eighth century BC). In the four centuries BC the Greek artists rediscovered the pebble technique art form, with precise geometric patterns and detailed natural scenes. Next specially manufactured pieces (tesserae) were being used to give extra detail and range of color to the work. Using small tesserae with different shapes and colors, the mosaics could imitate paintings. Many of the mosaics preserved at, for example, Pompeii were the work of Greek artists. Although not so famous it should be mentioned the important mosaic art in Mexico made by the ancient Mayan civilization.

The next evolution in the field was due to the expansion of the Roman Empire although the level of skill and artistry was diluted. Typically Roman subjects were scenes celebrating their gods, domestic themes, and geometric designs. Different characteristics were exploited with the rise of the Byzantine Empire from the 5th century onwards, centered on Byzantium (now Istanbul, Turkey). In particular the main novelties were involved in style (i.e., Eastern influences) and the use of special material (e.g., glass tesserae called *smalti*). Differently than before, the Byzantines specialized in covering walls and ceilings. The *smalti* were ungrouted, allowing light to reflect and refract within the glass. Also, they were set at slight angles to the wall, so that they caught the light in different ways. The gold tesserae sparkle as the viewer moves around within the building. Roman images were absorbed into the typical Christian themes of the Byzantine mosaics, although some work is decorative and some incorporates portraits of Emperors and Empresses. Another important aspect of the mosaic history is based on the west of Europe where the Moors brought Islamic mosaic and tile art into the Iberian peninsula in the 8th century, while elsewhere in the Muslim world, stone, glass, and ceramic were all used in mosaics. Islamic motifs are mainly geometric and mathematical [21].

In central Europe, mosaic went into general decline throughout the Middle Ages even if the tile industry led to mosaic tiling patterns in abbeys and other major religious buildings. During the Gothic Revival there was some influence of the medieval themes. Some famous artists like Antonio Salviati and Antoni Gaudì were able to give new emphasis to the mosaic world in the modern era.

15.3   The Digital Mosaic Problem

A first step toward the solution of the problem of digital mosaic creation is to state it within a mathematical framework. In particular the translation of a raster source image into a digital mosaic may take the form of a mathematical optimization problem as follows:

> Given a rectangular region I^2 in the plane R^2, a tile dataset and a set of constraints, find N sites $P_i(x_i, y_i)$ in I^2 and place N tiles, one at each P_i, such that all tiles are disjoint, the area they cover is maximized, and the constraints are verified as much as possible.

The above definition is general and is suitable for many applications beyond the computer graphics field. Indeed Harmon in 1973 [29] published the first results related with this kind of problems in the context of modeling human perception and automatic pattern recognition. Within this framework the problem can be viewed as a particular case of the cover problem or as a search and optimization problem. The mosaic construction as formulated above can also be regarded as a low-energy configuration of particles problem. In the specific case of mosaics four different definitions can be given to solve specific problems:

Crystallization Mosaic: *Given an image I^2 in the plane R^2 and a set of constraints (i.e., on edge features), find N sites $P_i\,(x_i, y_i)$ in I^2 and place N tiles, one at each P_i, such that all tiles are disjoint, the area they cover is maximized, and each tile is colored by a color which reproduces the image portion covered by the tile. In this case in order to allow a solution the requirements have to be relaxed asking only that the constraints are verified as much as possible.*

Ancient Mosaic: *Given an image I^2 in the plane R^2 and a vector field $\phi\,(x, y)$ defined on that region by the influence of the edges of I^2, find N sites $P_i\,(x_i, y_i)$ in I^2 and place N rectangles, one at each P_i, oriented with sides parallel to $\phi\,(x_i, y_i)$, such that all rectangles are disjoint, the area they cover is maximized, and each tile is colored by a color which reproduces the image portion covered by the tile [30].*

Photo-mosaic: *Given an image I^2 in the plane R^2, a dataset of small rectangular images and a regular rectangular grid of N cells, find N tile images in the dataset and place them in the grid such that each cell is covered by a tile that "resembles" the image portion covered by the tile.*

Puzzle Image Mosaic: *Given an image I^2 in the plane R^2, a dataset of small irregular images and an irregular grid of N cells, find N tile images in the dataset and place them in the grid such that the tiles are disjoint and each cell is covered by a tile that "resembles" the image portion covered by the tile.*

Different solutions to the problems above have been proposed. Most of these solutions are reviewed in this chapter within a unified framework, focusing mainly on the solutions proposed for the first two types of mosaics. More precisely we single out two different mosaic types:

1. crystallization mosaics (a.k.a. tessellation);

2. ancient mosaics.

These two types of mosaics decompose a source image into tiles (with different color, size, and rotation), reconstructing the image by properly painting the tiles. They may hence be grouped together under the denomination of *tile mosaics*. Many mosaic techniques may fit in more than a single class and it is likely that other new types of mosaics will appear in the future. Other classifications of the known techniques for digital mosaics may be chosen if other criteria are taken into account. For example mosaics could be classified into:

1. fixed tile and variable tile (picture) size;

2. Voronoi-based and non-Voronoi based approach;

3. deterministic and non-deterministic (probabilistic, random) algorithm;

4. iterative and one-step method;

5. interactive and batch system.

Finally, the relative performances with respect to the computational complexity have to be taken into account for an effective evaluation and classification. In the following most of all previously published algorithms are reviewed. The reader is provided with a general idea about the working of each technique. A discussion of the weakness and the strong points of each method is provided and, whenever possible, the relationships between algorithmic choices and visual qualities of the resulting images are emphasized.

15.4 The Crystallization Mosaics

Many sophisticated mosaic approaches adopt smart strategies using computational geometry (e.g., Voronoi diagrams) together with image processing. These techniques generally lead to mosaics that simulate the typical effect of some glass windows in the churches. A Voronoi diagram is a geometric structure that represents proximity information about a set of points or objects [37]. Given a set of sites or objects, the plane is partitioned by assigning to each point its nearest site. The points whose nearest site is not unique form the Voronoi diagram. That is, the points on the Voronoi diagram are equidistant to two or more sites. From a geometric point of view Voronoi cells can be considered convex polygons. Voronoi diagrams were first discussed by Lejeune-Dirichlet in 1850, but it was more than a half of a century later, in 1908, that these diagrams were written about in a paper by Voronoi, hence the name Voronoi diagrams. The Voronoi cells/polygons are sometimes also called Dirichlet Regions. There are a variety of algorithms available to construct Voronoi diagrams (see for example [16, 37]), but the most famous algorithm was presented by Fortune [26]; he developed a plane-sweep algorithm which is more efficient in time than any other incremental algorithm. The algorithm guarantees an $O(n \ln n)$ complexity in the worst case.

Haeberli [28] used Voronoi diagrams, placing the sites at random and filling each region with a color sampled from the image. This approach tessellates the image with tiles of variable shapes but it does not attempt to follow edge features; the result is a pattern of color having a cellular-like look (see Figure 15.2(b)). The effect may be efficiently implemented by z-buffering a group of colored cones onto a canvas. This allows one to make the best use of hardware acceleration provided by modern graphics cards. Although very simple, Haeberli's idea is a milestone in this field and a starting point for many subsequent techniques.

In [19] Dobashi et al. extended the original idea of Haeberli. Their results are aesthetically more pleasant because the technique that they propose integrates edge information with Voronoi tessellation. The method, however, suffers from the variability that Haeberli's algorithm produces on tile shapes (Figure 15.2(c)). The strategy to better approximate the source image is simple. An error function E for the output image is defined as:

$$E = \sum_{x,y,c} \left(P_{(x,y,c)}^{Input} - P_{(x,y,c)}^{Output} \right)^2 \tag{15.1}$$

The function E is minimized by iteratively moving the centers of the Voronoi's polygons; the movement is limited to the 8-pixel neighborhood. The authors introduce also a heuristic strategy to speed up the moving process. Faustino and Figueiredo [23] presented a technique similar to Dobashi's. The main difference is that the sizes of tiles vary along the image: they are smaller near image details and larger otherwise (Figure 15.2(d)). To obtain this result, the authors use centroidal Voronoi diagrams with a density function that depends on image features (edge magnitude in this case). Differently from Dobashi, they do not start from a hexagonal lattice, but the seeds of the first Voronoi diagram are found by sampling the image by using a quadtree [24]: the seeds are the centers of the leaf cells. A leaf is created when the color of its corresponding cell pixels are close to the average color of the cell. In particular, for each cell, they test if:

$$\max_{p \in C} d(I(p), c)^2 \le \epsilon \tag{15.2}$$

where $I(p)$ is the color of the pixel p in C, c is the average color in C, and ϵ is a user-selected tolerance value. The image in Figure 15.2(d) has been obtained by using 2,557 seed points, 10 iterations, $\epsilon=0.150$, and a minimum cell size equal to 36 pixels.

Figure 15.2 is provided to compare the above techniques in terms of visual appearance. Taking into account high-frequency details (edge features and their orientation) the general image structure is in some way preserved. Only Faustino and Figueiredo's approach makes use of tile size according to the different edge magnitude but without using the relative orientation.

A different approach is presented in [34] where Mould proposes a technique to reproduce medieval stained-glass windows. He presents an unsupervised method for transforming an arbitrary image into a stained-glass version (see Figure 15.3). The key issues in designing a stained-glass window are tile boundaries and colors. He uses erosion and dilation operators to manipulate and smooth an initial region segmentation into a tiling. The algorithm chooses tile colors from the palette of heraldic tinctures and renders a displacement-mapped plane to obtain the final image. The algorithm can be summarized as follows:

1. Obtain an initial segmentation of the image by using an image-processing algorithm (for example, he uses EDISON: [11, 14, 33]);

2. Evolve the segmentation to obtain an appropriate tiling having smooth boundaries and approximately convex pieces and lacking excessively large or excessively small pieces; the smoothing is obtained by the application of simple erosion and dilation operators from mathematical morphology;

3. Choose a color for each tile; in particular it is possible to adopt the "heraldic" palette: for a given tile, determine its average color in the original image and the distance of this color from each heraldic color; the tile is then colored with the nearest heraldic color;

4. Apply a displacement map to a plane, representing the leading and irregularities in the glass;

5. Render the result.

The method is able to reproduce stained-glass images in a very effective way. Only a few parameters are involved; the edge magnitude and relative orientation are implicitly considered by combining segmentation and morphological operators.

15.5 The Ancient Mosaics

Since ancient times the art of mosaic has been extensively used to decorate public and private places. Today it is still possible to see some of such artistic works realized first by Greeks and Romans and later during the Byzantine Empire. Different kinds of mosaics were produced also by pre-Columbian people. Finally also in the modern era several artists have continued to deal with artistic mosaics [1, 25].

Ancient mosaics are artworks constituted by cementing together small colored tiles. A smart and judicious use of orientation, shape, and size may allow one to convey much more information than the uniform or random distribution of N graphic primitives (like pixels, dots, etc.). For example, ancient mosaicists avoided lining up their tiles in rectangular grids, because such grids emphasize only horizontal and vertical lines. Such evidence may distract the observer from seeing the overall picture. To overcome this potential drawback, old masters placed tiles emphasizing the strong edges of the main subject to be represented. In our context we are not interested in physical design of a mosaic work (e.g., cementing materials, etc.), but in the way that the individual mosaic pieces - known as the *tesserae* - are laid down. By using different materials and/or combining the tesserae in various ways, many different artistic styles and effects can be obtained. The general "flow" of the mosaic is known as *andamento*. The typical ancient mosaics, today available in computer graphics, have a specific categorization in the field of cultural heritage. The Latin term "opus" is used to describe the overall look of the mosaic. In particular the implemented techniques until now are the *opus musivum* and the *opus vermiculatum* (see Figure 15.4).

The epithet of "opus musivum" means of "quality worthy of the muses," of great visual refinery and effect. "Opus vermiculatum" takes its name from the Latin for "worm." It refers to lines of tiles

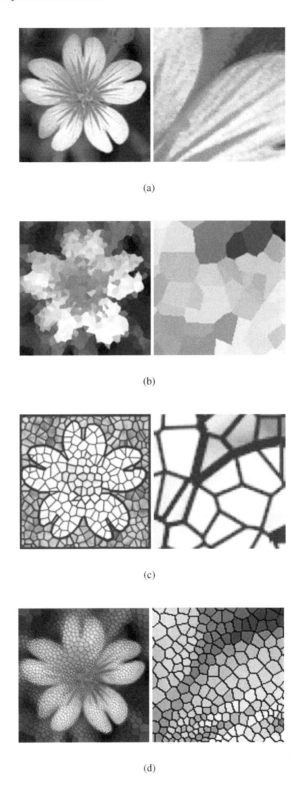

FIGURE 15.2
Crystallization mosaics: (a) the original image, (b) Haeberli [28], (c) Dobashi et al. [19], (d) Faustino and Figueiredo [23].

FIGURE 15.3
Medieval stained glass by Mould [34].

that snake around a feature in the mosaic. Often two or three rows of "opus vermiculatum" appear like a halo around something in a mosaic picture, helping it stand out from the background. The rendering of "opus vermiculatum" mosaics requires a clear separation between foreground and background because the two regions of the image have to be managed in different ways. The foreground region is covered as an "opus musivum," while the background region has to be covered by a regular grid of tiles (possibly perturbed by a random noise in size, position, and rotation).

The first attempt to reproduce a realistic ancient mosaic was presented by Hausner [30]. He proposed the mathematical formulation of the mosaic problem as described in Section 15.3. He obtained very good results using Centroidal Voronoi Diagrams (CVD), user-selected edge features, L_1 (Manhattan) distance and graphic hardware acceleration (Figure 15.5(b)). In particular the method uses CVDs (which normally arrange points in regular hexagonal grids) adapted to place tiles in curving square grids. The adaption is performed by an iterative process which measures distances

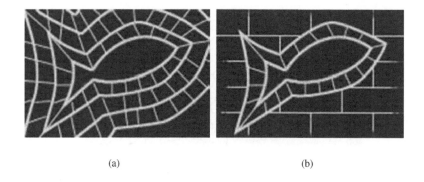

(a)   (b)

FIGURE 15.4
Examples of ancient mosaics styles [1]: (a) opus musivum, (b) opus vermiculatum.

with the Manhattan metric whose main axis is adjusted locally to follow a chosen direction field (coming from the edge features). Computing the CVD is made possible by leveraging the z-buffer algorithm available in many graphics cards. Hausner's algorithm can be outlined as follows:

1. S = list of random points on the input image;
2. while not converged:

   - for each p in S, place a square pyramid with apex at p;
   - rotate each pyramid about the z axis to align it to the field direction;
   - render the pyramid with an orthogonal projection onto the xy plane, producing a Voronoi diagram;
   - compute the centroid of each Voronoi region;
   - move each p to the centroid of its region.

The number of iterations to reach convergence is one of the main drawbacks of this technique, mainly when there is no direct access to the graphic acceleration engine.

Another algorithm for the creation of ancient mosaics is presented in [4, 18]; this approach is based on directional guidelines and distance transform and leads to very realistic results (Figures 15.5(c) and 15.7(b)). The algorithm uses some known image-processing techniques in order to obtain a precise tile placing that can be summarized as follows:

1. Segment the image by using the Statistical Region Merging algorithm [36];
2. Subdivide the image into background and foreground regions (optional);
3. For each pixel of the image evaluate the distance transform from the segmented region bounds;
4. Evaluate the gradient matrix and the level line matrix;
5. place the tile.

In particular point 5 can be described in more detail as follows:

1. While there are chains of pixels not yet processed in the level line matrix:

   - select a chain;
   - starting from an arbitrary pixel on it "follow" the chain;
   - place new tiles at regular distances along the path (the orientation of the tiles is assigned using the gradient information from matrix).

The authors, in order to obtain a high degree of similarity in terms of style with respect to ancient mosaics, also consider tile cutting (Figures 15.5(c) and 15.7(b)).

A novel technique for ancient mosaics generation has been presented in [39]. The authors present an approach for stroke-based rendering that exploits multi-agent systems; they call the agents *RenderBots*. RenderBots are individual agents representing in general one stroke. They form a multi-agent system and undergo a simulation to disseminate themselves in the environment. The environment consists of a source image and possibly additional G-buffer support images (edge image, luminance image, etc.). The final image is created when the simulation is stopped by having each RenderBot execute its painting function. The complete mosaic generation process can be described as follows:

1. Set up the environment (a number of RenderBots of a specific class are created and distributed in the environment);

2. Distribute the RenderBots (randomly or interactively by the user);

3. While the image is not finished:

   - simulate each bot (control of the bot physical behavior, computation of the new direction and velocity values and, possibly, change of the internal state of the RenderBot);

   - move each bot (perform the actual movement of the bot by computing a new position);

   - (eventually) paint each bot.

RenderBot classes differ in their physical behavior as well as their way of painting so that different styles can be created in a very flexible way. Thus they provide a unified approach for stroke-based rendering. Different styles such as stippling, hatching, painterly rendering, and mosaics can be created using the same framework. This is achieved by providing a specific class of RenderBots for each style. Figures 15.6(a) and 15.8(a) show two images obtained by RenderBot; before generating the mosaics, the images had been manually segmented. This is necessary because MosaicBots orient themselves using the nearest edges. The generation of NPR-images using RenderBots is an (iterative) interactive process, so production-times depend on the artist's requirements and the desired quality. It took about an hour to produce the results presented here. The number of MosaicBots involved was approximately 9,000 and 8,000 respectively for the images in Figures 15.6(a) and 15.8(a).

A very advanced approach to the rendering of traditional mosaics is presented in [22] (Figure 15.9(a)). This technique is based on offset curves that get trimmed-off the self intersecting segments with the guidance of Voronoi diagrams. The algorithm requires a mathematical description, as B-splines, of the edges and allows a very precise tile placement. Another point of this approach is the use of variable size tiles. Although the results are very good the technique seems limited to the case of a single, user-selected and closed edge curve.

A very interesting technique can be found in [27]; the authors present a new and efficient method to interactively create visually pleasing and impressive ancient mosaics. The algorithm is based again on the Lloyds method for CVT (Centroidal Voronoi Tessellation) computation and can be viewed as a smart extension and/or optimization of the technique proposed by Hausner [30]. They use a placement algorithm in an interactive fashion enabling the user to arrange tiles of various shapes and sizes. The user can easily control the distribution process by adding some other data such as contour lines and directional information. Tiles can be sized or shaped in order to better approximate the master image features. Additionally, this technique is less time consuming than using heuristic-controlled automatic methods. An interactive tool is preferred because "the proper arrangement of individual tiles is a highly artistic process" [27]. These authors claim that heuristic methods produce unwanted artifacts such as misaligned tiles. In more details this algorithm can be summarized as follows:

1. M = set of randomly distributed and randomly oriented polygons P;

2. While movement/rotation is above a threshold value:

   - for each P approximate its Voronoi region by using geometric modeled distance;

   - perform the Principal Component Analysis on polygons and their Voronoi regions;

   - compute the center of gravity CG of the polygons and of their Voronoi regions;

   - move the polygons in order to match the CG of the polygons and the CG of their Voronoi regions;

   - rotate polygons in order to align their principal component with the principal component of their Voronoi regions.

To create a mosaic a user must choose a master image and he has to define its feature lines and control polygons on the basis of the master image. He can hence choose the desired mosaic tile prototype (circles, quads, or user-defined n-edges) and input the rough number of tiles to be inserted. After a preliminary unsupervised insertion the tiles can be manually inserted/deleted by making use of interactive tools. Figure 15.9(b) shows a typical output image of this technique. This algorithm clearly outperforms all previously presented techniques in terms of aesthetic result. Unfortunately it requires a (crucial) user intervention and it is strictly dependent on the user's aesthetic skill and experience: two different users could obtain two totally different aesthetic results starting from the same input data (image, tile prototype, etc.).

A novel technique for ancient mosaics generation has been presented in [32]. The authors, using graph cuts optimization algorithm, are able to overcome some typical weaknesses of the artificial mosaics approaches: user interaction, fixed tiles number, explicit edge detection phase (see Figure 15.6(b)). First it generates tile orientation field and then packs the tiles. These two steps can be summarized as follows:

1. Tile orientation field is generated considering the strong edges of the underlying image. Moreover orientations are forced to vary smoothly in order to produce pleasing mosaics and reduce the gap between tiles. This field is obtained by using a global optimization approach (α-expansion algorithm) [10].

2. The packing of the tiles is performed in two steps:

   - A set of mosaic layers $(M_1, M_2, ..., M_n)$ is generated. Starting from a random pixel p each mosaic layer is created with a region-growing strategy based on a greedy assumption (the nearby pixel s that does not overlap and with the minimum gap space with respect to p is chosen).

   - The mosaic layers are stitched together taking into account gap space minimization, the absence of broken tiles, and the crossing of strong edge intensity. This task is performed through the graph cuts algorithm.

In [5, 6] the authors propose a novel approach based on Gradient Vector Flow (GVF) [43] computation together with some smart heuristics used to drive tiles positioning. Almost all previous approaches filter out high frequencies in order to simplify mosaic generation. On the contrary, GVF properties permit one to preserve edge information maintaining hence image details (see Figure 15.6(c) and 15.8(b)). Some heuristics are then used to follow principal edges and maximize the overall mosaic area covering. In particular the tiles positioning is not based only on gradient magnitude but makes use of local considerations to link together vectors that share the same logical edge. The algorithm can be summarized as follows:

1. Compute GVF from image I;

2. Put in a queue Q the pixels whose GVF magnitude is greater than a threshold t_h;

3. Sort Q with respect to GVF magnitude;

4. While Q is not empty:

   - extract a pixel (i, j) from Q and place a tile in according to the corresponding GVF phase, skipping tile positioning if there is overlap with previously placed tiles;

   - starting from (i, j) follow the chain of local maxima (GVF magnitude greater then a threshold t_l with $t_l \leq t_h$), placing a tile according to the GVF phase (skipping tile positioning if there is overlap with previously placed tiles);

5. Considering all the pixel, from top to bottom, from left to right, place in sequential order the tiles according to GVF phase (skipping tile positioning if there is overlap with previously placed tiles).

To summarize this section, the key of any technique aimed at the production of digital ancient mosaics is clearly the tile positioning and orientation. The methods presented in this section use different approaches to solve this problem, obtaining different visual results. Some techniques are based on a CVD approach [22, 27, 30] whereas other methods [5, 6, 32] compute a vector field by making use of different strategies (i.e., graph cuts minimization, gradient vector flow). Tile positioning is then performed with iterative strategies [22, 27, 30, 32] or reproducing the ancient artisan's style by using a "one-after-one" tile positioning [4–6, 18]. A different non-deterministic approach is used in [39].

15.6   The Ancient Mosaics in a 3D Environment

Recently [7, 8] some attempts have been made to reproduce digital mosaics in a 3D environment, extending in some sense the original ideas by a series of heuristics devoted to generalizing the rendering process; in particular in the three-dimensional representation the use of vectorial primitives, together with smart tricks to simplify the original cuts described in [18], allows for very effective results. Just adding some simple random changes on tile color, position, and tilt angle brings a strong realistic impact. Moreover, some features have been implemented, which attain a three-dimensional mapping of the mosaicized image to different 3D surfaces (domes, pyramids, cylinders, etc.). The overall way to access the mosaic changes given to the users provides the possibility to highlight the requested level of details by just navigating inside the 3D virtual space where the mosaic is rendered. Figure 15.10 shows some examples of such work obtained from an ancient mosaic realized by [5, 6] without tile cutting. Also we mention a method that simulates mosaic sculptures using tiles with irregular shapes, a method known by mosaicists as *opus palladium*, or simply crazy paving, due to the inherent freedom of mixing the tiles [20]. Special mosaic-like effects are obtained just by controlling the tile distribution and considering proper texture mapping strategy.

15.7   Final Discussions

In this section we briefly summarize the various characteristics of each presented method. Table 15.1 shows in a compact way, grouped by category, some specific details. The overall computational complexity is reported with respect to the number n of pixels of the input image and the number k of iterations for iterative methods. For iterative methods the parameter k is not always known a priori: usually a suitable tuning phase is required. Note that for the sake of clarity in some techniques we consider the grid size and/or granularity g proportional to the number of pixels n of the input image. To have a low computational complexity (almost linear in the number of involved pixels) is fundamental to being able to reproduce high-resolution mosaics without requiring expensive hardware resources. The remaining columns in Table 15.1 list other parameters and methodologies. One of the main effects of using fixed or variable tile size is the different emphasis given to specific details of the input image; the variability of the tile sizes produces a degradation in size along edges and textured areas especially if compared with classical mosaic appearance. Of course the particular partitioning strategy used by each method impacts the final result even if the tile positioning (orientation, cutting, etc.) is more crucial especially for unsupervised techniques (i.e., the key factor of each proposed strategy).

Although some criteria (e.g., covered area) have been used [5, 6] to compare different techniques, no evaluation is given in terms of aesthetic pleasure: no objective metrics are available to measure the effectiveness/accuracy of any digital mosaic. Aesthetic evaluation of any work produced by using supervised or unsupervised CA techniques is not so easy, because any objective metric is clearly inadequate [12]. Only an artist could give a more reliable, but subjective, judgment. The aesthetics of an output could be further evidenced by non-scientific and non-academic interests; for example the

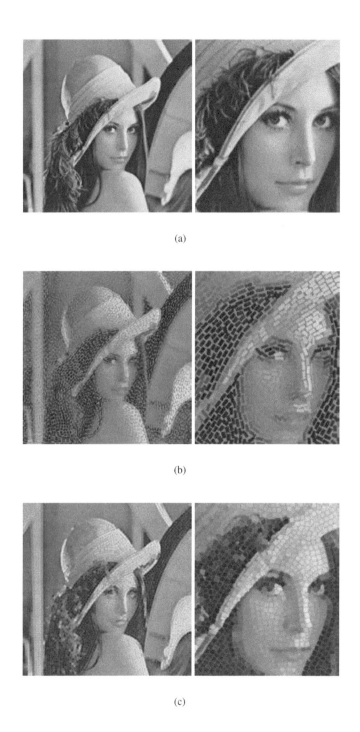

FIGURE 15.5
Examples of mosaics: (a) the original image, (b) Hausner [30], (c) Di Blasi and Gallo [18].

(a)

(b)

(c)

FIGURE 15.6 (SEE COLOR INSERT)
Examples of mosaics: (a) Schlechtweg et al. [39], (b) Liu et al. [32], (c) Battiato et al. [6].

(a)   (b)

FIGURE 15.7 (SEE COLOR INSERT)
Other examples of ancient mosaics: (a) the original image, (b) Battiato et al. [4].

(a)   (b)

FIGURE 15.8
Other examples of ancient mosaics: (a) Schlechtweg et al. [39], (b) Battiato et al. [6].

(a)   (b)

FIGURE 15.9
Other ancient mosaics: (a) Elber and Wolberg [22], (b) Fritzsche et al. [27].

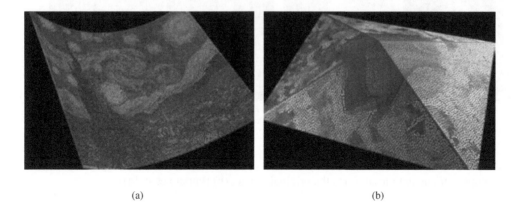

(a)   (b)

FIGURE 15.10
Some examples of 3D mosaics on cylinder (a) and pyramid (b).

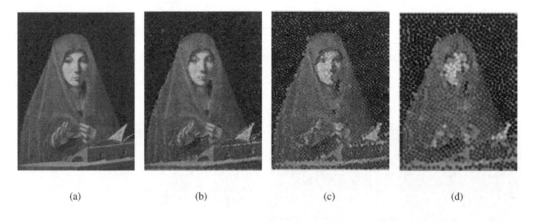

(a)   (b)   (c)   (d)

FIGURE 15.11
Mosaics generated with increasing tile size: (a) 3×3, (b) 6×6, (c) 10×10, (d) 14×14.

(a)   (b)

FIGURE 15.12
Example of a mosaic realized by an artist starting from a painting (a) and a detail (b).

software proposed by Di Blasi and Gallo [18] has received interest from many artists, companies, and fine arts academies.

There are several ways to improve the aesthetic of the results and several ideas started from these works. In particular:

- A different strategy for choosing, in case of overlapping tiles, which tile has to be cut; heuristic rules or, perhaps, randomized choices could produce different outcomes.

- Generalization of the "mosaicists' heuristic" to other kinds of primitive-based non-photorealistic image processing seems possible and quite promising.

- Some generalizations proposed in [22], such as variable size tiles and photo-mosaic, are also considered for future work and research; it is also interesting to explore the possibilities offered using different basic shapes other than rectangular tiles.

- Automatic optimized choices of tile scale relative to each input image is an open problem worth further investigation (Figure 15.11).

- The extension of mosaic techniques to other kinds of mosaics as proposed in [22].

- A different method to better find the directional guidelines is an important research investigation issue.

- Exploitation of hardware graphics primitives to accelerate the mosaic synthesis.

- The usage of methods able to infer from high-quality pictures of ancient mosaics the underlying structure or original schema [9].

- Inferring the tile distribution and novel aesthetic procedures to arrange the overall mosaic just considering the work of some artists that have tried to "copy" some painting in a mosaic (see Figure 15.12).

	Method	Computational Complexity (n = number of pixels) (k = number of iterations)	Size	Partitioning	Deterministic	Iterative	Interactive
Crystallization Mosaics	Haeberli [28]	$O\,(n\ln n)$	Fixed	Voronoi	No	No	No
	Dobashi et al. [19]	$O\,(k \cdot n\ln n)$	Fixed	Voronoi	No	Yes	No
	Mould [34]	$O\,(n)$	Variable	Segmentation	Yes	No	No
	Faustino and Figueiredo [23]	$O\,(k \cdot n\ln n)$	Variable	Voronoi	No	Yes	No
Ancient Mosaics	Hausner [30]	$O\,(k \cdot zBufExecTime)$	Variable	CVD	No	Yes	No
	Elber and Wolberg [22]	$O\,(k \cdot n\ln n)$	Variable	CVD	No	Yes	No
	Battiato et al. [4]	$O\,(n)$	Fixed	None	Yes	No	No
	RenderBot [39]	$O\,(k \cdot botExecTime)$	Fixed	None	No	Yes	Yes
	Fritzsche et al. [27]	$O\,(k \cdot n\ln n)$	Variable	CVD	Yes	Yes	Yes
	Liu et al. [32]	$O\,(k \cdot n)$	Fixed	None	No	Yes	No
	Battiato et al. [5,6]	$O\,(k \cdot n) + O\,(n\ln n)$	Fixed	None	Yes	Yes	No

TABLE 15.1
List of the presented digital mosaic approaches and related features.

15.7.1   Final Summary

In this chapter, we surveyed several approaches to non-photorealistic rendering of digital images in the field of mosaic generation. The various methods have been grouped together according to the main features. In particular, we singled out two different kinds of mosaics: crystallization and ancient mosaics. It is also possible to group the mosaic creation methods by using different criteria. The common and different ideas among the methods were reported and described. The various techniques were compared also with respect to the overall performances both in terms of achieved visual effects and computational complexity.

Bibliography

[1] Mosaic art in vitreous glass, millefiori, tesserae mosaics by shelby glass studio; http://www.mosaic-tile-art.com/mosaic.html, 2006.

[2] A. Hertzmann, C. Jacobs, N. Oliver, B. Curless, and D.H. Salesin. Image analogies. In *Proceedings of the 28th Annual Conference on Computer Graphics and Interactive Techniques (SIGGRAPH '01)*, pages 327–340, New York, NY, USA, 2001. ACM.

[3] S. Battiato, G. Di Blasi, G. M. Farinella, and G. Gallo. Digital mosaic frameworks – an overview. *Computer Graphics Forum*, 26(4):794–812, 2007.

[4] S. Battiato, G. Di Blasi, G.M. Farinella, and G. Gallo. A novel technique for opus vermiculatum mosaic rendering. In *Proceedings of the 14th International Conference in Central Europe on Computer Graphics, Visualization and Computer Vision (WSCG'06)*, pages 133–140, 2006.

[5] S. Battiato, G. Di Blasi, G. Gallo, G.C. Guarnera, and G. Puglisi. Artificial mosaics by gradient vector flow. In *Short Proceedings of EUROGRAPHICS*, 2008.

[6] S. Battiato, G. Di Blasi, G. Gallo, G.C. Guarnera, and G. Puglisi. A novel artificial mosaic generation technique driven by local gradient analysis. In *Proceedings of International Conference on Computational Science (ICCS 2008) - Seventh International Workshop on Computer Graphics and Geometric Modeling (CGGM 2008)*, Lecture Notes in Computer Science, volume 5102, pages 76–85, 2008.

[7] S. Battiato, G. Di Blasi, G. Gallo, and A. Milone. Realizzazione di mosaici artificiali in ambiente 3D. In *Proceedings of V Convegno AIAR Associazione Italiana di Archeometria*. Siracusa, February 2008.

[8] S. Battiato and G. Puglisi. 3D Ancient Mosaics. In *Proceedings of ACM Multimedia Technical Demos*, Florence, Italy, 2010.

[9] L. Benyoussef and S. Derrode. Tessella-oriented segmentation and guidelines estimation of ancient mosaic images. *Journal of Electronic Imaging*, 17(4), October 2008.

[10] Y. Boykov, O. Veksler, and R. Zabih. Fast approximate energy minimization via graph cuts. *IEEE Transactions on Pattern Analysis and Machine Intelligence*, 23(11):1222–1239, Nov 2001.

[11] C.M. Christoudias, B. Georgescu, and P. Meer. Synergism in low level vision. In *Proceedings of the 16th International Conference on Pattern Recognition (ICPR'02)*, volume 4, pages 150–155, Washington, DC, USA, 2002. IEEE Computer Society.

[12] J.P. Collomosse. *Higher Level Techniques for the Artistic Rendering of Images and Video*. PhD thesis, University of Bath, UK, May 2004.

[13] J.P. Collomosse and P.M. Hall. Cubist style rendering from photographs. *IEEE Transactions on Visualization and Computer Graphics*, 9(4):443–453, October 2003.

[14] D. Comaniciu and P. Meer. Mean shift: a robust approach toward feature space analysis. *IEEE Transactions on Pattern Analysis and Machine Intelligence*, 24(5):603–619, 2002.

[15] C.J. Curtis, S.E. Anderson, J.E. Seims, K.W. Fleischer, and D.H. Salesin. Computer-generated watercolor. In *Proceedings of the 24th Annual Conference on Computer Graphics and Interactive Techniques (SIGGRAPH'97)*, pages 421–430, New York, NY, USA, 1997. ACM Press/Addison-Wesley Publishing Co.

[16] M. de Berg, M. van Kreveld, M. Overmars, and O. Schwarzkopf. *Computational Geometry*. Springer-Verlag, Berlin, 1997.

[17] O. Deussen, S. Hiller, C. Overveld, and T. Strothotte. Floating points: A method for computing stipple drawings. *Computer Graphics Forum (EG'00)*, 19(3):40–51, 2000.

[18] G. Di Blasi and G. Gallo. Artificial mosaics. *The Visual Computer*, 21(6):373–383, July 2005.

[19] J. Dobashi, T. Haga, H. Johan, and T. Nishita. A method for creating mosaic images using Voronoi diagrams. *Computer Graphics Forum (EG'02)*, pages 341–348, 2002.

[20] V.A. Dos Passos and M. Walter. 3D virtual mosaics: Opus palladium and mixed styles. *The Visual Computer*, 25(10):939–946, 2009.

[21] M. Du Sautoy. *Symmetry: A Journey Into the Patterns of Nature*. HarperCollins, 2008.

[22] E. Elber and G. Wolberg. Rendering traditional mosaics. *The Visual Computer*, 19(1):67–78, 2003.

[23] G.M. Faustino and L.H. de Figueiredo. Simple adaptive mosaic effects. In *Proceedings of the 18th Brazilian Symposium on Computer Graphics and Image Processing (SIBGRAPI 2005)*, pages 315–322, Natal, RN, Brazil, 2005. IEEE Computer Society.

[24] R.A. Finkel and J.L. Bentley. Quad trees: A data structure for retrieval on composite keys. *Acta Informatica*, 4(1):1–9, March 1974.

[25] I. Fiorentini Roncuzzi and E. Fiorentini. *Mosaic: materials, techniques and history*. MWeV, 2002.

[26] S. Fortune. A sweep-line algorithm for Voronoi diagrams. *Algorithmica*, 2:153–174, 1987.

[27] L.P. Fritzsche, H.S. Hellwig, S. Hiller, and O. Deussen. Interactive design of authentic looking mosaics using Voronoi structures. In *Proceedings of the 2nd International Symposium on Voronoi Diagrams in Science and Engineering VD 2005 Conference*, pages 1–11, 2005.

[28] P. Haeberli. Paint by numbers: Abstract image representation. In *Proceedings of the 17th Annual Conference on Computer Graphics and Interactive Techniques (SIGGRAPH'90)*, pages 207–214, Dallas, TX, USA, 1990. ACM Press.

[29] L.D. Harmon. The recognition of faces. *Scientific American*, 229(5):71–82, 1973.

[30] A. Hausner. Simulating decorative mosaics. In *Proceedings of the 28th Annual Conference on Computer Graphics and Interactive Techniques (SIGGRAPH '01)*, pages 573–580, New York, NY, USA, 2001. ACM.

[31] S. Hiller, H. Hellwig, and O. Deussen. Beyond stippling - methods for distributing objects on the plane. *Computer Graphics Forum*, 22(3), September 2003.

[32] Y. Liu, O. Veksler, and O. Juan. Simulating classic mosaics with graph cuts. In *Proceedings of Energy Minimization Methods in Computer Vision and Pattern Recognition (EMMCVPR 2007)*, pages 55–70, 2007.

[33] P. Meer and B. Georgescu. Edge detection with embedded confidence. *IEEE Transactions on Pattern Analysis and Machine Intelligence*, 23(12):1351–1365, 2001.

[34] D. Mould. A stained glass image filter. In *Fourteenth Eurographics Workshop on Rendering*, pages 20–25, 2003.

[35] L. Neumann, M. Sbert, B. Gooch, and W. Purgathofer, editors. *Computational Aesthetics 2005: Eurographics Workshop on Computational Aesthetics in Graphics, Visualization and Imaging 2005, Girona, Spain, May 18-20, 2005*. Eurographics Association, 2005.

[36] R. Nock and F. Nielsen. Statistical region merging. *IEEE Transactions on Pattern Analysis and Machine Intelligence*, 26(11):1452–1458, 2004.

[37] F.P. Preparata and M.I. Shamos. *Computational Geometry: An Introduction*. Springer-Verlag, New York, 1985.

[38] T. Saito and T. Takahashi. Comprehensible rendering of 3-D shapes. In *Proceedings of the 17th Annual Conference on Computer Graphics and Interactive Techniques (SIGGRAPH'90)*, pages 197–206, Dallas, TX, USA, 1990. ACM Press.

[39] S. Schlechtweg, T. Germer, and T. Strothotte. Renderbots-multi-agent systems for direct image generation. *Computer Graphics Forum*, 24(2):137–148, June 2005.

[40] A. Secord. Weighted Voronoi stippling. In *Proceedings of the 2nd International Symposium on Non-Photorealistic Animation and Rendering (NPAR'02)*, pages 37–43, Annecy, France, June 3-5 2002.

[41] A. Secord, W. Heidrich, and L.M. Streit. Fast primitive distribution for illustration. In *Proceedings of the Thirteenth Eurographics Workshop on Rendering*, 2002.

[42] Norman Tellis. Ancient mosaics; http://www.classicalmosaics.com/photo_album.htm.

[43] C. Xu and L. Prince. Snakes, shapes, and gradient vector flow. *IEEE Transactions on Image Processing*, 7(3):359–369, 1998.

16

Pattern Discovery from Eroded Rock Art

Yang Cai

Carnegie Mellon University
Email: ycai@cmu.edu

CONTENTS

16.1 Introduction

Rock art is perhaps the oldest cultural heritage on earth. For more than ten thousand years, our ancestors have carved figures on smooth rock surfaces by pecking and scratching, generating textural patterns under the sun. Unfortunately, as time passes, most outdoor rock art surfaces have weathered and are disappearing. They have been cracked, colored, eroded, or accumulated lichen or moss, and dirt. Recovering the illegible art on the damaged surfaces has been a challenging task.

Prehistoric rock art sites are mainly on flat rock panels in remote areas. The contents are primitive figures carved by picking tools such as rods or stones. While most prehistoric sites are away from man-made pollution, they have been gradually eroded by nature. The Neolithic Baite Fles Saline Rock Art site in Val Camonica in Northern Italy is an example [16] (Figure 16.1). It is an inclined panel on a hillside where the closest driveway is about one kilometer away. The huge steep angled rock has been eroded by weather, cracked, broken away, or covered by the dirt and algae. Many figures are illegible to the naked eye. In North America, the Indian rock art in Western Pennsylvania carved by local Indians has been there for thousands of years [5]. Many of these sites are underwater for most of the year except summer. The stream and sand have smoothed the rock art until it appears illegible.

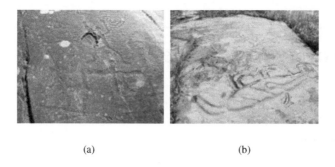

(a)   (b)

FIGURE 16.1
(a) The Neolithic rock art at Paspardo, Val Camonica, Italy. (b) The petroglyph at Parkers Landing, Western Pennsylvania, USA.

Modern rock art sites are normally less than a few hundred years old. They are mainly in habited areas such as on buildings, walls, and tombstones. In contrast to prehistoric rock art, modern art involves human languages. The processes involve modern tools such as cutters or drills. Human factors have also contributed to the erosion process, including selection of stones, installation angles, and air pollution. For example, the Old St. Luke Church, established in 1787 in southern Pennsylvania, has over 100 historic tombstones [11, 20, 45]. Each of them is a valuable source for genealogy, local heritage, and even epidemic studies. The causes of death for those buried here are largely unknown, but some of the deaths were from epidemic diseases that hit the region between 1873 and 1916. The dates on the tombstones bear witness. Sadly, due to weathering, the words on the stones are disappearing.

In this chapter, we investigate how to recover textural patterns such as figures and words from eroded outdoor rock surfaces. Some case studies including representative archeological sites worldwide, from prehistoric to modern eras, will be reported.

16.2 Surface Imaging Methods

Light is critical to observing the eroded rock art. Today, rock art archeologists commonly use a direct light or an indirect light (mirror) to reveal the surface structures, which is simple but unreliable [15]. They also use contact recording methods such as overlay rubbing and tracing to document the rock art. However, these methods involve subjective judgment and potential surface damage. Besides, those methods cannot record any depth information, which is important for revealing packing patterns.

To record a large surface of a sample of rock art, we cannot use contact probers, X-ray, or rotating platforms. Here we will investigate potential field survey methods in three categories: structured lighting, multiview, and image-based digitizing methods. Structured light methods normally need to project lines to the surface and calculate the 3D points from recorded 2D images. The methods include laser scan, pattern projection, and stick shadow. The multiview imaging methods use multiple cameras or multiple images taken from a single camera. The image-based methods do not build a truly 3D model. Instead, they synthesize an image based on interpolating multiple images. The typical methods include Polynomial Texture Maps (PTM).

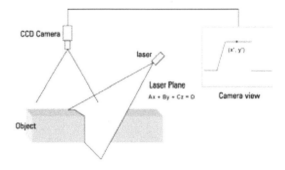

FIGURE 16.2
Triangulation of a laser light.

16.2.1 Laser Scan

A broad spectrum of laser scanners can be used for surface digitizing, ranging from 10 kilometer to sub-millimeter resolutions. The long-range laser scanners use the *time-of-fly* method to measure the distance between the laser and the surface to generate a coarse 3D point cloud. The close-range laser scanners use *structured light* and *triangulation* to generate a fine point cloud. For rock art surface scanning, the structured light and triangulation methods are desirable because the laser light can work on accumulated leaches, dirt, colored and smooth (featureless) surfaces. However, although laser scanning works with most rock surfaces, it often fails to work with highly reflective rock surfaces such as polished marbles.

A structured light laser scan projects a laser light across the surface of the object. It is a very precise version of structured light scanning. It is good for high-resolution 3D digitizing, but it needs many images and takes time. Laser scan can avoid problems due to correspondence point computation and surface colors.

By uniformly swiping laser lines across the surface, the camera captures laser lines in frames. The triangulation algorithm extracts 3D points from the frames. Figure 16.2 shows the triangulation process of the laser light.

The laser point light is transformed into a line through a cylindrical lens and the laser beam sweeps across the object. Assume the f is the camera focus length, the depth from the ray-plane can be found from the intersection of camera ray with the light plane [34].

$$x = x'z/f \tag{16.1}$$

$$y = y'z/f \tag{16.2}$$

$$z = \frac{Df}{Ax' + By' + Cf} \tag{16.3}$$

After the triangulation, we get a 3D point cloud. We then can generate mesh and rendering with artificial lights on the reconstructed surface. We can also transform the 3D data into a 2D image and proceed with pattern recognition manually or automatically.

To date, the laser scanner is perhaps the most accurate sensor for digitizing rock art surfaces. The greatest advantage of this method is that it can generate accurate depth information. For example, it can reach accuracy less than 0.01 mm. Typical laser scanners include the commercial products Cyberware [37], Minolta [32], and Next Engine 3D [35]. However, laser scanners also generate a large amount of data, which consumes a lot of time and power. For a fine-resolution scan, the normal scan size is very limited. For example, with off-the-shelf product NextEngine laser scanner [35], the approximate scan size is about 20 cm by 30 cm, at 1 mm resolution. To cover a large surface area, we need to tile the scanning patches together. In many cases, we can combine different kinds of scanners

FIGURE 16.3
The data flow of the laser scan process.

(a)   (b)

FIGURE 16.4
The BSF site in Val Camonica, Italy (a) and the Old St. Luke Church site in Pittsburgh, Pennsylvania in the United States (b).

for multiple resolutions, for example, the time-of-fly based scanner for large coarse regions and the triangulation-based scanner for fine details [38].

In field trials, we have used the laser scanner based surface imaging system, which contains the following components: (1) a laser projector with a step motor, (2) digital cameras, (3) a laptop with at least 1 GB RAM and 2 GB hard disk space, (3) the 3-D rendering and filtering software, (4) a support equipment such as a tripod or a customized box, (5) a black tent for shading the scanner, (6) a car battery, and (7) a DC to AC inverter if needed. The left image of Figure 16.4 shows a field setup at the BSF site in Val Camonica, Italy. Similar setups were used for the site in Indian Rock (on the right of Figure 16.4) and Old St. Luke Church, Western Pennsylvania (on the right of Figure 16.4). The structural line based 3-D scanning method is used to reconstruct the rock art surface. The scanning hardware contains eight laser diodes, of which four sets are used for micro-mode scanning and four are used for the wide-range scanning. Two sets of 2 MB pixel digital cameras are used to record the laser line images for both modes. The scanner works in two modes: micro mode at the depth resolution of 0.07 mm at a distance between 5.5 to 7.7 inches for a size of 4 inch by 6 inch area, and wide mode at the depth resolution of 0.01 mm at a distance between 16 and 18 inches for a size of 8 by 10 inches. In this study, we used wide scanning mode to cover the whole figure area. As the laser lines swipe across the object, the line deforms according to the surface profile. The laptop receives the image sequences from the camera and runs the triangulation algorithm to generate 3D point data. The data from the scanner can be exported in forms of the XYZ format or Wavefront's OBJ format. In this study, we use XYZ data for filtering and OBJ data for 3D rendering with artificial lighting. The main scheme is illustrated in Figure 16.3.

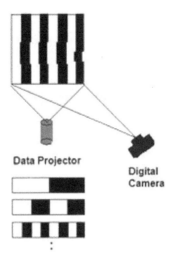

FIGURE 16.5
Binary pattern projection from a data projector.

16.2.2   Pattern Projection

Single structured light beam yields the most accurate resolution but it is rather slow. Projecting multiple beams simultaneously can speed up the digitizing process. However, there is a correspondence problem: which beam is which? To solve the problem, we have to assign each stripe with a unique code spatial pattern over time, for example, in forms of binary or color patterns.

The binary encoding light projects $2^n - 1$ stripes in n images. For example, three progressive binary encoded patterns: 01,0101,01010101 allow the measuring surface to be divided in eight sub-regions. The color encoding light needs few images (one or two). It could be used in real-time 3D imaging. But it needs a more complex correspondence algorithm to determine which color strip is which because the color stripes could be not unique. Pattern Projector is a low-cost improvised 3D imaging method. It started in Italy for digitizing the ancient sculptures [6]. The technology has been commercialized as FlexScan [44], in which users can integrate their own laptop, cameras and projector with the licensed software. The procedure includes: (1) calibrating the camera and projector with a chessboard, (2) taking multiple photos as the projected pattern changes, (3) detecting lines in the captured images, (4) reconstructing 3D points by triangulation, and (5) generating a mesh with a texture image.

The projector can make hundreds of line patterns in a single frame. The computer automatically synchronizes the projector and camera. It's fast and low-cost. Figure 16.6 shows a field setup for scanning gravestones at Old St. Luke Church, in Pennsylvania in the U. S. Unfortunately, a digital projector (2000 L) is power consuming. Ambient light results in noisy data on sunny days.

16.2.3   Stick Shadow

Stick shadow is a weakly structured light method [3, 4, 21, 22], based on a line of shadow from a distant parallel lighting source (e.g., the sun). In this process, the user first calibrates the angle of the shadow with a standing object and a chessboard. Then the user swipes a ruler above the surface so that it casts a black line in a video. By extracting dots from the black line at each frame, a three-dimensional point cloud can be calculated from a triangulation model. In summary, the stick shadow method has the following steps:

First, camera calibration. Most images from an uncalibrated camera are optically distorted,

FIGURE 16.6
Pattern projected onto the gravestone at Old St. Luke Church.

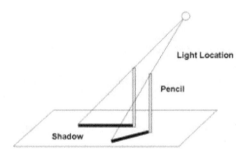

FIGURE 16.7
Light calibration.

especially around the edges. The goal of the camera calibration is to find the intrinsic parameters of the camera and make corrections with the camera calibration model [24]. To do so, we place a chessboard-like calibration board on the scan plane and take a few images with different angles. The calibration algorithm then automatically detects the edges of the black blocks. A Hough Transformation is used to find the straight lines and their intersections on the calibration board. The coordinates are then fed into the calibration matrix.

Second, light calibration, is used to estimate the location of the light source. To do so, we just place a standing pencil on the plane and take at least two images, including the pencil shadow. Since we already know the height of the pencil and the locations of the pencil base and the shadow, we can easily triangulate the location of the light source by finding the intersection of the light rays. When there are more than two input images, we use the mean of the light locations calculated from each unique pair. This reduces the error associated with human input and also compensates for the fact that the light is not a perfect point source. The output of the light calibration is a 3D coordinate of the light. The detailed procedure of the light calibration can be found in the reference [4, 21].

Third, find shadow shape. First we define a rectangular area as the base plane. We make sure that the object is smaller than the base plane so the shadow lines near the top and bottom edges are

FIGURE 16.8
Stick shadow in process.

straight and form a shadow chord (shadow base). We then find the shadow shape by searching for the dark pixels outside of the chord line. Therefore, we get the shape of the shadow.

Finally, extract 3D points. For each image, we have one shadow shape above the chord. From the top pixel of the shadow, we measure the length of the shadow from the tip of the shadow to the chord. Since we know the light location and the intrinsic parameters of the camera, we can calculate the 3D coordinate of the pixel of the tip of the shadow. We repeat this process for all pixels in the image to get a point cloud. This appeared to be a low-cost method for a field survey with excellent mobility. Stick shadow method can be viewed as a "poor man's 3-D scanner." However, the resolution and accuracy are poor due to the uncertainty of the angle of the light, manual swiping velocity, and height. In addition, it doesn't work well with dark colored surfaces.

16.2.4 Multiview Imaging

Multiview imaging is an approach used to reconstruct the 3D information of a scene from two or more photos taken from different viewpoints [14, 18, 24]. Having two or more cameras, we can configure a pre-calibrated stereo imaging system, such as ZScan [48]. However, synchronization of the stereo pair images needs an electronic device or a computer. Recently, handheld camera-based stereography has become an attractive option due to its simplicity. But the calibration and modeling are more challenging than stationary imaging systems. Here we briefly review the process for 3D imaging from a handheld camera.

First, we need to calibrate the camera. We place a calibration board inside the view and take multiple images. Then we use the camera calibration procedure discussed above to process the images and get the intrinsic camera parameters. Then we remove the calibration board and take multiple images with varying angles.

We then detect features and find the correspondent points in image pairs. Correspondence is used to determine which item in the left image corresponds to which item in the right image when two images are involved. In the case of more than two images, correspondence is used to find the points in all images which correspond to the same point in the real world [12]. The computational implementation of the key point (feature) detection and correspondence is based on scale-invariant feature transform (SIFT) [27–29, 46]. The procedures for detecting key points in SIFT are as follows: first the image is convolved with Gaussian filters at multiple scales. Then compare each pixel in the Difference of Gaussians (DoG) images to its neighbors at the same and neighboring scales. If the pixel value is the maximum or minimum among all compared pixels, it is selected as a candidate key point. After obtaining a group of possible key points, their localizations are checked in order to reject false detecting. Finally, a descriptor is used to ensure invariance to image location, scale, and rotation. The biggest advantage of SIFT is the invariance to some of the image properties such as

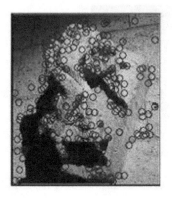

FIGURE 16.9
Example of correspondence points.

image scaling, translation and rotation. And it is partially invariant to illumination changes and affine or 3D projection, which is a practical advantage too. Then the essential matrix and the fundamental matrix [2] are computed to determine the 3D location and structure of the observed objects given corresponding parts of the images and information of the camera. In the practical application, the information given by point corresponding is not enough to construct a decent 3D model, since the 3D points recovered from the images are sparse. As a result, a dense corresponding approach is proposed for stereo matching in order to recover a continuous surface model [43]. Considering the weight and power supply for cameras and its mobility, this approach is excellent for rock art recovery. Two of the recent successful cases are [17] and [36]. However, it is still challenging to model the rock art surfaces because this approach depends on the quality of feature detection, where in many cases features are lost due to poor lighting, contrast, or resolution.

16.2.5 Polynomial Texture Maps (PTM)

Polynomial Texture Maps (PTM) is an image-based representation of the appearance of a surface under multiple lighting directions [30]. The basic idea is to record multiple photos under varying lighting directions and then use a polynomial function to reconstruct images in any lighting direction. PTM assumes that the chromaticity of a particular pixel is constant under varying light source direction; it is largely the luminance that varies. Therefore, for each pixel in the texture map, we have a luminance model that projects each pixel in a light direction space:

$$L(u, v; l_u, l_v) = a_0 l_u^2 + a_1 l_v^2 + a_2 l_u l_v + a_3 l_u + a_4 l_v + a_5 \tag{16.4}$$

where, u, v are texture coordinates, $a_0 - a_5$ are fitted coefficients stored in texture map, and l_u, l_v are projection of light direction into the texture plane.

Given $N + 1$ images, for each pixel, the best fit for $(a_0 - a_5)$ can be obtained using Singular Value Decomposition (SVD) [42]. Figure 16.10 illustrates a basic concept of PTM.

To construct PTM, we need to know the light position, including the lighting direction and distance between the light source and the rock surface. One simple approach is to build a light hemispheric cage with fixed-light positions. Automatic control of lights and camera can acquire a PTM with great speed, e.g., between 5 and 15 minutes. However, fixed-light position equipment has its disadvantages. The light distance from the subject limits the object diameter [33]. The bigger subjects require proportionally larger cage size and a more powerful lamp. To avoid using an elaborate light stage with a known light source position, the user may position a handheld light source at varying locations, and the software can recover the lighting direction from the specular highlights

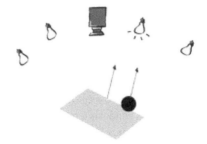

FIGURE 16.10
Illustration of the Polynomial Texture Maps method.

Method	Resolution (mm)	Depth	Mode	Power (W)
Laser Scan	< 0.01	yes	active	300
Pattern Projection	> 0.1	yes	active	500
Stick Shadow	> 2	yes	active	7
Multiview Imaging	> 5	yes	passive	7-14
PTM	< 0.1	no	active	320

TABLE 16.1
Comparison of 3D geometric imaging methods.

produced on a black sphere included in the field of view captured by the camera [33]. To measure and manage the light source radius, the "Egyptian Method" can be applied. This low-tech approach is to use one string with one end tied to the light source and another end tied to the center of the subject. Two people can hold each end straight. So the light distance is measured and the light held steady. A look-up table of distance-dependent light power values can help the field workers to change the string length and light power at the same time. The per-pixel surface normals are extracted from the representation to enhance the surface details. For example, specular enhancement. Simulation of specularity is particularly effective at enhancing the perception of surface shape. This is important to discover patterns in rock art. PTM requires only a single camera and a light for which the angle and distance can be measured. PTM is a low-energy technique that is desirable for field applications. This method has been used to discover a 3,000-year-old cuneiform tablet. The interactive texture map viewer is available online [19] and it has been popular in the archeologist community. Unlike photometric stereo or laser scans, PTM is implemented without the use of 3D geometry, eliminating 3D geometry's associated costs in terms of hardware and software. However, PTM does not provide any depth information which is often useful to study the packing patterns.

Mobility, scalability, and visibility are three important factors when designing a surface scanning system for surveying rock art sites. The scanner should be able to operate on a steep panel with intensive ambient light and heat. It should have enough digitizing resolution to reveal the pecking details within a whole figure area, and its post-processing software should be able to highlight the figures with digital filters.

For a large archeological site, researchers often use two kinds of scanners to cover both large backgrounds with the lower resolution time-of-fly laser range finder and small rock art areas with a fine-resolution scanner such as a laser structure line imaging system [38].

16.3   Pattern Discovery Methods

To discover patterns from a point cloud is a more challenging problem here. Given unorganized point clouds derived from laser scanner data or photogrammetric image measurements, the reconstruction of precise surfaces involves modeling and visualization in the case of incomplete, noisy, and sparse data. This process is beyond general mesh and rendering techniques [13,26]. For rock art, we are interested in surface feature analyses in particular. For example, line detection based on fitting elevated data [47], Hough Transform based vertical line and circle detection [41], Radon Transform based shape recognition [7,8], and shape similarity measurement [40]. It is found that most of the automatic pattern recognition algorithms assume ideal conditions such as an isolated object or a clean background, which is not always the case in rock art data. Recently, researchers proposed to combine modeling with visualization tools to improve the quality of the non-uniform data [13]. Nevertheless, human perception is still in a dominant position for the pattern discovery.

The great challenge to pattern discovery is to find truly novel facts that have never been reported before. In the later part of this chapter, we will test how the interactive system helped to discover interesting patterns in rock art. Obtaining 3D scanning data is just the first step toward feature detection and visualization. In this study, we used multiple filters to enhance the rock art features. One example from this study included a figure of a warrior on a horse at the bottom of the rock. Figure 16.11 shows the original photo of the scanning area. Obviously the figure is impossible to see without a mirror and light.

We use the wide scanning mode (17 inches from the surface) with a counterweighted support frame and sunshade tent (a huge black garbage bag). We exported the 3D data to OBJ and XYZ formats that are common data exchange formats for many 3D modeling software such as MAYA [1], Studio 3D Max [1], Mesh Lab [31], VTK (Visualization Toolkit) [23], and David 3D Fusion [9]. We can do interactive pattern discoveries with the following methods.

16.3.1   Simulated Lighting

Rendering the 3D data with meshes and artificial lights enables us to explore the textural features in the lab instead of in the field. This is a programmable and reproducible process. The mesh rendering software David [9] is used to simulate the lighting effect. We also input the 3D data to software MAYA [1] and use script language MEL to move the artificial lights around.

16.3.2   Data Transformation

The rendered mesh models contain optical illusions to highlight the surface contours, similar to the mirror and light approach. However, the information about the depth is not explicit. Here we have transformation models to map the depth data to a feature space: linear, exponential, sinusoidal, tangent, logarithms, and pseudo colors.

Assume n is the Z-coordinate of current point and m is number of levels, which is

$$m = \frac{Zmax - Zmin}{D} \tag{16.5}$$

where D is the interval along Z-axis, we have:

$$Linear : T = k \times \frac{n}{m} \tag{16.6}$$

$$Exponential : T = k \times \frac{e^{\frac{n}{m}}}{e}; \tag{16.7}$$

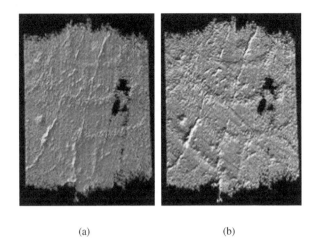

(a)   (b)

FIGURE 16.11
Simulated lighting conditions on the 3D mesh model: back light (a) and lights at 135 degrees and -45 degrees (b).

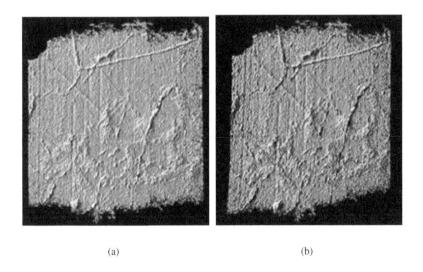

(a)   (b)

FIGURE 16.12
Simulated lighting conditions with different angles: front (a) and 20 degree (b).

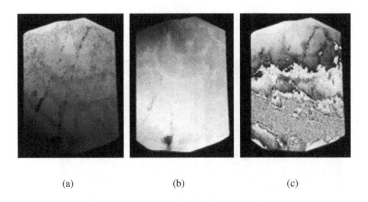

(a)   (b)   (c)

FIGURE 16.13
Linear (a), sinusoidal (b), and tangent (c) transformations of the depth map in grayscale.

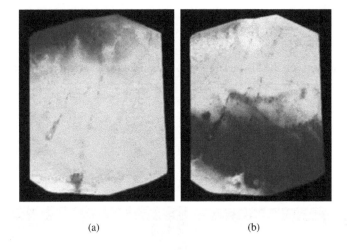

(a)   (b)

FIGURE 16.14
Pseudo colored linear transformation (a) and sinusoidal transformation (b).

$$Logarithmic : T = k \times log_2(\frac{n}{m}); \tag{16.8}$$

$$Sinusoidal : T = k \times sin(\pi\frac{n}{m}); \tag{16.9}$$

$$Tangential : T = k \times tan(\pi\frac{n}{m}); \tag{16.10}$$

where k is a constant. In linear transformation, the pixel intensive is proportional to the depths along the surface normal. If the ground panel is relatively flat, then the linear transformation can show rather uniform outlines of the object. But they are in a lower contrast. Furthermore, the sinusoidal transformation shows the top of outlines clearly.

We also transform the depth map to pseudo color images. Figure 16.14 shows the sample results of the pseudo colored linear and sinusoidal transformations.

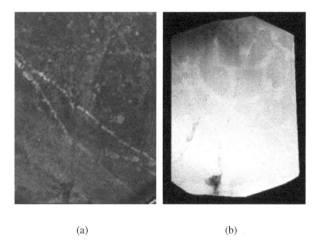

(a)   (b)

FIGURE 16.15
Original photo of the scanning area (a) and the reconstructed Neolithic rock art from BSF site in Val Camonica, Italy (b).

16.3.3 Transformation Examples

Based on interactive visualization and digital filtering, we are able to discover the rock art that was invisible to human eyes. Figure 16.15 shows a Neolithic rock art reconstruction result. Figure 16.16 shows the result of the reconstruction of a 2,000 to 3,000 year old Indian rock art sample.

Another site is the cemetery of Old St. Luke Church near Pittsburgh. Most of the gravestones are over 200 years old and heavily weathered. Using our method, we are able to read the words and figures on the stone surfaces. We have corrected errors in a historical book about the text on the stone. Our work has benefited genealogists and local historians who want to identify the details on the gravestones.

16.3.4 Off-Line Handwriting Recognition

To recognize words or symbols on rock, we begin with OCR (Optical Character Recognition) methods. Traditional OCR methods are quite mature and work well on printed text data. However, rock carving is not printing on paper. In fact, it is handwriting or drawing! Due to the noisy nature of these images and the fact that we have no contextual information, and since they contain mostly names which may be obsolete today or contain rare spellings, applying handwriting OCR detection to recognize individual characters is a more reasonable approach than using printed text OCR methods. Handwriting recognition has to be more robust to variations in the character representation whilst still being able to accurately classify each character and such methods would produce more reliable results with our data.

On-line recognition methods are very popular in mobile devices. They use gesture or writing sequences to achieve a high recognition rate. Unfortunately, we have no way to know the writing sequences of ancient rock art. Therefore, we have to use the off-line handwriting recognition methods. Off-line handwriting recognition involves the automatic conversion of text in an image into letter codes [25, 39]. The off-line handwriting recognition is comparatively difficult, as different people have different handwriting styles. Nevertheless, limiting the range of input can allow recognition to improve. For example, the ZIP code digits are generally read by computer to sort the incoming mail. The recognition of digits and letters in digital images is tackled here by extracting each separate

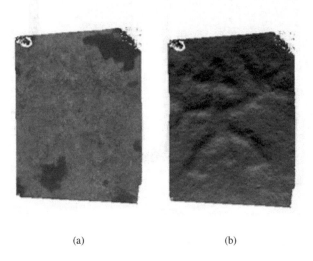

(a)   (b)

FIGURE 16.16
The original surface image of the Indian rock art (2,000 - 3,000 years old) (a) and the reconstructed human figure (b).

(a)   (b)

FIGURE 16.17
The original tombstone (218 years old) (a) and the reconstructed surface with artificial lighting (b).

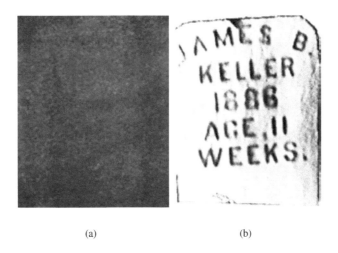

(a)   (b)

FIGURE 16.18
Original tombstone at Old St. Luke Church (a) and the linear filtered reconstruction (b).

character as a glyph and attempting to match this glyph to a prior known set of targets. Since our target database contains only the characters we want to identify and not the exact shapes or fonts, we implement a series of preprocessing algorithms to produce a character representation that allows for generic recognition across similar shapes.

For standard OCR using printed text data such as images of text, faxes, or scanned text accurate results can be achieved, however using images which contain non-printed text as part of the background adds further levels of complexity to the problem. In this case, the depth image of a gravestone (Figure 16.19(a)) is used and although the font seems quite generic we have no exact samples for it; along with this the image is quite noisy. Preprocessing algorithms can produce clean images of the data for an OCR algorithm to work on. The first problem is how to threshold the image to a binary color space, as parts of the background are grayed (Figure 16.19(b)). Simple thresholding is not dynamic and can result in joining of adjacent letters or the splitting of some letters into separate parts along weaker connections. We implemented histogram equalization to introduce greater separation between the regions; then we used a high threshold to cut away only the darkest parts (Figure 16.19(c)). A dilation and erosion algorithm is implemented on this data to try and recombine broken letters, without combining nearby characters (Figures 16.19(c)) and 16.19(d). A thinning algorithm is then implemented on this data to try to extract the basic character shape (Figure 16.20 top). While the thinning process can accentuate the splitting effect and can cause the loss of minor details in the image (such as the center lobe of the "E" character) the process has a normalization effect, allowing the basic shape of the character to be recognized without the excessive noise seen in the original image.

To recognize the words, a vector representation is calculated for each glyph; this vector representation is compared against those representations held in the database to determine the closest match. The vector can be calculated using any of a number of different descriptors or it can combine several. The simplest way is either a pixel based template matching or a comparison of the bounding box aspect ratio, pixel density, regional features and measures of compactness (circularity, rectangularity); more complex methods can take into account vector features such as the position and direction of "limbs" on the digits. It computes the correlation between an input image, and a sequence of templates gave us reasonable results (Figure 16.20 bottom). More advanced models include recurrent neural networks for estimating probabilities for the characters represented in the skeleton, and the hidden Markov model that calculates the best word in the lexicon [39].

FIGURE 16.19
Preprocessing for the engraved text.

FIGURE 16.20
Thinned text and the recognition result.

In our study, we have not included the semantic context information in pattern recognition. If we incorporate a large database of words or figures, it would increase the accuracy of the recognition, similar to our reading in daily life. We can recognize misspelled words. The reference [25] provides a survey about large vocabulary off-line handwriting recognition methods.

16.3.5   Interactive versus Automatic Discovery

Ideally, we want to have automated processes to discover hidden patterns in a large database of rock art worldwide. Several algorithms have been developed for 3D pattern recognition. For example, automated 3D line detection [47], circle detection [41], shape recognition [7, 8], and shape similarity measurement [40]. Unfortunately, millions of eroded rock art heritages have not been digitized. Many artifacts are not visible to cameras or human eyes. To identify the meaningful patterns requires both machine and human intelligence, especially archeological domain knowledge, field experience, and environmental awareness. Perhaps that's why archeologists haven't used robots to do the fieldwork yet. Normally, interactive processes are used for early stage discoveries in the field; automated processes are used for later stage analysis in a lab to test hypotheses, develop theories, and compare patterns. Diverse environments, styles, and carbon dates often destroy statistical assumptions; and the sparse samples prevent us from making sense out of data mining, because rock art pieces don't simply repeat themselves. At the end, manual adjustments are often needed in an automated process for pre-processing, post-processing, and model parameter tuning for selected data. In reality, all processes are in fact interactive. What is the true value of the computerized pattern recognition? In archeology of rock art, many theories and expertise are untested and unreliable. For example, how to distinguish a natural circle versus man-made? Experts found that naturally broken-away circular patterns have sharp edges, but man-made pecking patterns have smoother edges. A computerized model can verify the hypothesis, just like a lab experiment, which is repeatable and measurable. It takes the guesswork out of debates in the field. Furthermore, computer models can simulate physical and chemical impacts on rock art and can turn the clock back or forth. The Internet increases the information flow for rock art studies. The public involvement enables interactive discoveries on a grand scale. For example, genealogy researchers use the Internet to discover the family trees from online databases and gravestones. A stream of feedback was received after the author posted the reconstructed text on the gravestones at Old St. Luke Church [10]. A few errors were corrected based on the online feedback. The Internet has become a venue for public "debugging" and discovering. Nevertheless, the ultimate challenge here is to discover truly novel facts or knowledge that has never been reported before. The objective of this study is to build a semi-automatic process that maximizes the visibility of surface patterns and minimizes guessing and labor-intensive measurement. We have developed interactive surface modeling and visualization software based on triangulation-based laser and pattern projection data. In the system, we enhance the human-computer discovery process with data transformation to support humans' effective logic and visual reasoning.

16.4 From Reconstruction to Knowledge

In many cases, heavily eroded rock surfaces result in missing data. Modeling and visualization are not enough. Domain knowledge and logic reasoning must be included in the process to solve the puzzle. Taking the tombstone reconstruction as an example, we found that the horizontal tombstones are the most vulnerable to erosion. In this case, we can only reconstruct a portion of the words on the stone, for example, Captain David Steel's tombstone at the Old St. Luke Church. We reconstructed a few words shown in Figure 16.21. According to the local historian [11], the words came from John Gay's "The Beggar's Opera." See also the plaque on the exterior wall of Trinity Cathedral, Oliver Avenue, Pittsburgh. The poem in the historic book was [11]:

This world's a farce,
And all things show it;
Once I thought so,
But now I know it.

Based on our measurement of the gaps between words and characters, we found it is impossible to fit the word "I" in between the words "once" and "thought." We discovered that the text on the tombstone is actually like this:

This world's a farce,
And all things show it;
Once thought so,
But now I know it.

After we discussed this with the historian and the author of the book, we agreed to omit the word "I" in front of "thought".

In addition, reconstruction is also a learning process that crosses fertile visual analysis with semantic knowledge. For example, we learned the old English used "%" as "and" ("&") without prior knowledge. We found that based on the semantic reasoning, it must be an "and" semantically. Therefore, the symbol "%" must be a "&."

16.5 Interaction Design

Digital heritage is the property of humankind. The Internet enables people to view exquisite rock art remotely and in a timely manner. The 3D digitized models enhance the pattern recovery and add more dimensions to archeology, culture, and genealogy studies. Interaction design here serves

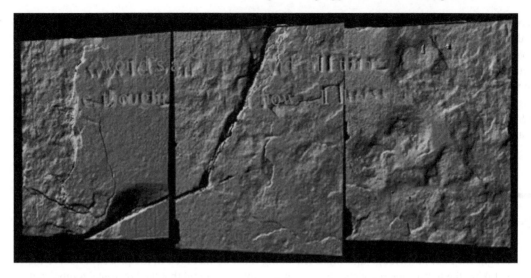

FIGURE 16.21
Partially reconstructed Captain David Steel tombstone (the red text is the recognized words).

not only as an education channel, but also as a collective intelligence vehicle throughout the world. After we posted our results of the identified tombstone text, our team received lots of feedback and corrections from genealogy researchers and family members of the buried. The dynamics encouraged us to integrate the virtual reality technologies with the digital heritage data. We view it as a digital surrealism, mixing the virtual world with the reality behind the surface.

Figure 16.22 shows a prototype of the digital cemetery for the Old St. Luke's Church, where we superimposed the reconstructed text on top of the virtual tombstones. The first-person point-of-view enables the system to dynamically collage the reconstructed digital heritage data and present it to the user in real-time. In addition, the system is able to integrate geographical information such as an ice age glacier into the rock art site, increasing the scientific ambient intelligence.

16.6   Conclusions

In conclusion, we found that structured lighting is the key for rock art reconstruction. Laser and pattern projection scanners are the most accurate 3D digitizing devices. Polynomial Texture Map (PTM) is an effective imaging and reconstruction tool without depth map. Besides, multiview imaging and stick shadow are opportunistic low-cost alternatives in the field. However, their accuracy is yet to be improved. For rock art applications, we need to balance the factors of mobility, power consumption, and accuracy. Digital transformations and artificial lights are critical for reconstructing visible surface patterns. An interactive process is applied to debug the reconstructed text. In many cases, multiple modal approaches are used to discover the hidden patterns. The meaningful discoveries cannot happen without an intimate understanding of the background knowledge, such as the linguistics, typology, and crafting. Using data transformation, researchers have indeed discovered something new and corrected an error in the historical book with the interactive pattern discovery toolkit. Rock carving is not printing on paper. It is handwriting. To automatically recognize words or symbols on rock, we use the off-line handwriting OCR detection technology to recognize individual characters. The results show that it is more robust than the OCR for printing text. Finally, the interaction design

FIGURE 16.22
Digital cemetery of Old St. Luke Church, where reconstructed text is superimposed on the tombstones.

methodology is presented to combine the virtual world with reconstructed data to form a surreal world that enables us to see something invisible to our naked eyes - a similar experience to viewing an x-ray image of the body.

Acknowledgments

The author would like to thank Samantha Stevick, Pat Sweeney, Cheong Kin Ng, Shizhe Liu, Iryna Pavlyshak, and Yingchung Xia from Carnegie Mellon University for their assistance. Furthermore, many thanks to Professor Angelo Fossati of Università Cattolica del S. Cuore, Italy, Rev. Canon Richard W. Davies of Old St. Luke Church, and Scott Township and Kenneth Burkett from Jefferson County History Center, Brookville, Pennsylvania, for contributing their expertise in archeology and field studies to this study.

Bibliography

[1] AutoDesk. http://usa.autodesk.com/. *captured on 26 April*, 2010.

[2] B.G. Baumgart. *Geometric modeling for computer vision*. Stanford University, 1974.

[3] J.V. Bouguet and P. Perona. 3D photography on your desk. In *Proceedings of the Sixth International Conference on Computer Vision, IEEE Computer Society*, 1998.

[4] J.Y. Bouguet and P. Perona. Closed-form camera calibration in dual-space geometry. *European Conference on Computer Vision*, 1998.

[5] K. Burkett and E. Kaufman. On the rocks at parkers landing. *Pennsylvania Archeologist, Vol. 75, No.1, Spring*, 2005.

[6] C. Montani P. Pingi C. Rocchini, P. Cignoni, and R. Scopigno. A low cost 3D scanner based on structured light. *EUROGRAPHICS*, 2001.

[7] P. Challenor, P. Cipollini, and D. Cromwell. Use of the 3D radon transform to examine the properties of oceanic rossby waves. *Journal of Atmospheric and Oceanic Technology*, 18, 2001.

[8] P. Daras, D. Zarpalas, D. Tzovaras, and M.G. Strintzis. Shape matching using the 3D radon transform. In *Proceedings of the 2nd International Symposium on 3D Data Processing, Visualization, and Transmission (3DPVT'04)*, 2004.

[9] David-Scanner. http://www.david-laserscanner.com/. *captured on 26 April*, 2010.

[10] R. Davis. http://www.oldsaintlukes.org/stones.htm. *captured on May 11*, 2010.

[11] R.W. Davis. *Rebellion and revelation: The history of Old St. Luke's Church*. Old St. Luke Church, 1996.

[12] E.Trucco and A.Verri. *Introductory techniques for 3D computer vision*. Prentice Hall, 1998.

[13] R. Fabio. From point cloud to surface: The modeling and visualization problem. *International Workshop on Visualization and Animation of Reality-Based 3D Models*, February 2003.

[14] O. Faugeras. *Three-dimensional computer vision*. The MIT Press, 1999.

[15] A. Fossati. Tracing the past. *http://www.rupestre.net/tracce/tracing.html*, 1997.

[16] A. Fossati. But they are only puppets. problems of management and educational programs in the rock art of Val Camonica and Valtellina, Lombardy, italy. *Rock Art Research,* 20(1): 25-30, 2003.

[17] Y. Furukawa and J. Ponce. Carved visual hulls for image-based modeling. *ECCV*, pages 564–577, 2006.

[18] R. Hartley and A. Zisserman. Multiple view geometry in computer vision. *Cambridge University Press*, 2003.

[19] http://www.hpl.hp.com/research/ptm/. *captured on May 11*, 2010.

[20] C. Jimenez. Scans reveal lost gravestone text. *BBC News, 3 October*, 2007.

[21] J.Y. Bouguet and P. Perona. Camera calibration from points and lines in dual-space geometry. *Proc. 5th European Conf. on Computer Vision*, 1998.

[22] J.Y. Bouguet and P. Perona. 3D photography using shadows in dual-space geometry. *International Journal of Computer Vision*, volume 35, number 2:564–577, 1999.

[23] Kitware. http://www.vtk.org/. *captured on 26 April*, 2010.

[24] R. Klette, K. Schluens, and A. Koschan. *Computer vision - three-dimensional data from images*. Springer, 1998.

[25] A.L. Koerich, R. Sabourin, and C.Y. Suen. Large vacabulary off-line handwriting recognition: A survey. *Pattern Anal Applic.* 6:97-121, 2003.

[26] M. Lemmens. 3D laser scanner software. *GIM International*, September 2006.

[27] D.G. Lowe. Object recognition from local scale-invariant features. *International Conference on Computer Vision,* Corfu, Greece (September 1999), pp. 1150-1157, 1999.

[28] D.G. Lowe. Local feature view clustering for 3D object recognition. *IEEE Conference on Computer Vision and Pattern Recognition,* Kauai, Hawaii (December 2001), pp. 682-688, 2001.

[29] D.G. Lowe. Distinctive image features from scale-invariant keypoints. *International Journal of Computer Vision,* 60, 2, 2004.

[30] T. Malzbender, D. Gelb, and H. Wolters. Polynomial texture maps. In *Proceedings of SIG-GRAPH 2001.*

[31] MeshLab. http://meshlab.sourceforge.net/b. *captured on 26 April,* 2010.

[32] Minolta. http://konicaminolta.com. *captured on April 30,* 2010.

[33] M. Mudge, T. Malzbender, C. Schroer, and M. Lum. New reflection transformation imaging methods for rock art and multiple-viewpoint display. In *The 7th International Symposium on Virtual Reality, Archeology and Cultural Heritage (VAST),* 2006.

[34] S. Narasimhan. http://www.cs.cmu.edu/afs/cs/academic/class/15385-s06/lectures/ppts/lec-17.ppt. *captured on May 11,* 2010.

[35] NextEngine. http://www.nextengine.com. *captured on April 30,* 2010.

[36] M. Pollefeys et al. Visual modeling with a hand-held camera. *International Journal of Computer Vision,* 59(3).

[37] E. Keogh Q. Zhu, X. Wang and S.H. Lee. Augmenting the generalized Hough transform to enable the mining of petroglyphs. *Proceedings of 15th ACM SIGKDD,* 2009.

[38] F. L. Van Scoy, J. Jarrell, and G. Wagaman. Cemetery preservation and laser scanning. In *Proceedings of the Seventh International Conference on Virtual Systems and Multimedia (VSMM'01),* IEEE Press, 2001.

[39] A. Senior and A. Robinson. An off-line cursive handwriting recognition system. *IEEE Trans. on Pattern Analysis and Machine Intelligence,* Vol. 20, No. 3, pp. 309-321, 1998.

[40] H.Y. Shum, M. Hebert, and K. Ikeuchi. On 3D shape similarity. *CMU-CS-95-212,* November 1995.

[41] P.K. Sinha and Q.H. Hong. Detection of vertical lines and circles in 3D space using Hough transform techniques.

[42] http://en.wikipedia.org/wiki/svd. *captured on April 30,* 2010.

[43] E. Tola, V. Lepetit, and P. Fua. A fast local descriptor for dense matching. *Proceedings of Computer Vision and Pattern Recognition (CVPR),* June 2008.

[44] T. Tong. Flexscan3d. *http://www.3d3solutions.com,* 2007.

[45] J. Vellucci. CMU team's scanner unlocks secrets from the past. *Pittsburgh Tribune-Review,* October 8, 2007.

[46] L.L. Williams. Reduced sift features for image retrieval and indoor localisation. *Australasian Conference on Robotics and Automation,* 2004.

[47] D.M. Woo. Generation of 3D building model using 3D line detection scheme based on line fitting of elevation data. *PCM 2005, part 1, LNCS 3767,* pages 559–569, 2005.

[48] htttp://www.menci.com/zscan

17

Copyright Protection of Digital Images of Cultural Heritage

Vito Cappellini, Roberto Caldelli, Andrea Del Mastio, Francesca Uccheddu

University of Florence
Email: *vito.cappellini@unifi.it, roberto.caldelli@unifi.it,
andrea.delmastio@unifi.it, francesca.uccheddu@unifi.it*

CONTENTS

This chapter discusses *digital watermarking* technology and how it can be used to protect digital documents (copyright protection). In the field of cultural heritage (CH), the protection of images is a crucial issue, stated in the importance of their diffusion, both with dissemination aims and in order to attract visitors to real museums. Digital marking techniques are a key application for the protection of Cultural Heritage digital images.

In the following, digital marking techniques are presented, with particular attention given to the "perceptual" invisibility of the mark (in order to not deteriorate the image itself) and its "robustness" (the strength of the mark, that is, its resistance to removal). Examples concerning artworks of the Polo Museale Fiorentino are shown.

The chapter is organized as follows: in the introductory section and subsections, a brief history of watermarking is presented; the basic elements of a digital watermarking system are then explained, from data embedding to data recovery. The given description is very general to avoid confusion with data-hiding applications. After that, some application scenarios for watermarking techniques will be provided in the third subsection of the introduction, which explains how such tools can be implemented in practice by paying particular attention to digital assets representing Cultural Heritage. Two sections follow, discussing some concepts about the fundamental features of watermarking for digital images (2D contents, Section 17.2) and for digital models (3D digital objects, Section 17.3).

17.1 Introduction

The aim of watermarking techniques [46] is to hide in a multimedia object some perceptually invisible codes (signatures), usable for carrying information related to legal or commercial properties of the data contents.

Such a code should be easily and reliably identifiable, its insertion should not perceptually deteriorate the content, and its identification should also be possible after some processing has been made on it [59]. With reference to this last aspect, lately the robustness of the watermark (that is its capacity to be resistent to manipulations) has become an important requirement [71] to be satisfied.

17.1.1 A Brief History of Watermarking

A first application of watermarking systems can be found in the *Histories* of Herodotus, where the following story, which took place around 480 B.C., is reported. Histiaeus wanted to secretly notify the regent of the Greek city of Miletus to start a revolt against the Persian occupier. Histaeus chose an ingenious, albeit rather slow, secret communication method: to shave the head of a slave, tattoo the message on his skull, allow the hair to grow back, and finally dispatch the slave to Miletus; there the slave was shaved again to disclose the secret message. In this way, the physical communication medium is kept out of plain sight.

The secret communication can also be carried in plain sight, for example, using the following method developed in ancient China. The message sender and recipient share identical copies of a paper mask with holes cut out at random locations. The sender places the mask over a sheet of paper, writes the secret message through the holes, removes the mask, and fills in the blanks with an arbitrary composed message, giving the appearance of an innocuous text. This method was reinvented 500 years ago by the Italian mathematician Cardan and has become known as the Cardan grille.

A commercial application in which information is camouflaged in a visible physical medium is logarithmic tables used in the 17th and 18th centuries. Errors were deliberately introduced in the least significant digits in order to assert intellectual property rights.

In the examples shown above, casual inspection of the message carrier fails to detect the presence of hidden information. Moreover, a secret code is used to embed the information, for example the location of the holes in the paper mask and the location of the numerical errors in logarithmic tables.

17.1.2 Watermarking Basics

A watermarking system, which can be considered a communication system, is composed of three main elements: a transmitter, a communication channel, and a receiver. The embedding of the to-be-hidden information within the host signal (e.g., a digital image) plays the role of data transmission; any processing, applied to the watermarked content along with the interaction between the hidden data and the host data itself, represents the transmission through a communication channel. The extraction of the hidden information from the host data constitutes the receiving part (see Figure 17.1). The to-be-hidden information represents the basic input of the system. Such information can be considered as a binary string $\mathbf{s}$ usually named with the term *watermark code*. The string $\mathbf{s}$ is inserted within a piece of data generally called host data or host signal I (e.g., an audio file, an image, a piece of video, etc.). The embedding module may optionally accept a secret key K as an additional input.

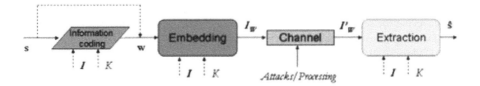

FIGURE 17.1
The watermarking model.

Such a key, whose main goal is to introduce some secrecy within the embedding step, is usually used to make the recovery of the watermark impossible for non-authorized users which do not have access to such key.

In many watermarking systems, the information message **s** is not introduced directly within the host signal (see Figure 17.1). On the contrary, before insertion such message is transformed into a *watermark signal* **w** which is more suitable for embedding. Similarly to what happens in digital communications, the watermark code may be used to modulate a much longer spread-spectrum sequence or it may be transformed into a bipolar signal where zeros are mapped in $+1$ and ones in -1. In addition to this, before transforming the watermark code into the watermark signal, **s** could be channel-coded to increase robustness performance of the watermarking system. After that, during watermark casting, an embedding function C takes the host document I, the watermark signal **w**, and, possibly, a key K, and generates the watermarked asset I_W. How C is defined usually depends on the selection of a set of features extracted from the digital data, named *host features*; such features are altered according to the watermark signal to convey information. These features can be extracted directly from the asset domain or can be obtained by applying a transformation (i.e., Discrete Fourier Transform [DFT] or Discrete Cosine Transform [DCT]) which necessarily has to grant invertibility to achieve the watermarked content I_W.

An important aspect to be taken into account during embedding is that the hiding operation has to be imperceptible. This can be accomplished either by properly choosing the set of host features and the embedding rule, or by introducing a concealment step after watermark embedding. To this aim, the properties of the human senses must be carefully studied, since imperceptibility ultimately relies on the imperfections of such senses. Thereby, still image and video watermarking will rely on the characteristics of the Human Visual System (HVS), whereas audio watermarking will exploit the properties of the Human Auditory System (HAS). After this watermarking phase, the marked asset enters the channel, that is it can undergo a series of manipulations. Modifications may explicitly aim at maliciously removing the watermark from the data, or may have been done simply for data compression or editing.

At the end of the whole system the receiver is located. Basically two kinds of receiver can be designed. The first one is a detector which takes as inputs the watermarked asset (optionally the original one in the non-blind-case), the watermark code **s** and the secret key K, and answers if **s** is contained or not in the document. Alternatively, the receiver may work as a decoder: in this case the watermark code **s** is not known in advance, but it will be the output of the watermark decoder. The two different schemes make a distinction between algorithms embedding a mark that can be read (*readable*) and those inserting a code that can only be detected (*detectable*). Note that in readable watermarking, the decoding process always results in a decoded bit stream; however, if the asset is not marked, decoded bits are meaningless. Detectable watermarking is also known as 1-bit watermarking (or 0-bit watermarking), since, given a watermark, the output of the detector is just yes or no.

Some notes concerning the main features of a watermarking framework have been given. Some references to better deepen the argument are also given in the following. An interesting overview of watermarking covering the second half of the 20th century can be found in [14], while in [17,18] it is possible to find some protocol issues that were brought to the attention of watermarking researchers by S. Craver, N. Memon, B. L. Yeo, and M. M. Yeung, demonstrating the problems derived from the adoption of a non-blind watermark detection strategy. As data hiding has grown, terminology and symbolism tended to get more and more uniform; however, even now after ten years since digital watermarking first came to the attention of researchers, a complete agreement on a common terminology has not been reached. A first attempt to define data-hiding terminology and symbolism can be found in [56], while in [15] an effort as been made to define a non-ambiguous terminology.

17.1.3 Watermarking Application Scenarios

Now that we have presented the basic watermarking model and its main components, let us see now which are the possible uses of such a technology in relation to some application scenarios. The main chosen framework, that is the more pertinent with the case of digital assets representing cultural heritage, is Intellectually Property Rights (IPR) protection, and, in particular, within this context, two specific scenarios are discussed: demonstration of rightful ownership and fingerprinting. Before going into detail, it is worth underling that with the term IPR the protection of the rights possessed by the creator, or the legitimate owner, of a multimedia piece of work is intended. Furthermore two other scenarios, which seem to be interesting to the debate are the use of watermarking for digital media authentication and for hiding embedded annotations.

Ownership of the rights. One of the most interesting scenarios in which watermarking can be called to operate is surely to assess the rights ownership of a specific digital document. In this case, the author of a work wishes to prove that he is the only legitimate owner of the work [31]. To do so, as soon as he creates the work, he also embeds within it a watermark identifying him unambiguously. The watermark could bring information regarding the name of the owner or an identifier of the digital archive where the asset is contained and so on. It is worth pointing out that the watermark may still be used by the rightful owner for his own purposes. For example, the author may wish to detect suspicious products existing in the distribution network. Such products could be individuated by an automated search engine looking for the watermark presence within all the works accessible through the network. Then, the author may rely on a more secure mechanisms to prove that he was the victim of a fraud, for example by depositing any new creation to a Registration Authority. A common way to confer the watermark verification procedure a legal value is to introduce the presence of a Trusted Third Party (TTP) in the watermarking protocol. For example, the watermark identifying the author may be assigned to the owner by a trusted Registration Authority. In this way, in fact, it would be by far more difficult to invert the watermarking operation, especially when blind watermarking is used, since pirates can not rely on the design of an ad hoc fake original work. As to the requirements a watermarking algorithm to be used for rightful ownership verification must satisfy, it is obvious that for any scheme to work, the watermark must be a secure one, given that pirates are obviously interested in removing the watermark, possibly by means of computationally intensive procedures. In addition, private watermarking is preferable, due to its inherently superior security. Finally, capacity requirements depend on the number of different author identification codes the system must accommodate for.

Fingerprinting. A second classical application of digital watermarking is copy protection [10]. In such a scenario, a so-called copy deterrence mechanism is adopted to discourage unauthorized duplication and distribution. Copy deterrence is usually achieved by providing a mechanism to trace unauthorized copies to the original owner of the work. In the most common case,

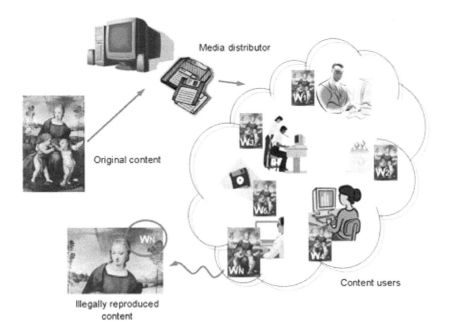

FIGURE 17.2

Schema of fingerprinting technique: the original content is marked with different watermarks (one for each of the users); if an illegal use of the fingerprinted media is made, the traceability is guaranteed by the mark itself (courtesy of Ministero per i Beni e le Attività Culturali, Soprintendenza per il Patrimonio Storico Artistico ed Etnoantropologico e per il Polo Museale della Città di Firenze [1]).

distribution tracing is made possible by letting the seller (owner) insert a distinct watermark, which in this case is called a fingerprint, identifying the buyer, or any other addressee of the work, within any copy of data which is distributed. If, later on, an unauthorized copy of the protected work is found, then its origin can be recovered by retrieving the unique watermark contained in it. Of course, the watermark must be secure, to prevent any attempt to remove it, and readable, to make its extraction easier. In Figure 17.2 it is possible to see a schema depicting how the fingerprinting technology acts: note that the readability requirement may be relaxed if the owner has the possibility to guess in advance the watermark content. A problem with the plain fingerprinting protocol described above is that it does not take into account buyers rights, since the watermark is inserted solely by the seller. Thus, a buyer whose watermark is found in an unauthorized copy can not be inculpated since he can claim that the unauthorized copy was created and distributed by the seller. The possibility exists, in fact, that the seller is interested in fooling the buyer. Let us consider, for example, the situation where the seller is not the original owner of the work, but an authorized reselling agent. The seller may distribute many copies of a work containing the fingerprint of buyer B without paying the due royalties to the author, and claim that such copies were illegally distributed or sold by B. As in the case of rightful ownership demonstration, a possible solution consists in resorting to a Trusted Third Party. The simplest way to exploit the presence of a TTP to confer a legal value to the fingerprint protocol is to let the TTP insert the watermark within the to-be-protected work, and retrieve it in case a dispute resolution protocol has to be run. Despite its simplicity, such an approach is not feasible in practical applications, mainly because the TTP must do too much work, then it may easily become the bottleneck of the whole system. In addition, the protected work must be transmitted from the seller to the TTP and from the TTP to the customer, or, in an even worse case, from the TTP to the seller and from the seller to the customer, thus generating very heavy traffic on the communication channel.

Authentication. Another important application scenario regards authentication of a digital document by means of *fragile* or *robust* watermarking [4, 47]. With the term fragile watermarking it is intended that the hidden information is lost or altered as soon as the host signal undergoes any modification: watermark loss or alteration is taken as an evidence that data has been tampered with, whereas the recovery of the information contained within the data is used to demonstrate data integrity and, if needed, to trace back to data origin. Additional functionalities could include the capability to localize tampering, or to discriminate between malicious and innocuous manipulations (e.g., moderate image compression). In the latter case, a semi-fragile watermarking scheme has to be used, since it is necessary that the hidden information survives only a certain kind of allowed manipulations. On the contrary, the use of robust watermarking for data authentication relies on a different mechanism: a summary of the host signal is computed and inserted within the signal itself by means of a robust watermark. Information about the data origin is embedded together with the summary. To prove data integrity, the information conveyed by the watermark is recovered and compared with the actual content of the sequence: a possible mismatch is considered as evidence of data non-authenticity. The capability to localize manipulations will depend on the accuracy of the embedded summary. If tampering is so heavy that the watermark is lost, watermark absence is simply taken as evidence that some manipulations occurred and the output of the authentication procedure is a negative one. Note that in this case watermark security is not a pressing requirement, since it is unlikely that someone is interested in intentionally removing the watermark. On the contrary, hackers would be interested in modifying the host data without leaving any trace of such alteration. Though the approaches to data authentication relying on fragile (semi-fragile) and robust watermarking may seem rather different, it is possible to demonstrate that both of them can lead to the same applicative framework.

Hidden notations. The last application scenario concerns the adoption of watermarking for the insertion of hidden notations within the host document [14]. Despite the fact that digital watermarking is usually looked at as a means to increase data security, such a data-hiding scheme can be simply regarded as the creation of a side transmission channel, associated to a piece of work. Interestingly, the capability of the watermark to survive digital to analog and analog to digital conversions led to the possibility of associating the side channel to the work itself, rather than to a particular digital instantiation of the work. This interpretation of digital watermarking involves different potential applications, in which the watermark is simply seen as annotation data, inserted within the host work to enhance its value. The range of possible applications of annotation watermarks is very large. Note that the requirements annotation watermarks must satisfy can not be given without carefully considering application details. In many cases, watermark capacity is the most important requirement; however, system performance such as speed or complexity may play a predominant role. As to robustness, the requirements for annotation watermarks are usually much less stringent that those raised by security or copyright protection applications.

17.2   2D Watermarking

The copyright protection of multimedia contents is a typical field where watermarking can be applied; however, since the protection of the rights of the creator, or the legal owner, of a multimedia masterpiece involves many different aspects, including also moral rights protection, e.g., the insurance that the integrity of the work is respected not to violate the moral beliefs of the owner/creator, watermarking applications also involve a number of related issues. In general, data-hiding technologies allow one to provide a communication channel multiplexed into original content [33], through which it is possible to transmit some information, depending on the application at hand, from a sender to a receiver.

As an example, let's say a watermarking has to be used to identify the owner of a multimedia document, in an unambiguous way: the owner can simply insert into the document a code (watermark) stating his identity. Thus, the watermark has to be as robust as possible, that is nobody except the owner is able to remove the code from the documents (for a more precise definition of *robustness* see the following); this prevents a hacker from removing the original code and substituting it with a personal code, thus "stealing" the Property Rights on the document. However, this is not the only way a hacker can harm the owner's rights: for example, he can also simply add his own watermark to the watermarked document, making it impossible to state who is the legitimate owner by simply reading the watermark (or watermarks) contained in the multimedia contents. To be more precise, let's say that Alice (A) is the owner of a document (D) she wants to protect; to do this, she adds to it a watermark with her identification code, thus obtaining a watermarked version of her document $D_{w_A} = D + w_A$. She is thus able to publish her document/work.

Let's suppose that a third party, say Bob (B), takes the watermarked media and adds his own watermark code on it, thus obtaining another version of the document ($D_{w_A w_B} = D + w_A + w_B$). Since in this last version of the original document D, Alice and Bob's watermarks are present, it is impossible to define the property of the rights on it: in fact, both of them are able to demonstrate that the document has their code on it. This is a way to confuse the ownership evidence provided by the watermark. A way to solve the matter is to ask both Alice and Bob to show the original asset, that is the document where no watermarking codes are present; actually, this is easy to be done by Alice (the owner), while it seems to be impossible for Bob (the hacker), who owns a copy of the document with the watermarking code by Alice. And now, the hacker can harm Alices property rights in a different way. Let's suppose that Alice's watermarking technique is not-blind, that is the detector needs the original documents to extract the code (see the following for further details), let's say, for sake of simplicity, by a difference between the watermarked document and the original version. Alice can use the true original asset to show that Bobs asset contains her watermark and that she possesses an asset copy, D_{w_A} containing w_A but not w_B:

$$D_{w_A w_B} - D = D + w_A + w_B - D = w_A + w_B; \qquad (17.1)$$

$$D_{w_A} - D = D + w_A - D = w_A \qquad (17.2)$$

But the problem is that Bob can also do the same thing, that is to demonstrate that he owns a document where only his code is present, while in another asset both Alice and Bob's codes are present: he can simply build a fake original by subtracting his watermarking code from the Alice-marked asset:

$$D_f = D_{w_A} - w_B = D + w_A - w_B \qquad (17.3)$$

In this way, the detector acts like this:

$$D_{w_A} - D_f = D + w_A - (D + w_A - w_B) = w_B \qquad (17.4)$$

thus demonstrating that Bob is able to show a original D_f, which is perceptually identical to the original document D, where Alice's watermarking code is not present. This is just a simple example to focus on the fact that watermarking can be useless to prevent some kind of abuses, unless a proper protection protocol is established. Similarly, a watermarking system has to satisfy a number of properties, which cannot be exactly defined unless focusing on the particular scenario in which the system has to be used; thus, the same algorithm can be decisive with respect to some application, while useless with respect to other fields.

In the following, watermarking system properties are briefly reviewed.

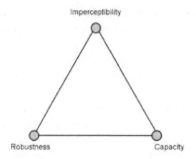

FIGURE 17.3
Imperceptibility, capacity, and robustness: a trade-off is needed depending on each application.

17.2.1   Capacity

With the term capacity we mean the amount of information bits that a watermarking is able to convey. Capacity is a fundamental property of watermarking algorithms, making techniques suitable or not for specific uses.

Capacity is not directly dependent on the particular algorithm used, while it depends on the host signal, the attack strength, etc.; it can be also focused on the application of the watermarking system which determines the requirements for the capacity. In the watermarking framework, security has a very precise meaning; among the various items that can be found in literature, we state that *security refers to the inability of unauthorized users to access the additional channel created with watermarking* [33].

From what is stated above, it is also clear that capacity requirements contend with robustness and imperceptibility requirements, since the more capacity one wants the less robust and imperceptible the mark is. In Figure 17.3 a scheme is depicted, showing that a trade-off is needed when dealing with watermarking algorithms and their applications.

17.2.2   Insertion of Multiple Watermarking Codes

The insertion of more than one code in a document or a content can sometimes be a request for the watermarking algorithm. This can be the instance for a copyright protection scheme, where the need can arise to insert both the author and the consumer names, the last being the legal user of the contents.

The requirements for such kinds of application are to not deteriorate the quality of the host data, as well as to be able to correctly extract all the inserted watermarks.

The presence of more than one watermark inserted in a host signal can lead to some ambiguities in the interpretation of the information hidden within the protected piece of work, thus suggesting that the system designer carefully consider this opportunity.

17.2.3   Robustness

The robustness of a watermarking signal refers to the capability of the hidden data to be not removed from the host asset by manipulations, both malicious and not-malicious. We consider as malicious all the manipulations explicitly aiming to remove the watermark and damage the hidden information; not-malicious are for example the manipulations not aimed at making the watermark unreadable (even if they are able to do it), such as for example the JPEG compression in images.

Robustness depends on the particular application we are dealing with; generally speaking, it is possible to define four qualitative robustness levels, able to embrace most of the situations:

Secure watermarking: This is the typical case of the copyright protection environment; the requirement on the watermark is to survive both non-malicious and malicious manipulations. The hidden data can be lost only if the host data is so degraded so as to be not useful for the application at hand. Considering malicious manipulations, it is assumed that attackers are able to conceive ad-hoc watermark removal strategy, knowing the watermarking algorithm. Non-malicious manipulations include instead a huge variety of processing tools, including lossy compression, linear and non-linear filtering, cropping, editing, scaling, noise addition; strictly speaking of images, we consider also zooming and shrinking, rotation, contrast enhancement, histogram manipulations, and row/column removal or exchange. Note that the need on the watermarking system is not to be completely unbreakable, but it has to be strong enough, that is watermark breaking does not need to be impossible (which probably will never be the case), but only difficult enough.

Robust watermarking: The requirements on the watermarking are less strict than in the previous case, demanding only that the watermark has to be resistant against non-malicious manipulations. The application field of it includes all the situations in which the involved actors are not interested in removing the mark, even if the normal use of the data themselves involve a number of manipulations, which have to not damage the hidden data.

Semi-fragile watermarking: In applications where robustness is not the principal requirement, semi-fragile watermarking is suitable to the aim. Examples can be fields where the host signal can suffer only some minor manipulations, such as data labeling for archival retrieval, where the only manipulation performed on the data is a compression of the watermarked content (archives are usually created by compressed data). More generally, in such a framework the watermark has to survive only to a limited, well-specified, set of manipulations leaving the quality of the host document virtually intact.

Fragile watermarking: This is a case which is completely opposite to the ones defined above. In such applications, hidden data are irremediably altered as soon as any modification is applied to the host signal, both on the whole content (global) or on a part of it (local). This kind of watermarking system is usually adopted to authenticate data (data authentication), where the presence of the hidden data ensures that no manipulation has been performed, while the absence of it demonstrates that some sort of processing has been performed.

Since many signal processing applications aim to modify perceptually insignificant parts of the signal itself (such as compression, whose aim is to not deteriorate the signal even if reducing its size), robustness is better achieved if the embedded signal is placed in perceptually significant parts, which are able to better survive to signal processing. Robustness against malicious and non-malicious manipulations is probably the most important requirement for watermarking algorithms, especially in certain application contexts such as for Intellectual Property Right (IPR) protection. For this reason, one of the main issues in Digital Watermarking of the last years has been the evaluation of the robustness of watermarking algorithms. In particular, a series of benchmarks has been studied and developed. One of the first benchmarks for image watermarking has been the StirMark [55] package, developed by Fabien A. F. Petitcolas [53, 54] while earning his PhD at Cambridge University in 1997. StirMark benchmark is one of the most used benchmarks for still image watermarking. Other contributors currently include researchers at INRIA, Eurécom, University of Magdeburg, and USTL-LIFL. Some of the image alterations implemented in StirMark are: cropping, flip, rotation, rotation+scaling, Gaussian filtering, sharpening filtering, linear transformations, random bending, aspect ratio changes, line removal, and color reduction. Jane Dittmann, Andreas Lang, and others are also working on an audio version of StirMark called AudioStirMark. Other benchmarks for still image watermarking are the Optimark software, developed by Artificial Intelligence and Information Analysis Laboratory of the Aristotle University of Thessaloniki, Greece, and the Checkmark, developed at the Computer Vision and Multimedia Laboratory of the University of Geneva. Optimark benchmark includes several image attacks plus some watermarking algorithm performance evaluation methods such as statistic to evaluate detector performances (e.g., Bit Error Rate (BER), probability of false detection, probability of missing detection), and the estimation of mean embedding and detection time. The CheckMark software is one of the most recent benchmarks for still images and includes some new classes of attacks such as non-linear line removal, collage attack, denoising, wavelet compression (JPEG2000), projective transformations, copy attack, etc. The CheckMark package is primarily developed by Shelby Pereira [52].

17.2.4   Blind and Non-Blind Techniques

Watermarking algorithms differ also from the way they are able to recover hidden data. We call *blind* (or *oblivious*) a technique which doesn't need a comparison between the original and marked data to extract the watermark, while we call *non-blind* the techniques which needs such a comparison. One can think, and even early works in digital watermarking insisted on it, that blind techniques are less robust than non-blind ones, since the original unknown data where the mark is hidden have to be treated as disturbing noise. This is not completely true: theoretically, it can be demonstrated that under some particular conditions blindness doesn't cause any loss of capacity and robustness (the host signal is known by the encoder); however, at a more practical level, some sort of loss in capacity and robustness is present for blind algorithms with respect to non-blind ones, even if the loss is lower than people may expect. Another consideration has to be made: in most of the application scenarios the original data is not present, since the marked version of it is the one distributed; thus, non-blind techniques are completely useless for this set of practical applications.

17.2.5   Private and Public Techniques

Concerning watermarking, *private* and *public* attributes refer to users allowed to recover the watermark; more precisely, a watermark is called private if only authorized users are able to recover it, for example by assigning to each user a different secret key, whose knowledge is necessary to extract the watermark from the host document. In contrast to private watermarking, techniques allowing anyone to read the watermark are referred to as public. Non-blind techniques are private by themselves, since only by owning the original data is possible to recover the mark; thus only the author or the document owner is able to. A famous principle known as Kerkhoff's principle [36] states that security can not be based on algorithm ignorance, but rather on the choice of a secret key; as a consequence it is possible to assert that private watermarking is likely to be significantly more robust than public watermarking, since once the embedded code is known, it is much easier for an attacker to remove it or to make it unreadable, for example by inverting the encoding process.

17.2.6   Readable and Detectable Watermarks

Once the watermark is inserted in the host document, it is later possible to read it again or simply to detect its presence; this distinction leads to two different classes of watermarking schemes. In readable watermarking, the bits composing the mark can be read with no knowledge of them in advance; the response of the watermarking algorithm is thus the watermark itself. On the contrary, in detectable watermarking it is only possible to verify if a given code is present in the document, thus the algorithm answers only with a present/non-present information. It is clear that detectable watermarking techniques are intrinsically private, since it is impossible for an attacker to guess the content of the watermark without knowing anything about it. The practical use of a watermark

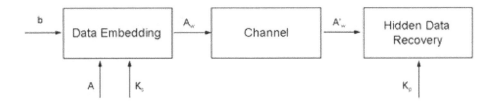

FIGURE 17.4
In asymmetric watermarking two distinct keys, K_s and K_p, are used for embedding and retrieving the watermark.

strongly depends on the readable/detectable characteristics of the hidden mark: even if an a priori knowledge of the inserted watermark is very difficult in many applications, detectable watermarks are, generally speaking, more robust than readable ones, because of their private nature and of the smaller payload they can convey.

Note that, given a readable watermarking scheme, it is easy to construct a detectable scheme, for example simply by comparing the retrieved information against the to-be-searched code.

17.2.7   Invertibility and Quasi-Invertibility

As previously described, the watermarking process can be inverted by reverse engineering, having thus the possibility of building a fake original document where to insert a fake watermark and resulting in a watermarked asset which is equal to the one obtained by the "real" original and watermark. This is a particular kind of attack, aiming to remove or modify the information the hidden data carry.

Starting from this general concept, it is possible to define more precisely the invertibility feature. Let A be a digital media, and let's assume a non-blind detectable watermarking scheme to ensure the property of the content. We also call $\mathcal{E}$ the embedding function, while $\mathcal{D}$ identifies the detection function; thus it is possible to say that:

The watermarking scheme is invertible if for any asset A it exists an inverse mapping $\mathcal{E}^{-1}$ such that $\mathcal{E}^{-1}(A) = \{A_f, w_f\}$ and $\mathcal{E}(\{A_f, w_f\}) = A$, where $\mathcal{E}^{-1}$ is a computationally feasible mapping, and the assets A and A_f are perceptually similar. Otherwise the watermarking scheme is said to be non-invertible.

The previously mentioned attack can be extended to a more sophisticated version, stating that the effectiveness of the attack doesn't depend on the need that the insertion of the fake watermark within the fake original asset produces an asset which is identical to the initial one, i.e. A, but it is only needed that the application of the detector $\mathcal{D}$ to A, by using the fake original asset as original non-marked document, results in revealing the fake watermark. Mathematically, this means that $\mathcal{D}(A, A_f, w_f) = yes$.

Thus, similarly to one given for the invertibility, it is possible to give a definition for the quasi-invertibility feature:

A non-blind watermarking scheme, characterized by an embedding function $\mathcal{E}$ and a detector function $\mathcal{D}$, is quasi-invertible if for any asset A it exists an inverse mapping $\mathcal{E}^{-1}$ such that $\mathcal{E}^{-1}(A) = \{A_f, w_f\}$ and $\mathcal{D}(A, A_f, w_f) = yes$, where $\mathcal{E}^{-1}$ is a computationally feasible mapping, and the assets A and A_f are perceptually similar. Otherwise the watermarking scheme is said to be non-quasi-invertible.

We call A_f and w_f respectively fake original asset and fake watermark. The above mentioned

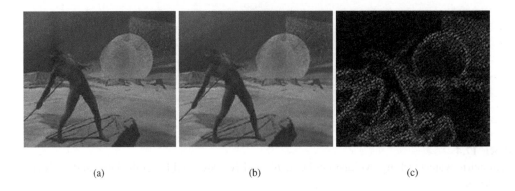

(a)   (b)   (c)

FIGURE 17.5
Riccardo Saldarelli, "Verso il 2000" (courtesy of the author): (a) the original image, (b) the water-marked image, (c) difference between (a) and (b) multiplied by a factor of 32 (Tuscany & Gifu Art Gallery - http://lci.det.unifi.it/Projects/ArtGallery/index.html).

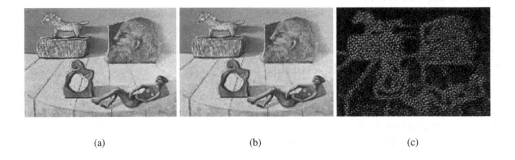

(a)   (b)   (c)

FIGURE 17.6
Pirzio, "Composizione uomini di sempre" (courtesy of the author): (a) the original image, (b) the watermarked image, (c) difference between (a) and (b) multiplied by a factor of 32 (Tuscany & Gifu Art Gallery - http://lci.det.unifi.it/Projects/ArtGallery/index.html).

definitions are related to detectable non-blind techniques; the concept of watermarking invertibility can anyway be easily extended to readable or blind techniques. Great attention should be given to watermark invertibility as well. In [18] it was stated that for a watermarking scheme to be successfully used to demonstrate rights ownership, non-invertibility of the watermark has to be granted.

17.2.8 Reversibility

The *reversibility* deals with the possibility to *extract* (which is different with respect to *detect*) a watermark from the host data. In particular, we can speak of a *strict-sense reversibility* when the mark can be completely removed from the host asset, and the original media can be exactly recovered with no degradations applied on it. We can similarly speak of a *wide-sense reversibility*, when the watermark can be extracted and the recovered original asset doesn't present any perceptible distortion. From the above stated definitions, it is clear that the strict-sense reversibility implies the wide-sense reversibility, while the opposite is not true.

17.2.9 Asymmetry

We define as *symmetric* watermarking algorithms the ones using the same set of parameters for the embedding and detection/decoding phase; such parameters include for example a secret key purposely introduced to bring in some secrecy in watermark embedding, and all the parameters defining the embedding process, e.g., the number and position of host features, all of them inserted in a secret word used for the embedding/decoding. The usage of the same secret word for both the insertion and detection phase means that both the owner and the final user share it, and this can lead to security problems, in particular for applications involving consumer scenarios, for which a widespread diffusion is forecasted: the knowledge of the secret word can in fact be sufficient for pirates to have enough information concerning the watermarking technique, thus being able to force it and to recover the original asset; even if the recovery key could be kept secret, it is possible to effectively estimate the detection boundary in a reasonably simple way.

To overcome this problem, asymmetric techniques have been developed [23] (see Figure 17.4). These kinds of methods allow public watermark recovery without the need to disclose enough information for watermark removal; in practice, asymmetric schemes use a private key (that is kept secret) to perform watermark embedding, and a public key (available to anybody) to perform watermark recovery, in such a way that:

- It should be computationally impossible to estimate the private (embedding) key from the public (recovery) key.

- The knowledge of the public key should not help the attacker to effectively remove (or make unreadable) the watermark.

In general, asymmetric schemes show a degradation of robustness with respect to symmetric techniques, making them more sensitive to all the manipulations which have been mentioned above; practically, the amount of degradation to be introduced for removing or making unrecoverable the watermark is much lower than for symmetric methods. Moreover, with reference to the two characteristics that should be exhibited by asymmetric algorithms listed above, and generally speaking, it is possible to make another consideration. While it is impossible or at least very difficult to recover the embedding key from the recovery parameters (which is the first constraint), it is not too difficult to remove the watermark based solely on the recovery key, that is, they fail to satisfy the second requirement. More details about the importance of asymmetric watermarking in security-oriented applications may be found in [2]. The interest in asymmetric watermarking was triggered by the works in [21, 22, 65]. Since then researchers have investigated the potentiality offered by asymmetric schemes [16, 48]; however, an ultimate answer on whether asymmetric watermarking will permit one to overcome some of the limitations of conventional methods has not yet been given.

17.2.10 Examples

Some examples of the use of watermarking for copyright protection are shown in Figure 17.5 and Figure 17.6, where the original, the watermarked, and the difference (between the original and marked) images are shown. As it can be seen, the original and marked images are perceptually identical; to better appreciate the changes occurred when applying the watermark, some specific processing is applied.

The watermarking technique is also applicable in the field of Cultural Heritage; to prove this, in Figure 17.7, Figure 17.8, and Figure 17.9 some images concerning the "Annunciazione" by Leonardo da Vinci are shown. The imperceptibility of the watermark is a fundamental requirement in this field, and figures show how watermark acts.

(a)

(b)

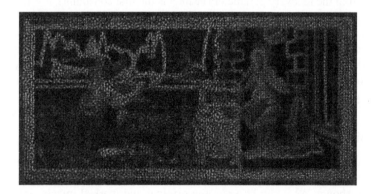

(c)

FIGURE 17.7 (SEE COLOR INSERT)
Leonardo da Vinci, "Annunciazione": (a) the original image, (b) the watermarked image, (c) difference between (a) and (b) multiplied by a factor of 32 (courtesy of Ministero per i Beni e le Attività Culturali, Soprintendenza per il Patrimonio Storico Artistico ed Etnoantropologico e per il Polo Museale della Città di Firenze [1]).

(a)   (b)   (c)

FIGURE 17.8

Leonardo da Vinci, "Annunciazione," detail of the Archangel Gabriel: (a) the original image, (b) the watermarked image, (c) difference between (a) and (b) multiplied by a factor of 32 (courtesy of Ministero per i Beni e le Attività Culturali, Soprintendenza per il Patrimonio Storico Artistico ed Etnoantropologico e per il Polo Museale della Città di Firenze [1]).

(a)   (b)   (c)

FIGURE 17.9

Leonardo da Vinci, "Annunciazione," detail of the Saint Mary: (a) the original image, (b) the watermarked image, (c) difference between (a) and (b) multiplied by a factor of 32 (courtesy of Ministero per i Beni e le Attività Culturali, Soprintendenza per il Patrimonio Storico Artistico ed Etnoantropologico e per il Polo Museale della Città di Firenze [1]).

17.3 3D Watermarking

The watermarking of 3D objects, usually referred to as *3D watermarking*, is a new frontier of digital watermarking concerning the embedding of information into 3D objects in an imperceptible way. The interest in 3D watermarking technology is constantly growing in these last years due to the large diffusion reached by 3D models. In fact, geometric data are widely used in a large number of contexts, for example in mechanical engineering for virtual prototyping and simulation, in architecture to preview the final impact of a project, in Cultural Heritage to realize virtual museums and archeological sites reproduction, in scientific visualization, in entertainment industries for movies and video games, in the Internet to build virtual worlds and for e-commerce applications, and so on. The explosion of this new multimedia data has been supported both by the tremendous graphics power reached by the modern consumer graphics accelerator board and by the effort spent by the Computer Graphics Research community to develop efficient algorithms to build, edit, and compress 3D objects. Now, this trend has reached the digital watermarking community and the number of works related to 3D watermarking is increasing every year. In the following, except where otherwise specified, when we refer to a 3D object we assume a triangular mesh representation because this is the most used representation for 3D models.

17.3.1 Requirements

Watermark *imperceptibility* is a crucial requirement for 3D watermarking. In fact, it is very important that a watermarking system for 3D objects guarantees that the watermarked model has the same visual appearance of the original one since the typical intended use of a 3D model is viewing. The problem is that any watermarking algorithm introduces distortions in the watermarked media. Such distortions are caused by the alterations affecting the host features during the embedding process. In the case of image and video watermarking the knowledge of the Human Visual System (HVS) has been widely applied to develop methods to make such distortions imperceptible [3,57,68]. The imperceptibility is achieved by processing the image or video content and evaluating, for each region of the image, how much the modifications will result in perceptible appearance. Then, the watermark is embedded only in those parts of the image, or of the video, such that the changes will be less perceptible. This process is called *visual masking*. From our knowledge, no analogous studies exist for 3D objects. In fact, to mask the visual distortions in a 3D model is very difficult for two reasons. The first one is that the visibility of the watermark depends on the particular rendering conditions used to visualize the mesh, i.e., by the shading algorithm used, by the surface properties, by the textures, and so on. The second reason is that, typically, the user interacts with the 3D model and hence the model can be observed from several viewpoints, thus making the perceptual analysis of the introduced visual artifacts more difficult than in the case of image and video. So, watermark imperceptibility for 3D watermarking requires new studies and approaches.

3D watermarking is a young technology, in fact no benchmark tools for 3D watermarking algorithms have been developed. So, there are no "standards" to evaluate the robustness of 3D watermarking algorithm. Typically the attacks used to test the robustness performances of 3D watermarking are extrapolated from the image attacks. Obviously, such attacks require one to be adapted to the meshes. In addition to these adapted image attacks a lot of other sophisticated and difficult-to-prevent attacks such as mesh optimization and remeshing are possible on a polygonal mesh. Table 17.1 shows the battery of mesh attacks we propose and the corresponding image attacks. Note that some mesh attacks have no corresponding image attack. In the following we describe in detail such attacks.

Translation/Rotation/Uniform Scaling. These simple geometric transformations are commonly used in Computer Graphics to position a 3D model inside a scene. In fact, usually a virtual scene

Image Attacks	Mesh Attacks
Translation/Rotation/Scaling	Translation/Rotation/Scaling
Additive Noise	Additive Noise
Down-sampling	Simplification
Up-sampling	Refinement
Compression (e.g., JPEG)	Mesh Compression
Low-pass Filtering	Smoothing
Cropping	Cropping
Bi-dimensional Free-Form Deformation	Three-dimensional Free-Form Deformation
Second watermark insertion	Second watermark insertion
⋆ ⋆ ⋆	Vertices and/or faces re-ordering
⋆ ⋆ ⋆	Re-triangulation
⋆ ⋆ ⋆	Remeshing

TABLE 17.1
Comparison between image and mesh attacks.

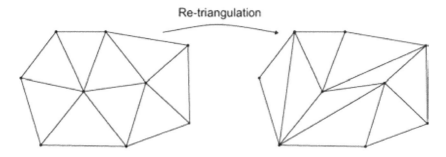

FIGURE 17.10
Re-triangulation attack.

is composed by hundreds of objects that must be rotated, scaled, and translated to obtain the final results. So, this is a basic non-malicious attack to consider. Other more complex geometric transformations such as affine and projective transformations may be used to attack the model, even if they are less common than plain translations, rotations, and isotropic scaling.

Noise. Noise attack consists in the addition of uniform or Gaussian noise to the digital media in order to compromise the watermark. Several ways to add noise to a geometric surface are possible. One way is to perturb the mesh vertices position by adding them random displacement vectors. If the maximum modulus of such vectors is small with respect to the mesh dimensions the overall shape of the model is preserved. Figure 17.11 shows an example of this process.

Re-triangulation. This attack regards the topology of the mesh. It is performed by changing the connections between mesh vertices leaving their position unaltered (Figure 17.10). One important characteristic of this attack is that it is very easy to implement and very fast to apply, so a lot of pirates may use it to attack the watermarked mesh.

Vertices and/or faces re-ordering. This attack consists in re-ordering the vertices, or the faces, of the mesh preserving the aspect of the mesh exactly. For example the vertex v_i can be swapped with the vertex v_j by taking care to modify opportunely the indices of those triangles v_i and v_j belong to. Further, the list of the triangles of the mesh can be ordered in any way. This attack is specific of 3D objects since, as previously noticed, the samples of other media such as audio, image, and video, have an implicit order that preserves them to the intensive use of this attack.

The conclusion is that it is not possible to assign to a vertex a particular meaning during the embedding or the decoding/detection phase unless such vertex is identified without the use of its indices.

Mesh smoothing. Mesh smoothing is a geometric operation that is usually performed on the meshes obtained through three-dimensional scanning. In fact, usually 3D scanners produce "noisy" surfaces due to the approximation errors introduced during the surface reconstruction process. Such errors could be compensated by fairing mesh techniques [19, 32, 38, 39, 61, 62, 64]. In Figure 17.11 an example of the Bunny model after the application of Taubin filter is showed. This filter acts on a mesh as a low-pass filter attenuating the roughness of the surface.

Cropping. Cropping concerns the disjunction of a part of the model. Pirates can discard the pieces of the model that they do not need (e.g., the hand of a statue). This simple attack presents a lot of complications since often it can cause synchronization problems during the extraction phase. Figure 17.11 shows the Bunny model cropped with a plane.

Mesh compression. Despite many years of compression research a widely supported compression standard for 3D models, such as JPEG for images and MPEG for videos, does not exist in practice. The attempts to establish similar compression standards (e.g., MPEG-4, compressed-binary VRML) did not succeed. The main reason for this is that a single compressed format suitable for all cases is difficult to achieve. In fact, 3D models can either have or not have texture information, there can be pre-computed colors (rarely), material attributes present high variability in their data form, and so on. Furthermore, it is difficult to define a widely accepted standard and, consequently, the robustness performance against compression is difficult to evaluate. As just pointed out the compression schemes for polygonal meshes are typically lossless for the connectivity and near-lossless for the geometry. Hence, in general, to take into account the effects of the compression of the mesh on the watermark it should be sufficient to consider the effect of the quantization of the vertex coordinates. We remember that typically each vertex coordinate is quantized with 12 bits instead of the 32 bits of the standard float representation of the uncompressed case.

Simplification. *Mesh simplification*, also referred to as *decimation*, regards the reduction of the number of vertices and triangles of a polygonal mesh model while preserving its shape. Typically, an atomic simplification operation is defined and applied several times to obtain the simplified mesh. An example of atomic simplification is the removing of a vertex with a successive retriangulation of the produced hole. The sequence of atomic simplification steps to apply to the model is built using a similarity metric. The similarity metrics measure the impact of each atomic step and determine the quality of the simplified output mesh. Such metrics can be of two types: geometric and image-based. Some examples of algorithms that use geometric metrics are [13, 25], and image-based metrics are [42, 45, 67]. This kind of processing can be considered as a non-malicious attack when it is applied to optimize the mesh, i.e., to eliminate over-sampling of the model's surface.

Refinement. The refinement operation regards the insertion of new vertices, edges, and faces into a triangular mesh. This operation could be performed for different reasons. For example, a refinement could be done to improve the quality of the mesh, i.e., the uniformity of the area and angles of its triangles. Another application context is finite element analysis of a certain model and/or surface, for example the evaluation of the stress suffered by a mechanical piece during its work condition. Even in this case the mesh may be refined to build the basic elements to perform the finite element analysis. So, this attack belongs to the category of non-malicious attacks. Obviously, the insertion of new vertices, edges, and faces with malicious intentions is always possible.

Remeshing. Remeshing is used to regularize a mesh converting an irregular mesh into a semi-regular [20, 27, 41] or a completely-regular [26, 43, 60] one. In short, this operation can be described as a geometric re-sampling of the shape of the model followed by a re-definition of the vertices connections in order to give the mesh vertices the desired valence. From our knowledge there are no experimental results about the resistance of 3D watermarking algorithms against this attack. Our feeling is that to make a watermarking system robust against remeshing is very difficult. Fortunately, the use of this attack is limited by its characteristics since it is very complex to implement and very computationally expensive.

17.3.2 3D Objects Representation

Basically, a 3D model is composed of a collection of geometric data, representing the shape of the model, plus other data defining the visual appearance of the model, such as surface properties, textures, and colors. For other kinds of digital media such as audio, image, and video, 3D objects data can be represented in a lot of different ways. In this section we focus on geometric data representation, providing a panoramic of the most used representations and presenting the peculiarities of each. The visual appearance of the model will be discussed in the third part, where the perceptual issues related to 3D watermarking will be debated.

One first categorization of 3D objects representation can be done by considering whether the surface or the volume of the object is represented:

Boundary-based: the surface of the 3D object is represented. This representation is also called *b-rep*. Polygonal meshes and implicit and parametric surfaces are common *b-rep* representations.

Volume-based: the volume of the 3D object is represented. Voxels and Constructive Solid Geometry (CSG) are commonly used to represent volumetric data.

Usually the representation of a 3D model depends on the way the model has been created and by its application context. The typical applications of each representation will be provided in the following. Concerning 3D model generation a lot of different ways are possible; to give some examples, a 3D model can be generated by scanning a real 3D object, can be designed by hand using geometric modeling software, can be automatically obtained from the analysis of medical images, can be reconstructed from a set of images through the use of Computer Vision techniques, etc.

Polygonal Meshes

Intuitively, a 3D *polygonal mesh* is a collection of vertices, edges, and faces in 3D space joined together to form the shape of the 3D object. Generic meshes are rarely used in Computer Graphics, the most used meshes are triangular and quadrilateral ones. A *triangular mesh* is a mesh composed by triangles only, in the same manner a *quadrilateral mesh* is composed only by quadrilaterals. In the following, when we refer to a mesh, we always intend a triangular mesh. This assumption implies no loss of generality since every polygon of a non-triangular mesh can be triangulated to obtain a triangular mesh. A mesh is *orientable* if each of its faces is oriented in the same way. The *orientation* of a face is given by the order of the vertices that form it; in particular two faces f_1 and f_2 have consistent orientation if for each shared edge its vertices appears in opposite order in the description of f_1 and f_2 (see Figure 17.12). Conventionally, the mesh orientation is called clockwise or anti-clockwise. After assuming that a face has a clockwise (anti-clockwise) orientation, the face oriented in the opposite way has an anti-clockwise (clockwise) orientation. Another important characterization of meshes is their *regularity*. A mesh is called *irregular* if its vertices can have any valence, *completely-regular* if all vertices have the same valence, and *semi-regular* if most of its vertices have the same valence except a small number that can have any valence. This last definition arises because a semi-regular mesh is obtained by repeatedly subdividing an irregular mesh. During the subdivision process the irregular vertices of the initial mesh remain irregular

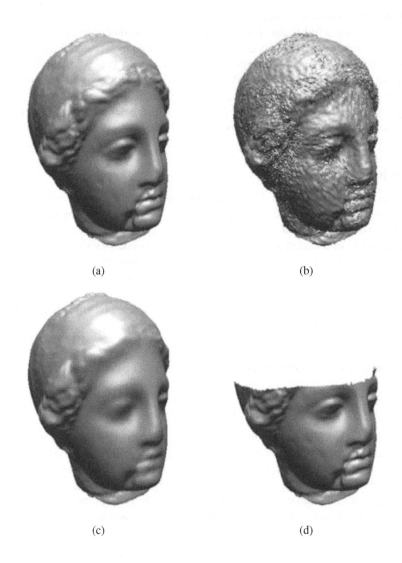

(a)   (b)

(c)   (d)

FIGURE 17.11
Mesh attacks. (a) Original model. (b) Additive noise attack. (c) Smoothing. (d) Cropping.

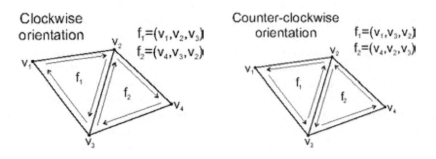

FIGURE 17.12
Mesh orientation.

while most of the newly inserted vertices converge to valence six (for a triangular semi-regular mesh). This classification is very important since regular meshes are at the basis of *digital geometry processing*. For this reason in the following we will give a detailed description of semi-regular meshes and other strictly related topics such as subdivision surface, mesh parameterization, and remeshing. In particular, we will see how the properties of semi-regular meshes can be used to build a multi-resolution framework based on wavelet analysis; our 3D watermarking method is based on such a framework.

Polygonal meshes present several limitations. First of all, since they are a *discrete* representation, curved surfaces can only be approximated; in fact each vertex can be seen as a sample of the curved surface the mesh represents. So, this representation is not compact, i.e., a highly detailed model requires a huge amount of data to be represented. Another problem is that direct editing is not easy, in fact, for this purpose, other representations are typically used, in particular parametric surfaces such as NURBS. Finally, there is no natural parameterization. Despite all of these problems the Computer Graphics research community has put a lot of effort in mesh processing and a huge number of applications use this kind of representation. One of the main reasons for this is that meshes are the common denominator of the other representations, i.e., it is easy to convert other representations to this one. Another motivation is that modern graphics hardware is able to render thousands of millions of triangles every second, in other words, graphics accelerated board, are designed to work with triangular meshes.

17.3.3 State of the Art

Digital watermarking of 3D objects is far from the level of maturity of other watermarking technologies such as the ones for audio, images, and video.

One of the first works about 3D watermarking was done in 1997 by Ryutarou Ohbuchi and other researchers of the IBM Tokyo Research Laboratories [49]. In 1998 the first multiresolution approach was proposed by Kanai et al. [34] and after one year a non-blind technique with very good robustness properties was developed by Praun et al. [58]. At the middle of 2000 Oliver Benedens from the Fraunhöfer Institute of Technology, Germany, in his work "Towards Blind Detection of Robust Watermarks in Polygonal Models," presented the first paper about blind and robust watermarking for 3D objects [7]. More recent is the work of Kim et al. [37] that presents a novel robust to topological attacks and blind watermarking schema. Until now, many other techniques have been proposed by many other researchers and the effort to develop 3D watermarking techniques is constantly increasing. The development of a blind and robust 3D watermarking algorithm yet is still a difficult task.

In the following, a detailed review of the most important 3D watermarking algorithms for polygonal meshes will be presented. The algorithms have been categorized according to the host features used for the embedding. Usually watermarking techniques are divided into four main categories: those operating in the *asset domain*, be it the spatial domain or the time domain depending on the asset; those operating in the *transformed domain*, for example in the DCT or DFT domain; those operating in the *hybrid domain*, retaining both spatial/temporal and frequency characterization of the host asset; and, finally, those operating in the *compressed domain*, i.e., working directly on the compressed version of the host asset.

Asset Domain

The most straightforward way to embed information within a digital host asset is to modify the asset in its original signal space, in other words the host features correspond to the signal samples. For audio signal this means that the watermark is embedded in the *time domain*; for still images this means the *spatial domain*, and so on. The main peculiarity of embedding in the asset domain is that in this way the temporal/spatial localization of the watermark is trivial, thus permitting a good control of the local distortions introduced by the watermark. For 3D watermarking in the asset domain several options are available; the watermark can be embedded in the topology, in the geometry, or in the appearance attributes of the 3D model.

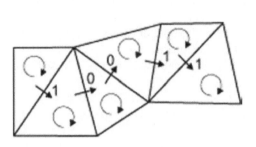

FIGURE 17.13
Example of information encoding in a trian-gle strip (TSPS).

FIGURE 17.14
Example of Macro Density Pattern (MDP) embedding (from [49]).

Topological features. For *topological embedding*, or embedding using the topology of the model, we intend both those algorithms that encode the information in the connectivity of the model and those algorithms that use the topology of the model to drive the alterations of the geometry of the mesh. In other words all the algorithms for which the topology plays a fundamental rule in watermark casting. Some remarkable examples of this kind of algorithm can be found in the first work by Ohbuchi et al. [49], where four watermarking algorithms for polygonal meshes based on topological features were presented. Another example of topology-driven mesh alterations embedding is the Tetrahedral Volume Ratio (TVR) algorithm. This algorithm has been designed to be robust against affine transformation. In fact, its host feature, i.e., the ratio of two tetrahedral volumes, it is invariant against affine transformation. In short, the embedding consists of the identification of a sequence of tetrahedron on the input mesh and then to embed the information by modifying the ratio of the volume of these tetrahedrons. Like the TSQ algorithm, TVR is a blind, readable watermarking scheme. The connectivity of a strip of triangles encodes the information in the Triangle Strip Peeling Sequences (TSPS) algorithm. The base idea is to peel off a sequence of triangles, to be more specific a strip of triangles, and then encode the information in the strip. After the "peeling" operation the strip is connected to the rest of the mesh only by its first edge. By assuming that the information to embed has the form of a string of bits, the strip is created in the following way: starting with an edge of a triangle, the two opposite edges identify two corresponding triangles. If the mesh to watermark is orientable these two triangles can be ordered in a clockwise or counter-clockwise way. According to the selected orientation, a "0" or "1" is assigned to the first and the second triangle. Obviously, some strings cannot be fitted by the topology of the input mesh, for example the strip can hit a boundary of the mesh or circle-back to itself. To avoid such problems the shape of the triangle strip is manipulated by alternating data symbols with *steering symbols*, i.e., symbols that steer the direction of growth of the triangle strip. An example of such kind of connectivity-based information encoding is shown in Figure 17.13 (left). The decoding of the watermark is achieved by identifying the peeled-off strip into the watermarked mesh and then directly reading the bits by analyzing the strip's connections.

The Macro Density Pattern (MDP) is a *visual watermarking* method, in other words the watermark is retrieved by visual inspection and not by the computer, likely as a logo. The idea is to modulate the density of the triangles by a proper tessellation of the input models. If this tessellation respects the local curvature of the mesh the density changes are invisible when the model is rendered with common shading algorithms, but visible when the edges of the mesh are visualized. Figure 17.14 (right) shows an example of mesh watermarked with this technique. Another topology-driven watermarking scheme is the Triangle Flood Algorithm (TFA) [5] developed by Oliver Benedens. This algorithm uses both topological and geometric information to generate a unique path of

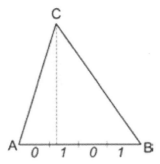

FIGURE 17.15
Watermark embedding in modified TSPS.

triangles on the mesh. Vertices of triangles along this path are modified both to embed bits (by modifying triangle height) and to order the triangles yielding a unique traversal path. Another remarkable example of a 3D watermarking algorithm that operates in the topological domain is the recent blind method for readable watermarking proposed by François Cayre and Benoit Macq [12]. This algorithm takes the basic ideas of TSPS and improves it by adding security properties. In fact the major drawback of the original TSPS implementation is that it is easy for an opponent to locate the payload and hence to change or discard it. In this implementation the triangle strip does not encode the watermark but it is built according to a sequence of bits given by a secret key K (private watermarking). After being built, each triangle of the strip is used to embed 1 bit of the watermark by modifying its geometric properties. In particular the state of the triangle ("0" or "1") depends on the position of the orthogonal projection of the triangle summit C on the "entry edge" AB (see Figure 17.15). This way of proceeding can be seen as a quantization index modulation (QIM) scheme extended on a discrete partition of a physical measurement. The starting triangle of the strip is determined on the basis of specific geometric characteristics. In this way, the peel-off operation is avoided and it is impossible to locate the embedded information if K is unknown, thus resulting in a secure topological embedding scheme. The main problem of watermarking algorithms based on topological embedding is their robustness. In fact, a simple re-triangulation attack destroys the watermark. An exception to this is the MDP that does not suffer re-triangulations attacks; nevertheless, the watermark is known and it is very simple to modify or remove it by using a simple geometric editing software. In conclusion, these kinds of algorithms are suitable for annotation or similar applications, where the robustness requirements are relaxed.

Geometric features. The most intuitive way to embed information into a 3D model is through modification of its geometry, e.g., vertex coordinates. Several host geometric features are possible but, at the end, the most used features for watermark embedding are basically two: the position of the mesh vertices and the vertex (or face) normals. Concerning vertex and face normals some clarifications are opportune. The direction of the face normal must be consistent with the orientation of the face. Typically, once computed, the direction of each normal is adjusted such that all the face normals point outward toward the 3D object. This adjustment is trivial if all the faces of the mesh have the same orientation. T. Harte and G. Bors [28] developed a blind and readable watermarking scheme to embed the bits of the watermark in the position of the vertices. In this scheme each vertex encodes 1 bit. A vertex encodes 1 or 0 if its position is outside or inside a bounding volume defined by its neighborhoods vertices. This algorithm is robust against geometrical transformation such as translation, rotation, and scaling. No evaluation tests about the robustness of the technique against other attacks such as noise addition, simplification, and so on, has been led. Yeo and Yeung [70] developed an authentication algorithm for 3D

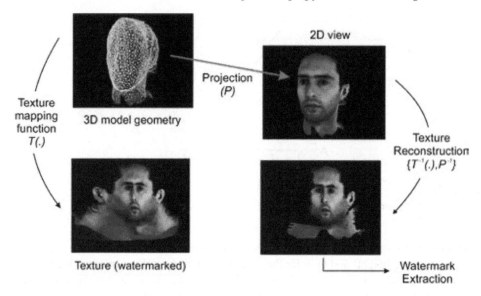

FIGURE 17.16
Texture-based 3D watermarking.

objects based on small perturbations of mesh vertices. Embedding is achieved by adjusting vertex coordinates; this technique allows quick localization and visualization of slight modifications of the watermarked model. One of the most interesting algorithms that exploits face normal features is the Normal Bin Encoding (NBE) [6] of Oliver Benedens. As previously stated face normals are directly related to the shape of the models, in fact this non-blind algorithm for readable watermarking is robust against attack like mesh simplification, i.e., all those attacks that alter the mesh considerably but that preserve the overall shape of the model. Another interesting algorithm based on shape-related geometric features is the one by Wagner [66]. Such an algorithm uses as host features an approximation of the surface normals; as demonstrated by Wagner, this algorithm can be made robust against affine transformation, by replacing the Euclidean norm with an affine-invariant vector norm.

Thus we can conclude that techniques based on geometric features can be used to develop 3D watermarking algorithms with different characteristics in terms of capacity and robustness. Hence, it is possible to conclude that geometric-based techniques are more flexible than those based on topological embedding. Robustness is achieved at the expense of imperceptibility, i.e., the greater the robust requirements the greater the visual impact of the distortions introduced by the watermark. The techniques based on shape-related geometric features, such as face normal, have good robustness against complex attack such as mesh simplification. On the contrary, they could suffer attacks like noise addition. Following these considerations further studies about the embedding by the combination of different geometric features would be desirable. One last consideration about robustness of geometry-based techniques is that, often, the impact of re-triangulation attack is negligible.

Surface attributes. The last set of host features belonging to the asset domain that can be used to embed information are the surfaces attributes. Typical surface attributes are colors, textures, and other additional data that can be used to model the material of the surface such as density, transparency, and so on. Since now, this kind of feature has been rarely used to embed watermarks in 3D objects. In the next section we illustrate the basic principles of texture-based 3D watermarking following recent work of Garcia and Dugelay [24]. The basic idea of texture-based watermarking of 3D objects is to embed the watermark into the texture of the 3D model. A texture is an image associated to the geometry of the model to enhance the visual content of the model itself. *Texture mapping* is the name of the process that maps the vertices to the texture. Since the texture is an image it is possible to watermark it by any image watermarking algorithm. So, the embedding phase is brought back to an image watermarking problem. The decoding/detection phase consists of two steps: in the first the texture is reconstructed from a visual representation of the model, for example a single or a set of images depicting the model; in the second step the watermark is extracted from the reconstructed texture. The first step of the extraction phase concerns the reconstruction of the texture from images of the model since the main goal of texture-based 3D watermarking is the protection of the visual representation of the object and not the protection of the 3D object itself. In fact, the crucial point of this approach is the reconstruction of the watermarked texture. The overall process is schematized in Figure 17.16. Garcia and Dugelay [24] propose a 2D/3D registration scheme to estimate the projection matrix and then invert the texture map process. To do this the geometry of the original model is required, hence this kind of technique is not blind. Sometimes the projection parameters are not sufficient to reconstruct the texture but even other parameters that influence the visualization of the model, for example the position of the light, have to be estimated to correctly reconstruct the original texture. The robustness performances tested by Garcia and Dugelay demonstrate that small errors in projection parameters estimation have the same effect of severe image attacks, such as JPEG compression at 75% level of quality, confirming the importance of the reconstruction phase.

Transformed Domain

These kinds of techniques perform the insertion of the watermark into the coefficients of a mathematical transformation of the host asset. The most used transformation are the ones related to the frequency domain, typically the Discrete Fourier Transform (DFT) or the Discrete Cosine Transform (DCT). Usually, these techniques exhibit good robustness against attacks. For example, by spreading the watermark over the whole asset, they are intrinsically more resistant to cropping than the techniques operating in the asset domain. Another example is robustness against shifting and scaling, that it is easily achieved in transformed domain. For instance, in the DFT domain, the magnitude of the DFT coefficients is invariant to shifting. One disadvantage of these methods is that the localization of the distortions introduced by the watermarking process is more difficult with respect to the techniques that operate in the asset domain. Another drawback of transformed domain is the high computational cost of such techniques. In fact, it is necessary to pass from the asset domain to the transformed domain and, after embedding, to apply the inverse transformation to obtain the watermarked asset. As previously mentioned, mathematical tools such as DFT or DCT are not available for geometric data, so today few techniques for mesh watermarking work in the transformed domain. One recent extension of the Fourier analysis to mesh, called *mesh spectral analysis*, has been employed for the first time by Karni and Gotsman [35] for lossy compression of polygonal meshes. Mesh spectral analysis is based on the extension of the classical spectral graph theory [8,9,69]. Ohbuchi et al. have developed a 3D watermarking algorithm based on mesh spectral analysis [51]. The watermark is inserted in mesh spectral domain by modifying the mesh spectral coefficients and then the inverse transformation is used to obtain the watermarked mesh. If the mesh has n vertices the computational complexity to calculate the eigenvectors is $O(n^3)$, that is prohibitive for mesh with thousands of vertices. In order to reduce the complexity, the input mesh is portioned in sub-meshes [35], then each sub-mesh is watermarked separately. The watermark extraction requires the original mesh spectral coefficients to be achieved. This non-blind algorithm is robust against additive vertex noise, partial resection (cropping) of the mesh, similarity transformation (translation, rotation and uniform scaling), and mesh smoothing. An improved version of this algorithm [50] uses a resampling phase to obtain robustness against remeshing and mesh simplification. Another 3D watermarking algorithm based on mesh spectral decomposition has been proposed by Cayre et al. [11]. In this work they propose some improvements to reduce the computational burden of mesh spectral decomposition, like the use of a fixed spectral basis, and a new partitioning scheme of the input mesh characterized by overlapping between the sub-meshes to improve the quality of the reconstructed mesh and the performance of the watermarking algorithm. The robustness performance against several geometric attacks such as noise addition, mesh smoothing, geometric transformation, and so on, are very good. The proposed method is also able to resist compression of the spectral mesh coefficients. No connectivity-based attacks have been tested. Recently Lavoue et al. [40] have presented a robust watermarking scheme for subdivision surfaces, based on the modulation of spectral coefficients of the subdivision control mesh. The algorithm optimizes the trade-off between watermarking redundancy and imperceptibility by modulating coefficients and by using error-correcting codes. The algorithm provides efficient robustness against remeshing or simplification attacks; unfortunately, is a non-blind schema, thus not suitable for IPR protection aims.

As previously stated, typically, these kinds of techniques are more robust than the ones operating in the asset domain. From this point of view 3D watermarking does not present exceptions. All the techniques based on mesh spectral domain, from our knowledge the only transformed domain for polygonal meshes used for 3D watermarking, present very good robustness against several attacks.

Hybrid Domain
The development of hybrid techniques has been inspired by the attempt to obtain both the main advantages of the techniques operating in the asset domain, i.e., the localization of the watermarking disturbance, and the good robustness properties typical of the transformed domain techniques. Hybrid techniques keep track of the temporal/spatial characterization of the host signal and, at the same time, use frequencies interpretation to achieve better information analysis and processing. The hybrid techniques that have received more attention recently are the ones based on block DCT/DFT and those relying on wavelet decomposition.

Wavelet decomposition was the first multiresolution analysis tool used in 3D watermarking. In 1998, Kanai et al. [34] proposed one of the first multiresolution approaches to 3D watermarking based on wavelet decomposition of meshes developed some years before, in 1995, by Lounsbery et al. [44]. In short, the wavelet coefficients computed by this analysis are vectors (to be more specific, these wavelets are *lazy wavelets* since the mathematical framework is not wholly applied). Embedding is performed by changing the modulus of such vectors. In particular, the least significant bits of the modulus of each wavelet are used to insert the watermark. Not all the wavelet coefficients are watermarked in this way but a selection policy based on geometric thresholds is used to minimize the visual impact of the wavelets' modifications. The watermark is extracted in a non-blind way by applying the wavelet decomposition and comparing the original and the watermarked wavelet coefficients. This technique is robust against affine transformation, additive noise, and partial resection of the mesh. Praun et al. [58] developed, in 1999, one of the most robust algorithms for mesh watermarking. In fact, this technique is robust against several attacks such as translation/rotation/uniform-scaling, additive noise, simplifications, mesh smoothing, re-triangulation and re-meshing. The main idea of this algorithm is to displace the vertices of the mesh according to a set of radial basis functions defined over the surface of the model. These basis functions are built on salient features of the mesh by a multiresolution approach performed with a progressive mesh [30] encoding of the input model. The watermark is extracted by considering the linear correlation between the original and the watermarked vertices' position. The watermark recovery phase is performed on a registered and re-sampled version of the watermarked mesh. For resampling, here, we intend that both the vertices' position and the connectivity of the original mesh is reconstructed. Such registration/resampling phase is the kernel of the excellent robustness properties of this algorithm. Nevertheless, it needs the original model to be performed, thus resulting in a non-blind method with more limitations in its practical applicability.

A completely different approach is used by Song et al. [63]. In this work a virtual 3D scanner is simulated to create several *range images* of the model. A range image is an image whose pixels value corresponds to depth distances. So, a range image contains the information about the shape of the model. The obtained range image is watermarked using a standard image watermarking method based on DCT [29]. Then, the vertices of the model are moved accordingly to watermarked range image. This technique is not hybrid in the strict sense of the term but it uses the spatial localization given by the range image and the frequency analysis for the embedding. For this reason we decided to classify it as a hybrid technique.

Compressed Domain

The watermarking techniques that embed the information directly in the compressed domain present several advantages, in particular if a compression standard is chosen for the embedding (e.g., MPEG2). For example it is possible to avoid all the problems relative to watermark coefficients that are typically discarded or coarsely quantized after the compression. Additionally, to work directly in the compressed domain makes the computational burden low. This is especially true for video applications, where the videos are typically compressed using some video compression standard such as MPEG2, MPEG4, H.263, or H.264. In fact, in this way, the watermarking can be achieved without the necessity of decompressing the video, saving a considerable amount of time. A disadvantage of such techniques is that often they are sensitive to *transcoding*, i.e., to the change of compressed domain (e.g., convert an MPEG2 bitstream to an MPEG4 bitstream).

This review about the current state-of-the-art in 3D watermarking poses in evidence one of the main drawbacks of the current watermarking technology for 3D objects (i.e., the existing techniques with good robustness properties for 3D watermarking are not blind). To be more specific, most of the algorithms that work in the asset domain, in particular those based on the topology of the model, are blind, readable, and capable of achieving high payload, but not robust, making them suitable only for all those applications where the robustness requirements are relaxed. Geometric-based methods are the more flexible, in term of characteristics, in the asset domain. All the blind techniques of this type are not robust against both malicious and non-malicious manipulations. In general, such algorithms achieve robustness at the expense of imperceptibility. The methods based on shape-related features seem to be intrinsically robust to certain complex attacks, such as mesh simplification. Watermarking through surface's attributes is still an unexplored topic except for the texture-based 3D watermarking method. It is important to remark that these kinds of watermarking algorithms appear to be more suitable to embed data into the visual representation of a 3D object than into the 3D object itself. The algorithms that work in the transformed domain exhibit good robust properties against several and complex attacks. The main drawback of this kind of technique is the difficulty of controlling the distortions caused by the watermarking process and hence the difficulty of obtaining a watermarked model with high visual quality. Those algorithms that work in the hybrid domain have good robustness properties.

17.4   Conclusions

In the Cultural Heritage area the information technologies have made very significant advances in recent years. Efficient digital acquisition systems have been developed for 2D objects (e.g., paintings) and 3D objects (e.g., sculptures or archaeological items). In particular, very high quality digital images, having also color calibration, can be obtained, creating new digital galleries or virtual museums.

The need of copyright protection of the 2D and 3D acquired data is of high importance, for digital archives and in particular for digital content distribution and remote user access. Digital watermarking technologies have been presented in this chapter for copyright protection of 2D and 3D data, outlining the main general properties and still open research lines (especially for 3D watermarking). Some examples have been reported, showing how in practice watermarking technologies can be applied to real 2D and 3D objects.

At the present state of the art, watermarking of digital images appears a quite well developed area, to be actually used for museum and gallery artwork copyright protection.

Bibliography

[1] C. Acidini and V. Cappellini. *Reale e Virtuale nei Musei - Due Visioni a Confronto*. Pitagora Editrice Bologna, Bologna, 2008.

[2] M. Barni, F. Bartolini, and T. Furon. A General Framework for Robust Watermarking Security. *Signal Processing*, 83(10):2069–2084, 2003.

[3] F. Bartolini, M. Barni, V. Cappellini, and A. Piva. Mask Building for Perceptually Hiding Frequency Embedded Watermarks. In *Proceedings of the 5th IEEE International Conference on Image Processing, ICIP98*, volume I, pages 450–454, Chicago, IL, USA, October 1998.

[4] F. Bartolini, A. Tefas, M. Barni, and I. Pitas. Image Authentication Techniques for Surveillance Applications. In *Proceedings of IEEE*, volume 89, pages 1403-1418, October 2001.

[5] O. Benedens. Two High Capacity Methods for Embedding Public Watermarks into 3D Polygonal Models. In *Proceedings of the Multimedia and Security-Workshop at ACM Multimedia 99*, pages 95–99, Orlando, Florida, 1999.

[6] O. Benedens. Watermarking of 3D Polygon Based Models with Robustness against Mesh Simplification. In *Proceedings of SPIE: Security and Watermarking of Multimedia Contents*, volume 3657, pages 329–340, 1999.

[7] O. Benedens and C. Busch. Towards Blind Detection of Robust Watermarks in Polygonal Models. In *Proceedings of the EUROGRAPHICS '2000. Computer Graphics Forum*, volume 3, pages C199–C208, Interlaken, Switzerland, August 2000. Blackwell.

[8] N. L. Biggs. *Algebraic Graph Theory (2nd) edition*. Cambridge University Press, 1993.

[9] B. Bollobás. *Modern Graph Theory*. Springer, 1998.

[10] D. Boneh and J. Shaw. Collusion-Secure Fingerprinting for Digital Data. In *IEEE Transactions on Information Theory*, volume 44, pages 1897-1905, 1998.

[11] F. Cayre, P. R. Alface, F. Schmitt, B. Macq, and H. Matre. Application of Spectral Decomposition to Compression and Watermarking of 3D Triangle Mesh Geometry. *Image Communications – Special Issue on Image Security*, 18:309–319, April 2003.

[12] F. Cayre and B. Macq. Data hiding on 3-D Triangle Meshes. 51(4):939–949, 2003.

[13] J. Cohen, A. Varshney, D. Manocha, G. Turk, H. Weber, P. Agarwal, F. Brooks, and W. Wright. Simplification Envelopes. In *SIGGRAPH '96: Proceedings of the 23rd Annual Conference on Computer Graphics and Interactive Techniques*, pages 119–128. ACM Press, 1996.

[14] I. J. Cox and M. L. Miller. The First 50 Years of Electronic Watermarking. *EURASIP Journal on Applied Signal Processing, 2002*, (2):126-132, 2002.

[15] I. J. Cox, M. L. Miller, and J. A. Bloom. *Digital Watermarking*. Morgan Kaufmann, 2001.

[16] S. Craver and S. Katzenbeisser. Security Analysis of Public-Key Watermarking Schemes. In M. S. Schmalz, editor, *Mathematics of Data/Image Coding, Compression, and Encryption IV, Proc. SPIE, San Diego, CA, USA*, volume 4475, pages 172-182, July 2001.

[17] S. Craver, N. Memon, B. L. Yeo, and M. M. Yeung. On the Invertibility of Invisible Watermarking Techniques. In *Proc. 4th IEEE Int. Conf. on Image Processing, ICIP97, Santa Barbara, CA, USA*, volume I, pages 540-543, October 1997.

[18] S. Craver, B. Yeo, and M. Yeung. Technical Trials and Legal Tribulations. *Communications of the ACM*, 41(7):44–54, 1998.

[19] M. Desbrun, M. Meyer, P. Schröder, and A. H. Barr. Implicit Fairing of Irregular Meshes Using Diffusion and Curvature Flow. In *Proceedings of the 26th Annual Conference on Computer Graphics and Interactive Techniques*, pages 317–324. ACM Press/Addison-Wesley Publishing Co., 1999.

[20] M. Eck, T. De Rose, T. Duchamp, H. Hoppe, M. Lounsbery, and W. Stuetzle. Multiresolution Analysis of Arbitrary Meshes. In *Proceedings of the 22nd Annual Conference on Computer Graphics and Interactive Techniques*, pages 173–182. ACM Press, 1995.

[21] J. J. Eggers, J. K. Su, and B. Girod. Public Key Watermarking by Eigenvectors of Linear Transforms. In *Proceedings of X Europ. Signal Processing Conf., EUSIPCO00, Tampere, Finland*, volume III, pages 1685-1688, September 2000.

[22] T. Furon and P. Duhamel. Robustness of an Asymmetric Watermarking Technique. In *Proceedings of 7th IEEE International Conference on Image Processing, ICIP00, Vancouver, Canada*, volume III, pages 21-24, September 2000.

[23] T. Furon, I. Venturini, and P. Duhamel. A Unified Approach of Asymmetric Watermarking Schemes. In E.J. Delp and P.W. Wong, editors, *Security and Watermarking of Multimedia Contents III, Proceedings of SPIE, San Jose, CA*, volume 4314, pages 269–279, 2001.

[24] E. Garcia and J. L. Dugelay. Texture-Based Watermarking of 3D Video Objects. *IEEE Transaction on Circuits and Systems for Video Technology*, 13(8):853–866, 2003.

[25] M. Garland and P. S. Heckbert. Simplifying Surfaces with Color and Texture using Quadric Error Metrics. In *VIS '98: Proceedings of the Conference on Visualization '98*, pages 263–269, Research Triangle Park, North Carolina, United States, 1998. IEEE Computer Society Press.

[26] X. Gu, S. J. Gortler, and H. Hoppe. Geometry Images. In *Proceedings of the 29th Annual Conference on Computer Graphics and Interactive Techniques*, pages 355–361, San Antonio, Texas, 2002. ACM Press.

[27] I. Guskov, K. Vidim, W. Sweldens, and P. Schröder. Normal Meshes. In *Proceedings of the 27th Annual Conference on Computer Graphics and Interactive Techniques*, pages 95–102. ACM Press/Addison-Wesley Publishing Co., 2000.

[28] T. Harte and A. G. Bors. Watermarking 3D models. In *Proceedings of IEEE International Conference on Image Processing 2002*, volume III, pages 661–664, Rochester, NY, USA, 2002.

[29] J. R. Hernandez and F. Perez-Gonzalez. Statistical Analysis of Watermarking Schemes for Copyright Protection of Images. 87(7):1142–1166, 1997.

[30] H. Hoppe. Progressive Meshes. In *Proceedings of the 23rd Annual Conference on Computer Graphics and Interactive Techniques*, pages 99–108. ACM Press, 1996.

[31] I. J. Cox J. A. Bloom, T. Kalker, J.-P. M. G. Linnartz, M. L. Miller, and C. B. S. Traw. Copy Protection for Digital Video. In *IEEE – Special Issue on Identification and Protection of Multimedia Information*, volume 87, pages 1267-1276, July 1999.

[32] T. R. Jones, F. Durand, and M. Desbrun. Non-Iterative, Feature-Preserving Mesh Smoothing. *ACM Transactions on Graphics*, 22(3):943–949, 2003.

[33] T. Kalker. Considerations on Watermarking Security. In *IEEE Multimedia Signal Processing, MMSP'01 Workshop, Cannes, France*, pages 201–206, 2001.

[34] S. Kanai, H. Date, and T. Kishinami. Digital Watermarking for 3D Polygons using Multiresolution Wavelet Decomposition. In *Proceedings of the Sixth IFIP WG 5.2 International Workshop on Geometric Modeling: Fundamentals and Applications (GEO-6)*, pages 296–307, Tokyo, Japan, 1998.

[35] Z. Karni and C. Gotsman. Spectral Compression of Mesh Geometry. In *Proceedings of the 27th Annual Conference on Computer Graphics and Interactive Techniques*, pages 279–286. ACM Press/Addison-Wesley Publishing Co., 2000.

[36] A. Kerckhoffs. La Cryptographie Militaire. *Journal des Sciences Militaires*, IX:5–83, 1883.

[37] M. S. Kim, J. W. Cho, H. Y. Jung, and R. Prost. A Robust Blind Watermarking for 3D Meshes using Distribution of Scale Coefficients in Irregular Wavelet Analysis. *Acoustics, Speech and Signal Processing 2006*, 5:477–480, 2006.

[38] L. Kobbelt. Discrete Fairing. In *Proceedings of the Seventh IMA Conference on the Mathematics of Surfaces '97*, pages 101–131, 1997.

[39] L. Kobbelt, S. Campagna, J. Vorsatz, and H. P. Seidel. Interactive Multi-Resolution Modeling on Arbitrary Meshes. In *Proceedings of the 25th Annual Conference on Computer Graphics and Interactive Techniques*, pages 105–114. ACM Press, 1998.

[40] G. Lavou, F. Denis, and F. Dupont. Subdivision Surface Watermarking. Technical Report RR-LIRIS-2006-011, LIRIS UMR 5205 CNRS/INSA de Lyon/Universit Claude Bernard Lyon 1/Universit Lumire Lyon 2/Ecole Centrale de Lyon, June 2006.

[41] A. W. F. Lee, W. Sweldens, P. Schröder, L. Cowsar, and D. Dobkin. MAPS: Multiresolution Adaptive Parameterization of Surfaces. In *Proceedings of the 25th Annual Conference on Computer Graphics and Interactive Techniques*, pages 95–104. ACM Press, 1998.

[42] P. Lindstrom and G. Turk. Image-Driven Simplification. *ACM Transaction on Graphics*, 19(3):204–241, 2000.

[43] F. Losasso, H. Hoppe, S. Schaefer, and J. Warren. Smooth Geometry Images. In *SGP '03: Proceedings of the Eurographics/ACM SIGGRAPH Symposium on Geometry Processing*, pages 138–145, Aachen, Germany, 2003. Eurographics Association.

[44] M. Lounsbery, T. D. DeRose, and J. Warren. Multiresolution Analysis for Surfaces of Arbitrary Topological Type. *ACM Transactions on Graphics*, 16(1):34–73, 1997.

[45] D. Luebke and C. Erikson. View-Dependent Simplification of Arbitrary Polygonal Environments. In *SIGGRAPH '97: Proceedings of the 24th Annual Conference on Computer Graphics and Interactive Techniques*, pages 199–208. ACM Press/Addison-Wesley Publishing Co., 1997.

[46] B. M. Macq and J. J. Quisquater. Cryptology for Digital TV Brodacasting. *Proceedings of the IEEE*, 83(6):944–957, June 1995.

[47] E. Martinian and G. W. Wornell. Authentication with Distortion Constraints. In *Proceedings of IEEE International Conference on Image Processing*, volume II, pages 17–20, 2002.

[48] M. L. Miller. Is Asymmetric Watermarking Necessary or Sufficient? In *Proceedings of XI European Signal Processing Conference, EUSIPCO02, Toulose, France*, volume I, pages 291-294, September 2002.

[49] R. Ohbuchi, H. Masuda, and M. Aono. Watermaking Three-Dimensional Polygonal Models. In *Proceedings of the Fifth ACM International Conference on Multimedia*, pages 261–272, Seattle, Washington, United States, 1997. ACM Press.

[50] R. Ohbuchi, A. Mukaiyama, and S. Takahashi. A Frequency-Domain Approach to Watermarking 3D Shapes. In *Proceedings of EUROGRAPHICS 2002*, Saarbrucken, Germany, September 2002.

[51] R. Ohbuchi, S. Takahashi, T. Miyazawa, and A. Mukaiyama. Watermarking 3D Polygonal Meshes in the Mesh Spectral Domain. In *Proceedings of Graphics Interface 2001*, pages 9–17, Ottawa, Ontario, Canada, 2001. Canadian Information Processing Society.

[52] S. Pereira, S. Voloshynovskiy, M. Madueño, S. Marchand-Maillet, and T. Pun. Second Generation Benchmarking and Application Oriented Evaluation. In *Proceedings of Information Hiding Workshop*, Pittsburgh, PA, USA, april 2001.

[53] F. A. P. Petitcolas. Watermarking Schemes Evaluation. 17:58–64, September 2000.

[54] F. A. P. Petitcolas, R. J. Anderson, and M. G. Kuhn. Attacks on Copyright Marking Systems. In David Aucsmith, editor, *Proceedings of the 2nd International Workshop on Information Hiding, IH98*, pages 219–239, Portland, Oregon, USA, April 1998. Springer-Verlag.

[55] Fabien Petitcolas and other contributors. Stirmark benchmark 4.0.

[56] B. Pfitzman. Information Hiding Terminology. In R. Anderson, editor, *Proceedings of 1st International Workshop on Information Hiding, IH96, of Lecture Notes in Computer Science, Cambridge, UK*, volume 1174, pages 347-350. Springer Verlag, May/June 1996.

[57] C. I. Podilchuk and W. Zeng. Image-Adaptive Watermarking using Visual Models. 16:525–539, April 1998.

[58] E. Praun, H. Hoppe, and A. Finkelstein. Robust Mesh Watermarking. In *Proceedings of the 26th Annual Conference on Computer Graphics and Interactive Techniques*, pages 49–56. ACM Press/Addison-Wesley Publishing Co., 1999.

[59] J. J. K. Ó Ruanaidh, W. J. Dowling, and F. M. Boland. Watermarking Digital Images for Copyright Protection. *Proceedings of the IEEE – Visual Image Signal Processing*, 143(4):250–256, August 1996.

[60] P. V. Sander, Z. J. Wood, S. J. Gortler, J. Snyder, and H. Hoppe. Multi-Chart Geometry Images. In *SGP '03: Proceedings of the Eurographics/ACM SIGGRAPH Symposium on Geometry Processing*, pages 146–155, Aachen, Germany, 2003. Eurographics Association.

[61] R. Schneider and L. Kobbelt. Geometric Fairing of Irregular Meshes for Free-Form Surface Design. *Computer Aided Geometric Design*, 18(4):359–379, 2001.

[62] R. Schneider and L. Kobbelt. Mesh Fairing Based on an Intrinsic PDE Approach. *Computer-Aided Design*, 33:767–777(11), 2001.

[63] H. S. Song, N. I. Cho, and J. W. Kim. Robust Watermarking of 3D Mesh Models. In *Proceedings of Multimedia Signal Processing Workshop 2002*, pages 332–335, Seoul, South Korea, 2002.

[64] G. Taubin. A Signal Processing Approach to Fair Surface Design. In *Proceedings of the 22nd Annual Conference on Computer Graphics and Interactive Techniques*, pages 351–358. ACM Press, 1995.

[65] R. G. van Schyndel, A. Z. Tirkel, and I. Svalbe. Key Independent Watermark Detection. In *Proceedings of IEEE International Conference on Multimedia Computing and Systems, ICMCS 99, Florence, Italy*, volume I, pages 580-585, June 1999.

[66] M. G. Wagner. Robust Watermarking of Polygonal Meshes. In *Proceedings of Geometric Modeling and Processing 2000. Theory and Applications*, pages 201–208, Hong Kong, China, 2000.

[67] N. Williams, D. Luebke, J. D. Cohen, M. Kelley, and B. Schubert. Perceptually Guided Simplification of Lit, Textured Meshes. In *Proceedings of the 2003 Symposium on Interactive 3D Graphics*, pages 113–121, Monterey, California, 2003. ACM Press.

[68] R. B. Wolfgang, C. I. Podilchuk, and E. J. Delp III. Perceptual Watermarks for Digital Images and Video. In Ping W. Wong and Edward J. Delp III, editors, *Security and Watermarking of Multimedia Contents*, volume 3657, pages 40–51, San Jose, CA, USA, 1999. SPIE.

[69] J. Wu and L. Kobbelt. Efficient Spectral Watermarking of Large Meshes with Orthogonal Basis Functions. *The Visual Computer*, 21(8-10):848–857, 2005.

[70] B. L. Yeo and M. M. Yeung. Watermarking 3D Objects for Verification. 19(1):36–45, 1999.

[71] W. Zeng and B. Liu. On Resolving Rightful Ownerships of Digital Images by Invisible Watermarks. In *Proceedings of IEEE International Conference on Image Processing '97*, pages 552–555, Santa Barbara, CA, October 26-29 1997.

Index

T - #0312 - 101024 - C0 - 254/178/28 [30] - CB - 9781439821732 - Gloss Lamination